Tropical Grassland Husbandry

TROPICAL AGRICULTURE SERIES

The Tropical Agriculture Series, of which this volume
forms part, is published under the editorship of
G. Wrigley

ALREADY PUBLISHED

Tobacco *B. C. Akehurst*
Tropical Grassland Husbandry *L. V. Crowder and H. R. Chheda*
Tea *T. Eden*
Rice *D. H. Grist*
The Oil Palm *C. W. S. Hartley*
Cattle Production in the Tropics Volume 1
   W. J. A. Payne
Spices Vols 1 & 2 *J. W. Purseglove* et al.
Tropical Fruits *J. A. Samson*
Tropical Pulses *J. Smartt*
Agriculture in the Tropics *C. C. Webster and P. N. Wilson*
An Introduction to Animal Husbandry in the Tropics
   G. Williamson and W. J. A. Payne
Cocoa *G. A. R. Wood*

Tropical Grassland Husbandry

L. V. Crowder
Professor of Agriculture
The Rockefeller Foundation, Indonesia

H. R. Chheda
Professor of Agronomy
University of Ibadan, Nigeria

Longman
London and New York

Longman Group Limited
Longman House
Burnt Mill, Harlow, Essex, UK

*Published in the United States of America
by Longman Inc., New York*

First published 1982

British Library Cataloguing in Publication Data

Crowder, L.V.
    Tropical grassland husbandry. — (Tropical agriculture series)
    1. Pastures — Tropics
    I. Title   II. Chheda, H.R.   III. Series
    633.2'02'00913   SB199

    ISBN 0-582-46677-6

Library of Congress Cataloging in Publication Data

Crowder, Loy V.
    Tropical grassland husbandry.

    (Tropical agriculture series)
    Bibliography: p.
    Includes index.
    1. Range management — Tropics.   2. Pastures — Tropics.   3. Forage
plants — Tropics.   4. Grazing — Tropics.   I. Chheda, H. R.   II. Title.
III. Series.
SB193.3.T76C76 1983   633.2   82-15239
                                        AACR2

*Printed in Singapore by
Tien Mah Litho Printing Co (Pte) Ltd*

Contents

To Eloise and Sati

Preface

Our experience in carrying out research with tropical pastures and forage crops and teaching grassland husbandry has extended over many years and into several countries. In developing teaching materials, we have drawn from this research and experience, as well as from that of colleagues throughout the tropics and subtropics. In addition, an effort has been made to maintain awareness of current literature. Our lecture notes, and hand-outs have formed the basis of this publication. The book was written primarily as a text for students but will be useful as a reference for them, and for teachers, researchers and lay persons. Information found herein is too extensive and comprehensive for a single course, but we have found the contents divisible into introductory and advanced course study. Subject matter contained in some chapters can be utilized in a modular form for team teaching. We expect and anticipate that instructors will select and modify the information to fit their individual needs.

We made an effort to integrate the soil–plant–animal relationships, particularly from the point of view of pasture agronomy and livestock husbandry. Attention was also given to historical aspects, environmental effects, botany and systematics, the interface with farming systems, pasture improvement through plant-breeding, and seed production of superior cultivars.

Yield records and other measurements reported in the literature have been recorded in the metric and avoirdupois systems. Recalculation of the latter into metric for this publication has sometimes led to rounding of figures, so that slight differences may be noted. Latin scientific names have been used for the most part. In some instances they appear interchangeably with names that are widely accepted in the vernacular.

A major portion of grassland research and pasture development has taken place during the past 50 years. The Australian grassland scientists must be recognized for their leadership role in plant exploration, evaluation and improvement of pasture crops, utilization of these species and cultivars in the improvement of native and naturalized grazinglands and in the development of sown pastures.

Recognition must also be given to pasture and forage researchers at the International Centre for Tropical Agriculture (CIAT) in Colombia, and at various agricultural experimental stations in the Andean zone, Brazil, Puerto Rico and Mexico, East and West African countries, and in the Philippines.

Grazinglands in which grasses predominate still provide the basic feed requirements for growth and development of livestock, their maintenance and the production of their products. In some grazinglands, legumes and various herbaceous species contribute to overall available feedstuff, but for the most part tropical legumes continue to remain an elusive element. Crop residues are of vast importance in many regions of the tropics, especially where smallholdings prevail. Despite research efforts, sown pastures occupy a small portion of the total grazinglands of the tropics and subtropics, below 5 per cent even in countries where pasture research is well advanced.

Animal output from fertilized grass exceeds that from grass–legume mixtures in the tropics. The increasing cost of nitrogen fertilizer and the need to conserve energy, however, has created a critical need to intensify tropical legume investigation. Unfortunately, there is no 'queen' among the tropical legumes to compare with lucerne in the temperate zone. A wider array of warm-season legumes exists than cool-season legumes, but as yet attention, research efforts and financial support have not focused on a single species, as was the case with lucerne. Pasture scientists have placed emphasis on the herbaceous types for the most part and tended to overlook the woody species, especially the taller-growing forms. We urge grassland students and scientists to give keen attention to ecological systems in their examinations of tropical grassland husbandry and to integrate pasture development as a part of an overall farming system.

November 1980   L. V. Crowder
                                                       H. R. Chheda

Acknowledgements

We are grateful to the following for permission to reproduce copyright material:

The Australian Institute of Agricultural Science for our Fig. 12.1 from Fig. 1 (Norman & Stewart 1964); the author, R. C. Bruce for our Fig. 9.10 adapted from Fig. 1 (Bruce 1974); Cambridge University Press for our Fig. 5.5 adapted from Fig. 6 (Haggar 1970), our Fig. 10.3 adapted from Fig. 5 (Jones & Sandland 1974), our Fig. 18.4 adapted from Fig. 1 (Shelton & Humphreys 1971); Commonwealth Scientific and Industrial Research Organisation, Australia for our Fig. 6.3 (Hartley 1958b), our Fig. 6.4 (Hartley 1958a), our Fig. 10.7 from Fig. 1 (Hamilton et al 1970), our Fig. 10.10 from Fig. 1 (Stobbs 1973b), our Fig. 17.2 (Clements 1977), our Fig. 17.4 from Fig. 3 (McWhirter 1969); Crop Science Society of America for our Fig. 17.1 from Fig. 1 (Burton & Monson 1972), our Fig. 17.5 modified from Fig. 3 (Taliaferro & Bashaw 1967); W. H. Freeman & Co for our Fig. 10.9 from Fig. 14.3 © 1972 W. H. Freeman (McDowell 1972); the author, C. T. Gates for our Fig. 9.9 adapted from Fig. 1 (Gates 1970); Iowa State University Press for our Fig. 10.2 redrawn from Fig. 12.2 © 1973 by Iowa State University Press (Heath et al 1973); IPC Science and Technology Press Ltd for our Fig. 13.3 from Fig. 1 (Dradu & Harrington 1972); the author, Dr E. Q. Javier for our Fig. 18.2 from Fig. 3 (Javier 1970); McGraw Hill Book Company for our Fig. 6.1 adapted from (Koeppe & de Long 1958); the author, Professor M. J. T. Norman for our Fig. 10.8 (Norman 1970); Royal Dutch Geographical Society for our Fig. 12.2 from Fig. 2 (Raay & Leeuw 1970); Tropical Grassland Society of Australia and the author for our Fig. 16.5 (Evans 1968); University of Queensland Press and the author for our Fig. 16.4 (Evans 1970); Van Nostrand Reinhold Co for our Fig. 17.3 from Fig. 20.1 © Van Nostrand Reinhold Company 1967 (Briggs & Knowles 1967).

Whilst every effort has been made we are unable to trace the copyright holders of our Figs. 5.3, 5.4 and 18.6 and would appreciate any information which would enable us to do so.

Metric-Imperial Conversions

Metric units	Imperial equivalent	Imperial units	Metric equivalent
Length			
1 centimetre (cm)	0.39 inch (in)	1 in	2.54 cm
1 metre (m)	3.28 foot (ft)	1 ft	0.30 m
1 metre (m)	1.09 yard (yd)	1 yd	0.91 m
1 metre (m)	0.05 chain (ch)	1 ch	20.12 m
1 kilometre (km)	0.62 mile	1 mile	1.61 km
Area			
1 cm^2	0.155 in^2	1 in^2	6.452 cm^2
1 m^2	10.758 ft^2	1 ft^2	0.093 m^2
1 m^2	1.196 yd^2	1 yd^2	0.836 m^2
1 km^2	0.386 mile^2	1 mile^2	2.59 km^2
1 hectare (ha)	2.470 acre	1 acre	0.405 ha
Volume			
1 litre (l)	1.76 pint (pt)	1 pt	0.56 l
1 l	0.22 gallon (gal) (Imp.)	1 gal (Imp.)	4.54 l
1 l	0.27 gal (US)	1 gal (US)	3.76 l
1 kilolitre (kl)	28.4 bushel (bu) (US)	1 bu (US)	35.24 l
1 kl	27.4 bu (Imp.)	1 bu (Imp.)	36.37 l
Mass			
1 gram (g)	0.04 ounce (oz)	1 oz	28.35 g
1 kilogram (kg)	2.20 pound (lb)	1 lb	453.59 g
1 kg	0.02 hundredweight (cwt)	1 cwt	50.80 kg
1 tonne (t)	0.98 ton (long)	1 ton (long)	1.02 t
1 kg/ha	0.89 lb/acre	1 lb/acre	1.12 kg/ha^2
1 kl/ha	11.2 bu/acre^2	1 bu/acre	0.09 kl/ha

Temperature

°Celsius	0	10	20	30	40	50	60	70	80	90	100
°Fahrenheit	32	50	68	86	104	122	140	158	176	194	212

Rainfall

Inches	10	20	30	40	50	60	70
Millimetres	254	508	762	1016	1270	1524	1778

Chapter 1

The tropical environment

The extremes of wet and dry, hot and cold, daily downpours and extended drought, fertile flood plains and eroded overgrazed hills, steaming lowlands and permanent snowclad peaks, along with other inconsistencies of nature, make up the tropical environment. Three elements – climate, vegetation and soils – are its most essential components. Climate is the dominating factor and shapes the vegetation, modifies the soil and ultimately affects all forms of life. The type and distribution of tropical grasslands are largely determined by climate and its interaction with the soil. Total annual rainfall, and its distribution, regulate the adaptation, growth and production of grasses, legumes and browse plants, even though other factors, such as temperature, humidity, sunlight, elevation, slope and exposure of terrain exert a strong influence. Man's activities, however, may drastically alter the environment and thus change the composition and productivity of grasslands. Man, in fact, can completely destroy the natural grasslands and other vegetation, replace them with productive grazingland and then maintain a high level of herbage output over a long period of time.

Tropical boundaries

Solar or astronomical tropics

As the earth spins the tilt of the polar axis causes an annual shift in the path of the sun over the earth's surface. The angle of tilt is approximately 23° 27′ north and south of the equator. These points of latitude mark the limit of the sun's migration away from the equator. The area of the earth between the two latitudes receives direct vertical rays from the sun and is known as the tropics (from the Greek *tropikos*, meaning solstice).

This zone is bounded on the north by the Tropic of Cancer and on the south by the Tropic of Capricorn. On 21 June the constellation 'crab' (Cancer), in the northern hemisphere, rises and sets with the sun. On 21 January the constellation 'horned goat' (Capricorn), in the southern hemisphere, rises and sets with the sun. These dates and events coincide with the maximum wandering of the sun and mark its turning back towards the equator.

About one-fourth of the land surface and two-fifths of the earth's total surface lie between the tropics of Cancer and Capricorn. The region receives over one-half of the earth's total rainfall. Approximately one-third of the earth's population lives within the tropical zone.

The subtropics lie immediately north and south of the solar tropics. The limit has not been precisely defined but is taken to be about 35° north and south of the equator.

Isothermic tropics

Some geographers object to the artificial boundary of the solar tropics and have suggested that the mean isotherm (line of equal temperature) of 21.1 °C be taken as the limits of the tropics. The isotherms by and large run parallel to the latitudinal circles. They bulge poleward over some land masses, extending beyond the tropics of Cancer and Capricorn. Such variation reflects the amount and distribution of radiant energy received by the sun. The use of temperature lines has not gained recognition as boundary markers for the tropics. If temperature alone separates the tropical and temperate zones, then the isotherms probably would be valid limits. To those interested in tropical grassland husbandry, however, rainfall and soil moisture have as great, or greater, importance on plant adaptation and productivity than does temperature.

Biological tropics

It has been suggested that the 30° lines of latitude be considered as boundaries for the biological tropics since the transition of tropical and temperate vegetation occurs in this region (Davies, 1960; Mannetje, 1978). There could be no definite latitudinal line as the vegetation zones fluctuate above and below the 30° latitude lines. This is particularly noticeable for the type and composition of grasslands. Approximately one-half of the grazinglands of the world lie within the biological tropics. The area covers a wide range of environments, from very humid conditions to the arid lands of the desert fringes, and from sea level to very high mountains. Something like one-half of the earth's livestock are found within the biological tropics. These cattle produce about one-third of the total meat supply and one-sixth of the dairy products. The concept of a biological tropical boundary has not been widely used but it does hold merit for those interested in grassland husbandry.

Climate

Climate and weather are frequently used interchangeably, but a distinction should be made between them. Weather is a state of the atmosphere with respect to heat or cold, wet or dry, calm or storm, clear or cloudy. It changes from day to day and variation is influenced by geographical location, topography, distribution of land and water, mountain barriers, altitude, winds, ocean currents, vegetation. Climate is made up of a composite of the day-to-day weather conditions. It is an average of weather over time. Climate can be used to describe local or regional conditions, or applied to a few principal types (Garbell, 1947; Kendrew, 1953; Riehl, 1954). The major atmospheric elements making up climate in the tropics are moisture, temperature, light and air movement (Mohr and Van Baren, 1954; Pendelaborde, 1963).

Moisture

Precipitation is the most important climatic element, since temperature and light are less likely to be limiting for the growth of plants in the tropics. Rainfall furnishes the major portion of water in the tropics. Dew forms during cool nights when water vapour condenses on leaves and other surfaces. Fog occurs along coastal areas and at higher elevations. Both fog and dew are important factors of plant growth, since they reduce transpiration and evaporation but are insufficient to increase significantly herbage yields of fodder species. Snow and ice provide an appreciable amount of precipitation only in high mountains, but by feeding streams below they increase water available for irrigation. Hailstorms are common at higher elevations and are sporadically discharged in the lowlands, sometimes causing damage to field and forage crops.

Average rainfall and distribution

The total amount of rainfall fluctuates widely from one region or locale to another. In general, the equatorial belt between 5° N and 5° S latitude receives most rain, with more occurring in the southern latitudes than in the northern belts. The annual means range from about 250 mm to over 6 000 mm, but exceed 10 000 mm on the western coast of South America between 3° S and 8° N latitude, the coastal area of the Cameroons and the northeastern coast of the Gulf of Bengal (Mohr and Van Baren, 1954).

Average rainfall data are usually of limited value and distribution through the year is more meaningful for the agriculturist and the grassland husbandryman. Months with more than 100 mm are considered 'wet', 60–100 mm as 'moist' and less than 60 mm as 'dry' (Mohr and Van Baren, 1954). Below 60 mm, the soil rapidly dries, causing a water stress situation for many plant species.

On the average, rains in the tropical and subtropical regions follow recognised patterns.

In regions with high rainfall (1 000–2 500 mm annually), a two-peak season occupies most of the year with short intervening day periods. One dry interval, however, may extend over 3–5 months, to be followed by rain, then a short dry spell of about 1 month and a second rainy period of 1 or 2 months. In some regions two rainy and two dry seasons are prevalent, each of about 3 months' duration. The length of the dry period is extended to 6 months or more in areas with a single rainy period. In low rainfall areas (500 mm or less) the rains usually come during a period of 2–3 months. In the very arid and desert zones most rainfall is of the thunderstorm type. It comes in downpours with great water runoff and flooding because of little vegetative cover.

Grass and legume adaptation and production are largely determined by the amount and distribution of rainfall. Species such as *Pennisetum purpureum, Brachiaria mutica, Desmodium intortum* and *Calopogonium mucunoides* thrive best with more than 1 000 mm annual rainfall. Others have drought tolerance and survive dry periods, e.g. *Cenchrus ciliaris, Andropogon* and *Aristida* species.

Under all conditions, the distribution of rain determines the pattern of plant growth. Herbage yields of tropical grasses and legumes are seasonal and closely follow the cyclic pattern of rainfall. During the dry season growth declines. It may cease altogether with prolonged drought when the aerial portion of the plants become brown and parched and plants may even die. In some instances, duration of the dry period alters the life form of the plant. For example, *Stylosanthes humilis* at Townsville in northeast Queensland, Australia, is an annual under the prolonged dry period of 8 months or more. At other locations such as Ibadan, Nigeria, and the Cauca Valley, Colombia, with about the same total rainfall (1 200 mm annually), the species tends to be perennial under a shorter dry period of about 4 months.

Rainfall intensity

In many parts of the tropics a high proportion of the rainfall descends in heavy storms of short duration. Intensities of 50–100 mm/h are commonly recorded for periods of 5–40 min. (Webster and Wilson, 1966). Occasionally, up to 250 mm/h falls for a short period of time. Intensity of rain in the tropics generally exceeds that in the temperate zone. For example, 22 and 25 per cent of the total amount may be released in heavy showers (100 mm or more/h) in Indonesia (Mohr and Van Baren, 1954) and in Uganda (Webster and Wilson, 1966). These figures exceed those in Central Europe by 10–15 times.

High rainfall intensity may have several detrimental effects:

1. On an open soil surface, or with poor vegetational cover, the impact of raindrops breaks up the soil aggregates, or crumb structure. The finer particles fill up the soil pores and reduce infiltration of the rainwater. Runoff is increased, causing serious erosion.
2. With newly established seedings, the excess water moves a sufficient amount of soil so as to cover the small seeds of grasses and legumes too deeply, or washes them from place.
3. Where the soil surface is well covered, there may be excessive runoff over the vegetation, with less percolation into the soil.
4. If seed production is important, severe losses may occur because of shattering caused by the heavy rainfall.
5. Fodder crops and plants grown for daily cutting and green feeding fall over under heavy downpours, especially with accompanying winds.
6. High daily rainfall masks the monthly total and may come at infrequent intervals, causing periods of severe water stress. On certain soil types this stress is noticeable unless rains fall each 5–6 days.

Rainfall reliability

Total rainfall fluctuates widely at a given locality. Frequently the beginning of the rainy season and the onset of the dry season are changeable and vary several weeks or even months. Thus, it becomes difficult to predict accurately or to anticipate rainfall patterns from year to year.

The reliability of annual rainfall can be calculated, however, by knowing the average annual rainfall and the range over a given number of years. Thus, it might be determined that once in 5 years a fluctuation of 200 mm from the mean might be expected. For the cattleman this suggests that he would need supplemental feeding or should reduce the stocking rate of his grazingland. Such information might also allow him to judge the optimum time of sowing, applying fertilizers or burning rangeland. The risk, of course, lies in which year to expect the deviation.

Effectiveness of precipitation

The percentage of rainwater made available for plant utilization is influenced by a number of variables:

Quantity of rainfall over a given period of time.

About 80 per cent of the rainwater reaches the soil surface under dense grass cover (Mohr and Van Baren, 1954). Light showers (4.0 mm or less) do not contribute materially to the supply of soil moisture. Some water, however, can be absorbed by the leaves and other plant parts. Droplets frequently form on the leaves and trickle toward the leaf base. As the quantity enlarges there is downward movement along the stem. In this way soil at the base of the plant is moistened to a much greater extent than the surrounding soil surface.

The time interval over which rain falls contributes to the disposition of the water. Whether 20 mm descends in 10 h or in a 10 min. downpour determines in part the amount of runoff or the quantity entering the soil.

Evapotranspiration

Water is lost to the atmosphere from the soil and plant surfaces through evaporation and is returned to the air via plant transpiration. Together they are called evapotranspiration (Thornthwaite, 1948). If rainfall is the same as potential evapotranspiration there is no water surplus or deficit. Thus, the climate is neither wet nor dry and perennial grasses and legumes in the tropics should grow continuously, provided other factors are favourable. As water becomes deficient, the evapotranspiration potential exceeds the moisture supply and the climate in more arid. Under such conditions species adaptation is different and plant growth is variable.

Surface runoff of rainwater

Water movement into the soil depends on the intensity of rainfall, amount of plant cover, surface litter and soil permeability (conduct of water downward into the soil profile). On well-established grazinglands little runoff occurs, thus benefiting the ground water available for plant utilization.

Drainage of rainwater

Whether drainage occurs depends primarily on the amount of rainfall within a given period of time and the internal structure of the soil at the lower levels of the soil profile. Free movement of water beyond the root zone means that less is available for plants. The roots of certain plant species, however, grow to greater depths during the dry season and are in a more favourable position to utilize the ground water.

Amount of stored water

The water-holding capacity of the soil is of great practical significance with regard to water available for plants. Soil texture, structure and depth are the most important factors affecting the capacity of the soil to hold and store water.

Humidity

Moisture in the atmosphere is usually expressed as relative humidity, i.e. the percentage of water vapour present given as a percentage of the amount which could be held at saturation (Wilsie, 1962). It is usually uniformly high in the tropical humid climate, but seasonal in the dry and humid mesothermal climates.

In the more humid regions the daily relative humidity may approach 100 per cent during the morning hours. This condition favours development of fungal diseases. It also hinders the formation and production of high-quality seeds. For instance, many of the tropical legumes such as *Glycine wightii*, *Macroptilium atropurpureum* and *Centrosema pubescens* form seed pods among the mass of herbage which remains green into or throughout the dry season. Generally, the pods are excessively wet in the early morning but dry out in the afternoon. Under such conditions seeds frequently do not develop completely and maturity is not uniform, making harvest difficult.

Temperature

The tropics and subtropics have temperatures that permit plant growth throughout the year except at higher elevations. Seasonal differences occur with greater ranges in the wet–dry and arid climates than in the equatorial humid regions. In the latter, temperatures rarely exceed 38–40 °C with monthly means varying no more than one or two degrees throughout the year. In the drier areas maxima of over 45 °C are common with monthly means and daily ranges varying 5–10 °C.

Many grass and legume species are adapted to a broad range of temperatures in the tropics and subtropics but grow more luxuriantly within an optimal range. For example, *Pennisetum purpureum* and *Panicum maximum* thrive well in the hot, lowland tropics but when taken to more northerly latitudes and higher altitides herbage yields are reduced. Species such as *Chloris gayana*, *Setaria anceps*, *Desmodium uncinatum* and *D. intortum* flourish at medium elevations where nights are cooler but are less productive at lower elevations. *Chloris gayana* plants develop stolons and a solid stand at medium elevations but where higher temperatures prevail, the plants of this species tend to grow in bunches and send out relatively few creeping stems.

Tropical grasses and legumes have relatively high chilling temperatures and do not tolerate freezing weather. *Pennisetum clandestinum*, which thrives best at the higher and cooler elevations, withstands frosting. The leaves are burned, turn brown and wither but the stolons are unaffected. *Glycine wightii* has some frost tolerance and can be extended into the subtropics. Growth of most species, however, is restricted under sustained cool temperatures.

Some species withstand higher temperatures than others, but this may be partly a matter of drought tolerance. The effects of high temperature and low relative humidity are commonly seen as manifestations of water stress. Temporary wilting frequently takes place during the heat of the day because of excessive transpiration but the condition is usually transitory unless soil water becomes limiting.

Light

Light is of basic importance as the source of energy for the photosynthetic process. The intensity and quality of light varies with the angle of the sun's rays, duration of the light period and atmospheric conditions (Wilsie, 1962). For example, water vapour absorbs light so that intensity is greater in arid than in humid belts; dust, gasses and smoke scatter the light and may absorb 20 per cent or more; clouds and fog reduce the total sunshine and light intensity. All of these factors alter plant development. The total radiation received per hour and per year is higher in the tropics than in the temperate zone. The amount received during the day, however, may be no greater, or even slightly less, because of the shorter light period. Thus, dry matter production per unit of time in the tropics may be less than in the temperate zone.

The duration of daylight (length of day) varies with latitude and season. At the equator day and night are equal, being 12 h the year around. At 17° latitude the longest possible day is 13 h and at 30° latitude just more than 14 h. Differences in day length regulate the time of flowering of some plants. That is to say, certain species are photosensitive (respond to photoperiod). Some require long days to induce floral formation (e.g. *Bromus inermis*), others short days (*Brachiaria mutica*), and some are insensitive (*Desmodium sandwicense* and selections of *Panicum maximum*).

Air movement

High temperatures near the equator give rise to upward movement of heated and moisture-laden air. Because of lowered pressure, air currents flow from the north and south to replace it, their direction being deflected eastward due to the rotation of the earth (Blumenstock, 1958). These are the north and south trade winds. They never reach the equator, but rise as a part of the equatorial upward circulation of air, leaving a belt of equatorial calm. In this belt, and in the zones north and south of it, the rains vary with the location and size of land masses and the direction of prevailing winds. Movements of air are determined by differences in pressures which are linked with temperature phenomena. The airflow patterns are also modified by friction produced by the earth's surfaces, especially mountain ranges.

The desert regions of the globe strongly modify the pattern of equatorial airflow. During certain periods of the year they become heated to high temperatures, producing updraughts of dry air and indraughts of winds from adjacent areas. In the Asian monsoon region the updraughts occur over the deserts of central Asia. This causes inland air movement from the southwest traversing the Indian Ocean from June to October (Pendelaborde, 1963). Its effect extends over a large expanse of sea and extends across the equator. As the winds move over the ocean the air picks up moisture, bringing on the monsoon season over the countries of southern and southeastern Asia. During the winter months the desert becomes cooler than the tropical areas. This gives rise to very dry winds blowing from the desert regions and again causes a reversal of the movement of the air mass. Eventually, as the season advances, the process is repeated and the monsoon sets in again.

In West Africa the annual rainfall cycle is similar to that of the Indian monsoon (Whyte, 1968). Since the airflow does not cross the equator, however, it is not considered a true monsoon. There are two opposing air masses: the equatorial maritime air blowing from the southwest and the tropical continental air from the northeast. Where the two meet an intertropical front develops and lies roughly east–west. The air masses move seasonally twice each year and are responsible for the seasonal distribution of rainfall. As the intertropical front moves northward in April and May line squalls occur, ushering in the rainy season. Near the Sahara the line squalls result in dry dust storms. Later in the season the process is reversed so that the front moves southward. This brings about a second rainfall in the southern areas. As the front traverses southward, winds move from the northeast in October and November, bringing the dry, desiccating winds, known as the *harmattan*. This season dries up plants and coats everything with a layer of dust particles moved in from the Sahara. The dust haze fluctuates in duration, extent and intensity. At times, it extends for some distance over the Gulf of Guinea.

Climate and vegetation

A close relationship exists between vegetation and climate as a consequence of plant evolvement and adaptation over ages of time.

Because of this interaction dominant natural vegetation groups have become associated with a particular climate (Blumenstock and Thornthwaite, 1941; Trewartha, 1954). Members of the plant formation making up the natural vegetation is referred to as the 'climax' formation. Thus, we associate rainforests and lush, dense vegetation with wet and humid climates; tall grass and deciduous trees with subhumid climates; short grasses and bush or shrubs with semi-arid climate; and drought-tolerant, tufted and wiry grasses and sparse thorn bushes with very arid climates. A climax plant formation is also called an association and consists of several species. Locally, however, in many areas it may be difficult or impossible to recognize a climax vegetation, especially where man has interfered over a period of time.

Classification of climates

Attempts at classifying climate did not begin with elements of the climate itself, but rather with vegetation (Critchfield, 1966; Trewartha, 1954–68). Since natural vegetation integrates the effects of the climate it can be taken as an index of climatic conditions. Climatologists and geographers started with major plant associations and tried to relate their formation with the effects of climatic factors. A difficulty in classifying climates, however, is the gradual merging of one type into another. Fixed boundaries do not exist in the atmosphere, nor between vegetational belts on the ground. Thus a line on the map represents a transition zone. In most instances it would be nearly impossible to recognize where one region ends and the other begins.

World climates by Köeppen

In 1900 the German biologist Wladimir Köeppen devised a classification based largely on vegetation zones but used climatic elements to delineate boundaries (Critchfield, 1966; Rumney, 1968; Trewartha, 1954–68). It was revised in 1918 with greater attention given to temperature, rainfall and their seasonal characteristics.

The Köeppen system included five major categories as follows:
A – Tropical forest climates, hot all seasons.
B – Dry climates, cool seasons.
C – Warm temperate rainy climates; mild winters.
D – Cold forest climates, severe winters.
E – Polar climates.

Capital letters represented the main climatic types with additional symbols for further subdivision. A second letter referred to the rainfall regime, the third to temperature characteristics and the fourth to special features of the climate. The symbols were combined into sets to show seasonal variations. By this means, a wide variety of climatic conditions could be expressed and mapped. The sets of symbols representing climatic types make the system relatively simple and easy to follow since each has a specific meaning. For example, B Sh indicates a tropical steppe, semi-arid and hot; Cwb a tropical upland, having a mild, dry winter and a short, warm summer. This system has been used as a model for most of the schemes of climatic classification.

Thornthwaite's classification

A climatic classification applicable to North America was published in 1931 and to the world in 1933. It followed Köeppen's system in that climatic boundaries were identified by values of temperature and rainfall, vegetation groups recognized and symbols used to designate climatic types. A departure was made by use of the expressions of 'precipitation effectiveness' (P-E), the ratio of monthly precipitation and monthly evaporation, and 'temperature effectiveness' (T-E), computed on the basis of monthly temperatures. A weakness of the system was the lack of data for evaporation at many weather-recording stations. Without this information mapping of moisture effectiveness became difficult.

A modification was introduced in 1948 with the concept of 'potential evaporation', that is, the amount of moisture which could be evaporated from the soil and transpired from plants if it were available (evapotranspiration). A moisture index showed the amount of water needed to meet the demands of evapotranspiration after deducting that supplied by rainfall and the stored soil moisture. The T-E index was also calculated in relation to the potential evaporation. Boundaries were determined by mathematical analyses of climatic data.

Vegetation was no longer the criterion for separation of major zones. A climatic map for the USA was drawn, but global maps based on this system are not available, due, in part, to the lack of adequate evapotranspiration data at sufficient numbers of weather stations so as to allow accurate calculations.

Trewartha's modified Köeppen system

One of the best-known revisions of Köeppen's classification is that of Trewartha (1954) who simplified the world climatic map and gave emphasis to cores of climatic regions. The map was redrawn in 1968 with minor changes in major types and subtypes. Six major types were described with a seventh group of undifferentiated highlands. The latter was not considered divisible by climatic interpretation. The seven types are briefly described below with only subtypes corresponding to the tropics and subtropics.

A. Tropical humid climate – constantly hot or warm, killing frost absent; rainfall rarely lower than 750–1000 mm; intense sunlight;

occupies about 20 per cent of the earth's land surface and 43 per cent the ocean surface.

Ar. Tropical wet (rainforest) – astride the equator and extending 5–10° latitude on either side; uniformly high temperatures; heavy rainfall for 10–12 months.

Aw. Tropical wet and dry (savanna) – extending beyond 20° latitude; rainfall lower, less well distributed and less

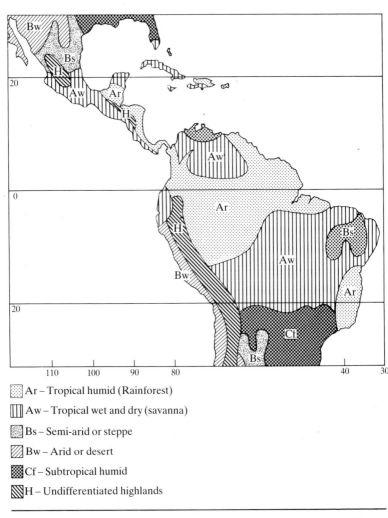

Ar – Tropical humid (Rainforest)

Aw – Tropical wet and dry (savanna)

Bs – Semi-arid or steppe

Bw – Arid or desert

Cf – Subtropical humid

H – Undifferentiated highlands

Fig. 1.1   Köeppen's world climates as modified by Trewartha (redrawn from Trewartha's 1954 and 1968 maps).

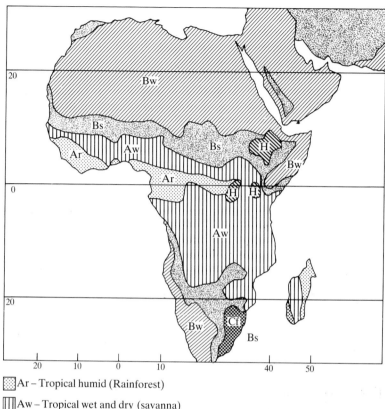

Ar – Tropical humid (Rainforest)

Aw – Tropical wet and dry (savanna)

Bs – Semi-arid or steppe

Bw – Arid or desert

Cf – Subtropical humid

H – Undifferentiated highlands

Fig. 1.1 (cont'd)

reliable than in Ar.; shorter wet and longer dry season, latter may extend to 9 months.

B. Dry climate – exclusively developed over land surface and boundary fixed by rainfall; annual water loss through evapotranspiration exceeds precipitation; extends from the equator through the tropics and into the middle latitudes; occupies about 26 per cent of continental area.

Bs. Semi-arid or steppe – rainfall meagre, highly variable and

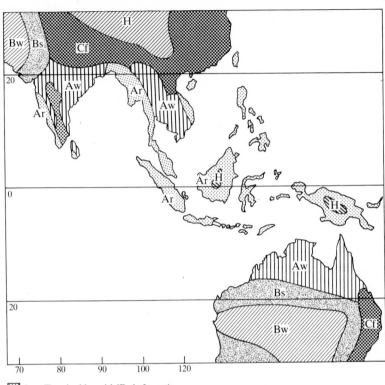

Fig. 1.1 (cont'd)

erratic; a transitional belt between the desert and humid tropics, boundary arbitrary.

Bw. Arid or desert – nearly rainless; precipitation in violent downpours over a limited area; hostile to life.

C. Subtropical climate – 8–12 months over 10 °C; occurring largely in the middle latitudes, except in tropical mountains; seasons distinguished as wet and dry since temperature contrasts are small.

Cs. Subtropical dry summer (Mediterranean) – seasonal rainfall of dry summers and humid winters; warm to hot summers; mild winters; occupies about 1.7 per cent of the earth's surface.

Cf. Subtropical humid – hot summers with rainfall, relatively mild winters but with frost and occasional snow; located on eastern side of a continent.

D. Temperate climate – mostly beyond 40° latitude; divided into milder oceanic and more severe continental climates.

E. Boreal (subarctic) climate – extending poleward beyond 50° latitude.

F. Polar climate – found in the higher latitudes of tundra and ice cap.

G. Highlands – scattered over all continents except Australia; elevation of dominant importance; identification of climatic types difficult.

Note that types are identified with capital letters and subtypes by small letters. In the 1954 maps, sub-subtypes were designated by an additional small letter, but in the 1968 map they were largely eliminated. The major climatic types and subtypes are illustrated in Fig. 1.1.

Seasonal climates of the earth

Troll (1965) separated the seasonal climate into (1) Polar and subpolar zones, (2) cool temperate zones, (3) warm temperate zones and (4) tropical zones. He then made subdivisions within each zone. This resulted in a more detailed map than that of Trewarth, but boundaries of major types were similar. General regions of tropical Africa and South America were separated on the basis of precipitation of the wet–dry periods as follows:

Humid months	Description	Arid months
12–9½	Belt of tropical rainforest and transitional woodland	0–2½
9½–7	Humid savanna belt	2½–5
7–4½	Dry savanna belt	5–7½
4½–2	Thorn savanna belt	7½–10
2–1	Semi-desert belt	10–11
1–0	Desert belt	11–12

This is a simplified classification satisfactory for the distinction of vegetational zones and has practical use for those interested in grassland husbandry and species adaptation.

Holdridge life zone system

In 1947 Holdridge proposed a system whereby the world plant formations could be determined from simple climatic data of temperature, precipitation and evaporation. A later modification divided the earth into more than 100 life zones, or groupings of natural vegetation associations (Holdridge, 1967; Holdridge *et al.*, 1971). These were arranged by latitudinal regions, altitudinal belts and humidity provinces. The latitudinal regions extend from the tropical belt of the equator and proceed poleward through the subtropics and warm temperate regions to the polar caps. Similar climatic boundaries also occur in the tropics where they are separated by elevations. The tropical boundary, whether separated by latitude or altitude, is fixed by a 24 °C limit. The subtropical and warm temperate zones lie between 12 °C and 24 °C. A critical temperature of about 12 °C prevails at the frost line.

Humidity effects are separated into provinces as divisional units and cover moisture conditions, their boundaries being based on annual precipitation and potential evapotranspiration. They range from semi-parched and superarid (receiving 250 to 500 mm annual rainfall), humid (about 4 000 mm rainfall) to superhumid and semi-saturated (more than 8 000 mm rainfall). Deserts are found at the dry end of the scale where annual rainfall is never sufficient to meet the potential deficit. Rainforests predominate in the more humid regions and available moisture exceeds evapotranspiration throughout the year.

The life zones are diagrammatically arranged in a triangle divided into hexagonals representing vegetational zones. These zones can be calculated from records of nearby weather stations or from field observations of the vegetational physiognomy (nature of its appearance) with little or no supporting climatic data (Holdridge *et al.*, 1971).

The Holdridge system appears to be complicated and greatly detailed at first sight. Once a map is prepared however, it can be easily followed. A prominent feature of the classification is that the life zones can be localized. This means that a 'tropical dry forest, moist province transition' shown on a map of Colombia would be comparable to the same classification in the Ivory Coast. Even though species are not used in the derivation of life zones, their adaptation should be similar in the two localities. This has great implication and significance to the agrostologist for the acquisition of grasses and legumes for assessment.

Vegetational maps

Since the climatic classifications are based on vegetation associations, they are essentially bioclimatic by nature (relationship of climate and living matter). Frequently, maps prepared to show such classifications are referred to as vegetation or bioclimatic maps.

Such a map has been compiled for Africa by Keay (1959) to show the effect of rainfall on vegetation associations (Fig. 1.2).

Rainfall along the coastal area of West Africa exceeds 2 500 mm annually, except for one region near the median of 0° longitude (Church, 1968). The total declines steadily as one moves inland and northward to the Sahara. The changing rainfall patterns coincide fairly well with boundaries of the vegetation zones, but transitional belts correspond to the length of the dry season.

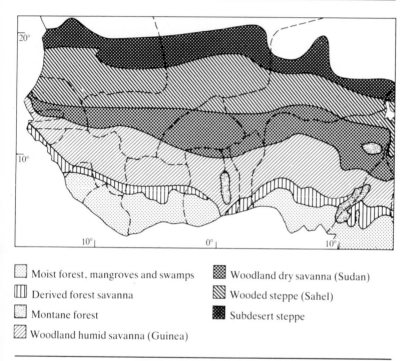

⬚ Moist forest, mangroves and swamps	▨ Woodland dry savanna (Sudan)
⠿ Derived forest savanna	▧ Wooded steppe (Sahel)
⬚ Montane forest	▩ Subdesert steppe
▨ Woodland humid savanna (Guinea)	

Fig. 1.2   Vegetational zones of West Africa (Keay, 1959): (1)

The vegetation map (Fig. 1.2) and the rainfall map (Fig. 1.3) of West Africa are closely related. Near the coast there is no distinguishable dry period and the region is classified as tropical rainforest and swamps. Over much of the zone, two peaks of rainfall alternate with two dry seasons. A longer drought period prevails from October or November to March or April and a shorter one occurs in July or August.

The forests are broken vertically with two or three layers of trees, the tallest storey emerging at more than 30 m. Tall, coarse grasses appear in the more open lands of the forests. A semi-

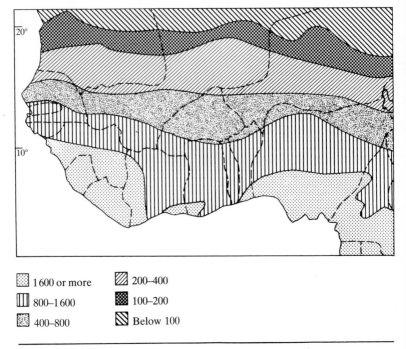

Fig. 1.3 Rainfall distribution (mm/year) in West Africa (Thompson, 1965).

deciduous forest exists at the outer fringes of the rainforests and comprise the 'derived savanna' where the trees have been cleared for cultivation and where the dry season ranges from 3–5 months.

The derived savanna merges into two 'woodland savannas' which are separated on the basis of moisture and vegetational cover. The 'humid savanna' is characterized by 5–7 months of dry season, usually continuous. Rainfall varies from about 1 000 to just less than 1 500 mm. The area is largely woodland with fire-resistant, broad-leaved, deciduous trees. The canopy may be full or open at 15 to 20 m. Shrubs occur as undergrowth but are dominant in some localities. Tall perennial, tufted grasses grow up to 3 m beneath the scattered trees and up to 5 m open places. Since the cattle population is relatively sparse, a heavy growth of grass accumulates by the end of the rainy season. During the dry season the herbage becomes very parched. The land surface of the entire zone is burned each season and widespread fires rage beyond control.

The 'dry woodland savanna' receives from 500 to 800 mm annual rainfall with 7–9 months having less than 100 mm total. The region is wooded but many single trees occur and display wide, spreading crowns and small leaves. The trees grow from 10 to 15 m in height and are shorter than in the humid savanna. There are many lower-growing shrubs and bushes in the southern areas. Thorn bush is prevalent in the northern part of the dry savanna. Grass cover is shorter than in the humid savanna, from 1.5 to just over 3.0 m in height when mature, less tufted, more feathery with finer leaves and stems, and fewer perennials. Much of the area is burned annually, but fires are less severe than in the derived savanna.

In the 'wooded steppe' a water deficit exists for most of the year and many areas receive less than 200 mm of rainfall. The rain usually occurs in downpours scattered over a 2 or 3 month period. The original climax was probably thorn woodland. This has opened up with scattered dwarfed trees of 5–10 m height. They are deciduous with deeply penetrating and spreading roots. The many acacias are characteristic of the region. Thorn shrubs of 2–3 m height with short conical bases and divided stems are common. Grasses are short, discontinuous, wiry and tufted. In some areas the land surface is barren. Fires are less serious than further south. There is not the density or cover of trees and no great accumulation of grass for excessive burning.

The southern Sahara is fringed with a 'subdesert steppe'. In some places a dispersed, permanent vegetation prevails, being composed of small shrubby plants and bushes, with acacias, other trees and shrubs. These are usually more dwarfed than in the wooded steppe. Rainfall seldom exceeds 150 mm and is extremely unreliable. After summer rains annual grasses and herbs appear and soon mature. Grasses grow as isolated tufts. A few plants are sometimes maintained by winter dew. Altitude modifies the vegetation due to increased humidity and cloudiness, lower temperatures and less evaporation. In the Guinea and Cameroon highlands a tall, canopied, montane forest exists, with woodland at the higher elevations where less precipitation occurs. The latter is easily destroyed by fire and gives rise to grassland.

Relative humidity has a marked effect on the vegetation association in the different regions. The coastal areas have a mean monthly relative humidity of 95 per cent at 06.00. It may drop below 60 per cent at noon in the driest months. In the north the morning relative humidity climbs to 90 per cent during the rainy season. In the dry season it seldom reaches 30 per cent but drops to less than 10 per cent before noon.

A temperature gradient extends from the coastal forest zone to the Sahara. Temperature lines run east and west, as do the vegetational zones. The gradient effects on vegetational associations is less visible than those of rainfall. A gentle rise in elevation occurs from south to north. This also causes a change in temperature and has some influence on vegetation type.

The vegetation zones in West Africa are so paramount that some have been given distinctive names: Guinea savanna = humid savanna; Sudan = dry savanna; Sahel = wooded steppe. Agriculture and cattle-keeping are fairly well defined within each zone. In travelling from one zone to another it would be difficult to recognize boundaries because of the broad transitional belts. This is not true, however, for the division between the semi-arid rainforest and the derived woodland savanna. This man-made line is one of the earth's clearest marked vegetational boundaries.

Other detailed information on climatic classification in the tropics (Blumenstock, 1958; Budowski, 1968; Carter, 1954; Garbell, 1947; Howe, 1953; Kendrew, 1953) and vegetation maps (Phillips, 1959; Van Steenis, 1958; Williams, 1955) are available for continents and regions. None demonstrate as vividly, however, the close relationship of water and vegetational association as that of West Africa.

Altitude and vegetation

Altitude duplicates, in some respects, the effects of latitude on temperature and changes in vegetation. Temperature decreases at a rate of about 0.3 °C for every 300 m elevation (3.3 °F per 1 000 ft). Precipitation usually increases with altitude. At higher elevations, the lower temperatures are more likely to affect plant growth than lack of moisture.

In the Andean zone of tropical Latin America, four vertical vegetational zones are commonly recognized in the highlands. From sea level to about 1 000 m is the *tierra caliente* (hot land) which has a tropical lowland type of vegetation. Within this zone is a preponderance of the tropical grasses and legumes. The *tierra templada* (temperate land) rises to about 2 000 m and embraces a subtropical vegetation. Mean temperatures range between 18.3 and 24 °C (65–75 °F). The *tierra fria* (cold land, but cool in many places) extends from 2 000 to about 3 000 m. The upper limit closely coincides with the range of cultivated crops. Cool-season grasses and legumes are adapted to this altitudinal zone. At the higher levels many grasses are short, tufted and wiry. Above 3 000 m, and extending to permanent frost (snow on high mountains), is the *tierra helada* (frost or frigid land). In the lower areas of this region are alpine meadows. Trees are dwarfed and frequently gnarled. Low-growing bushes and herbs extend beyond the tree line, but eventually give way to bare ground.

Tropical highlands in other parts of the earth contain vegetational zones somewhat similar to those of Latin America. Vegetation associations are highly variable, as the altitudinal belts are composed of many microclimates. The types of vegetation are strongly influenced by slope, exposure, soil condition (especially drainage), wind, cloud cover and other mountain effects.

Climatic zones, grasses and legumes

Tropical grasses and legumes usually exhibit a wide range of adaptability, grass species usually more so than the legumes. It is not unusual to find that species thrive from the lowland equatorial regions, across several latitudes and beyond the tropics of Cancer and Capricorn. A similar extent of adaptation can be found for the altitudinal zones. For example, *Panicum maximum*, *Paspalum notatum*, *Melinis minutiflora* and *Cenchrus ciliaris* flourish in the tropical wet climate. They also push beyond the 30° latitude line in the tropical wet and dry climates and into the humid subtropical climate. Others, such as *Cynodon dactylon*, *Hyparrhenia rufa*, *H. subplumosa* and *Andropogon gayanus* prosper in the forest zones of West Africa and extend across the Guinea and Sudan savannas (semi-arid climate and steppe). Among the legumes, *Glycine wightii*, *Stylosanthes guianensis*, *Macroptilium atropurpureum* and *Leucaena leucocephala* show broad adaptation.

Soils

The varying climates in the tropics have profoundly affected development of soils and their differentiation into complex patterns. Parent materials of many tropical soils are older than in the temperate zones, so that mineral reserves are small for additional weathering (Imp. Bur. Soil Sci., 1934). Leaching and weathering patterns are also different in the tropics. Constantly high temperatures accelerate microbiological activities and increase chemical processes under conditions of favourable moisture. With excessive rainfall heavy leaching takes place. In regions with high rainfall and long rainy periods, the amount of leaching becomes more severe than in regions with less rain and extended dry periods. During the prolonged dry season, the vegetation removes all or most of the water within the root zone. Thus, the next season begins with little or no reserve of moisture. When the rains begin the regeneration of chemical processes, the renewal of microbial activities and the development of plant roots are different than under conditions of more favourable moisture supply through the year. These differences result in variability of soils from locality to locality. In fact, extreme differences can be encountered within a distance of a few metres as brought on by micro-environmental effects.

Classification

A number of classification systems have been developed for soils of the world and for intertropical application (Ahn, 1970; Aubert and Tavenier, 1972; d'Hoore, 1968). The principles of classification embody the characteristics of the soil, the processes and regimes under which they evolved and the factors which influenced their development. Five major factors largely control the kinds of soils that developed (1) climate (particularly precipitation and temperature), (2) living organisms (especially native vegetation), (3) nature of parent material (texture and structure, chemical and mineral composition), (4) topography of the land and (5) time over which parent material has weathered.

A most striking characteristic of many tropical soils is their red colour, due primarily to a high degree of oxidation of iron compounds into ferric oxides. In addition, silicates are frequently split into their components parts and the silica is leached, leaving an excess of free alumina and ferric oxide in the soil. This further intensifies the red colour. Tropical soils for the most part do not show the distinctive horizontal zones which are typical of soil profiles in the temperature regions. The lack of well-defined horizons is related to intensity and distribution of rainfall, excessive leaching, and rapid decomposition of organic matter. Types of vegetational cover are related to soil development in both the tropics and humid temperate regions. In the tropics the living plant is the more directly active soil-forming agent, but in the temperate climates the plants affect soil development mainly after death and during subsequent humification. Plant residues in the tropics decompose so rapidly that they do not persist in the soil in sufficient quantities and for the length of time to affect soil formation greatly.

Practical application of soil survey for classification

Research with pasture improvement, grazinglands and range management in most tropical and subtropical countries is related to overall land development. The regional approach of the Commonwealth Scientific and Industrial Research Organization (CSIRO) to obtain data and information for the development of grazinglands in northeastern Australia provides an excellent example of an integrated team accomplishment (Davies and Shaw, 1964). It also demonstrates how the soil survey was carried out for a practical agriculture purpose.

The area of northern Queensland is partly covered with open eucalyptus forest having a ground cover of semi-natural grasses and soils of medium fertility. Cattle density is low with one animal unit per 10–20 ha. Large regions with a potential for animal production were identified for survey. Teams consisting of an ecologist, soil surveyor, geologist and agrostologist covered a given region to define the natural boundaries, to describe the pattern of landscape and to map the land, using a system which included the major features of climate, geology, soils and vegetation (Coaldrake, 1964; Isbell and McCown, 1976; Moore and Russell, 1976). The soil survey was aimed at assessing the suitability of the soils for growing improved grasses and legumes (Hubbel *et al.*, 1964). A reconnaissance survey was made to determine the nature and distribution of taxonomic units of the great soil groups and subgroups. In defining the units, attention was given to major differences in soil morphology, parent material and the degree of development of profile features. If little factual information was available, the surveyor noted the general topography, nature of the vegetation (whether mostly grassland, woodland or forest), general soil characteristics such as sandy, clayey or stony, predominant colour and kind of soil, obvious geological features as granite, basalt or sandstone, and the general forms and intensities of land use. More precision was given to the reconnaissance survey by studying several representative areas in detail. Soil types were characterized and mapped as individual units or as complexes of two or more units. For subsequent interpretation various soil types were grouped into land classes. Sample pits dug at representative sites of the main soil types allowed for more detailed study of profiles and chemical properties.

As a complement to the chemical analyses, experiments were carried out to assess the nutrient status of the soils by growing *Macroptilium lathyroides* (syn. *Phaseolus lathyroides*) in pots under controlled glasshouse conditions. A combination of elements (nitrogen (N) phosphorous (P), potassium (K), calcium (Ca), etc.) was used as a control and the response to a given element measured by withholding it from the appropriate treatment. These data indicated nutrient deficiencies and the need for further laboratory or field trials. Based on the landscape and land system maps, the soil survey maps and data from the laboratory analysis and pot experiments, the agrostologist established grass and legume introduction nurseries for evaluation at representative sites, along with appropriate fertility trials.

Land capability classification

The taxonomic classification systems, based partly on soil genesis (profile development and horizon sequence or weathering stages of minerals), reflect to some extent the agricultural value of the soil. For more precise recommendations, however, it is useful to employ additional or alternative classifications based on local needs (Ahn,

1970; d'Hoore, 1968; Gwynne, 1977). A simple classification could be based on soil texture, moisture storage capacity or drainage characteristics. A farm map showing this type of classification would be of far more use to a farmer than one showing whether the soils are Oxisols or Inseptisols. The 'land capability' classification devised in the USA illustrates such a system (Steele, 1951). Eight broad classes of soils are separated on the basis of factors which limit their use, such as susceptibility to erosion, factors which interfere with cultivation (e.g. stoniness), slope gradient, natural soil productivity and climatic limitations. Classes I–IV describe soils suitable for cultivated crops; Classes V–VII soils suitable for permanent vegetation as tree crops, forestry and grazinglands; and Class VIII rough or steep land, extremely sandy, wet or arid that might have value for wildlife or recreation purposes. The soil suitability classification is useful in regions where agriculture is to be intensified, where new crops are to be introduced or where new land is to be opened up for development. It is particularly helpful in delineating areas for development of potential grazinglands. In tropical regions the basic agricultural experimentation, especially on lands for development, is often inadequate for predicting soil behaviour. Suitability maps in the tropics have been prepared for parts of Central and South America, Africa and Southeast Asia.

Organic matter and humus

Organic matter in the soil is important because (1) it is a source of nutrients for plants and for micro-organisms, (2) it helps to reduce the surface runoff of water, (3) it increases the infiltration rate, (4) it improves the water-holding capacity, (5) it increases cation-exchange capacity, and (6) it improves soil structure (Ahn, 1970; Buckman and Brady, 1969; Nye and Greenland, 1960; Russell, 1961).

The primary source of organic matter is plant tissue: leaves, twigs, branches, stems, roots and fruits. In the forested zone these are derived from trees, vines and the understorey growth of bushes, shrubs and herbs. In the natural and semi-natural grasslands and savannas the residues come largely from the grasses and herbs, but also from bushes, shrubs and trees depending on the type of vegetational cover. In grazinglands animal droppings contribute to the organic matter but in limited amount, unless the pastures are intensively managed. Animals are considered a secondary source of organic matter. As they consume the plant parts and residues, they contribute waste products and leave their own bodies when the life cycle is completed. The lower plant forms of fungi, actinomyces and bacteria also contribute to the organic matter complex.

In the forest zones of Africa the production of plant residue is about 15 000 kg/ha annually (Nye and Greenland, 1960). Something like 10 000 kg comes from the litter (leaves, branches, decaying trunks) and the remainder from dead roots, root slough and exudates. In the savanna regions the growth of grasses is dependent on rainfall. In the high grass Andropogoneae region seasonal yields of about 8 000 kg of dry leaves and stems can be expected. Unfortunately, much of this is lost because of annual burn. Root material adds something like one-third of the aerial growth. Thus, it is estimated that the grassland savanna containing bush provides around 2 700 kg of dry material annually as a source of organic matter.

Plant residues deposited on the soil surface and those left in the soil medium are readily attacked by insects, fungi and other organisms. There is a rapid increased of the microbial population which is synthesizing and decomposing at the same time. The organic matter is converted to carbon dioxide and a great variety of substances, ranging from the lignins and cellulose to microbial cells. At some stages of decomposition the microbial tissue may account for as much as one-half of the organic fraction in the soil. These organisms are, of course, subject to decay and attack by other living organisms. As the raw organic supply diminishes, microbial activity declines, causing a release of nitrates and sulphates. The organic residues from all sources is a dark, incoherent and heterogeneous colloidal mass referred to as 'humus' (Ahn, 1970; Buckman and Brady, 1969; Russell, 1961; Vine, 1968).

Humus is not a stable end-product, but remains in the soil in variable amounts, continuously being decomposed and supplemented. It is the main source of mineral N for plant growth and supplies a proportion of the phosphate, sulphate and micro-nutrients.

The amount of humus in a soil depends on the balance of its rates of formation and decomposition. The quantity of plant residue, of course, is of prime importance, as previously discussed. Rainfall has a direct effect on the type of vegetation and on the production of aerial and root growth, and thus effects the quantity of organic matter available for formation of humus.

Higher temperatures favour an increased rate of organic matter decomposition, but under conditions of equal rainfall the rate of accumulated humus does not change (Ochse *et al.*, 1961; Vine, 1968; Webster and Wilson, 1966). A good forest soil contains from 5–7 per cent of humus. Savanna soils generally consist of 1–3 per cent humus. The percentage increases in soils at higher elevations (up to 15 and 20%) due to the reduced rate of decomposition. When soils are brought under cultivation, humus content can be markedly decreased until a new equilibrium is reached. During the cropping period the quantity of fresh organic matter received by the soil is decreased, and the rate of mineralization is increased because of exposure to climatic elements.

References

Ahn, P. M. (1970) *West African Soils*, Vol. I, Oxford Univ. Press: London, p. 331.

Aubert, G. and **R. Tavenier** (1972) Soil survey. In *Soils of the Humid Tropics*, National Academy Sciences: Washington, D.C., pp. 17–44. III. map, Soils of Tropics.

Blumenstock, D. I. (1958) Distribution and characteristics of tropical climates, *Proc. 9th Pacif. Sci. Cong.* **20**, 3–21.

Blumenstock, D. I. and **C. W. Thornthwaite** (1941) Climate and the world pattern. In *Climate and Man, Yearbook of Agriculture*. U.S. Dept. Agr., pp. 98–127.

Buckman, H. O. and **N. C. Brady** (1969) *The Nature and Property of Soils*, Macmillan: New York.

Budowski, G. (1968) Climatological data and natural vegetation. In *Agri-coclimatological Methods* (*Natural Resources Series*, VII), UNESCO: Paris.

Carter, D. B. (1954) Climates of Africa and India according to Thornthwaite's 1948 classification, *Pub. in Climatology* **7**, 453–74, Johns Hopkins Univ.

Church, R. J. (1968) *West Africa. A Study of the Environment and of Man's Use of it* (6th edn), Longman: London, Ch. 3 and 4.

Coaldrake, J. E. (1964) Plant ecology. In *Some Concepts and Methods in Subtropical Pasture Research*, Commonw. Bur. Past. Fld Crops, Bull. 47, Hurley, England, Ch. 4.

Critchfield, H. J. (1966) *General Climatology* (2nd edn), Prentice Hall: New Jersey, Ch. 6 and 10.

Davies, W. (1960) Temperate and tropical grasslands, *Proc. 8th Int. Grassld Congr.*, pp. 1–7.

Davies, J. G. and **N. H. Shaw** (1964) General objectives and concepts. In *Some Concepts and Methods in Subtropical Pasture Research*, Commonw. Bur. Past. Fld Crops, Bull. 47, Hurley, England, Ch. 1.

d'Hoore, J. L. (1968) The classification of tropical soils. In R. P. Moss (ed.) *The Soil Resources of Tropical Africa*, Cambridge Univ. Press: London, Ch. 1.

Garbell, M. A. (1947) *Tropical and Equatorial Meteorology*, Pitman: London.

Gwynne, M. D. (1977) Some ecological criteria for land evaluation of different types of rangeland, *Misc. Soil Paper, Kenya Soil Survey No. N-11*, Min. Agric.: Nairobi, pp. 49–60.

Holdridge, L. R. (1967) *Life Zone Ecology*, Tropical Science Center: San José, Costa Rica. 206 pp.

Holdridge, L. R., W. C. Grenke, W. H. Hatheway, T. Liang and **G. A. Tosi Jr** (1971) *Forest Environments in Tropical Life Zones*, Pergamon Press: New York, Ch. 1 and 2.

Howe, G. M. (1953) The climates of Rhodesia and Nyasaland according to Thornthwaite's classification, *Geogr. Rev.* **43**, 525.

Hubbel, G. D., A. E. Martin, R. F. Isbell, R. S. Beckwith and **G. B. Stirk** (1964) Soil survey and soil assessment. In *Some Concepts and Methods of Subtropical Pasture Research*, Commonw. Bur. Past. Fld Crops, Bull. 47, Hurley, England, Ch. 5.

Imperial Bureau Soil Science (1934) *Soil, Vegetation and Climate*, Tech. Comm. No. 29, Imp. Bur. Soil Sci., England.

Isbell, R. F. and **R. L. McCown** (1976) Land. In N. H. Shaw and W. W. Bryan (eds) *Tropical Pasture Research*, Commonw. Bur. Past. Fld Crops, Bull. 51, Hurley, England, Ch. 3.

Keay, R. W. J. (1959) *Vegetation Map of Africa South of the Tropic of Cancer*, Oxford Univ. Press: London.

Kendrew, W. G. (1953) *The Climates of the Continents*, Oxford Univ. Press: London.

Mannetje, L. 't (1978) The role of improved pastures for beef production in the tropics, *Trop. Grassld* **12**, 1–9.

Mohr, E. C. J. and **F. A. Van Baren** (1954) *Tropical Soils: A Critical Study of Soil Genesis as Related to Climate, Rock and Vegetation*, Royal Tropical Institute: Amsterdam.

Moore, A. W. and **J. D. Russell** (1976) Climate. In N. H. Shaw and W. W. Bryan (eds), *Tropical Pasture Research*, Commonw. Bur. Past. Fld Crops, Bull. 51, Hurley, England, Ch. 2.

Nye, P. H. and **D. J. Greenland** (1960) *The Soil Under Shifting Cultivation*, Tech. Comm. No. 51, Commonw. Bur. Soils, England.

Ochse, J. J., M. J. Soule, Jr., M. J. Dijkman and **C. Wehlburg** (1961) *Tropical and Subtropical Agriculture*, vol. I, Macmillan: New York, Ch. 1 and 2.

Pendelaborde, P. (1963) *The Monsoon*, Transl. by M. J. Clegg. Methuen and Co.: London.

Phillips, J. (1959) *Agriculture and Ecology in Africa*, Faber and Faber: London.

Riehl, H. (1954) *Tropical Meteorology*, McGraw-Hill: New York.

Rumney, G. R. (1968) *Climatology and the World's Climates*, Macmillan: New York.

Russell, E. W. (1961) *Soil Conditions and Plant Growth* (9th edn), Longman: London.

Steele, J. G. (1951) *The Measure of our Land*, Soil Cons. Serv., Govt. Print. Off.: Washington, D.C.

Thompson, B. W. (1965) *The Climates of Africa*, Oxford Univ. Press: London.

Thornthwaite, C. W. (1948) An approach toward a rational classification climate, *Geogr. Rev.* **38**, 55–94.

Trewartha, G. T. (1954–68) *An Introduction to Climate*, McGraw-Hill: New York.

Troll, C. (1965) Seasonal climates of the earth. In E. Rodenwaldt and H. J. Justaz (eds.), *World Maps of Climatology*, Springer-Verlag: New York. (With map.)

Van Steenis, C. G. J. (1958) *Vegetation Map of Malaysia*, UNESCO: Paris.

Vine, H. (1968) Developments in the study of soils and shifting agriculture in tropical Africa. In R. P. Moss (ed.), *The Soil Resources of Tropical Africa*, Cambridge Univ. Press: London, Ch. 5.

Webster, C. C. and P. N. Wilson (1966) *Agriculture in the Tropics*, Longman: London, Ch. 1, 2, 3.

Whyte, R. O. (1968) *Grasslands of the Monsoon*, Faber and Faber: London, Ch. 2.

Williams, R. J. (1955) Vegetation regions. In *Atlas of Australian Resources*, Canberra, Commw. Dept. Natl Dev.

Wilsie, C. P. (1962) *Crop Adaptation and Distribution*, W. H. Freeman: San Francisco.

Chapter 2

Grass, grazers and man

The meaning of grass in a broad sense goes beyond the limits of the family name. It sometimes refers to any plant constituent used as roughage for livestock, including members of the legume family.

The term grass is also applied to the general vegetational cover of grasslands. Grasses proper may comprise the dominant component, but usually occur in a mixture with other herbs, shrubs and trees.

Origin and movement of grasses

The origin of grass as a distinct plant form has not been precisely dated. Fossil records indicate that the grasses emerged as a part of the Angiosperms about 130 million years ago during the Cretaceous period (Barnard and Frankel, 1964). It is believed that bamboo-like plants appeared during this time. Separation of different types occurred within the Tertiary era (Semple, 1970). Thus, speciation began some 70 million years ago and has remained in continuous progress. Prior to the Miocene era, dating some 20 million years ago, grasses became important components of the earth's vegetation. Their movement accelerated with action of the Pleistocene ice age, about 1 million years ago. They spread into every continent of the globe and evolved as the third largest family of the plant kingdom (Barnard and Frankel, 1964), after the composites and orchids. The wider dispersal of grasses favoured intermingling and intercrossing of types, creating greater variation. Movement into diverse environments led to further differentiation and refinement of species. Within isolated valleys and on mountain slopes, variants were naturally selected as ecotypes.

In recent times man recognized the variability and similarity of plant types and devised systems of classification. It was observed that distributional patterns followed climate relationships, but regional groupings of genera and species were apparent. Within the territorial limitation of recognized types, it was assumed they were indigenous or endemic to that area. Among the cultivated grasses, three main regions of origin were important: Eurasia, East Africa and subtropical South America (Hartley and Williams, 1956). The temperate species arose in Central Eurasia and the Mediterranean. Important tropical grasses with African origin include species of *Andropogon, Brachiaria, Cenchrus, Chloris, Cynodon, Dichanthium, Digitaria, Eragrostis, Hyparrhenia, Melinis, Panicum, Pennisetum, Setaria, Sorghum,* and *Urochloa*; those with South American origin *Axonopus, Paspalum, Tripsacum, Zea.*

Origin of legumes

Fossil records indicate that legumes were distinct plant types in the upper Cretaceous period. Their evolution apparently followed a process similar to grasses (Semple, 1970), the basal form emerging as a tropical tree. Structural modifications led to the evolution of shrubs, woody climbers, perennial herbs and finally annual herbs. They were widely dispersed as the earth's surface changed with the uplifting of mountain ranges and as climate altered environments. Natural or semi-natural legumelands have not developed as have grasslands. Leguminous trees and shrubs exist as the predominant vegetational cover in many of the African savannas, but they are not called legumelands. Herbaceous types seldom occur as dominant species *en masse* as do the grasses. They are associated with grasses and other herbs in grasslands, but their presence seldom exceeds 10 per cent of the flora.

If we consider cultivated legumes for animal utilization, more attention has been given to the temperate species of *Medicago, Trifolium, Vicia* and *Melilotus* than to tropical species. These cool season species originated in the Near-Eastern region of Turkestan or present-day Afghanistan, southern Russia and Turkey and parts of Iran. It is only within the past two decades that tropical forage legumes have begun to assume importance as pasture and forage crops. Those of African origin include *Glycine, Vigna, Indigofera, Dolichos* and *Alysicarpus*. The greater majority are indigenous of tropical America: *Calopogonium, Centrosema, Desmodium, Leucaena, Phaseolus, Stylosanthes, Teramnus.*

Origin of grazing mammals

The history of animal evolution is more complete than that of grasses and legumes. Fossil records show that a group of carnivorous and omnivorous mammals existed in all continents except Australia during the early Tertiary era (Simpson, 1950). These gave rise to the ungulate (hoofed) orders of herbivorous mammals: the perissodactyls (e.g. horses, rhinoceros and tapirs, having odd numbers of toes or toes unevenly disposed in relation to the axis of the foot), the non-ruminant artiodactyls (e.g. cattle, buffalo, sheep, goats, deer and antelopes – the artiodactyls have even numbers of functional toes). Non-ruminant mammals emerged in early Tertiary times. Ruminants appeared later during the Miocene, some 20 million years ago.

The early ungulates were browsing rather than grazing animals. The change to grazing is best recorded in the evolution of the horse (Barnard and Frankel, 1964). The ancestor of the present-day horse was small, with five toes and lived in Europe and America. Through evolution during the Tertiary period the body became larger, hoofs developed and the mouth changed. The lips became more mobile and the teeth higher-crowned (Simpson, 1950). These characteristics were more suited to grazing grass.

The present types of domestic cattle apparently arose from one ancestral strain, *Bos primigenis* Bojanus (wild urus) (Issac, 1962). This beast ranged through Asia, Europe, Eurasia and into North Africa from the Pleistocene era to the seventeenth century. Some authorities point to North Africa as being a centre of origin. Zoological studies indicate, however, that *B. nomadicus* Falconer and Cautley (relics found in Asia) and *B. opisthonomus* Ponel (relics found in North Africa) were the same animal and a descendant of the wild urus. *Bos taurus longiforms* appeared at a later date and seems to have been the first bovine type economically exploited. It emerged in Mesopotamia. *Bos indicus* probably evolved from *B. nomadicus* in south central Asia.

Interaction of grass and grazers

As grasses spread over the earth's surface, vast areas of grasslands developed where soil and climate were favourable. Today, grasslands cover approximately 45.9 million km^2, and make up about 24 per cent of the total vegetational cover (Shantz, 1954). The larger natural and semi-natural grasslands of the steppes of Asia, plains of North America, pampas of South America and velds of Africa are believed to have existed from ancient times (Barnard and Frankel, 1964). The North American plains and prairies probably date back to the Miocene period (Weaver, 1954). Some ancient grasslands formed in Europe. The Philippines (Mindanao), Southeast Asia (Indonesia) and areas of West Africa are of more recent origin. They emerged after the destruction of forests by cutting and firing and have persisted through the interference of man.

Grasslands developed as edaphic, climatic and biotic conditions permitted. Usually they were established in interaction with browsing and grazing animals, and during the past 20 million years grasses and animals have evolved together with reciprocal interacting factors altering the natural selection of both. Structural changes in mammals modified their feeding habits on herbs and grasses. This, in turn, influenced the evolutionary development of morphological features of the grasses. The trampling of animals must have had some effect on modifying characteristics. The short basal internodes, basal tillering and the development of rhizomes of present-day grasses provide some resistance to trampling and tolerance to close grazing. The tufted, prostrate and creeping growth forms help in withstanding grazing. Regrowth from basal buds and new stem shoots from rhizomes and stolons support the regeneration of herbage after grazing and cutting. Expansion of leaves from the basal meristem is a feature suitable for grazing. This assures continued growth, especially in the vegetative period of development. The fact that stem apices remain near the ground for some period of time, and are enclosed by the rolled or folded leaf tips, reduces damage from close cropping and trampling. Rapid stem elongation and short flowering and seeding periods reduce the time of exposure to the grazing animal. Appendages of many seeds provide easy transport for distribution by the grazer. Ingestion of seeds by the grazing animal is another means of spread. Furthermore, passage of seeds through the digestive tract reduces the hard seed coat which is characteristic of some species. In this way germination is enhanced.

Of some 10 000 species of grass, approximately 40 are actually sown for pastures and fodder crops (Hartley and Williams, 1956). The great majority of these are indigenous to Eurasia and Africa and have evolved under grazing conditions of the bovine.

Influence of man

Grass and grazers were around for several million years before the appearance and interference of man. Man's separation from other primates by the use of articulate sounds, stone tools and fire has not been specifically dated (Table 2.1). It may have taken place during the Pleistocene period, probably 1 million years ago (Stewart, 1956). Early man gathered food from wild plants. Grasses became a part of his sustenance. His diet was supplemented by hunting and eating the browsing and grazing animals. Thus began the interaction of grass, grazers and man.

The very presence of man prompted modification of both grass and grazers. But the use of fire brought about more notable and significant changes than any other factor. Ignition of vegetation from camp fires and the deliberate firing of vegetation to flush out game must have caused widespread conflagrations more devastating than

Table 2.1 *A chronology of events in the evolution of grass, grazers and man*

Era	Period	Epoch	Time (*years*)	Event
Mesozoic	Jurassic	Upper	160 million	Primitive mammals, dinosaurs, reptiles
	Cretaceous	Neocomian	130 million	Emergence of legumes and grasses; other angiosperms
Cenozoic	Tertiary	Palaeocene	70 million	Speciation of grasses; first horse Emergence of non-ruminant mammals Transformation from browsers to grazers
		Miocene	20 million	Grasslands became important component of earth's vegetation Emergence of ruminants
	Quaternary	Pleistocene	1 million	Rapid spread of grasses Ancestor of domestic cattle, wild urus (*Bos primigenis*) Probable separation of man from other primates
	Recent		13 000	Domestication of food crop
			12 000	Nomadic man settled into villages Domestication of livestock
			9 400	Domestication of sheep and goats in Libya
			7 200	Cattle used for traction
			5 000	Domestic cattle in Africa

any previous 'natural' burning. Ecological processes were wholly altered and readjusted. Grasslands were not destroyed, but in fact increased. Forest climaxes were replaced by grassland savannas and wooded savannas. The elasticity of grasslands, i.e. the wide array of grass species and their broad adaptability, made possible the invasion of newly created environments.

Browsers and grazers were not so fortunate. Their numbers and productivity fluctuated with the modified ecological communities, seasonal variations and the changing demands of man for food, clothing and shelter. Mass migrations took place. Many species did not survive and were replaced by others.

Man migrated and took with him the mighty sword of fire: a flame to cook his food and light the darkness; a blanket of warmth against the chill; a tool for clearing land; a weapon to rouse and track down game; an armour with which to rout the enemy.

Man also collected and carried seeds of plants used as food. Many of these were grasses. Some intercrossed with those of the new environments, thus adding to the evolving pool of plant variability. With selection of the more appealing types, man brought under domestication (some 11 000 years ago) such staple food crops such as sorghum (*Sorghum*), bulrush millet (*Pennisetum americanum*), finger millet (*Eleusine coracana*), teff (*Eragrostis abyssinia*) in Africa; rice (*Oryza sativa*) in Asia; wheat (*Triticum* spp.), rye (*Secale cereale*), and barley (*Hordeum* spp.) in Eurasia; maize (*Zea mays*) in America (Flannery, 1969). The growing of these crops favoured the establishment and spread of other grasses. After harvesting grain the land was abandoned for a new site. Annual grass species rapidly invaded the area to be followed by perennial grasses and woody plants.

The transition of man's nomadic life into semi-permanent villages took place between 11 000 and 10 000 years ago (Reed, 1969). Domestication of livestock came at about the same time (Flannery, 1969). A sedentary way of life with an established food source may have been a requisite for this domestication. Grazers in the wild state were considered crop-robbers by early man. After domestication they could be kept away from fields until after the grain harvest. There was no competition for food between man and the ruminants. The diet of the latter consisted of mature grasses, crop residues and leaves, branches and twigs of trees and shrubs. In fact, they were converters of cellulose materials, not digestible by man, into products for his use – milk, meat, blood, hides, hair and wool. They also served as surpluses in live storage for lean years.

Cattle were the first domestic animals used for pulling a plough. The earliest representation of ploughs is placed about 7200 BC, but they were probably used earlier (Issac, 1962). Domestic cattle became beasts of burden, thus further lightening the load of man. In time, dung was used as fertilizer for crops and as fuel for cooking. Man continued to prey upon wild game for supplemental food and by-products. As his numbers increased, theirs decreased, often to the point of extinction.

Distribution of domestic cattle in the tropics

Domestic animals probably appeared in Africa during the late Stone Age. Identification of fossil bones showed that sheep and goats were living in Libya before 7000 BC, and possibly as early as 9400 BC (Higgs, 1967). Both were well known in southern Asia at the earlier date. It is likely that they were herded across the Sinai (not a desert then as we know it today) into northern Egypt and then west into the mountains of Libya.

Cattle arrived some time later. They probably were brought into northeastern Africa from southwestern Asia between 5000 and 2500 BC by the proto-Hamites (Payne, 1964). Rock engravings depict cattle in Tibesti (Chad), even though mammalian life does not exist there today. At that time, alternate wet and dry periods would have allowed cattle migration across the present desert.

Three major cattle types were imported or migrated with nomadic peoples into northeastern Africa: Hamitic longhorn, humpless Shorthorn and humped Zebu. The Hamitic longhorns and the Shorthorns belonged to the ancestral *Bos taurus* type and the Zebus to the *B. indicus* type.

Possible migration routes have been outlined and are shown in Fig. 2.1. From the northeastern sector of the continent, they moved west along the Mediterranean coastal region, then south and east across the Sahara into Chad. A westward trek circumscribed the eastern Sahara perimeter via the Tibesti and Tassili highlands. The southern migratory route passed through present-day Ethiopia and spilled into East Africa. From here the route continued south, with a trail leading west above the forested zone. The migration of cattle-owning people was restricted by available feed and water and by the location of the tsetse fly (*Glossinia* spp.). It has been pointed out that the savanna fly (*G. moristans* West) does not penetrate the forest (Leeson, 1953). Neither does the riverine fly (*G. tachnoides*

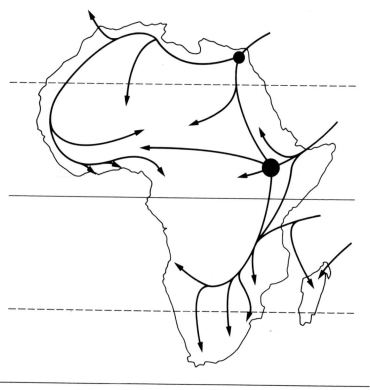

Fig. 2.1   Possible migration routes of cattle in Africa (after Payne, 1964).

West) survive in the savanna. Thus, relatively fly-free corridors probably existed between the two (Fig. 2.2). These, no doubt, coincided with some of the migratory routes.

Hamitic longhorns blazed the migratory trials (Payne, 1964). Their southwest movement along the highlands of the present Sahara was recorded by rock paintings and stone engravings. The drive southward is less clearly defined. But early figures preserved in stone show that longhorn cattle passed through Ethiopia and present-day Somalia. These paintings predate those of game animals, camels, horses and other types. By 1000 BC, Hamitic people with their longhorned cattle pushed from North Africa toward the south and then into West Africa. The indigenous inhabitants were not cattle owners.

No Hamitic longhorns exist today. The N'dama is descended from the longhorn and still exists in West Africa. Its smaller constitution may be attributed to a lower nutritional plane. The N'dama

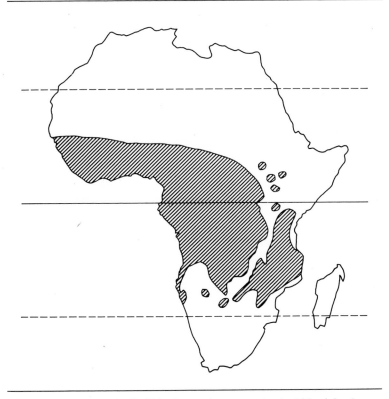

Fig. 2.2 Regions of tsetse fly (*Glossina* spp.) concentration in Africa (after Leeson, 1953). Their presence no doubt affected cattle migration routes.

penetrated the forest zone and have persisted because of tolerance to trypanosomiasis carried by the tsetse fly. A second descendant of the Hamitic longhorn in West Africa is the Buduma or Kuri. Larger than the N'dama, it has survived due to isolation in Chad. The influence of Hamitic longhorns may also be seen in other races, such as the Ankole of East Africa and the Binga of Zimbabwe. The latter possesses tolerance to trypanosomiasis, probably derived from the longhorn ancestor.

Shorthorns were present in Asia by 6000 BC. Records have placed them in Egypt between 2750 and 2625 BC (Payne, 1964). By 1700–1500 BC they emerged as the dominant type in lower Egypt. Their movements closely followed the route of the Hamitic longhorn. Present-day Egyptian cattle show strong Shorthorn characteristics and existing breeds in North Africa are largely of the Shorthorn type. They penetrated the forest zone and developed some tolerance to trypanosomiasis. The West African Shorthorn evolved and contributed to the development of other breeds and races. Shorthorns moved from Egypt into East Africa. It is also likely that they were brought into this region and into South Africa by other independent routes.

Zebus entered Africa about 2000–1700 BC. Evidence exists for admixture with the Hamitic longhorn and the Shorthorn between 2000 and 1500 BC. These crosses gave rise to the Sanga type. With successive waves of Zebus brought into East Africa continuous intercrossing took place. From these emerged types with more or less of the chest or neck-hump. In many regions the Zebu apparently showed outstanding local adaptation. Their progenies replaced derivatives of the Hamitic longhorns and the Shorthorns, except in North Africa. They apparently had greater tolerance to rinderpest than the humpless or the Sanga types. They became widespread in East Africa and were important in the ancestry of the Boran, East African Shorthorn and Ankole. Arab invaders took the Zebu westward on a route south of the Sahara and north of the rainforest in recent times. In fact, the White Fulani was first introduced into Nigeria during the fifteenth century.

The Sanga type of East Africa were moved southward probably first by the Hottentots and later by the Bantus (Payne, 1964). In South Africa they formed the base for the present-day Afrikander cattle.

Cattle did not exist in America or in Australia when the continents were opened up by the Europeans (Esminger, 1969; Williamson and Payne, 1978). The first cattle in America were taken by Columbus on his second voyage in 1493 to be used as work oxen. Spanish and Portuguese explorers made additional introductions into the American coastal areas. They soon spread over the Caribbean and inland regions through tropical and subtropical zones on the mainland. Cortés took cattle from Spain to Mexico in 1519. Other Spanish cattle were brought over about 1600 for work and as providers of milk by the Christian missionaries. During the nineteenth century British and other European breeds came into various parts of the American tropics. More recently, tropical types from Africa and Asia have been imported. Much of the European stock degenerated because of poor adaptation. Zebu were successful and increased rapidly. With intercrossing, many new races or breeds evolved in a number of locations. These came to be known as 'criollo' cattle and were given local names.

The first cattle importations into Australia were made less than 200 years ago (Williamson and Payne, 1978). Many of the early in troductions came from Africa and some from India. They did not thrive due to poor type selection and unsuitable relocation sites. Most were soon replaced by British breeds. These performed well in the more temperate regions but not in the tropical zones. Recently, American Brahmin, Santa Gertrudis and Red Sindhi have been introduced.

Livestock-keeping

The intensity of stock-keeping in the tropics largely depends on the herbage yield of grasslands and grazinglands, and this is directly affected by the amount of rainfall (Table 2.2).

The different types of livestock-keeping may be classified according to the permanence of the cattle owner or herder. Five main groups have been identified (Ruthenberg, 1976).
1. Total nomadism – no permanent residences, family moves with herd, temporary cultivated crops or none.
2. Semi-nomadism – quasi-permanent housing, family moves to distant grazing areas, supplementary cultivation practised.
3. Transhumance – quasi-permanent housing or permanent residences, seasonal migration to distant grazing, supplementary cultivation.
4. Partial nomadism – farmers live continuously in permanent settlements, herds remain in vicinity for most of the year, arable farming practised.
5. Sedentary (settled) animal husbandry – livestock remain on the holding or in the village the entire year, mixed farming or ranching practised.

With less than 50 mm average rainfall per annum (none in some years), the herbage yield from grasses and browse plants may not exceed 50 kg of dry matter per hectare. In such areas of Africa nomadic camel herders pass occasionally. In the semi-desert and steppes having up to 200 mm of rain, periodic grazing of camels is

Table 2.2 *Types of indigenous stock-keeping in the tropics* (*Ruthenberg, 1980*)

Rainfall (*mm/yr*)		Animals
Under 50	Occasional nomadism	Camel
50–200	Nomadism, long migrations	Camel
200–400	All types of nomadism, transhumance, supplemental arable farming	Cattle, goats, sheep
400–600	Semi-nomadism, transhumance; nomadism; stronger emphasis on arable farming	Cattle, goats, sheep
600–1 000	Transhumance and partial nomadism; semi-nomadism an ethnic tradition	Cattle
more than 1 000	Partial nomadism, permanent stock-keeping; semi-nomadism an ethnic tradition	Cattle

found, e.g. in the lower Sahara and northern Kenya. Semi-nomadism and partial nomadism are practised in regions having 200—1 000 mm annual rainfall. Seasonal grazing patterns generally follow established routes in the search of herbage and water. This is the common system in the African savannas. With more than 1 000 mm of rain per year livestock are kept on permanently settled farms. Semi-nomadism may be found but as an ethnic tradition rather than an imposed system of providing feedstuff for cattle.

Total nomadism

Traditional nomadism is found in East and West Africa where stock-keeping is the sole operational enterprise of the total nomad (Ruthenberg, 1980). It encompasses a way of life and involves the entire family unit. Total nomadism represents an agricultural system not suitable for technical investment. It exists under marginal production conditions and where natural climatic and soil forces hinder the application of cultivation techniques. The system operates at a minimal level of input with a low return of animal product.

The nomad is wholly dependent on his herd for sustenance. A primary aim of stock-keeping is the production of a daily supply of food (milk, milk fats, blood) and other products (hides, hair, wool, dung for fuel) (Dyson-Hudson, 1969). Large numbers of animals are associated with a large number of people. This contrasts with intensified stock-keeping, ranching and dairying, where large numbers are associated with relatively few people. Stock-keeping represents a variable degree of wealth, social status, community influence and man's legacy to his sons. Animals provide a means to formalize contracts, friendships and mutual assistance. They are offered in exchange or payment for crop foods and non-agricultural consumer goods. They afford transportation as mounts and pack animals. Sacrifice of cattle and use of terminal products (meat and blood) can both mark special occasions and stave off famine. Cattle are considered objects of man's affection and personal satisfaction of accumulated property.

The nomad, his family and livestock keep on the move with frequent changes of camps. Water and herbage are usually insufficient for a lengthy stay in any one location. Being dependent on native grasses and herbs for grazing and browsing, the migration follows available vegetation. Great distances are covered as herbage is usually sparse and watering places widely separated. Survival is of primary importance. Thus, the nomad has regard for hardy animals, which tolerate water stress, heat, cold, disease and periodic shortages of feed. He gives scant attention to a high yield of product per animal. Because of the vagaries of the environment and the extreme production risks, the nomad attempts to accumulate large herds to ensure a continuity of food supply and to increase social status.

Extremely dry climates favour camel- and sheep-keeping as they tolerate heat and can move long distances with restricted water. The camel only needs water at intervals of 2 or 3 days in the hot season and 2 or 3 weeks in the cool season. They seldom stray and can fend off small predators. Thus, they are left to range up to 80 miles unattended, browsing herbage not available to other domestic animals because of its distance from water. Camels provide milk and milk products and serve as pack animals. The nomad may also possess horses, mules, and donkeys for transport. He frequently keeps chickens, sheep and goats, and in areas of more rainfall, cattle for diversified food supply (Ruthenberg, 1980).

Cattle-keeping requires more attention than camel-keeping. Cattle need more water and herbage in greater abundance. They graze freely, but must be herded to and from watering sites. Water should be offered at least every second day, so that grazing is restricted within a range of 4–8 km. Goats and sheep graze more closely and browse on plants not acceptable to cattle. They tend to stray and are defenceless against predators. For this type of stock-keeping lax herding is needed during the day and kraaling at night.

Nomadic stock-keeping can be related to shifting agriculture in terms of agricultural evolution. It seldom leads to an improved way of life since the environment is not suitable for transfer to semi- or partial nomadism. Efforts to settle the total nomad have generally been unsuccessful. Thus, as the population of nomadic tribes increases, their plight worsens.

Partial nomadism

Animal keepers who operate from a quasi- or semi-permanent residence fall into the categories of semi-nomadism, partial nomadism and transhumance. They comprise the largest and most important groups of cattle keepers or pastoralists in tropical Africa. Those well known in West Africa are the Fulbe (or Fulani) and in East Africa the Masai, Suk, Turkana and Karamajong. Further north the Borana and Somali keep mixed herds of cattle and dromedaries (Webster and Wilson, 1980).

Animal owners may be keepers (herders) and travel with the herd. Families sometimes join the trek, especially during the dry season. Herders are sometimes employed by animal owners, with payment being in calves or animal products. In some instances, permanent arable farmers are owners but have no knowledge of cattle-keeping. They rent out and loan cattle or employ herders. Owners lend stock to relatives, poorer members of ethnic groups, or to friends.

The stock-keeping aims of the semi-nomad are similar to those of the total nomad, but vary with the extent of nomadism. Cattle are property and represent variable degrees of wealth, social status and community influence. They may be kept as the basis and sustenance of life or to cover the risk of crop failure or sickness. If land is not privately owned, they provide a means of support in old age. The factor of ox-plough cultivation is of importance in some systems. Frequently, manure collected from the kraal is the primary or only source of fertilizer for arable crops.

Communal and seasonal grazing are prominent and common characteristics of the semi-nomadic groups. The main feed consists of natural or semi-natural grasslands, fallow lands, harvest residues, open bush, loppings from edible trees and coppice. Herds graze freely over tribal lands. They generally remain in the vicinity of the village during the rainy season when grass is more abundant. After harvests of grain food crops, herders are permitted to enter the fields where cattle clean up the crop residues. With the onset of the dry season, grass around the dwellings and villages becomes scarce. Overgrazing is a serious problem since all herds have access to the communal grazinglands. Herd sizes are not regulated so that animal densities generally exceed the optimal stocking rate. The tendency is to acquire large herds and hoard animals against the risks of poor production. Where grassland is ample and water sufficient, the herds consists of a high proportion of males. The retention of older stock is also a practice and provides some insurance against recurrent diseases, as these animals have usually built up resistance by previous infections.

As grazing becomes more scarce, especially during dry years, and as cattle densities increase, the male cattle are reduced in number. With increasing cattle numbers and shortage of grazing around the villages, conflicts often arise between the pastoralist and the arable farmer, especially as the dry season progresses.

During the dry season, herders begin migrating in search of feed and water. The movement flows from regions of low to increasing amounts of rainfall. For instance, in West Africa, trails lead from north to south as rain declines, as herbage is utilized or becomes less nutritious, and as the tsetse fly recedes to wetter and cooler sites. This is reversed when the rains are renewed. The nomadic herders generally follow familiar grazing routes ranging over a distance of 100 miles or more during the season. Often, several groups or families migrate together. Temporary kraals and residences may be constructed near seasonal watering places. Animals are rotated around the various watering sites. They may be driven from 4 to 8 km away from water one day (overnight) and graze back toward the water the following day.

Herdsmen often recognize the feeding value of different graze and browse species (Allan, 1967). They have common names for many, recognize ecological associations and base stocking rates on the potential capacity of the vegetational cover. The nomads know that burning removes old grass, stimulates regrowth during the dry season and affords easier movement through the bush. They may even be accused of starting fires. They have knowledge of the position, quality and capacity of watering points, and the time required to move from one to another. The standard of animal husbandry practised by semi-nomads is highly variable. In some regions they subscribe to effective vaccination of their animals against diseases. In others, neither preventative nor veterinary methods are employed (Ruthenberg, 1968).

Most types of semi-nomadism operate under conditions of low returns per man-equivalent, per animal and per unit of land. The low productivity often arises from the desire to accumulate large herds, from communal land rights and from overgrazing rather than unfavourable natural conditions. Animal growth rates are seasonal and follow a stop-and-grow pattern in a cycle corresponding with the available herbage as influenced by rainfall. Cattle usually do not reach marketable size until they are 6 years old or more. Because of insufficient grazing and the low nutritional level of grass and crop residues, cows may not calve more than once in 3 years. Calf mortality is extremely high; 20 per cent may not survive the first 6 months and another 20 per cent fail to reach maturity (Webster and Wilson, 1980). The annual take-off rate (sale and slaughter) generally ranges between 6 and 10 per cent of the stock (Makings, 1967). This compares to 20 per cent on well-operated ranches.

Ruthenberg (1980) characterized the malpractices of the semi and partial nomad as follows:
1. High animal densities, overgrazing, and soil erosion damage, especially near watering points.
2. Insufficient herbage reserves and inadequate distribution over the year.

3. Herding of mature and infertile animals.
4. The need for kraaling because of thieves and predators, imposing daytime grazing.
5. Long treks between the kraals, the grazinglands and the watering places.
6. Low levels of calving and high calf mortality.
7. Lack of veterinary services and improved animal husbandry practices.
8. Lack of systematic breeding.

Semi- and partial nomadism may be likened to semi-permanent farming and in fact is a part of the system in some agricultural schemes. As land pressures increase, arable farming takes over more of the grasslands and savannas. This forces the nomads to move on to less desirable grazinglands or to modify their way of earning a living. In some areas they have taken the step to semi-permanent and mixed farming.

Sedentary animal husbandry

This is a type of mixed farming with animals kept on the farm. The primary objective may be stock-keeping or arable cropping. In either event, cattle remain on the land-holding or in the village the year around. Such types of integrated cattle and crop farms are seen among (1) smallholders – e.g. dairy farmers of East Africa and the higher elevations of Mexico, Central and South America; cotton farmers of Tanzania; maize farmers of Kenya; millet growers in Uganda; and groundnut–millet growers of West Africa and (2) large land holders – e.g. tobacco farms in the drier savannas of Zambia and Zimbabwe; wheat–cattle farms of the Kenya highlands; potato–cereal farmers of the Andean zone of South America; and irrigated rice of farmers of South America and Australia.

Depending on the size of holding and farming operation, long- or short-term fallows alternate with a period of arable cropping as part of a rotational scheme. This pattern of alternative cropping is referred to as a ley system. Two kinds can be distinguished; the unregulated and regulated ley (Ruthenberg, 1980). The unregulated ley is characterized by a fallow of natural or semi-natural grass cover and herbaceous plants. Usually communal grazing is allowed. There is little or no grazing management. The unregulated ley prevails in Africa. Cotton holdings of the savanna areas of East and West Africa and the groundnut–millet holdings of West Africa are representative of the unregulated ley. They are also found in the mountainous parts of the Andes of South America.

With the regulated ley system, fences are erected and grazing management imposed. Grasses may be sown or planted and fertilized. Other improved practices, as mowing for weed control, rotational grazing, separation of cows and calves and supplemental feeding may be incorporated into the system. Regulated leys are widespread in the temperate zones and subtropics. They are less prevalent in the tropics and more often found in the higher elevations and on larger farms. In the Kenya highlands for example, the wheat, pyrethrum, sheep, dairy and cattle operations represent regulated ley farming.

Ranching

The term 'ranch' generally refers to an expansive area of land legally owned or leased, with defined boundaries (fenced or unfenced) and improvements made by the owner or lessee (Ruthenberg, 1980). Ranching implies a large-scale operation of several thousand hectares and large permanent herds. Ranches are mostly located on marginal lands with little opportunity for alternative types of farming. Production is limited to stock-keeping, generally sheep and cattle. Sheep are better adapted to drier areas (150 mm rainfall annually) and more adverse regions. Cattle have less mobility, need a greater quantity of water and more frequently. They also demand larger quantities of herbage as found under conditions of 400 mm or more of annual rain. Stock numbers, capital investments and returns on ranches are large per man, giving a high output per man equivalent. Stocking rates are low, fluctuating between 2 and 30 ha per animal unit. Animal densities must remain minimal to provide a year-round supply of herbage. A supply of water is of utmost importance. Grazinglands are of little value without watering points at intervals of 4–8 km. Ranching is a long-term commitment with a high production risk because of the unreliability of rainfall and constant threat of extended drought.

Large-scale ranching is found in the low and high elevations of tropical America, in parts of East and Central Africa, in northern Australia and in areas of Asia. Ranching did not develop in the tropics, but is a system brought from Europe and modified to fit tropical conditions (Webster and Wilson, 1980). It was practised in Spain for many centuries and spread into the colonized countries during the fifteenth and sixteenth centuries. In many parts of South America, and to some extent in other tropical regions, some of the more successful ranches are owned by large companies which supply the capital. Managers with technical knowledge are employed to oversee the operation. In many countries ranch enterprises are located on government farms for experimental purposes and used to upgrade local herds by selling improved stock.

Development of grassland research

As grasslands developed a natural selection of adapted types occurred. Man eventually imposed selection when he picked out appealing forms. Improvement of grazinglands came about only recently. It was probably not until the sixteenth century that farmers in England began sowing selected grasses for meadows and grazing. Associations of grasses and legumes were made in the seventeenth century. Ley farming, i.e. alternate arable cropping with grass fallow, was introduced about 100 years later. Bones and bone dust had been used as fertilizer before the 1880s when basic slag became available from steel mills. This phosphatic fertilizer was applied to grass–clover pastures in England and Europe in the late nineteenth and early twentieth centuries.

Grassland and pasture research began with the establishment of the first agricultural research station at Rothamsted in 1843. Soil fertility and rotation plots were laid down and have been continued to the present. Within a few years, plots on a permanent meadow had been fertilized, the production of herbage recorded and the botanical and chemical composition determined. Fertilizer experiments with pastures started in Germany during the latter part of the nineteenth century. In 1902 at Cockle Park, England, sheep on fertilized pastures were weighed; this ushered in the beginning of experimental grazing studies. Investigation on the nutrition of grazing animals was begun at the Rowett Research Institute in Scotland in 1914.

Man carried seeds of food and feed crops in his early migrations. It was customary to do so when emigrants moved from one country to another. During the nineteenth century explorers, ambassadors and governmental emissaries were charged to collect and return seeds of promising crops, including herbage grasses and legumes. Breeding of pasture plants started with the establishment of the Welsh Plant Breeding Station at Aberystwyth in 1919. Comparable developments and investigations of grassland research were carried out in Europe and America.

Scant attention was given to tropical grasses and legumes until after the turn of the present century. During the 1930s grassland surveys were made in East and West Africa, with early descriptions coming from Kenya, Madagascar and Nigeria. Local grasses and legumes were collected and evaluated at a number of government farms and experimental stations. Intensive studies and selections among ecotypes at the National Agricultural Research Station at Kitale, Kenya, led to the development of improved cultivars of such species as *Chloris, Setaria, Panicum, Cenchrus, Cynodon* and *Glycine*. Agronomic studies of pasture and range management have now been conducted in all tropical countries. Grazing experiments and animal nutrition investigations have been carried out in a number of locations.

Many herbage species introductions were made into Australia by the Commonwealth Scientific and Industrial Research Organization (CSIRO) between 1930 and 1950. Improvements were largely confined to the sorting out of ecotypes. Since the early 1950s, the application of plant breeding methods at the Cunningham Laboratory in Brisbane, Queensland, have brought about the release of cultivars of the same species as studied in Africa, plus forage types of *Sorghum, Desmodium, Macroptilium, Glycine* and *Stylosanthes.*

Grasses serve mankind

Grasslands provide a major portion of the livestock feed for domestic and game animals. They furnish over one-half of the total feed supply in many countries and up to 85 and 90 per cent in others. Grasslands cover about 24 per cent of the earth's land surface and occupy a greater number of hectares than all other crops combined.

Under certain conditions other attributes of grasslands may be of more importance than providing livestock feed. Grass and legume cover retards water runoff, thus increasing the rate of water infiltration into the soil and reduces or prevents soil erosion by wind or water. When used in rotation with arable crops, and when properly managed, certain grasses (e.g. *Pennisetum purpureum*) can maintain or improve the fertility status of the soil. Under specialized conditions, grasses and legumes are used for maintenance of highway shoulders and embankments. They provide protective cover for roadways, for fire breaks, waterways, airstrips and general service areas. Grasses form an essential feature for recreational areas such as golf courses sports fields, parks and lawns. In some parts of the world, they are used for building homes and making furniture, thatching roofs, matting for beds or floor covers, making brooms and fans, ropes and fishing nets, baskets and boxes, ornaments such as earrings and necklaces, rosaries and beadwork, musical instruments, beverages and medicinal purposes, insect repellant and other conveniences and necessities of life.

References

Allan, W. (1967) *The African Husbandman*, Oliver and Boyd: London.

Barnard, C. and **O. H. Frankel** (1964) Grass, grazing animals, and man in historic perspective. In C. Barnard (ed.), *Grass and Grasslands*, Macmillan: London, pp. 1–12.

Dyson-Hudson, Rada and **Neville** (1969) Subsistence herding in Uganda, *Scientific Am.* **220(2)**, 76–89.

Ensminger, M. E. (1969) *Animal Science*, The Interstate Printers and Publishers: Danville, Illinois.

Flannery, K. V. (1969) Origin and ecological effects of early domestication in Iran and the Near East. In P. J. Ucko and G. W. Dimbleby (eds), *The Domestication and Exploitation of Plants and Animals*, Gerald Duckworth: London, pp. 12–100.

Hartley, W. and **R. J. Williams** (1956) Centres of distribution of cultivated pasture species and their significance for plant introduction, *Proc. 7th Int. Grassld Cong.*, pp. 190–201.

Higgs, E. S. (1967) Early domesticated animals in Libya. In W. W. Bishop and J. D. Clark (eds), *Background to Evolution in Africa*, Univ. Chicago Press, pp. 165–73.

Issac, E. (1962) On domestication of cattle, *Science* **137**, 195–204.

Leeson, H. S. (1953) The recorded distribution of certain tsetse flies and of human trypanosomiasis in Africa, *Trans. R. Soc. Trop. Med. Hyg.* **47**, 130–3.

Makings, S. M. (1967) *Agricultural Problems of Developing Countries in Africa*, Lusaka Oxford Univ. Press: Nairobi.

Payne, W. J. A. (1964) The origin of domestic cattle in Africa, *Emp. J. Exptl Agric.* **32**, 97–113.

Reed, C. A. (1969) The pattern of animal domestication in the prehistoric Near East. In P. J. Ucko and G. W. Dimbleby (eds), *The Domestication and Exploitation of Plants and Animals*, Gerald Duckworth: London, pp. 261–380.

Ruthenberg, H. (ed.) (1968) *Smallholder Farming and Smallholder Development in Tanzania, Africa-Studien No. 24*, IFO Institute: Munich.

Ruthenberg, H. (1980) *Farming Systems in the Tropics*, Clarendon Press: Oxford.

Semple, A. T. (1970) *Grassland Improvement*, Leonard Hill Books: London.

Shantz, H. L. (1954) The place of grasslands in the earth's cover of vegetation, *Ecology* **35**, 143–51.

Simpson, G. G. (1950) *Meaning of Evolution – A Study of the History of Life and of its Significance to Man*, Yale Univ. Press: New Haven.

Stewart, O. C. (1956) Fire as the first great force employed by man. In H. L. Thomas (ed.), *Man's Role in Changing the Face of the Earth*, Univ. Chicago Press, pp. 115–33.

Weaver, J. E. (1954) *North American Prairie*, Johnson: Lincoln, Nebraska.

Webster, C. C. and **P. N. Wilson** (1980) *Agriculture in the Tropics* (2nd edn), Longman: London.

Williamson, G. and **W. J. A. Payne** (1979) *An Introduction to Animal Husbandry in the Tropics*, (3rd end), Longman: London.

Chapter 3

Botany of grasses and legumes

The grasses

Grass is a universal term applied to all members of the family Gramineae. Grasses are monocotyledonous, i.e. the embryo of a grass seed has a single (mono) cotyledon, or seed leaf. The basic design of a grass plant is simple; it consists of: roots giving rise to cylindrical, jointed stems; leaf blades borne on sheaths which arise at the nodes and encircle the stem; and inflorescences consisting of several flowers from which seeds develop. Grasses may be 'annual' or 'perennial' in life form. An annual plant completes its life cycle in one growing season and dies. A perennial lives for more than two growing seasons. When a plant completes the life cycle in two seasons it is referred to as a 'biennial'. These terms are misleading in the tropics where wet and dry seasons influence growth patterns, rather than alternating hot (warm) and cold seasons. With ample moisture, a so-called annual of the temperate zone may persist over several seasons. For example, *Sorghum vulgare*, a strict annual where frost prevails, tends to be perennial in the tropics with plentiful soil moisture.

Almost all grass plants are herbaceous (non-woody) and are widely divergent in size, shape and habit of growth.

Vegetative organs

Subterranean parts

Primary roots develop after seed germination and during growth of the seedling. They persist only a short time and are replaced by secondary roots, which develop from the shortened basal nodes of the young stems. The secondary root system is fibrous and intertwines to form a dense mat just below ground level near the stem base.

Certain species, e.g. *Sorghum vulgare*, form aerial roots from the lower nodes which are known as 'buttress', 'brace' or 'prop' roots.

Trailing grasses, such as *Cynodon nlemfuensis* and *Digitaria decumbens*, develop secondary roots from the nodes of stolons. Those with partially or semi-dependent stems, for example *Brachiaria mutica*, send down roots from nodes which are in contact with the soil.

Rhizomes are underground stems present in certain species, with nodes and internodes covered with scale-like leaves and sheaths. They serve as a storage organ but may surface and form vertical stems with roots. These may separate to form plants that develop other rhizomes. Species such as *Cynodon dactylon* and *Imperata cylindrica* sometimes form long and fibrous rhizomes, making them difficult to eradicate. Others such as *Pennisetum purpureum* and *Sorghum almum* have short rhizomes.

Aerial parts

The *stems* of a grass plant, also called culms or haulms, are jointed; that is, they are made up of nodes separated by internodes. The nodes are solid, sometimes enlarged and serve to strengthen the stem. Elongation of stems takes place because of the presence of meristematic tissue just above the nodes. Internodes may be hollow (*Brachiaria mutica*), filled with a white pith (*Zea mays, Sorghum vulgare* and *Hyparrhenia* spp.), or solid (*Axonopus scoparius*). In a few grasses the basal internode is thickened and enlarged (*Panicum antidotale* and *Cenchrus ciliaris*), and serves as a storage organ.

Stems are usually cylindrical or rounded, but in some grasses there may be a groove, more prominent just above the node where axillary buds form. The lower part of the culm is sometimes compressed, as found in *Setaria splendida*. Stems are glabrous or pubescent, the latter form sometimes occurring as a ring below the nodes or below the inflorescence. Shoots develop from buds found at the nodes and produce side branches.

The basal portion of tufted grasses is sometimes referred to as the *crown*. As plants become older, and are repeatedly grazed or cut, the older stems die and new tillers develop, causing the crown diameter to enlarge.

Stolons are creeping stems that grow above the surface of the ground and develop roots and shoots at the nodes (Fig. 3.1*f*). It is easy to confuse stolons with procumbent stems that also root at the nodes. *Pennisetum clandestinum, Cynodon nlemfuensis* and *Digitaria pentzii* are examples of stoloniferous grasses.

The leaves consist of the sheath, the ligule and the leaf blade (Fig. 3.1*c*). In most grasses the edges of the sheath are free and overlap. All sheaths have chlorophyll and contribute to the production of photosynthate.

In the bud and growing tip, the young leaves are folded or rolled. As they emerge they normally expand and are flat, but may remain

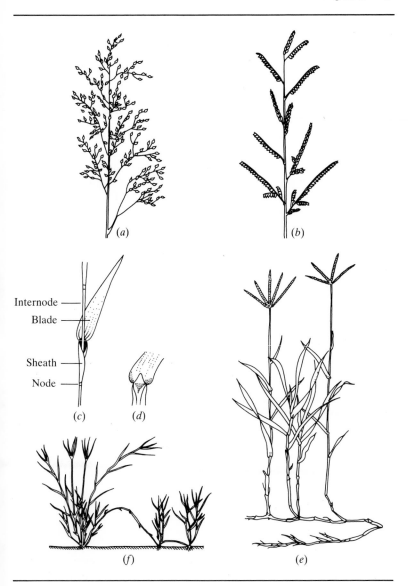

Fig. 3.1   Characteristics of the grass plant: (*a*) Spikelets in a panicle; (*b*) Spikelets in a raceme; (*c*) Culm, showing leaf attachment; (*d*) Ligule; (*e*) Rhizomes; (*f*) Stolons, and establishment of new plants.

more or less folded or rolled. When blades are wiry and bristle-like they are termed setaceous and when thread-like filiform. In some grasses the leaves are smooth to the touch, but in others rough because of minute, spiny hairs (e.g. *Rottboellia* and *Imperata* spp.).

The leaf midrib is usually prominent, but the lateral veins are often quite faint.

The *ligule* is an appendage found at the junction of the leaf blade and the sheath, usually closely adpressed to the culm (Fig. 3.1*d*). It may be a membrane, a fringe of hairs or a hardened ring, and varies in size, shape and texture. The structure is rather consistent within a species and considered a fairly reliable characteristic for identification.

Auricles are ear-like outgrowths at the leaf base of some species. They may be prominent and encircle the stem, minute and inconspicuous, or absent. They generally have no chlorophyll and appear to have no function but are used in taxonomic identification of species.

The *collar* marks the junction of the outer surface (upper region) of the sheath and leaf blade and is usually discoloured. When leaves are separated, they break easily at the collar.

The *prophyllum* is a two-keeled organ (a reduced leaf) covering the bud in the axil of the sheath. The branch shoot develops from this bud and the prophyllum is the first leaf of the shoot. It completely covers the bud but opens as the shoot develops.

Reproductive organs

The floral organs are modified shoots, consisting of stamens and pistils.

Inflorescence

The flowers or inflorescences may be terminal, arising at the end of the main stem, or axillary, i.e. the flower stalk developing from the axil of the leaf and stem. They are usually prominently displayed (*Panicum maximum, Pennisetum pedicellatum, Rhynchelytrum repens*), but may be partly concealed (*P. clandestinium*). The basic unit of the inflorescence is the spikelet, which consists of the flowers usually occurring in groups or clusters.

Inflorescence types are classified as spikes, racemes and panicles.
1. *Spike* – the spikelets are sessile (without stalks), or nearly so, on an unbranched axis (rachis). Species with typical spikes are *Lolium, Triticum, Secale, Hordeum, Agropyron*. The spike may be one-sided as in *Ctenium elegans*, digitate (finger-like) as in *Chloris* and *Cynodon* spp., or racemose on a central axis as in *Dactyloctenium* and *Leptochloa* spp.
2. *Raceme* (Fig. 3.1*b*) – the spikelets have pedicels along the axis, for example, species of *Digitaria, Paspalum*, and *Brachiaria*. Racemes are of more frequent occurrence than spikes.
3. *Panicle* (Fig. 3.1*a*) – the spikelets have short stalks on a branched inflorescence with a central axis and a number of side branches. The panicle may be open and loose (*Panicum maximum*),

contracted (species of *Sporobolus* and *Sorghum*), or spike-like and dense (*Cenchrus ciliaris* and *Setaria anceps*). The spike-like panicle is called a 'false-spike' when the branches are short enough to be concealed by the spikelets (*Pennisetum purpureum*).

Spikelet (Fig. 3.2a)

A typical spikelet consists of an axis (rachilla), two glumes and one to many florets. The glumes are borne at slightly different levels on the rachilla, i.e. an upper and a lower glume. They may be longer or shorter than the floret or florets. They vary in texture (thickness), shape, number of nerves and the presence or absence of hairs and other outgrowths.

The perfectly developed floret has a lemma and a palea (lower and upper bracts, respectively) which enclose the flower (Fig. 3.2b). The lemma may be shorter or longer than the legumes. The structure is such that it provides protection for the seeds and perhaps a

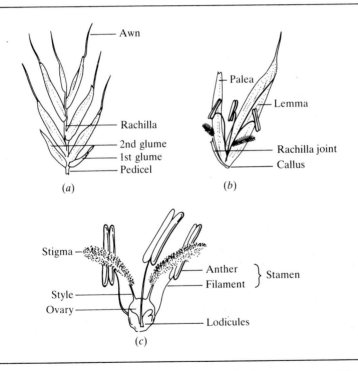

Fig. 3.2 Flowering of the grass plant: (*a*) Spikelet showing arrangement of florets; (*b*) Floret opening at blooming time; (*c*) Typical grass flower showing essential reproductive organs.

means of dispersal. In some grasses, it forms a hard seed cover and clasps the seed firmly. Sometimes it is thickened at the base to form a callus which may be blunt or sharp and barbed. The barbs may become attached to an animal and pierce the flesh. The lemma produces an awn in many grasses which may be round or flattened, straight or wavy, bent and twisted or feathery. Alternate wetting and drying of some awns may cause twisting and untwisting, an action which can drive the seed into the soil. Instead of an awn, the lemmas of some grasses bear hairs in tufts or fringes.

The palea is borne on the floral axis facing the lemma and embracing the flower. Generally, it is shorter than the lemma and thinner. It is mostly two-keeled and generally has two nerves. In some grasses various outgrowths develop, but are less prominent than on the lemma.

The flower

The floral organs consist of the gynoecium (female parts), androecium (male parts), and lodicules (Fig. 3.2c).

The gynoecium comprises the pistil and its parts – ovary, ovule, style and stigma. Stigmas form at the end of the styles or arise directly from the ovary. The stigmatic surface is papillose (having the nipple-like projections) or plumose (small feathery-like hairs). Pollen grains adhere to the sticky papillose surface and become caught among the hairs of the plumose surface, where they germinate and begin growth down the style.

The androecium consists of three, or one to six, stamens. Each stamen has a slender filament supporting a two-celled anther, which contains the pollen grains. Anthers are coloured yellow, purple, reddish or may be mottled. At 'anthesis' the anthers are thrust from the lemma and palea. Pollen is discharged through longitudinal slits of the anthers.

The lodicules are delicate, scale-like lobes found at the base of the flower, outside the stamens. They become turgid at anthesis. The swelling action forces the lemma and palea apart so that the stamens and stigmas may emerge.

The flowers of most grasses are perfect (hermaphroditic), i.e. the florets have both stamens and pistils, but some are imperfect, having one or the other. Members of the tribe Maydeae have '*monoecious*' inflorescence. In *Coix lagrima* and *Tripsacum laxum* the female spikelets form below the terminal raceme of male spikelets. In *Zea mays* the male and female inflorescences are separated on the same plant. Rarely are grasses 'dioecious', but *Spinifex littoreus* has the male and female flowers on different plants.

The grass seed

The fruit (seed) of most grasses is a caryopsis or kernel, sometimes called a grain. The single seed (mature ovule) is united to the peri-carp (ripened walls of the ovary). Sometimes the pericarp is not attached to the seed, for example species of *Sporobolus* and *Erag-rostis*. The pericarp of many grasses adheres to the seed and has the appearance of the seed coat. The seed coat, however, lies below the pericarp and is an ovular structure (integuments).

The embryo is found on the side of the caryopsis adjacent to the lemma. On the other side, the hilum marks the point of attachment to the ovary wall. Endosperm occupies a major portion of the grain.

Growth habit

The growth habits or patterns of grasses can be described as follows:
1. 'Tufted' (bunch-type, tussock) – a cluster, clump or hummock of vegetative shoots or culms, arising from a single crown, e.g. *Pani-cum maximum*. The culms of a tufted grass may grow erect, in a decumbent fashion (curving upward from a horizontal base), semi-erect or semi-decumbent. The stems may also lie flat on the ground for some length, then turn upward, in which case they are procumbent or prostrate (*Axonopus micay*).
2. 'Creeping' – the stems trail over or grow underneath the ground surface, e.g. *Cynodon* spp.
3. 'Scrambling' (climbing) – the plants are normally creepers, but the stems will grow upward and over upright objects, e.g. *Pennisetum clandestinum*.

The legumes

Legumes are dicotyledonous, i.e. the embryo consists of two cotyledons (seed leaves). The legume family is sometimes divided into three groups or subfamilies: Mimosaceae, woody plants and herbs with regular flowers; Caesalpinaceae, plants with irregular flowers; Papilionaceae, herbaceous and woody plants with a distinc-tive papilionate or butterfly-shaped flower. Most of the forage and economically important legumes come under the latter group. Legumes may be annuals, biennials or perennials, as with the grasses.

Vegetative Organs

There are distinct morphological differences among the legumes, but general characteristics of some plant parts are similar and rather uniform.

Subterranean parts

The root system of most legumes consists principally of an actively growing primary or tap root and its branches (secondary roots). The primary root of some legumes may penetrate the soil to a depth of 6–8 m, e.g. lucerne. In some it becomes slightly enlarged just below ground level, benefiting the plant during a prolonged drought, e.g. *Centrosema* spp. Secondary roots sometimes develop from nodes of stems resting on the ground.

The roots of many leguminous plants become infected by bacteria of the species *Rhizobium*, which grow and multiply, forming nodules. The nodules differ in size, shape and arrangement on the roots (see Ch. 16).

Aerial parts

The above-ground portion of the plant consists of a main stem with axillary branches, usually compound leaves, stipules and inflorescences. Tillers sometimes arise from the basal portion of the stem (crown) and stems also develop axillary branches. The stems are jointed, with nodes and internodes, and are usually hollow, except at the nodes. They may be covered with hairs or may be glabrous. Herbaceous stems contain chlorophyll.

The leaves are made up of a common leaf stalk (petiole), with three or more leaflets, each with its own stalk (petiolule). The leaves are 'palmately' compound, i.e. leaflets directly attached to the end of the petiole as in *Centrosema pubescens*. Or, they are 'pinnately' compound when the petiole extends into a long slender structure, the rachis, with the leaflets arising at intervals along this structure, e.g. *Clitoria ternatea*. In some species, such as *Vicia*, the upper leaflets are modified into tendrils, which may also arise directly from the node, as in *Lathyrus* species. A characteristic feature of the legume family is the presence of a pulvinus at the base of the leaflets and of the petiole (Fig. 3.3a). A mass of large, thin-walled cells surround the vascular strand and function in turgour movements. This tissue causes the terminal leaflet of some legumes to stand erect in bright sunshine. At the base of the main leaf stalk is a pair of leaf-like outgrowths, called stipules. They vary in shape and size and are useful in the identification of species. The leaflets and stipules may be smooth or possess hairs. The leaflet veins form a netted pattern as compared to the parallel venation of grasses.

Reproductive organs

Inflorescence

The Mimosaceae produces flowers in dense heads or small globular, spike-like inflorescences, and commonly has the floral parts arranged in sets of four. They are rendered conspicuous by the long, coloured filaments of the numerous stamens. *Leucaena leucocephala* and the *Acacia* spp. belong to this subfamily.

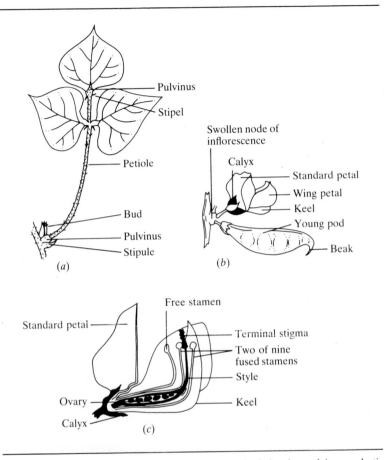

Fig. 3.3 Characteristics of a legume: (*a*) Trifoliate leaf showing pulvinus and stipule; (*b*) Flower and young seed pod; (*c*) Section showing essential organs of flower.

The Caesalpinaceae flowers appear in clusters or racemes, with overlapping petals. The stamens are usually separated. *Cassia* spp., *Ceratonia* spp. and *Gleditschia* spp. are members of this group.

The flowers of Papilionaceae are arranged in racemes as in *Desmodium* spp., in heads as in *Trifolium* spp., or spike-like racemes as in *Medicago sativa*. There is a central axis, along which the individual flowers develop. Each flower has its own short stalk or peduncle. The inflorescence may be terminal or axillary.

The flower

The *corolla* consists of five petals of three distinct kinds: 'standard' or 'banner', uppermost or outer petal, largest and most showy; two *wing petals*, with slender stalks called the claw, and an expanded portion; keel, two petals folded together, partially concealed by the wing petals, the expanded portions more or less united at the outer margin into a boat-shaped structure (Fig. 3.3*b*). When the standard is extended the flower somewhat resembles a butterfly, hence, the flower is termed papilionaceous.

The *calyx* with five teeth forms a tube at the base of the corolla, the length varying widely in different species. Usually a distinct vein runs down the centre of each tooth.

The keel encloses the stamens and pistil (Fig. 3.3*c*). The androecium consists of ten stamens, the filaments of which may be united ('monodelphous'). The staminal tube surrounds a superior ovary, an elongated structure comprising one carpel with one ovule or a single row of several ovules. A bent style surmounts the ovary and the stylar tip broadens into the stigmatic surface.

The *nectary* resides at the bottom of the corolla tube. The length of the corolla affects the activity of insects in search of nectar and thus influences pollination.

Fruit and seed

The ripened ovary forms a fruit of variable shape, called a *legume* or pod. In some species the pod is long and flattened, with two distinct junction lines running the distance of the two sides. In some cases, when the ripened pod dries out, it splits with a violent force, throwing the seeds some distance (e.g. *Centrosema pubescens* and *Macroptilium atropurpureum*). These pods are said to be *dehiscent*. The pod may be cylindrical or rounded as in *Crotolaria* species and non-dehiscent. The pods of some species have indentations between the enclosed seeds (*Glycine wightii*). They may be coiled, as found in *Medicago sativa*. Other forms are characteristic of different species. The pods are glabrous or covered with hairs.

Each seed is enclosed in the *testa* or *seedcoat*. A large oval scar, called the *hilum*, marks the place of attachment to the ovary wall. The legume seed has no endosperm, rather the reserve food is stored in the cotyledons.

Growth habit

Legume plants grow in the following ways:
1. 'Bush-type' – a central stalk with side branches appearing along the main stem and with axillary branches developing, e.g. *Cajanus cajan*, *Desmodium tortuosum*.
2. 'Bunch-type' – a single crown from which several stems and new tillers arise, making it difficult to identify a main stem. Stems may be erect or decumbent, e.g. *Stylosanthes guianensis*, *Medicago sativa*.

3. 'Creeping' – the stems trail over the ground surface, e.g. *Calopogonium mucunoides, Macroptilium atropurpureum*, some *Vigna* species.
4. 'Scrambling' – many of the creeping plants climb on to and grow over upright objects. Some are also 'twining' and encircle upright objects, e.g. *Centrosema pubescens* (Fig. 2.2), *Pueraria phaseoloides*.
5. 'Rosette' – a vegetative form of some perennials developed after flowering or with the onset of cool weather, e.g. *Medicago sativa* and *Trifolium pratense*, at the higher elevations of the tropics.

References

Arber, A. (1965) *The Gramineae*, Wheldon and Wesley (reprint by J. Cramer – Weinheim), London.
Bor, N. L. (1960) *The Grasses of Burma, Ceylon, India and Pakistan*, Pergamon: London, Ch. 1.
Brook, A. J. (1965) *The Living Plant*, Edinburgh Univ. Press.
Burkart, A. (1943) *Las Leguminosas Argentinas, Silvestres y Cultivades*, Acme Agency: Buenos Aires.
Chase, A. (1959) *First Book of Grasses*, Smithsonian Institute: Washington, D.C.
Chippindall, L. K. A. (1955) A guide to the identification of grasses in South Africa. In D. Meredith (ed.), *The Grasses and Pastures of South Africa*, Central News Agency: Cape Town. Part 1.
Cobley, L. S. (1976) *An Introduction to the Botany of Tropical Crops* (2nd edn revised by W. M. Steele), Longman: London.
Robinson, D. H. (1937) *Leguminous Forage Plants*, Edward Arnold: London, Ch. 1.
Strasburger's (1976) *Textbook of Botany*, New English edition. Translated from 30th German edition by P. Bell and D. Coombe, Longman: London.

Chapter 4

Systematics of grasses and legumes

The science of plant classification is a highly specialized field of study. Even the discipline of grass systematics is too large and comprehensive to master without years of study. The student or practitioner of grassland husbandry is not expected to be a grass or legume botanist, but should appreciate that certain groups of grasses and legumes possess common traits. Knowing the identity of grass and legume species, and recognizing likenesses and diversities as compared to other species, brings about a better understanding of their adaptability and potentiality as forage and fodder crops.

Separation into categories

The Gramineae (Grass family) and Leguminoseae (Legume family) are divided into lower or minor categories of tribes, genera and species (Lawrence, 1951). It is within these groups that characters can be assessed and described in more detail to facilitate orderly arrangement and plant identification.

Tribes are groups with natural affinities and certain phylogenetic (evolutionary) relationships. These groupings are of value to the botanist in plant systematics. They are not particularly relevant to one who is chiefly concerned with applied grassland husbandry.

Plants having several reproductives structures in common comprise a genus. They have other characters distinguishable from those of other genera. These similarities extend beyond morphological likeness to include genetic, cytologic, physiologic, ecologic and geographic relationships. The genus holds special interest for those interested in grassland husbandry. When opening up new grazinglands, or when searching for improved grass and legume types, selections which represent different genera are frequently brought together for evaluation. The generic name of a plant is the first of the two words making up the botanical binominal, for example *Chloris gayana*. Sometimes the generic name is used in discussing information related to grassland husbandry. For instance, we speak of *Chloris*, *Cenchrus*, *Glycine*, *Centrosema* without reference to the specific name.

The precise definition of a species is elusive and taxonomists do not always agree on its description or separation as a botanical unit. In general, the species consists of a natural population of plants with common morphological characteristics (phenotypically similar), having a common ancestry and capable of reproducing like types (Lawrence, 1951). It is identified taxonomically by the second name of the binominal used as a scientific name. This group is the most important botanical unit for the pasture agronomist and the cattleman, since plants of a species may have broad adaptability to diverse soil and climatic conditions.

The species is of particular interest to the plant breeder. He brings together introductions of the same species from many different sources and searches for types which excel in adaptation, herbage yield, persistence, etc. These may be released as cultivars, e.g. *Chloris gayana* 'Masaba', *Glycine wightii* 'Clarence', *Setaria anceps* 'Nandi'. In addition, interspecific crosses are made in the improvement of fodder and pasture types. For example, a number of the commercially grown selections of elephant grass came from interspecific crosses of *Pennisetum purpureum* and *P. americanum*. *Sorghum almum* arose as an interspecific cross between two *Sorghum* species (Parodi, 1943).

'Varieties' are morphological variants and subdivisions of the species. In botanical literature they will be written with the specific name, e.g. *Imperata cylindrica* var. *africana*, to distinguish this from types found in other areas or on other continents. The variety as used by the layman refers to a plant population which differs from another one of the same species in one or more recognizable inherited characters. These are the cultivated sorts (strains, selections) that the farmer talks about when he discusses fodder crops or pasture types. They are the names used when making orders from the seedsman. The pasture agronomist uses varietal names in making recommendations for seeds in the establishment of sown pastures.

'Cultivar' is the internationally recognized term for the agricultural variety. The term can be used in all languages and prevents confusion with the taxonomic term variety. A cultivar is a cultivated population of plants with recognized morphological, physiological, chemical or other differences. Cultivar names are written with a capital initial letter. They may follow generic, specific or common names, e.g. *Desmodium* 'Greenleaf', *D. intortum* cv. Greenleaf or *D. intortum* 'Greenleaf'.

Ways and means of classification

The earliest systems of classification were based on preselected characters, such as form or habit, e.g. trees, shrubs, herbs, annuals, biennials, perennials. Carolus Linnaeus (1707–78) was the Swedish naturalist and physician who revolutionized plant classification. Using a so-called sexual system of floral characters, in particular the stamens and pistils, he separated 24 classes of plants and divided them into orders. In *Hortus uplandicus* (published in 1732), his 'Klass 17' included legumes such as *Trifolium* and *Lathyrus*. His classification involved the grasses and the term *Gramina* formed the basis of the family Gramineae. In the first edition of *Species Planatatum* (published in 1753), Linnaeus listed grass species which included *Panicum, Cenchrus, Andropogon, Holcus (Sorghum), Ischaemum, Anthoxanthum, Phalaris, Hordeum, Oryza* and others (Burbidge, 1964).

It was during the early 1800s that the spikelet was recognized as a modified inflorescence and not a flower (Burbidge, 1964). The observation of fundamental differences in the spikelet led to the separation of two subdivisions of the grass family:

1. Paniceae – spikelet with two florets, lower one imperfect, majority of genera in warm or tropical regions.
2. Poaceae – spikelet of one or many florets, imperfect floret terminal, majority of genera in temperature regions.

This primary arrangement has been used in the identification of new tribes and new genera of grasses. Some taxonomists have used a major subdivision Fectucoideae in place of Poaceae, but retained Pancoideae.

Early botanists employed gross morphological features of plants and their flowers. Attention in recent years has been given to microscopic or submicroscopic features of spines, hairs, spores, pollen grains, starch grains, cellular inclusions and other characteristics as diagnostic features in the separation of species and genera (Heyward, 1967; Ornduff, 1967). By using light microscopy to study the leaf epidermis of cells, it was found that the grasses of Festucoideae and Panicoideae have different shaped siliceous (containing silicon) cells in the epidemis. Differences of embryo morphology in grasses had provided characters for regrouping genera immediately below the family level (Reeder, 1957). The shape and sculpture of pollen grains allowed the separation of *Macroptyloma* from *Lablab* species in the Leguminoseae (Verdcourt, 1970).

The use of cytology to determine chromosome numbers, their size and morphology, as well as their behaviour and structure at meiosis, aided the plant systematist in establishing cytotaxonomic relationships (Heyward, 1967). The basic number of chromosomes frequently occurs as exact multiples in different members of a group to give a polyploid series. For instance, a polyploid series exists for *Panicum maximum*, there being diploids, tetraploids, hexaploids and octaploids. This feature is sometimes used to separate species; e.g. *Brachiaria ruziziensis* is a diploid ($2n = 18$) and *B. mutica* is a tetraploid ($2n = 36$).

Biochemical properties of plants have been studied and used by taxomonists for a long time. Examples of plant substances include essential oils, pigments, alkaloids, flavonoids, glycosides, and non-protein free amino acids (Heyward, 1967). An example of chemotaxonomy was the chemical separation of essential oils which confirmed the morphological distinction of two species of *Cymbopogon*. In India, *C. travancorensis* and *C. flexuosus* had been separated by the minor variation in swelling of the pedicelled spikelet. Some years later a chemist found a difference in the oil secreted by the two species. In addition, the oil of *C. travancorensis* contained a chemical compound not found in other Indian *Cymbopogons*. In contrast, the species *C. martinii* exists in India as two similar forms (Bor, 1960). They can be distinguished by certain vegetative characters and distinctive odours when crushed. Both contain palmarose oil (used to perfume soap). It is known that the oil constituents are different. A chromosome count showed that one form of the species is a diploid and the other a tetraploid. They are identified by local names but have not been given separate species ranking.

Immunological studies of proteins yields measures of taxonomic relationships (Heyward, 1967; Leone, 1964; Swingle, 1946). In the technique of electrophoresis free amino acids in the plant extracts contained in a prepared agar gel diffuse towards one or both charged poles and form bands, that can be appropriately stained. The similarity of bands gives an indication of homology and natural affinity of plants. By use of electrophoresis, revisions were made in the Leguminosae, subfamily Papilionideae (Verdcourt, 1970). For example, a number of *Phaseolus* spp. were moved to *Macroptilium*.

Implications of the new systematics

Information from various biological fields when applied to systematics (sometimes called 'new systematics') has provided the taxonomist with new tools and has modified approaches to plant classification. As a consequence, revisions have been made in tribes, genera, species and subspecies. In some instances species have been transferred from one genera to another, necessitating a change in name. To the student of grassland husbandry this can create confusion. Reclassifying a species does not alter plant adaptability and response to management practices. Keeping abreast of the changing names, however, sometimes becomes an exercise in mental gymnastics.

Nomenclature

At first sight, 'scientific' names (sometimes called 'botanical' or

'Latin' names) seem formidable. Often they are long and unfamiliar, and their pronunciation may be difficult. The use of scientific names, however, has a number of advantages:
1. The same names are used in all languages.
2. They are uniformly bionomial (two names).
3. The binomials are exact in delineating a species.
4. They are descriptive (for those versed in Greek and Latin).
5. The choice of a name is governed by the International Rules of Botanical Nomenclature.

'Common' names (also called vernacular, colloquial, folk names) were applied to plants by nations and tribes, each in its own language, before classification systems were designed. The common name is sometimes descriptive (e.g. sword bean, *Canavalia ensiformis*, and lemon grass, *Cymbopogon flexuosus*), may bear the name of a person (e.g. Rhodes grass, *Chloris gayana*) or that of a location (e.g. Townsville stylo, *Stylosantheses humilis*), or may be associated with a habitat (e.g. beachgrass, *Ammophila* spp.). The term is sometimes learned in childhood as handed down from generation to generation (e.g. wire glass, species of *Cynodon*, *Aristida* and others; foxtail, species of *Setaria*, *Andropogon* and others; beggarweed, species of *Desmodium*). A common name may or may not represent a species. It may include everything within a genus and can be the generic name, as with *Desmodium*. The use of common names can be confusing and misleading:
1. They are restricted to one language or dialect, and perhaps to one locality.
2. The same term may be used for several species.
3. The names are indefinite.
4. They are not regulated by a constituted authority.

References

Bor, N. L. (1960) *The Grasses of Burma, Ceylon, India and Pakistan*, Pergamon Press: London.
Burbidge, N. L. (1964) Grass systematics. In Bernard (ed.) *Grasses and Grasslands*, Macmillan: London, pp. 13–28.
Lawrence, C. H. M. (1951) *Taxonomy of Vascular Plants*, Macmillan: New York.
Leone, C. A. (1964) *Taxonomy Biochemistry and Serology*, Ronald Press: New York.
Ornduff, R. (ed.) (1967) *Papers on Plant Systematics*, Little, Brown and Co.: Boston.
Parodi, L. R. (1943) Una nueva especie de Sorghum cultivada en la Argentina, *Rev. Argent. Agron.* **10**, 361–72.
Reeder, J. R. (1957) The embryo in grass systematics, *Amer. J. Bot.* **44**, 756–68.
Swingle, D. B. (1946) *A Textbook of Systematic Botany*, McGraw-Hill: New York.
Verdcourt, B. (1970) *Studies in the Leguminoseae – Papilionoideae for the 'Flora of Tropical East Africa': III and IV, Kew Bull.* **24**, 379–447, 507.

Chapter 5

Plant growth and development

Perennial tropical grasses and legumes have the potential for year-round growth and production of herbage. Their development is influenced by the energy supply (light), concentration of CO_2, available nutrients and water, and prevailing temperature. Light is the primary factor affecting their growth potential through its relationship to photosynthesis, but the input of solar energy may be limited by low temperature, water stress, shortage of soil nutrients and improper management of the pasture or fodder crop. The climatic limitations cannot be changed. The restrictions of water and soil nutrients can be offset by irrigation and fertilizer application, though it may not be economical to do so.

Plant metabolism

Grazingland production depends on the conversion of CO_2, nutrients and water into plant tissue (herbage) for animal utilization. Carbohydrates are the first tangible products of photosynthesis. They may be stored in the plant to serve as a source of energy or transformed by the plant into other food products (proteins and fatty substances). In tropical grasses, the production of 1 g of dry matter corresponds to a fixation of between 4 150 calories (*Tripsacum laxum*) to 5 020 calories (*Digitaria decumbens*) of bound energy (Butterworth, 1964). This represents about one-sixteenth of the incoming radiation available for photosynthesis.

Photosynthetic efficiency

In general, the photosynthetic rate of leaves is proportional to the light intensity. For example, the photosynthetic rate of *Paspalum dilatatum* and *Cynodon dactylon*, like those of sugar-cane and maize, increases with rising light intensity up to and exceeding 6 000 foot-candles, thus achieving conversion rates of 5–6 per cent of the incoming radiation (Cooper and Tainton, 1968). In contrast, the light saturation of temperate grasses such as *Lolium perenne* and *Dactylis glomerata* is reached at 2 000–3 000 foot-candles, giving a conversion rate of 2–3 per cent. At low light intensities, the photosynthetic efficiency of individual leaves is greater than at higher intensities, so that 12–15 per cent of the incoming light energy may be fixed (Cooper and Tainton, 1968). Even so, leaves of *P. dilatatum* give higher levels of net photosynthesis under low light conditions than *Lolium* species (Mitchell, 1956).

The higher net photosynthesis, greater energy conversion efficiency and increased growth potential of tropical grasses over temperate grasses can probably be contributed to one or several of the following factors:

1. The tropical grasses, including species of *Paspalum, Digitaria, Sorghum, Zea, Axonopus, Chloris* and *Eragrostis*, have a photosynthetic carbon pathway which differs from the classical Calvin cycle typical of temperate grasses (Cooper and Tainton, 1968; Downes and Hesketh, 1968; Hatch and Slack, 1970). Carbon is first fixed as C_4 rather than C_3 compounds, which may represent a more efficient process for the manufacture of photosynthates.

2. Temperate grasses photorespire when exposed to light, producing appreciable quantities of CO_2. When the O_2 content of the atmosphere is reduced, however, net photosynthesis is increased. Tropical grasses do not photorespire, and a decrease of atmospheric O_2 does not show a corresponding increase in net photosynthesis (Cooper and Tainton, 1968).

3. Tropical grasses can reduce the CO_2 content of the surrounding atmosphere to a lower level than temperate grasses; i.e. they have a lower CO_2 compensation point (Cooper and Tainton, 1968).

4. In the sheath cells surrounding the vascular bundles of most tropical grasses are found specialized plastids concerned with starch formation and storage (Barnard, 1964). They may also facilitate transportation of photosynthates from the leaves. Plastids of this type do not occur in temperate grasses.

In a study of 20 tropical pasture plants, Ludlow and Wilson (1970) noted that species differed in their photosynthetic capacity, with grasses having a higher potential than legumes, about double in some instances. Heslehurst and Wilson (1971) demonstrated that the superior growth rate of the tropical grass *Setaria anceps*, as compared to the legume *Desmodium intortum*, was due to increased photosynthesis rather than to canopy structure, i.e. arrangement of leaves. Tow (1967) found that *Panicum maximum* var. *trichoglume* produced more dry matter per unit of intercepted light than *Glycine wightii* cv. Tinaroo. Within species of temperate grasses, differences in light saturation level, and hence maximum photosynthesis, also occurred. In *Lolium perenne* and *L. multiflorum* variation of more than 50 per cent in maximum photosynthetic rate was noted among individual genotypes, being associated with variation in mesophyll structure of the leaf (Wilson and Cooper, 1967). This suggested that the plant breeder should be able to develop cultivars that are more efficient in conversion of light energy.

Products of photosynthesis

Storage products

The principal storage materials are sucrose and fructose, which are water-soluble and occur in the cell sap. In some species, starch, which is not water-soluble, occurs largely in storage organs but may accumulate temporarily in leaves.

The grasses have been roughly divided into two types (a) those which store the fructose form of sugar (temperate season types – *Lolium*, *Bromus* and *Dactylis*), and (b) those which store sucrose and starch (warm season types – *Sorghum*, *Zea*, *Saccharum*, *Cynodon*, *Chloris*, *Cenchrus*, and *Panicum*) (Wilson and Ford, 1973). In general, fructose forms are more readily digested than sucrose.

Fats are stored as glycerides of fatty acids, which can be broken down into simpler compounds and used by the plant or animal. They may occur as waxes (cutins) on the epidermis of the leaves and stems or as phospholipids in protoplasmic membranes.

Proteins are primarily constructed of amino acids, being synthesized from raw materials by plants. The mature ruminant animal can produce all of the amino acids required for its metabolism from relatively simple nitrogen compounds. The young ruminant, however, must depend on ingested feed for its source of proteins. Other animals cannot produce certain amino acids in sufficient quantity for normal growth and maintenance. These 'essential' amino acids must be supplied in the diet.

Structural products

Cellulose is the most abundant carbohydrate comprising the structural framework of the cell wall, but hemicelluloses are also present. The latter are composed of various polysaccharides which are easily hydrolysed to simple sugars and other products. Thus, they are more readily digested than cellulose. Some cell walls contain fatty substances such as suberin and waxes. Lignin is a non-carbohydrate complex organic compound deposited in the interstices of cell walls of leaves and stems. It is extremely resistant to chemical or enzymatic activity and its presence markedly decreases herbage digestibility. Tropical legumes generally contain more lignin than grasses during early growth. Their lignin content, however, remains fairly constant, while that of grasses increases with age in stems and leaves.

In animal production it is the storage materials that supply the major nutrients for production of meat, milk and wool. As forage agronomists, however, we are concerned with structural materials because the cell wall constituents are highly indigestible. Their presence does not hinder the digestion of cellular products, but in large quantities they reduce the total percentage digestibility of the herbage ingested.

Growth patterns

In the tropics grasses and legumes develop as annuals or as perennials. Pattern of growth of the latter is cyclic and much or all of the herbage dies because of maturity and insufficient water supply in the dry season. There are no true grass or legume biennials in the tropics.

With some species, such as *Chloris gayana*, *Cenchrus ciliaris*, *Hyparrhenia rufa* and *Panicum maximum*, a short vegetative period of 4–6 weeks is followed by stem elongation and emergence of floral parts. Unless plants are cut, new tillers develop and produce inflorescences throughout most of the growing season. A number of species, such as *Pennisetum purpureum*, *Tripsacum laxum*, *Andropogon gayanus*, *Axonopus scoparius*, *Melinis minutiflora* and *Brachiaria ruziziensis*, remain vegetative over a longer period of time. Stem elongation occurs during the latter part of the rainy season in these species, with a fairly uniform development of inflorescences and seed maturity in the dry season.

Perennial tropical legumes are more drought tolerant than grasses and many persist well into, or throughout, the dry season, depending on the amount of soil moisture. Growth is arrested and may cease where the dry period exceeds 6 months. Under grazing or cutting, upright species such as *Stylosanthes guianensis*, *Cajanus cajan* and *Leucaena leucocephala* produce new shoots near ground level or along the main stem. Lucerne is the only species which sends out many crown tillers, but under cutting or grazing *Pueraria phaseoloides*, *Clitoria ternatea*, *Macroptilium atropurpureum*, *Glycine wightii* and some *Desmodium* species develop sparse crown tillers. Some species, such as *Desmodium sandwicense*, are light-insensitive and develop flowers profusely throughout the growing season. Most perennial species produce flowers during the latter part of the rainy season or early dry season. Generally, flowering is non-uniform and spread over several weeks, or even months, making seed harvest difficult.

The growth pattern of grasses and legumes is different, and adds to the complexity of compounding mixtures for sowing, as well as to the difficulty in management of mixed swards. Furthermore, the imposition of any grazing or cutting treatment during one phase of the growth cycle of any given mixture may indirectly affect subsequent cycles to a great extent.

Stages of growth and development

The following general stages of growth can be identified:
1. Seedling – time from emergence to tiller formation or axillary branching.
2. Establishmental – a transitional period when young plants are producing leaves and tillers, nodal and secondary roots. The point in time or stage of development that a single plant becomes established is not easy to identify. With a solid stand, or sward, it is the time that the tiller number remains fairly constant.
3. Vegetative – production of leaves, shoots, stolons and rhizomes with no visible, or relatively few, floral stems. This stage includes the establishmental period for newly sown herbage crops, but also represents the time from renewed growth after the dry season until large numbers of floral stems begin to emerge for established swards.
4. Floral stem elongation – shoots having flower primordia increase in height, or length, as internodes lengthen. This is frequently a response to changing day length but may be influenced by temperature, water stress or some physical damage to the plant.
5. Reproductive – floral and seed production.

Seedling and young plant growth

Under favourable conditions, seedlings of some grass species emerge in 5 or 6 days after sowing but others may be delayed from 10 to 14 days. As soon as light strikes the coleoptile, photosynthesis begins in the foliage leaf but endospermic reserves are available for the first 7–10 days of growth. Leaf appearance generally occurs in a linear rate with time and tillering, beginning during exfoliation of the fourth to eighth leaves. Stolons and rhizomes develop during the third or fourth week, but vary with the species and time of sowing. The growing tip is found just below the soil surface, depending on depth of sowing, until three or four leaves are formed, then remains near the ground until stem elongation takes place. In the early stage of growth, grasses are more competitive than legumes because of increased tillering and development of fibrous roots.

Seedlings of legumes such as lucerne, the *Desmodiums* and *Macroptilium* emerge from the soil in 3–5 days but others may be delayed 2 weeks or more. Hard seed coats and degree of sacrification influence time of germination and emergence. Some cotyledons become green and begin to photosynthesize. Shortly after emergence the seedling leaf unfolds and development of the true leaf occurs in 10–12 days for those types which emerge early. Lateral branching usually takes place in another 2 weeks.

Dry weights of shoots and roots during seedling growth may be used as criteria of legume seedling vigour (Black, 1959; Chow, 1972; Whiteman, 1968). Seed size confers a definite advantage on seedling shoot and root growth. Seedlings from large-seeded species develop more rapidly than those of small-seeded species. The more rapid growth continues throughout the development of the young plant so that shoot and root weight are higher for a period of several weeks. The greater seedling vigour provides a competitive advantage in the early stage of plant development but may not determine final establishment. After 8–10 weeks the difference in growth rate begins to change but may still be visible at the end of the first season. If plants are cut during the floral bud or early flowering stages, the relative order of plant size will be different by the time that plants approach flowering a second time.

Some tropical grasses and legumes are established by transplanting crown splits or stem pieces rather than by sowing seeds. If the growing tip has not been damaged it usually continues growth. Otherwise, new shoots arise from crown buds in 10–14 days. When stem pieces with three to five nodes are set in the soil, it is usually the uppermost node that produces a bud because of apical dominance. This generally occurs in 8–14 days, to be followed by new shoots from the lower nodes. If the stem piece is rather juvenile, the upper portion may die back so that shoots form at the lower nodes. They may even come from nodes placed below ground level. Even though the upper nodes produce the first shoots, the lower ones usually grow more rapidly, because of proximity to the soil water and nutrient supply. When long stems are layered in furrows and covered with soil, or left exposed on the soil surface, the first nodal buds develop near the stem base rather than the tip and new shoots do not develop from all nodes. In fact, a portion of the upper stem usually deteriorates. Regeneration of growth is slower than setting stem pieces in the soil.

Growth curves

When established pastures are closely grazed or mowed the herbage regrowth curve is roughly sigmoid in form, as shown in Fig. 5.1. After an initial lag in recovery of growth, the exponential phase begins but its extent and duration depend on the degree of plant defoliation. Studies with *Cynodon nlemfuensis* in Nigeria showed that the phase lasted for about 25 days after defoliation to about 3.0 cm and when fertilized with 75 kg/ha of N (Mohamed Saleem, 1972). With

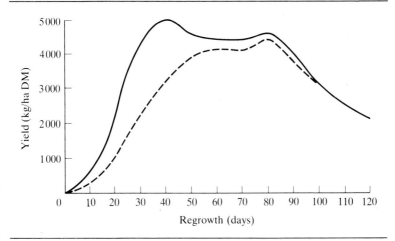

Fig. 5.1 Uninterrupted growth of *Cynodon nlemfuensis*, Ibadan, Nigeria (Mohamed Saleem, 1972). Solid line – 75 kg/ha N fertilizer applied; dashed line – no fertilizer.

less severe defoliation the period of rapid growth was reduced. The slope of the regrowth curve was modified by available soil nutrients. The *Cynodon* regrowth curve flattened out without applied N and the length of the exponential phase was extended over a period of about 45 days. The effect of the added N was lost after about 35 days from cutting and application of fertilizer, however, and subsequent growth of the two treatments was essentially the same. Following the period of rapid regrowth, herbage production declined because of senescence of lower leaves and reduced rate of root development.

The natural uninterrupted growth curves are obtained by cutting an established sward at a predetermined height above ground level. Then at given sampling dates different areas are cut at the same height and dry matter production determined. Similar studies were made with *Andropogon gayanus* in northern Nigeria (Haggar, 1970), *Cenchrus ciliaris* in Australia (Burt, 1968), and several grass species in Kenya (Taerum, 1970a, 1970b), and *Pennisetum purpureum* in Australia (Ferraris and Sinclair 1980).

Use of growth curves

Regrowth curves can be used as guides to establish rest periods for accumulation of dry matter in providing animals optimal quantities of herbage. This is illustrated in Fig. 5.2 where two growth curves of *Cynodon nlemfuensis* (Mohamed Saleem, 1972) have been reconstructed for uninterrupted regrowth periods of 70 days. After cutting the established sward to ground level and topdressing with

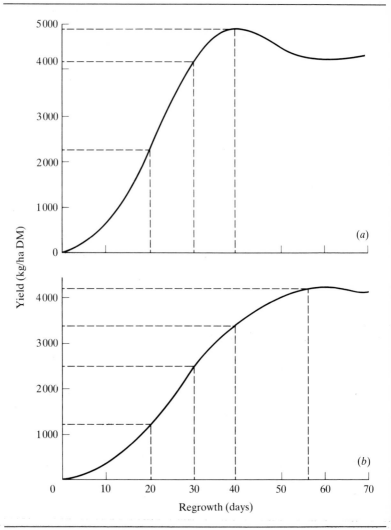

Fig. 5.2 Use of *Cynodon nlemfuensis* growth curves to determine optimal rest periods in rotational grazing: (*a*) 75 kg/ha of N fertilizer applied; (*b*) no N applied.

75 kg/ha of N, about 4 100 kg of dry matter were produced in 30 days. With no N fertilizer applied, almost 60 days were needed to produce the same amount of herbage.

In choosing a rest period the grazier is interested in herbage quality as well as quantity, and in the seasonal as well as daily production. In addition to N fertilization he will need to consider available moisture, seasonal effects, height of cutting or intensity of grazing and species composition of the sward. Regrowth of grass–legume mixtures will be different than that of species growing alone.

Leaf : stem ratio

In the early stages of growth, the herbage consists entirely of leaves. As grasses age, stems comprise a greater percentage of the bulk forage. This is illustrated for non-defoliated single plants of *Chloris gayana* and *Panicum maximum* in Fig. 5.3. Before plants elongated, the ratio exceeded 1.0, at flowering fell to about 0.5 and later dropped below 0.25 (Taerum, 1970b).

Under a sward condition of closely spaced plants and a rather constant rate of tiller appearance, the ratio would change less drastically. As the season progresses, however, it is difficult to maintain a ratio greater than 0.5, even under close grazing. Young stems have about the same percentage digestibility as leaves, so that the leaf : stem ratio is less critical in immature plants than in older plants with elongated flowering stems.

The changing ratio of upright growing legumes such as stylo and lucerne would follow a pattern similar to that of grasses, but in trailing types it would be less divergent.

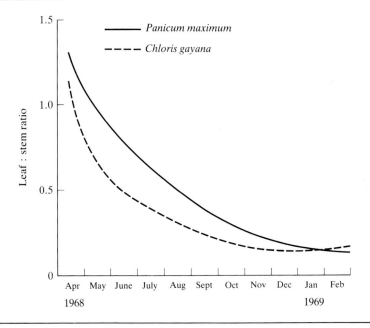

Fig. 5.3 Change in leaf : stem ratios during uninterrupted single plant growth (Taerum, 1970b).

Root growth

As the above-ground plant parts increase in size the root system also enlarges. The data in Table 5.1 show shoot : root ratios for *Cenchrus ciliaris, Chloris gayanus* and *Panicum maximum*. Their root dry matter yields, as percentage of the total plant, are given in Fig. 5.4. It is readily seen that:

1. Roots generally comprise less of the total plant weight than the shoots.
2. Roots usually constitute a greater proportion of the total plant weight in the juvenile stage than in the more mature plant stage, but exceptions occur as noted by the small variation in root percentage of *C. ciliaris*.
3. Species differ in their shoot : root ratios.

The information given for *Panicum maximum* and *Chloris gayana* (Table 5.1) portrays the usual pattern of seasonal change in shoot : root ratios, and more specifically that which occurs as grass plants mature. In the Kenya study ratios were small in early stages of plant development, became progressively larger as the plants grew and then showed a tendency to decline or fluctuate (Taerum, 1970b). This pattern of change was also noted in temperate species where increased ratios coincided with flower formation, decreased temperatures and advanced plant maturity (Troughton, 1961; Ozanne *et al.*, 1965; MacColl and Cooper, 1967). In Kenya, however, the higher shoot : root ratios occurred during a period of increased flowering when nutrients were diverted to seed development. Temperatures did not fluctuate widely but soil moisture decreased markedly because of the dry season.

Variation among tropical and subtropical species in the total root production and distribution has been noted (Laird, 1930; Burton, 1943). Burton (1943) found that *Paspalum notatum* produced three

Table 5.1 *Shoot : root ratios of grasses grown at Muguga, Kenya (Taerum, 1970b)*[*]

Sampling data[†]	Cenchrus ciliaris	Chloris gayana	Panicum maximum
18 April 1968	2.99	3.14	4.26
6 May	3.43	3.74	5.99
2 June	4.30	5.92	7.66
14 July	4.41	6.16	8.80
19 August	3.77	6.98	9.18
24 September	3.60	8.30	14.16
15 November	2.43	7.91	14.20
6 January 1969	2.81	4.77	8.33
1 March	2.54	7.72	10.94

[*] Single plants started in pots and transplanted in the field at the beginning of the rainy season in April.
[†] Species significantly different at 0.01 level; dates different for *C. gayana* and *P. maximum*.

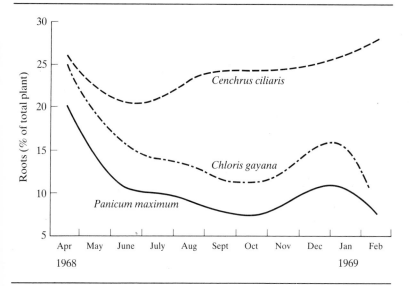

Fig. 5.4   Root dry matter yields as percentage of total plant (Taerum, 1970b).

times more total roots in a 6-foot soil profile than other *Paspalum* species and seven times more than *Axonopus afinis*. In Nigeria, *Cynodon* IB.8 showed greater root development as compared to several other selections of *C. nlemfuensis* (Mackenzie and Chheda, 1970). This appeared as a thicker mat of fine roots in the upper 15 cm of soil and a larger number of roots in the soil profile, particularly in the top metre.

Data from Kenya (Taerum, 1970b) illustrate that differences among species in depth of root penetration into the soil can be expected (Table 5.2). More than 50 per cent of the total root system occurred within the upper 20 cm of the soil, being typical of most grasses that possess a fibrous root system. The maximum rooting depth was attained in about 9 months from the time of sowing. *Cenchus ciliaris* produced a more extensive root system in terms of accumulated dry weight than *Panicum maximum* and *Chloris gayana*, even though the percentage distribution within the various soil depths did not greatly differ. When the grasses had grown undisturbed for 28 months, excavations to 3 m showed that *Cenchrus* yielded approximately 7 000 kg/ha of dry roots compared to 3 300 kg/ha for *Panicum*. The sward was then cut to ground level. After 4 months of plant regrowth, *Cenchrus* produced 19 500 kg/ha and *Panicum* 16 600 kg/ha of herbage. These data are not indicative of the seasonal herbage yields, since total production over time of *P. maximum* exceeds that of *Cenchrus*. They do, however, illustrate

Table 5.2 *Root dry weight distribution of three tropical grasses grown in Muguga, Kenya (Taerum, 1970b)**

Soil depth (cm)	Cenchrus ciliaris (%)	Chloris gayana (%)	Panicum defoliated (%)	P. maximum non-defoliated (%)
0–20	53.3	57.3	51.7	49.3
20–40	15.1	18.8	16.3	18.0
40–60	8.3	9.3	9.8	11.6
60–100	10.8	9.3	10.3	12.7
100–140	6.2	4.6	6.1	6.5
140–180	3.6	2.4	3.4	1.7
180–220	1.8	1.2	1.8	0.2
200–260	0.5	0.1	0.6	–
260–300	0.1	–	–	–

* Nine samplings made from April 1968 to March 1969.

the large reservoir of nutrient reserve which can be accumulated by some species.

Once roots are established in the deeper soil regions they can be maintained under proper grazing management, even though their depth fluctuates with the water-table. Rooting patterns of this type are of great importance in areas that have alternating wet and dry seasons. During the prolonged periods of drought the deeper roots play a significant role in the movement of water and nutrients from the lower soil regions.

One method of measuring the regrowth potential of sod crops where nutrient reserves are located in roots, rhizomes, stolons, crowns and stubble is that of removing sod plugs and allowing growth to take place in the dark (Burton and Jackson, 1962). The etiolated herbage is cut and weighed at intervals until no further re-growth occurs. Since photosynthesis is prevented, the regrowth can be related to the carbohydrates and other energy reserve materials stored in the plant organs. Akinola *et al.* (1971) used this method in Nigeria to measure the effects of cutting height and N application on several selections of *Cynodon nlemfuensis*. They clearly demon-strated that selections differed in regrowth potential, showing that the method could be used by the grass breeder in evaluating germ-plasm and comparing potentially new cultivars.

Legumes produce a primary or tap-root that develops vertically and reaches depths of 8 m or more, depending on the species and soil type. In Australia the tap-root elongation of *Stylosanthes humilis* averaged 1.5 cm growth per day over the first 40 days after sow-ing (Torssel *et al.*, 1968). No lateral roots developed until the tap root had reached a length of 12–13 cm. During the main growing season 80 per cent of the root length and 70 per cent of the root sur-face were found in the top 1.0 cm of soil.

Components of yield

The above-ground herbage of grasses can be separated into stems, leaves, leaf sheaths and inflorescences, the legumes into stems, leaves and inflorescences. After stems elongate and side branches develop in grasses, the stem fraction exceeds the leaf fraction of the whole plant (Fig. 5.5). The time at which stem weight exceeds leaf weight varies with species and with the season. For example, *Andropogon gayanus* remains leafy until rather late in the season in West Africa, where growth begins with the rains in May and continues into November (Haggar, 1970). *Cenchrus ciliaris* shows a different pattern of development. The stems rapidly surpass the weight of the leaves because of early stem elongation which becomes more pronounced as the season progresses (Burt, 1968). Although flower heads emerge early, they make up less than 10 per cent of the total plant dry matter. With other species, such as *Sorghum vulgare*, the floral organs constitute as much as 25 per cent or more of the yield components because they are larger with heavier seeds.

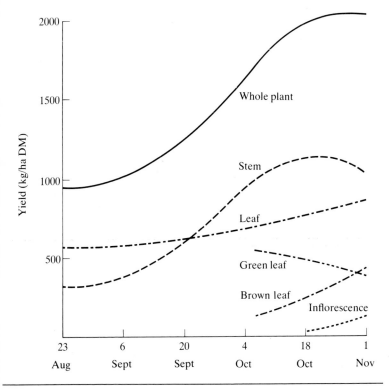

Fig. 5.5 Yield components of *Andropogon gayanus*, Zaria, Nigeria (Haggar, 1970).

Regulation of growth

A number of environmental factors affect plant growth and the effects of one may markedly depend on others. In addition, a number of conditions imposed by man and animals alter plant development.

Soil moisture

Rainfall is the greatest single factor affecting plant growth and herbage dry matter production in most of the tropics and subtropics. Its seasonal nature, variability and erratic incidence, and the high evaporation potential in many areas result in periods of short water stress and prolonged droughts.

The plant form itself increases the effectiveness of water availability. Leaves of many grasses tend to grow upright and flex upward from the midrib, thus forming a leaf catchment. In addition, the cuticle surface of the leaves is covered with minute wax platelets. If a leaf expands in a dry environment and in full sunlight, the platelets are more dense than when plants grow under high moisture or in the shade (Gwynne, 1966). A well-developed platelet formation increases the contact angle of water droplets falling on the leaf surface and promotes runoff. As droplets coalesce, they move towards the leaf base and join with others, creating a stem flow to the plant base. This results in water penetration of the soil and wetting around the plant roots. With 175 mm or less of rainfall at a given time, the spot wetting at the plant base will be considerably higher than the same amount falling on an open surface. With more than 175 mm in short duration, plant runoff and water delivery at the plant base exceeds the absorptive capacity of the soil so that the water spreads over the surface rather than penetrating close to the plant (Glover *et al.*, 1962).

Species differ in their utilization of water. This can be related to root extension and to the extraction of water from the soil interstices, but is also a genetic characteristic of the species or cultivar in the photosynthetic transformation of raw materials into plant tissue.

Temperature

The dry matter yield of plant tops and roots of tropical grasses increases markedly with an increase in temperature to an optimum which lies between 30 °C and 35 °C (Burt, 1968; Chudleigh *et al.*, 1977; Deinum and Dirven, 1972; Lovvorn, 1945; Mannetje and Pritchard, 1974; McCloud, 1963; Wilson and Ford, 1971, 1973; Ferraris, 1978). Most species grow vigorously at 35 °C and some up to 38 °C. The temperate grasses have an optimum between 20 °C and

25 °C, making considerable growth at 10–12 °C, but are usually killed at 38 °C.

Subjecting tropical or subtropical species to colder temperatures, such as occur at night in the higher elevations and upper latitudes, adversely affects plant growth (Ingle and Rogers, 1961; Youngner, 1959). In Australia, for example, plants of *Cenchrus ciliaris* were grown for 2 weeks at 25 °C, subjected to 10 °C for only 1 week and then returned to the original temperature (Burt, 1968). During the cold period growth rate declined, especially rhizome appearance and root development. During exposure to the low temperature, the distribution pattern of photosynthates was altered, with a higher proportion being used for stem development and less moving to the roots. On returning the cold-treated plants to the higher temperature, the photosynthate distribution pattern returned to normal, i.e. the same as the control plants, but the total dry matter production of the 6-week-old plants was reduced by about 30 per cent.

Tropical legumes have a lower optimal growth temperature than tropical grasses. Whiteman (1968) in Australia studied the effects of temperature on the growth of six legumes: *Desmodium intortum, D. sandwicense, D. uncinatum, Macroptilium atropurpureum, M. lathyroides* and *Glycine wightii*. Optimum temperature was 30/25 (day/night) ± 3 °C. Above 33/28 °C growth rate declined; it was also reduced at 18/13 °C, and became abnormal at 15/10 °C. Differences among species have been noted, e.g. *Stylosanthes humilis* produced higher herbage yields at 33/28 °C than at lower temperatures in northern Australia (Bryant and Humphreys, 1976; Cameron and Mannetje, 1977; Humphreys, 1967).

Temperature also influences growth of legumes by its effect on symbiotic N fixation (Mes, 1959). In *Stizolobium deeringianum* a low night temperature of 10 °C was more harmful to growth when the plant was dependent on the associated *Rhizobium* than when N was supplied by a nutrient medium. Nitrogen fixation was poor with night temperatures below 18 °C and increased with rising temperatures. This inferred that the low temperatures had an adverse effect on bacterial activity as well as on plant processes.

Light Intensity

It is not uncommon to hear livestockmen say that fattening cattle make less daily gain during long periods of cloudy weather than when skies are clear, even though being grazed on the same or comparable pastures. Ambient temperature may be lower under cloud cover but the difference is probably insufficient to affect animal output in the tropics. Thus, one suspects that some change might occur in the herbage when grown under reduced light to account for the difference in animal response.

Plant growth increases as light intensity increases, up to the point of light saturation of the leaves in a canopy exposed to full sunlight (Blackman and Templeman, 1938; Pritchett and Nelson, 1951; Wilson, 1962; Cooper and Tainton, 1968). This is illustrated by the *Cenchrus ciliaris* data given in Table 5.3, where it is shown that production of plant parts was enhanced by higher solar radiation (Burt, 1968). One could expect, then, that shading would reduce plant growth and development (Benedict, 1940; Mitchell, 1955).

Table 5.3 *Growth and development of buffel grass (Cenchrus ciliaris) plants in Australia (Burt, 1968)*

Sowing date	Day length (h)	Temper-ature (°C)	Daily radiation (cal/cm²)	Leaves* (no.)	Tillers* (no.)	Plant wt (g)	Root wt (g)	Rhizome wt (g)
June	9.48	10.0	350	0.119	0.170	4.7	2.4	0.2
July	10.06	9.0	250	0.166	0.219	5.2	4.2	0.3
August	11.26	11.5	410	0.214	0.222	6.4	5.4	0.2
November	14.03	19.5	690	0.468	0.524	6.4	4.5	0.4
January	14.27	22.0	700	0.492	0.508	4.6	2.8	0.2

* Daily rate of leaf appearance on primary shoot and tiller appearance on entire plant; both significantly different among sowing dates at P = 0.05 level.

In southeastern USA, Burton *et al.* (1959) showed that reduced light decreased herbage yields, production of roots and rhizomes, nutrient reserves for regrowth and total available carbohydrates in the herbage of *Cynodon dactylon* cv. Coastal (Table 5.4). The effect was even more dramatic with high rates of applied N, with forage yields being decreased proportionately as light was reduced, i.e. a reduction of light by 30 per cent decreased yields by about 30 per cent. In full sunlight applied N consistently increased plant density and leaf area. Both declined with shade and within 2 years many plants had died after receiving only 28.8 per cent sunlight.

Of most significant importance from the animal nutrition viewpoint was the reduction in total available carbohydrates in the herbage, especially when less than 50 per cent sunlight reached the grass canopy. The lower energy value of grass growing under reduced light conditions could limit rumen flora activity and affect animal output. Shade significantly increased the lignin content of the herbage which would decrease digestibility. Thus animals consuming forage produced under cloudy climatic conditions could be expected to make less liveweight gains.

Most tropical grasses show a reduced yield with lower light intensity, especially C_4 plants which are strongly depressed in the shade (Ludlow, 1978). Wong and Wilson (1980), however, noted that green panic gave a 27–30 per cent increase in shoot weight when grown under 68 per cent shade and defoliated each 8 weeks. This was reflected in total biomass and not just reallocation of

Table 5.4 *Influence of light intensity on the growth and production of* Cynodon dactylon *cv. Coastal (Burton et al., 1959)*

Available light* (%)	Seasonal dry matter (t/ha)	Roots and rhizomes[†] (t/ha)	Reserve index[‡] (g)	TAC[§] (%)	Lignin (%)
100.0	15.5	5.17	2.2	15.8	9.2
64.3	14.1	3.51	1.6	14.0	9.7
42.8	10.6	3.44	0.8	10.5	10.2
28.8	8.1	2.39	0.1	9.0	10.4

* Light reduction obtained by placing different thicknesses of cheesecloth over plots.
[†] Dry matter production in the 0–30.5 cm soil zone.
[‡] Fresh weight of herbage obtained from soil cores 15 cm in diameter and 30.5 cm deep; grown for 33 days in the dark.
[§] Total available carbohydrates in the herbage.

assimilates from roots to shoots. The increased herbage yield may have been related to greater uptake of nitrogen rather than a reduced light effect.

Photoperiod

The influence of day length on plant growth is usually overshadowed by its conspicuous effect on flowering in many grasses and legumes. When short-day plants, such as *Stylosanthes humilis* and *S. guianensis* (Mannetje, 1965) and *Hyparrhenia rufa* (Agregeda and Cuany, 1962), are grown in long days, many or all plants remain vegetative and accumulate dry matter. Increased tillering occurs in some grasses with longer days, as was shown for *Paspalum dilatatum* and *P. notatum* by Knight and Bennett (1953). Hutton (1970) reported that 'Siratro' yielded 30 per cent more herbage when grown under a 16-hour as compared to an 8-hour day. The increased yields came from continued growth of the elongating stems and development of axillary branches which did not produce flower primordia.

Leaf area index

The ratio of leaf surface to soil surface was termed leaf area index (LAI) by Watson (1947). The critical LAI, point of maximal growth and interception of about 95 per cent of the incident light, varies greatly with species, being about 4.0 for *Panicum maximum* var. *trichoglume* (Humphreys, 1966) and for *Cynodon dactylon* (Alexander and McCloud, 1962), 7.1 for *Lolium perenne* and 3.5 for *Trifolium repens* (Broughman, 1958). It is determined by the amount of intercepted light, size and shape of leaves, their angle of vertical orientation, the distance between them and their arrangement in the horizontal plane. With more upright leaf orientation a higher light intensity is needed for saturation since more light penetrates into the canopy.

At the critical LAI, photosynthesis reaches maximum efficiency and the rate of dry matter accumulation usually levels off or declines. The LAI *per se* does not stabilize at this point but may increase as the plants continue to develop and values up to 26 have been recorded (Alexander and McCloud, 1962). The leaf respiration rate may be reduced in the relatively cool microenvironment within the canopy so that growth rate beyond the optimal LAI declines gradually, thus allowing dry matter to accumulate. The optimum LAI usually increases with the approach of flowering because of the wider vertical spacing of leaves and the diminishing leaf size on the flowering stem. This also permits greater penetration of light into the canopy (Evans *et al.*, 1964).

Defoliation to soil level immediately reduces the LAI below the critical value and has a strong influence on plant regrowth potential. Under grazing or cutting conditions the sward should be managed in such a way so as to leave an ample residual LAI for continued or rapid renewal of growth (Alexander and McCloud, 1962; Brown and Blaser, 1968; Humphreys and Robinson, 1966; Broughman, 1958, Ludlow and Charles-Edwards, 1980).

Plant competition

As the sward of perennial grasses and legumes ages there is increased competition among plants for soil nutrients, soil moisture and solar radiation. Over time, a steady decline in plant growth and herbage production occurs. This has frequently been attributed to overcrowding of plants or excessive root mass accumulation, leading to a so-called 'sod-bound' condition. Often the basic problem is due to a progressive loss of soil fertility and aeration. Generally speaking, annual topdressing of fertilizer and proper pasture management will maintain a productive sward.

References

Agregeda, O. and **R. L. Cuany** (1962) Effectos fotoperiodicos y fecha de floración en jaragua (Hyparrhenia rufa), *Rev. Interam. Ciencias Agric. (Turrialba)* **12**, 146–9.

Akinola, J. O., J. A. Mackenzie, and **H. R. Chheda** (1971) Effects of cutting frequency and level of applied nitrogen on productivity, chemical composition, growth components and regrowth potential of three Cynodon strains, *W. Afr. J. Biol. Appl. Chem.* **14**, 7–12.

Alexander, C. W. and **D. E. McCloud** (1962) CO_2 uptake (net photosynthesis) as influenced by light intensity of isolated bermudagrass leaves contrasted to that of swards under various clipping regimes, *Crop Sci.* **2**, 132–5.

Barnard, C. (1964) Form and structure. In C. Benard (ed.), *Grasses and Grasslands*, Macmillan: London, Ch. 4.

enedict, H. M. (1940) Growth of some range grasses in reduced light intensities at Cheyenne, Wyoming, *Bot. Gaz.* **102**, 582–9.

lack, J. N. (1959) Seed size in herbage legumes, *Herb. Abst.* **29**, 235–41.

lackman, G. and W. G. Templeman (1938) The interaction of light intensity and nitrogen supply in the growth and metabolism of grasses and clover (Trifolium repens), *Ann. Bot. N. S.* **2**, 765–91.

roughman, R. W. (1958) Interception of light by the foliage of pure and mixed stands of pasture plants, *Aust. J. Agr. Res.* **9**, 39–52.

rown, R. H. and R. E. Blaser (1968) Leaf area index in pasture growth, *Herb. Abst.* **38**, 1–9.

ryant, P. M. and L. R. Humphreys (1976) Photoperiod and temperature effects on the flowering of Stylosanthes guyanensis, *Aust. J. Exptl Agric. Anim. Husb.* **16**, 506–13.

urt, R. L. (1968) Growth and development of buffel grass (Cenchrus ciliaris), *Aust. J. Exptl Agric. Anim. Husb.* **8**, 712–19.

urton, G. W. (1943) A comparison of the first year's root production of seven southern grasses established from seed, *J. Amer. Soc. Agron.* **35**, 192–6.

urton, G. W. and J. E. Jackson (1962) A method of measuring sod reserves, *Agron. J.* **54**, 53–5.

urton, G. W., J. E. Jackson and F. E. Knox (1959) The influence of light reduction upon the production, persistence and chemical composition of Coastal bermudagrass, Cynodon dactylon, *Agron. J.* **51**, 537–42.

utterworth, M. H. (1964) The digestible energy content of some tropical forages, *J. Agric. Sci., Camb.* **63**, 319–21.

ameron, D. F. and L. 't Mannetje (1977) Effects of photoperiod and temperature on the flowering of twelve Stylosanthes species, *Aust. J. Exptl Agric. Anim. Husb.* **17**, 417–24.

how, K. H. (1972) Morphological variation and breeding behaviour of some tropical and subtropical legumes, Ph. D. Thesis, Cornell University, Ithaca, New York.

hudleigh, P. D., J. G. Boonman and P. J. Cooper (1977) Environmental factors affecting herbage yields of Rhodes grass (Chloris gayana) at Kitale, Kenya, *Trop. Agric. (Trin.)* **54**, 193–204.

ooper, J. P. and N. M. Tainton (1968) Light and temperature requirements for the growth of tropical and temperate grasses, *Herb. Abst.* **38**, 167–76.

einum, B. and J. G. P. Dirven (1972) Climate, nitrogen and grass. 5. Influence of age, light intensity and temperature on the production and chemical composition of Congo grass (Brachiaria ruziziensis Germain and Everard), *Neth. J. Agric. Sci.* **20**, 125–32.

ownes, R. W. and J. D. Hesketh (1968) Enhanced photosynthesis at low oxygen concentration: differential response of temperate and tropical grasses, *Planta* **78**, 79–84.

vans, L. T., I. F. Wardlaw and C. N. Williams (1964) Environmental control of growth. In C. Bernard (ed.), *Grasses and Grasslands*, Macmillan: London, Ch. 7.

erraris, R. (1978) The effect of photoperiod and temperature on first crop and ratoon growth of Pennisetum purpureum Schum., *Aust. J. Agric. Res.* **29**, 941–50.

erraris, R. and D. F. Sinclair (1980) Factors affecting the growth of Pennisetum purpureum in the wet tropics, *Aust. J. Agric. Res.* **31**, 899–925.

lover, P. E., J. Glover and M. D. Gwynne (1962) Light rainfall and plant survival in East Africa. II. Dry grassland vegetation, *J. Ecol.* **50**, 199.

wynne, G. W. (1966) Plant physiology and the future. In W. Davies and C. L. Skidmore (eds), *Pastures*, Faber and Faber: London, Ch. 4.

aggar, R. J. (1970) Seasonal production of Andropogon gayanus. I. Seasonal changes in yield components and chemical composition, *J. Agric. Sci. (Camb.)* **74**, 487–94.

atch, M. D. and C. R. Slack (1970) Photosynthetic CO_2 fixation pathways, *Rev. Pl. Physiol.* **21**, 141–62.

Heslehurst, M. R. and **G. L. Wilson** (1971) Studies on the productivity of tropica pasture plants. III. Stand structure, light penetration and photosynthesis in fiel swards of Setaria and greenleaf Desmodium, *Aust. J. Agric. Res.* **22**, 865–78.

Humphreys, L. R. (1966) Subtropical plant growth. 2. Effects of variation in leaf are index in the field, *Queensl J. Agric. Anim. Sci.* **23**, 337–58.

Humphreys, L. R. (1967) Townsville lucerne: history and prospect, *J. Aust. Ins Agric. Sci.* **33**, 3–13.

Humphreys, L. R. and **A. R. Robinson** (1966) Subtropical grass growth. 1. Rela tionship between carbohydrate accumulation and leaf area in growth, *Queensld . Agric. Anim. Sci.* **23**, 211–59.

Hutton, E. M. (1970) Tropical pastures, *Adv. Agron.* **22**, 1–73.

Ingle, M. and **B. J. Rogers** (1961) The growth of midwestern strains of *Sorghu halepense* under controlled conditions, *Amer. J. Bot.* **48**, 392–96.

Knight, W. E. and **H. W. Bennett** (1953) Preliminary report on the flowering an growth of several southern grasses, *Agron. J.* **45**, 268–9.

Laird, A. S. (1930) A study of the root systems of some important sod formin grasses, *Florida Expt. Sta. Bull.* No. 211.

Lovvorn, R. L. (1945) The effect of defoliation, soil fertility, temperature and leng of day on the growth of some perennial grasses, *J. Amer. Soc. Agron.* **37**, 570–82

Ludlow, M. N. (1978) Light relations to pasture plants. In J. R. Wilson (ed.), *Plan Relations in Pastures*, CSIRO; Melbourne, pp. 35–49.

Ludlow, M. N. and **D. A. Charles-Edwards** (1980) Analysis of the growth of a tropi al grass/legume sward subjected to different frequencies and intensities of defoli tion, *Aust. J. Agric. Res.* **31**, 673–92.

Ludlow, M. N. and **G. L. Wilson** (1970) Studies on the productivity of tropical pa ture plants. II. Growth analysis, photosynthesis, and respiration of 20 species grasses and legumes in a controlled environment, *Aust. J. Agric. Res.* **21**, 183–9

MacColl, D. and **J. P. Cooper** (1967) Climatic variation in forage species. III. Seaso al changes in growth and assimilation in climatic races of Lolium, Dactylis and Fe tuca, *J. Appl. Ecol.* **4**, 113–27.

Mackenzie, J. A. and **H. R. Chheda** (1970) Comparative root growth studies Cynodon IB. 8. An improved variety of Cynodon forage grass suitable for sout ern Nigeria and two other Cynodon varieties, *Niger. Agric. J.* **7**, 91–7.

Mannetje, L.'t (1965) The effect of photoperiod on flowering, growth habit and d matter production in four species of the genus Stylosanthes, *Aust. J. Agric. Re* **16**, 767–71.

Mannetje, L.'t and **A. J. Pritchard** (1974) The effect of daylength and temperatu on introduced legumes and grasses for the tropics and subtropics of coastal Austr lia. 1. Dry matter production, tillering and leaf area, *Aust. J. Exptl Agric. Anir Husb.* **14**, 173–81.

McCloud, D. E. (1963) Temperature responses of some subtropical forage grasse *FAO Working Party Past. and Fodder Dev. in Trop. Amer.*, Sâo Paulo, Brazil.

Mes, M. G. (1959) Influence of temperature on the symbiotic nitrogen fixation legumes, *Nature* (*Lond.*) **84**, 2032.

Mitchell, K. J. (1955) Growth of pasture species. II. Perennial ryegrass, cocksfoc and paspalum, *N. Z. J. Sci. Tech.* **37A**, 8–26.

Mitchell, K. J. (1956) Growth of Lolium perenne and Paspalum dilatatum over range of constant temperatures, *N. Z. J. Sci. Tech.* **38A**, 203–15.

Mohamed Saleem, M. A. (1972) Productivity and chemical composition of Cynod IB. 8 as influenced by level of fertilization, soil pH and height of cutting Ph. Thesis, Univ. Ibadan, Nigeria.

Ozanne, P. G., C. J. Asher and **D. J. Kirton** (1965) Root distribution in a deep sa and its relationship to the uptake of added potassium by pasture plants, *Aust. Agric. Res.* **16**, 785–800.

Pritchett, W. L. and **L. B. Nelson** (1951) The effect of light intensity on the grow characteristics of alfalfa and bromegrass, *Agron. J.* **43**, 172–7.

Taerum, R. (1970a) Comparative shoot and root growth studies on six grasses in Kenya, *E. Afr. Agric. For. J.* **36**, 94–113.

Taerum, R. (1970b) A study of root and shoot growth in three grass species in Kenya, *E. Afr. Agric. For. J.* **36**, 155–70.

Torssel, B. W. R., J. E. Begg, C. W. Rose and **G. F. Byrne** (1968) Stand morphology and root development of Townsville lucerne (Stylosanthes humilis): seasonal growth and root development, *Aust. J. Exptl Agric. Anim. Husb.* **8**, 533–43.

Tow, P. G. (1967) Controlled climate comparisons of a tropical grass and legume, *Neth. J. Agric. Sci.* **15**, 141–54.

Troughton, A. (1961) The effect of photoperiod and temperature on the relationship between the root and the shoot system of Lolium perenne, *J. Brit. Grassld Soc.* **16**, 291–5.

Watson, D. J. (1947) Comparative physiological studies on the growth of field crops. I. Variation in net assimilation rate and leaf area between species and varieties and between years, *Ann. Bot. Lond.* **11**, 41–76.

Whiteman, P. C. (1968) The effect of temperature on the vegetative growth of six tropical legume pastures, *Aust. J. Exptl Agric. Anim. Husb.* **8**, 528–32.

Wilson, D. B. (1962) Effects of light intensity and clipping on herbage yields, *Canad. J. Pl. Sci.* **42**, 270–5.

Wilson, D. B. and **J. P. Cooper** (1967) Assimilation of Lolium in relation to leaf mesophyll, *Nature (Lond.)* **214**, 989–92.

Wilson, J. R. and **C. W. Ford** (1971) Temperature influences on the growth, digestibility and carbohydrate composition of two tropical grasses Panicum maximum var. trichoglume and Setaria sphacelata and two cultivars of the temperate grass Lolium perenne, *Aust. J. Agric. Res.* **22**, 563–71.

Wilson, J. R. and **C. W. Ford** (1973) Temperature influences on the *in vitro* digestibility and soluble carbohydrate accumulation of tropical and temperate grasses, *Aust. J. Agric. Res.* **24**, 187–98.

Wong, C. C. and **J. R. Wilson** (1980) Effects of shading on the growth and nitrogen content of green panic and Siratro in pure and mixed swards-defoliated at two frequencies, *Aust. J. Agric. Res.* **31**, 269–85.

Youngner, V. B. (1959) Growth of U.3 Bermuda grass under various day and night temperatures and light intensities, *Agron. J.* **51**, 557–559.

Chapter 6

Distribution of grasslands, grasses and legumes

Grasses make up an essential component of the vegetational cover of more than half the land surface of the tropics and subtropics. This characteristic is not a feature of the legumes, however, since they are not as widely dispersed.

Grasses constitute a part of the vegetation of the following plant communities (Rattray, 1960):
1. Grassland – dense grass cover, woody growth absent or confined to isolated clumps or to watercourses.
2. Savanna – conspicuous grass cover, woody growth of varying density.
3. Steppe – open and sparse grass cover, scattered woody growth, including desert conditions where grasses are short-lived.
4. Woodland – sparse grass cover, usually dense woody growth.
5. Forest – virtually no grass cover, dense woody growth.
6. Undifferentiated – variable grass cover and woody growth.

Grasslands

A typical grassland is an open plain or tract of land having a dense cover of tall or short grasses and associated herbaceous species. Shrubs and trees are absent or widely scattered, usually clumped into low-lying moist areas and spread along watercourses. The grasses may be endemic or native of a given region, being referred to as 'natural' components. They may have migrated into the region under the influence of man or grazing animals and are completely integrated into the natural flora, in which case they are said to be 'indigenous' or 'semi-natural'.

Terminology
Dry, highveld and montane grasslands are differentiated in South Africa from low to high elevations (Adamson, 1938). Highland grasslands are recognized in Kenya between 2 200 and 3 000 m with

a minimum of 1 000 mm rainfall and frequent mist (Phillips, 1959). Montane open grasslands occur in Ethiopia at about 2 500 m and upward, under 750 to 1 275 mm annual rainfall (Keay, 1959). The same types are described as cool, mountain grasslands in Latin America (Roseveare, 1948). Wet and dry grasslands have been identified in Ceylon (Holmes, 1946). These two have been subdivided into arid, subarid, mild arid, subhumid and humid in South Africa (Phillips, 1959). Tussock and hummock grasslands occupy parts of the steppe and subdesert regions of Australia (Wood and Williams, 1960). In Papua New Guinea grasslands are identified on the basis of plant height (Heyligers, 1965). Other adjectives attached to the term grassland include open, secondary open, mixed, seasonal, derived, edaphic, scattered tree, savanna, steppe, hill, plateau, watershed, flood plain, tall grass, short grass, tropical, subtropical, temperate and alpine.

In many regions open grassland is replaced by grass–woody plant associations. A proposal was made in East Africa to use canopy cover as a criterion in measuring the contribution of trees and shrubs to grazinglands (Pratt *et al.*, 1966). Areas dominated by grasses, but with widely scattered or grouped trees and shrubs having a canopy cover no greater than 2 per cent, were called grasslands. Bushed and wooded grasslands had scattered or grouped shrubs or trees, respectively, with less than 20 per cent canopy cover. Areas of arid or infertile land sparsely covered by grasses and dwarf shrubs not exceeding 1 m in height, but sometimes with widely scattered larger shrubs or stunted trees, were called dwarf-shrub grassland. Subtypes suggested for these categories were manifold in describing associations on a local level.

Grassland regions

Extensive areas of grasslands, and grass–woody associations that provide grazinglands, occur in all continents and on many larger islands. The map in Fig. 6.1 shows the distribution of plant associations in which grasses comprise a prominent component of the vegetation. These include (1) grasslands, savannas, woodlands and shrub associations forming natural communities, and (2) areas modified to produce grazinglands (Moore, 1964). Large tracts of grazinglands also exist within the regions not included in these two categories. Furthermore, much of the land indicated as grasslands or modified grazinglands is used for arable cropping, sometimes in rotation with pasturelands.

Africa

Almost 60 per cent of Africa is covered with grasslands, wooded, and low-tree and shrub savanna (Harlan, 1956). Extending from the

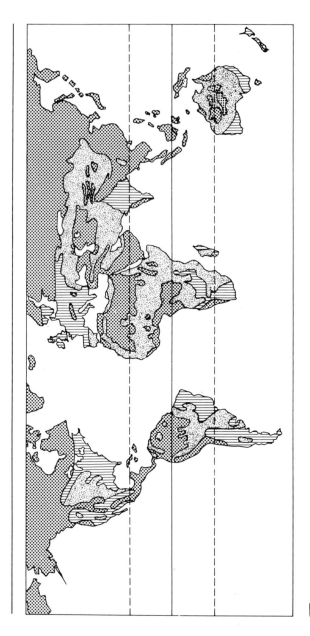

Fig. 6.1 Major grazing lands of the earth (redrawn from Moore, 1966: adapted from Koeppe & de Long, 1958).

Grasslands or grasses associated with natural vegetation which can be grazed with little or no modification

Modified grazing lands

Other vegetation or desert

humid forest area of the Congo basin, and along the coast of West Africa, are the woodland savannas with sparse grass cover, but in which the derived savannas with tall grasses have developed after cleaning of trees. Beyond are the low-tree and shrub savannas partially forming concentric belts around the humid forest of the Congo basin and extending westward above the woodland savanna and below the subdesert region of the Sahara. The *Hyparrhenia, Cenchrus, Andropogon* and *Loudetia* genera shown in Fig. 6.2 occupy much of these regions. Fringing the Sahara is the arid and semi-arid subdesert steppe with a predominance of *Aristida* species (Rattray, 1960). In some areas continuous heavy grazing and trampling by nomadic livestock have reduced the sparse vegetative cover to a condition somewhat comparable to the desert itself. In the highlands of East Africa, extending from the escarpments of Ethiopia south and westward across parts of Kenya, Uganda and Tanzania, are found mountain grasslands or temperate-type vegetation cover with associated grasses. The more important grasses include species of *Hyparrhenia, Sporobolus, Ctenium, Eragrostis* and *Themeda*. At the lower elevations species of *Setaria* and *Pennisetum* are associated with savanna or woodland vegetation. Further east, towards the Indian Ocean, annual rainfall drops rapidly and the region becomes desertic north of the equator. *Cenchrus, Chloris* and *Chrysopogon* are genera representative of these arid and semi-arid regions, but also associated with species of *Eragrostis, Andropogon, Aristida, Cymbopogon, Sporobolus*, etc. Just south of the equator, and extending to about 12° S, lies a moist, coastal belt with *Panicum* as the type genera, but species of *Pennisetum, Hyparrhenia, Brachiaria, Digitaria, Setaria, Cenchrus* and *Chloris* occur. Much of the southern part of the continent consists of arid lands having a predominance of *Eragrostis* and *Aristida*, along with desert shrub vegetation in which plants are prominent. To the east, rainfall becomes heavier and the vegetation grades into arid and semi-arid grasslands, then into shrub savanna and tall-grass savanna. *Themeda trianda* is an important species, growing under different soil and climatic conditions and ranging from short, tufted types of less than 1.0 m at the higher altitudes to tall, woody types reaching 4.0 m or more at the lower and warmer altitudes. Species of *Hyparrhenia, Cymbopogon, Andropogon, Eragrostis* and others are associated with the *Themeda* and contribute valuable forage.

South America

A great natural grassland extends from Patagonia in southern Argentina, through Uruguay, to southern and central Brazil (James, 1941; Roseveare, 1948). In the south the dominant grasses are cool-season types but in the north they grade into warm-season types such as *Andropogon, Paspalum, Panicum* and *Eragrostis*. An exten-

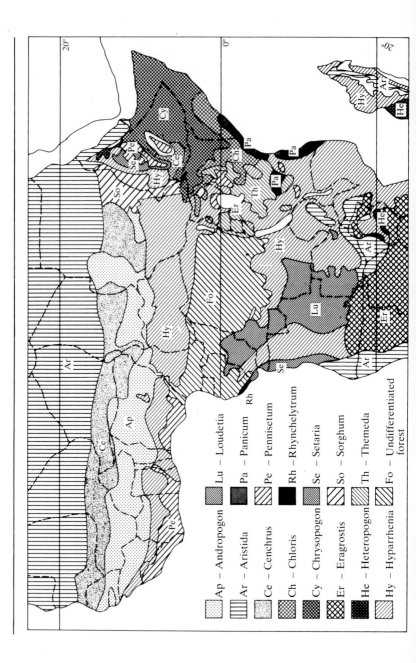

sive grassland, known as the *campos*, covers enormous areas of the elevated plains, often undulating, in the central, eastern and southern states of Brazil (Domingues, 1941; James, 1941; Roseveare, 1948). The *campos* are composed of a treeless savanna and a subdivision, *campo cerrado*, having scattered tree and bush savanna. While there is no uniform formation, the grass communities preponderate at one extreme but merge into scrub woodland at the other. From the earliest days of European colonization the *campos* have been used for stock raising, being improved in some areas by substituting *Panicum maximum*, *Melinis minutiflora*, *Hyparrhenia rufa*, *Chloris gayana* and *Pennisetum clandestinum* for the native or naturalized species. A feature of much of the *campos* is the shortage of fodder during the dry season which corresponds to the cool winter months. Herbage is usually present but quality deteriorates with a deficiency of protein, phosphorus and calcium. Northeastern Brazil is hot and for the most part semi-arid due to uneven distribution of rainfall. Coastal grazinglands are of relatively good quality and used for dairying where there is abundant and regular rain. The inland region consists mainly of *caatinga* (scrub forest or thorn-woodland) comprised of spinous woody plants, trees and shrubs, chiefly Mimoseae, with Cactaceae, Bromeliaceae and other xerophilous plants. The available grazing is composed of grass species such as *Aristida*, *Digitaria*, *Paspalum*, *Gymnopogon*, *Andropogon* and browse plants as *Cassia*, *Bauhinia*, *Ruellia*, *Cordia*, *Caesalpinia* and *Zizyphus*. In some areas herbage legumes such as *Desmodium*, *Centrosema*, *Stylosanthes* and *Zornia* are found.

The great Amazon basin has generally been regarded as one vast tropical rainforest, but numerous grassy savannas interrupt the tree cover. Enormous areas of wet and dry savanna occur in the state of Mato Grosso, being a southern extension of the Amazon. Important grasses include *Panicum* and *Paspalum* with various species of *Eragrostis*, *Andropogon*, *Chloris*, *Manisuris*, *Sporobolus*, and wide areas of *Tristachya* in the uplands. Of the legumes, *Desmodium* and *Stylosanthes* appear in greatest abundance. The Gran Chaco is a region of scrub forest interspersed with patches of savanna extending from Mato Grosso into Paraguay and eastern Bolivia. An extensive grassland savanna covers much of the department of Santa Cruz and part of the department of Cochabamba, with the expansive wet, grass plains of Yacuma and Mojos lying in the northern riverine lands of Bolivia.

A vast grassland, the *llanos*, spreads over about two-thirds of eastern Colombia, sloping gradually from the Andes and extending over eastern and northern Venezuela to the Atlantic Ocean (James, 1941; Pittier, 1926; Roseveare, 1948). In many places the impression is that of an ocean of grass from horizon to horizon. Scattered tree savannas (especially palms), however, emerge in some areas. Grasses of the *llanos* vary from a mere carpet to a height equal to a man on horseback. The principal genera consist of *Paspalum, Andropogon, Panicum, Eragrostis* and *Sporobolus* along with *Trachypogon, Cymbopogon, Axonopus, Aristida, Chloris, Cenchrus* and others. Legumes include species of *Desmodium, Vigna, Phaseolus, Cassia, Mimosa* and *Aeschynomene*. The upland savannas of the Guianas are an eastward extension of the *llanos* of Venezuela and the northern region of Brazil, occupying an extensive area of southwestern Guyana and continuing in a broken form across southern Surinam and French Guiana. Masses of heavily forested areas emerge with bunch-type grasses covering the intervening open sites, comprising species of *Andropogon, Cymbopogon, Trachypogon, Elyonurus, Paspalum, Arundinella* and *Heteropogon*.

North America

An enormous mid-continental prairie stretches from the central highlands of Mexico (Weaver, 1954). Arable cropping is practised wherever water permits, but much of the region only affords sparse grazing. Grasses encountered include species of *Euchlaena, Tripsacum, Andropogon, Cenchrus, Setaria, Panicum, Agrostis, Eragrostis, Aristida, Stipa, Sporobolus, Buchloe, Bouteloua* and *Muehlenbergia*. Much of northern and western Mexico is desert savanna where a general depletion of short grasses has occurred, causing increased woody vegetation. Brush encroachment and shrub invasion consisting of *Artemisia, Prosopis, Agave, Acacia, Mimosa* spp., along with various species of *Opuntia* (cactus) and *Yucca*, have modified the environment and created serious problems of range management (Harlan, 1956).

Europe

Western Europe has a forest climax modified by man for agricultural land and intensive grazingland management, the latter largely composed of improved temperate grasses and legumes. Around the Mediterranean a shrub-dwarf tree climax (chapparal) has been altered by heavy grazing and fire and is generally cropped where water is available. The extensive European grasslands at one time extended across Hungary, Romania and Russia including the productive farmlands of the Ukraine, being a tall-grass prairie.

Asia

Central mainland Asia consists of semi-arid and arid cool-temperate lands with desert shrub and steppe grasses such as *Agropyron, Elymus* and *Stipa*. Modified grasslands comprise parts of India and Pakistan, having been derived from humid forest, dry forest and some form of savanna from north to south. Southeast Asia, including the island nations, is for the most part rainforest and humid

forest with occasional grassland savannas. These have been described as follows:

1. Wet- and dry-zone grasslands in Sri Lanka (Ceylon) (Holmes,
   1946) with species of *Chrysopogon, Cymbopogon, Ischaemum,
   Themeda, Chloris, Cynodon, Digitaria* and *Aristida.*
2. Open and mixed grasslands and scattered tree savannas in
   Malaysia, Indonesia, Philippines, Celebes, the Moluccas and
   other East Indies islands, with species of *Andropogon, Heteropogon, Dichanthium, Bothriochloa, Themeda, Hyparrhenia,
   Ischaemum, Chloris, Sorghum, Saccharum, Pennisetum* (Van
   Steenis, 1958; Whyte, 1968).
3. Low grassland (up to 60 cm), mid-height (up to 1.5 m) and tall
   grassland (more than 1.5 m) in Papua and New Guinea (Heyligers, 1965; Whyte, 1968).
   Much of the grassland, however, is of the cogonal type (*Imperata
   cylindrica*) as a result of shifting agriculture and burning.

Australia

The central land mass of Australia is comprised of desert but has extensive ranching around the fringes where grasses and browse plants
are ample to support livestock. Immense areas of short and tufted
grasses and grassy–woodland associations project south, east and
northeasterly from the desertic region with arid scrub and *Spinifex*
formation to the north and northwest (Christian *et al.*, 1960). Rainfall increases toward the coastal regions, excepting the northwestern
part of the continent. Well-defined grass and grass–woody formations exist in the north and northeast tropical and subtropical zones.
Vast areas of the native and naturalized vegetation, however, have
been replaced by introduced species which are sown and fertilized
for intensive and semi-intensive management. To the south and
southeast of the desert lie grasslands that receive considerably more
rainfall and have been modified and developed by use of phosphates
and introduced temperate-zone species.

Grassland distribution maps

A detailed and informative map which portrays the grass cover of
Africa was prepared by Rattray (1960) to illustrate the major grass
associations on a floristic basis. A portion of the map, redrawn on a
broad basis, and limited to the tropics of Cancer and Capricorn, is
given in Fig. 6.2 (p. 88). Over a given region a grass genus emerged
as the predominating type on the basis of percentage frequency and
was selected to distinguish an association. Many other genera occur
within the region or zone, and in fact, in some localities no single
one appeared as the dominant grass. Furthermore, within many
plant communities a woody genus could have been chosen and used

in the descriptive terminology instead of a grass. In flying over, or driving through, many regions one would describe the vegetation as forest or as tree or shrub savanna. In many areas the associated woody species change within a broad vegetational unit, but the grass constituents extend over a wide geographical zone so that it is often difficult to decide where one type ends and another begins. Within the major grass associations a number of communities have developed because of local edaphic and biotic influences, so that the grass cover appears to be a mosaic of isolated plant colonies.

A grassland distribution map of Australia recognizes a number of geographical zones which are separated by grass associations (Christian *et al.*, 1960). In the tropical zone, comprising parts of Queensland, Northern and Western Territories, three major grassland regions are important as grazinglands:

1. 'Mitchell–Flinders grass' (*Astrebla* and *Iseilema* spp., respectively) – mainly open tussock grassland with associated herbs, shrubs and trees and various grass species.
2. 'Tropical tall grass' – a variety of open forests or woodlands, usually dominated by species of *Eucalyptus*, with the ground flora of tussock grasses from 1.0 to more than 2.5 m in height; major species include *Sorghum*, *Themeda*, *Heteropogon*, *Aristida*, *Chrysopogon* and *Imperata*.
3. 'Bunch spear grass' – open forest with dominant perennial grasses such as *Heteropogon contortus*, *Themeda australis* and *Bothriochloa ewartiana*.

Natural legumes form only a small proportion of the flora, but *Stylosanthes humilis*, introduced into the Townsville area of Queensland, has become rather widespread.

In India development of the grasslands has resulted from shifting agriculture, lopping of woody species, burning and grazing, and it is likely that none evolved as climax formations (Whyte, 1964). There are four major types of grass cover in the tropical and subtropical boundaries:

1. '*Sehima–Dichanthium*' – composed of 24 grass communities, each with a dominant species representing different stages of development within the type.
2. '*Dichanthium–Cenchrus–Lasiurus*' – covering much of central India.
3. '*Phragmites–Sacchrum–Imperata*' – found in the Ganges Plain and westward into the Punjab Plain.
4. '*Themeda–Arundinella*' – northern and northwestern montane regions below 2 100 m and between 1 800 and 2 100 m, a transitional zone extending into a 'temperate alpine' type at higher elevations.

A distribution map proposed for the types of grass cover of Pakistan (formerly West Pakistan) is somewhat similar to the one drawn for India (Johnston and Hussain, 1963). Two large group associations, '*Dichanthium–Cenchrus–Elyonurus*' and '*Chrysopogon*' types cover most of the country, dividing it into western and eastern regions, respectively. A '*Themeda–Arundinella*' association occurs in the northwestern mountain region and extends into an 'alpine' type.

Savannas

A savanna comprises woody-plant–grass vegetation complex in which the density of trees, bushes and shrubs is highly variable and the grass cover is well developed. The term was first used in 1535 by Oviedo in its Spanish form *zavana* to describe the open, grassy plains (*llanos*) of Venezuela (Roseveare, 1948). It exists in English as 'savannah' or 'savanna' and has been applied singly or with descriptive affixes to vegetation ranging from open grassland to woodland. The descriptions of the many types of savannas have led to a complexity and synonymy of terms and nomenclature in attempts to separate categories and classifications of plant formations. The following list illustrates the diversity of savanna types which have been employed: open, grassland, woodland, tree, low tree, shrub, thorn, mixed, closed, canopy, derived, subhumid, mild subhumid, subarid, arid, tropical, tall or high grass, low grass, *Acacia* desert grass, tiger savanna, arrow grass savanna. A proposed classification made by Cole (1963) in South Africa (Table 6.1) is broadly descriptive, encompasses the types of vegetation with which grasses are associated and has global application. It is unlikely, however, that the mass of terms can be readily and quickly replaced, because of diverse opinions as to the composition of a typical savanna and the historical concepts attached to the present understanding the different types.

The distinction between savanna and steppe has not been clearly defined since the classification of both is based on the nature of the herbaceous layer and on the density of the woody vegetation (Cole, 1963). In mapping the vegetation of Africa south of the Tropic of Cancer, the term savanna was used for vegetation in which mesophytic (medium conditions of moisture) perennial grasses were at least 80 cm tall; whereas steppe was used for vegetation in which xerophytic (limited supply of moisture) perennial grasses were less than 80 cm and might be interspersed with annual plants (Keay, 1959). These definitions led to difficulties in classifying the more arid grass–woody associations. It has even been suggested that the two terms be completely rejected in East Africa, since some grasslands could be placed in either category and others would be classified as savanna when ungrazed and as steppe when grazed or burned (Pratt *et al.*, 1966). On a global basis, however, the two

Table 6.1 *Proposed classification of savanna vegetation for unified nomenclature applicable on a world scale (Cole, 1963).*

1. *Savanna woodland*
   Deciduous and semi-deciduous woodland of tall trees (more than 8 m high) and tall mesophytic grasses (more than 80 cm high); the spacing of trees more than the diameter of the canopy
2. *Savanna parkland*
   Tall mesophytic grassland (grasses more than 80 cm high) with scattered deciduous trees (less than 8 m high)
3. *Savanna grassland*
   Tall tropical grassland without trees or shrubs
4. *Low tree and shrub savanna*
   Communities of widely spaced, low-growing perennial grasses (less than 80 cm high) with abundant annuals, and studded with widely spaced, low-growing trees and shrubs often less than 2 m high
5. *Thicket and scrub*
   Communities of trees and shrubs without stratification

terms have been historically employed and it is unlikely that their usage will be completely rejected.

In West Africa four of the main vegetational zones are commonly called savannas and are recognized as climatic regions representing different agricultural interests. Their boundaries are in part related to those delimited by the type genera of the grass distribution map of Africa (Fig. 6.2) (1) *Pennisetum* type – derived savanna, (2) *Hyparrhenia* type – southern Guinea and part of derived savanna, (3) *Andropogon* type – northern Guinea and part of Sudan savanna, (4) *Cenchrus* type – Sahel and part of Sudan savanna (Rattray, 1960).

In tropical America the term savanna frequently occurs in the description of grasslands. Within the *llanos* of Colombia and Venezuela three types have been recognized (1) open savanna with dominant grass cover, (2) orchard savanna with scattered gnarled bushes, (3) palm savanna in moist conditions (Beard, 1944; Pittier, 1942). The Bolivar savannas extend over the undulating and lowland plains of the upper Magdalena River in northern Colombia. In the lowland savannas of Costa Rica the Guanacaste grasslands spread over the province of the same name and extend into other low-lying areas bordering on the Pacific. Savanna regions have been referred to in the Amazon basin, uplands of the Guianas (Surinam, Guyana and French Guiana), Bolivia and the Gran Chaco (Roseveare, 1948).

Prairie is a grassland type somewhat comparable to savanna, and originally referred to level or rolling fertile lands in the temperate zone having a cover of tall coarse grasses and few trees. It has also been used in reference to open grasslands of the tropics and subtropics. In the original application, the tall-grass prairies were characteristic of some of the globe's most productive land areas, e.g. the so-called maize and wheat belts of the USA, the Ukraine of Russia, the Hungarian plains, the pampas of Argentina, the better agricultural lands of eastern South Africa and parts of Australia. Relatively little of the climax vegetation remains since most of the land is now used for diversified farming.

Steppe

The term steppe was originally used for open and treeless plains of xerophilous vegetation in Russia and Asia, i.e. short, wiry, tufted perennial grasses that developed under cool, temperate, low-rainfall or arid conditions. A warm steppe exists in tropical and subtropical regions where xerophytic plants developed under low-rainfall or arid conditions. Steppe vegetation was described in Africa (Keay, 1959; Rattray, 1960) and Australia (Wood and Williams, 1960) but has less common usage in South America and Southeast Asia.

The warm steppe refers to vegetation cover of low-growing trees or shrubs with widely spaced annual or perennial grasses, as well as treeless grass–herb subdesert formations. The grass cover map of Africa (Fig. 6.2) shows various grass genera associated with steppe formations (1) open – *Chloris, Cenchrus, Stipa* and *Aristida*, (2) tree – *Eragrostis, Chloris* and *Chrysopogon*, (3) shrub – *Eragrostis* and *Stipa* and (4) dwarf shrub – *Aristida* (Rattray, 1960).

In the steppe climates of Australia, bunch grasses predominate and are usually associated with scrub trees on thorn bushes and annuals (Moore, 1964). Tussock and hummock grasslands comprise a part of the important vegetational forms of steppes in the north where annual rainfall is 200 mm or less, while a shrub steppe occurs in parts of the south (Wood and Williams, 1960).

A cool grass steppe described in the eastern region of La Pampa (plain or flatland) in Argentina is an extension of the humid pampa to the north and a transitional zone abutting on a xerophilous region of woodland to the west (Monticelli, 1938; Parodi, 1930). The Patagonian steppe has been recognized in southern Argentina and is characterized by cool temperatures, low rainfall and xerophytic vegetation (Hauman, 1925).

Woodland

The woodlands with which grasses are associated are open forests with deciduous and semi-deciduous trees, often having their crowns touching, and with a sparse undercover of tall grasses that thicken when the trees are removed. In areas of high rainfall, grasses may be absent because of the closed tree canopy. When trees are cleared, tall grasses develop and become fibrous and highly lignified.

Burning is usually practised to remove the old accumulated material and to maintain an open woody plant formation, otherwise the derived grassland reverts to woodland.

Forest

This plant formation is comprised of evergreen or deciduous trees and bush or shrubby vegetation, occurring at a range of altitudes in the tropics and subtropics, and having extremely sparse grass or none. In regions of high rainfall, derived woodlands with tall grass undercover develop under cultivation and burning. At the higher elevations open grasslands are attained by clearing and maintained by frequent burning. In areas of low or medium rainfall grasslands and savanna types are derived after clearing of thickets and shrubby growth. A large percentage of the present grasslands developed from forests and are considered fire subclimaxes.

Distribution of grasses and legumes

Grasses

A floristic study of 87 regions in the northern temperate latitudes showed that the grass components ranged from 4.9 to 26.1 per cent (Hartley, 1964). Regions with fewer than 100 species of flowering plants had more than 14 per cent grasses. A steady decline was noted as the total flora increased to more than 4 000 species. The number of grass species in the flora of forests may be equal to that of grasslands but comprise less than 0.5 per cent of the total ground cover. In a grassland savanna, on the other hand, the same number of grasses would predominate and comprise more than 95 per cent of the vegetational cover.

The large grass tribes have a wide distribution over the earth's surface and exhibit definite climatic limits. In a study of the five largest and most widely distributed tribes between 60° N and 60° S latitudes, Hartley (1950, 1954) recorded the percentage contribution of species and obtained a global average (Table 6.2).

Tribes

The Paniceae is the largest tribe of the grass family and contains more than 1 460 species, or almost one-fourth of the total within the grass family. The world distribution map (Fig. 6.3) shows a complex pattern of dispersal (Hartley, 1958b). Conspicuous features of the map are as follows (1) predominantly tropical and subtropical distribution, (2) region of highest concentration in northeastern South

Table 6.2 *Global distribution of the major grass tribes within the grass flora of widely separated localities (Hartley, 1954)*

Tribe*	Per cent
Agrosteae	8.2
Andropogoneae (e)	17.2
Andropogoneae (w)	8.2
Eragrosteae	8.1
Festuceae	16.5
Paniceae (e)	18.3
Paniceae (w)	33.3

* Andropogoneae and Paniceae subdivided into eastern (e) and western (w) hemispheres.

America, (3) greater density in the western than in the eastern hemisphere and (4) complex distribution patterns, especially in southeast Africa, North and South America, and eastern Australia.

The unusual distributional patterns are influenced by local conditions, e.g. low frequencies in the high elevations of South America, altitudinal effect on lower densities in the tablelands of eastern Australia, reduced species percentages in Ecuador and Sri Lanka because of extremely high temperature. The regions having high percentages of species frequencies are closely associated with an extended growing season with favourable moisture and temperature for continued growth. The Paniceae are of greater importance in the forest and humid savanna regions than in the drier zones. Forms of the tribe, however, thrive under relative dry conditions and appear as frequent constituents of the more arid parts of India, the savannas of West Africa, and the Sahara.

The Andropogoneae comprises one of the larger grass tribes with a broad array of species represented in various types of grassland vegetation. A world distribution map of the Andropogoneae (Hartley, 1958a) prepared from lists of grass flora of 300 regions, showed (1) a concentration of the tribe in the tropics and subtropics, (2) the region of maximum abundance in Indo-Malaysia, and (3) a lower density in the western than in the eastern hemisphere (Fig. 6.4). Temperature appeared to be the most important climatic factor affecting distribution, and a close relationship was noted between latitude and frequency of species. There was no apparent correlation between distribution and type of dominant vegetation.

Genera and species

The distribution patterns of grass genera and species are less clearly defined than those of tribes. The limits of genera are strongly influenced by climatic conditions, but historical factors (especially time of evolutionary development) and geographical barriers are also important (Hartley, 1964). Species developed under more localized

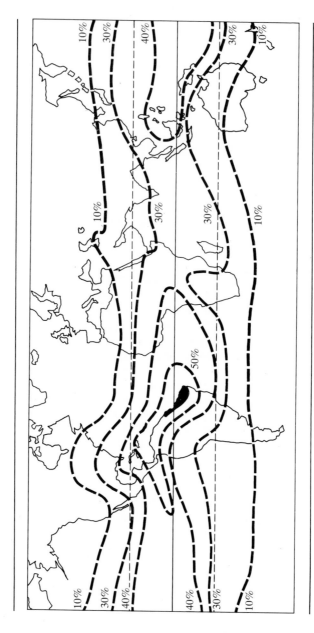

Fig. 6.3 World distribution of the grass tribe Paniceae, expressed as percentage of the total flora, shaded portion in South America over 60 per cent (redrawn from Hartley, 1958b).

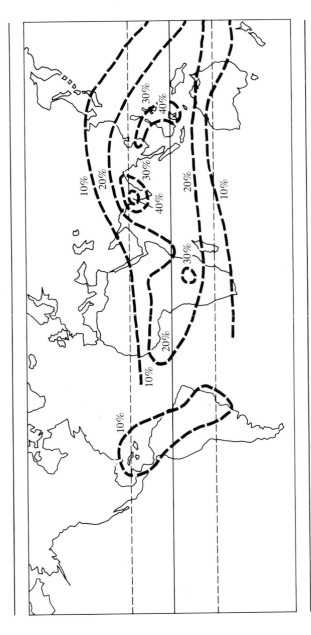

Fig. 6.4   World distribution of the tribe Andropogoneae, expressed as percentage of the total flora (redrawn from Hartley, 1958a).

conditions where biotic and edaphic factors played important roles in their distribution.

Many species have been widely distributed by man and grazing animals, e.g. *Cynodon dactylon*. This pantropic tetraploid was derived from diploid species which hybridized in Asia (Harlan, 1970; Harlan *et al.*, 1970a). Interest in revision of the genus led to detailed mapping studies of eight major species (Harlan *et al.*, 1970b). They fell clearly into four groups according to geographical distribution (1) *C. aracuatus* and *C. barberi* in regions of Southeast Asia and the Indian Ocean and Pacific islands, (2) *C. plectostachyus*, *C. aethiopicus* and *C. nlemfuensis* in East Africa, (3) *C. incompletus* and *C. transvaalensis* in South Africa and (4) endemic varieties of *C. dactylon* that are cosmopolitan. The fragmented geographic patterns, with limited distribution and complete genetic isolation of some species, and the narrow adaptation of some varieties of *C. dactylon* imply antiquity and long evolutionary history.

The dispersal of a species may occur relatively rapidly and cover a wide range of habitats. It can proceed without a noticeable change in the frequency and distribution of the genus or tribe. On the other hand, extension may be limited to a localized zone because of geographical barriers or to some restrictive environmental effect. In this way ecotypes develop that are specifically suited to the given locality. The displacement of a species takes place with the evolvement of a new species, development of hybrid swarms having a more pliable adaptability or the movement of an older species which invades and occupies a slightly modified environment.

Legumes

Distribution of legumes and factors affecting their dispersal have not been as well documented as the grasses. Legumes such as *Stylosanthes*, *Desmodium*, *Phaseolus*, *Centrosema*, *Leucaena*, *Calopogonium* and *Teramnus* are indigenous to tropical America (Bermudez Garcia, 1960; Whyte *et al.*, 1953). Centres of density exist in Central America, the lower elevations of the Andean zone of South America, eastern and southern Brazil and Paraguay.

Tropical Africa and parts of Asia contain species of *Glycine*, *Indigofera*, *Alysicarpus*, *Clitoria* and *Dolichos*. Of 229 genera occurring in tropical Africa, 94 are endemic. Four of the non-endemic genera are restricted to Africa and India, but at least 31 are distributed over Africa, Southeast Asia, Oceania and Australia (Brennan, 1965). Migrations have occurred to and from Africa, and among locations within other continents. A stronger similarity exists among species occurring in the drier regions of Africa and Asia than among those of the forest zones. This suggests that migrations were severed at an earlier time between forest areas than between drier areas.

Factors affecting grassland distribution

Climate

Climax grasslands developed when climatic conditions favoured emergence of grasses as the predominant vegetation. The tall-grass prairies of North America, the plains (*llanos*) of Colombia and Venezuela and part of the open grasslands of South Africa have been called climax formations. Some authorities, however, consider these to be fire subclimax grasslands, having been subjected to periodic burning (Pittier, 1942; Rattray, 1960). The proponents of climatic climax argue that the effects of fire have been of minor importance and would have evolved in response to the climate in the absence of burning (Borchert, 1950; Weaver, 1954). It is maintained that burning caused by lightning played a role in vegetational development prior to the destructive fires set by man and should be considered a factor of the natural environment.

It is common to describe grasses as cool-season (temperate zone) and warm-season types (tropical and subtropical zones). Genera such as *Poa, Dactylis, Festuca, Bromus, Agrostis, Stipa, Hordeum, Agropyron, Elymus, Deschampia, Danthonia* belong to the cool-temperate types. Those belonging to the warm-season types include *Panicum, Paspalum, Andropogon, Digitaria, Pennisetum, Chloris, Cenchrus, Setaria, Eragrostis, Bouteloua*. There is a general overlapping and comingling of species within each group so that clearly defined limits of adaptation are not distinguishable.

Edaphic

The effects of climate, especially rainfall, may be modified by the soil through its effect on the availability of water. Edaphic grasslands developed under conditions of inadequate drainage of the soil, seasonal flooding or waterlogging, increased water stress during dry periods and influences of soil type. Parts of the *llanos* of Colombia and Venezuela are under edaphic control because of annual flooding and poor surface and internal drainage of the soil. During the dry season hundreds of kilometres can be traversed without difficulty, but the same area in the rainy season may be obstructed by water channels and marshes (Pittier, 1942). Large areas remain waterlogged throughout the year but may provide fresh grazing during the dry season as cattle trample winding paths around grass hummocks. Two distinct forms of grasslands have been recognized on soils of different water-holding capacities in New South Wales, Australia (Costin, 1954):

1. Dry tussock grassland composed of *Stipa* spp. and *Themeda australis* on heavy soils in which moisture stress occurs during the dry season.

2. Wet tussock grassland having *Poa caespitosa* and *T. australis* associated with species of *Juncus* and *Carex* on soils with greater water retention.

Along major streams and waterways that drain grassland areas, one can observe abrupt changes in species adaptation due to the effects of flooding, soil texture and water retention. For example, along the Magdalena River, flowing through the Bolivar savanna of northern Colombia, are found vast areas of *Brachiaria mutica* in mixture with *Eriochloa polystacha*, both of which tolerate periodic flooding and remain lush during the early dry season when the waters recede. Away from the flood plains and on the sloping hillsides occur zones of *Panicum maximum* and further upward *Hyparrhenia rufa*. A similar situation exists in many parts of Africa where *Pennisetum purpureum* occupies the swales along water channels, with *Hyparrhenia*, *Andropogon* and more drought-tolerant *Pennisetum* species on the nearby higher ground.

Topography

The physiographic effects on grass and other vegetal associations are related to slope, exposure and elevation. In mountainous areas where slopes have been cleared, and in hillside regions, the soil and nutrients have been moved downward so that a gradation exists from fertile colluvium in the valley floor to thin, shallow, non-productive subsoil in the upper zones of the gradient. As these areas revert to grass cover the effects of soil fertility on the type of grass association and productivity are clearly visible. Topographic influences are frequently more noticeable at higher elevations and latitudes. In some instances northerly slopes are forested while southerly slopes support grass associations. High plateaus are sometimes cool, arid and windswept and have become covered with short grasses.

Biotic and fire

The influence of man on the development and maintenance of grasslands has been more widespread and significant than all other biotic influences. Man has wrought such modifications of grasslands and other vegetation communities that they now bear little or no resemblance to what might have developed had man not interfered (Moore, 1964). He has become a controlling factor in the balance of nature to the extent that he is capable of developing grazinglands in almost any place that grass will grow, and by suitable management can maintain such grazinglands, or by mismanagement cause the deterioration of presently established grasslands, which subsequently leads to the invasion of woody and weedy species.

Grasslands have been altered by man in the following ways (1) clearing of forest, (2) shifting agriculture, (3) continued arable cropping, (4) grazing management of domestic livestock, (5) influence on the population of wild game and interference with their habitat and (6) use of fire.

The most ancient and important influence of man on grasslands and other vegetation associations has been through the use of fire (Stewart, 1956). Relatively few grasslands exist which do not reflect the effects of burning. In fact, most present-day grasslands are fire subclimax associations and are maintained as such because of periodic burning. The effects and occurrences of the primeval natural fire mosaic has been changed by the activities of man but he has not destroyed or abolished them (Komarek, 1968, 1972). When these activities are removed, the patterns of lightning fires begin to be reasserted as plant succession recovers and fuel is accumulated by the natural process of maturity.

Natural grasslands, woodlands, tree and shrub savannas have been subjected to grazing by native fauna and influenced by aboriginal populations. Large elephant and hippopotamus populations in parts of Africa, especially overcrowded game preserves, have destroyed vast tracts of woodland savanna by debarking and breaking trees, allowing an increase in the distribution and percentage frequency of grasses (Olindo, 1972; West, 1972). In recent times changes in plant associations by exotic animals has been accentuated by continuous grazing and over- or understocking, distribution of watering sites in the more arid regions, and confinement of livestock in fenced areas. In many areas a balance between burning and grazing is highly important to maintain good grass cover. Reduction in fire frequency or time of burning and overstocking often result in the deterioration of the grassland and cause an invasion of woody species. For example, fire protection of derived savanna in Nigeria enabled the vegetation to change rapidly to forest climax (Charter and Keay, 1960). After 6 years a considerable difference existed between annual burning and protection from burning. An increase in the total number of tree species was observed but of most significance was the transition to fire-tender species. After 28 years the fire-protected areas contained virtually no grass and had a thick cover of trees and shrubs.

References

Adamson, R. S. (1938) *The Vegetation of South Africa*, British Empire Vegetation Committee; Kew, London.

Beard, J. S. (1944) Climax vegetation in tropical America, *Ecology* **25**, 127–58.

Bermudez Garcia, L. A. (1960) Leguminosas espontaneas del Valle de Cauca, Colombia, *Rev. Acad. Colombiano Cienc. Exactas, Fisicas y Natur.* **11(42)**, 51–83.

Borchert, J. R. (1950) The climate of the central North American grassland, *Ann. Assoc. Amer. Geographers* **40**, 1–39.

Brennan, J. P. M. (1965) The geographical relationships of the genera of *Leguminosae* in Tropical Africa, *Webbia* **19**, 545–78.

Charter, J. R. and **R. W. J. Keay** (1960) Assessment of the Olokemeji fire-control experiment (Investigations 254) Twenty-eight years after institution, *Niger. For. Inf. Bull.* (New Series) **3**, Fed. Govt. Printer: Lagos.

Christian, C. S., C. M. Donald and **R. A. Perry** (1960) In *The Australian Environment* (3rd edn), CSRIO: Melbourne, pp. 85–104.

Cole, M. M. (1963) Vegetation nomenclature and classification with particular reference to the savannas, *So. Afr. Geog. J.* **45**, 3–14.

Costin, A. B. (1954) *A Study of the Ecosystems of the Monaro Region of New South Wales, with Special Reference to Soil Erosion*, Govt Printer: Sydney.

Domingues, O. (1941) Nota preliminar sobre as regioes pastoris do Brasil, *Rev. Agric. Piracicaba* **16**, 325–40.

Harlan, J. R. (1956) *Theory and Dynamics of Grassland Agriculture*, Van Nostrand: Princeton, New Jersey, Ch. 3.

Harlan, J. R. (1970) Cynodon species and their value for grazing and hay, *Herb. Abst.* **40**, 233–8.

Harlan, J. R., J. M. J. de Wet and **K. M. Rawal** (1970a) Origin and distribution of the seleucidus race of Cynodon dactylon (L.) Pers. var. dactylon (Gramineae), *Euphytica* **19**, 465–9.

Harlan, J. R., J. M. J. de Wet and **K. M. Rawal** (1970b) Origin and distribution of the species of Cynodon L. C. Rich (Gramineae), *E. Afr. Agric. For. J.* **36**, 220–6.

Hartley, W. (1950) The global distribution of tribes of the Gramineae in relation to historical and environmental factors, *Aust. J. Agr. Res.* **1**, 355–73.

Hartley, W. (1954) The agrostological index: A phytogeographical approach to the problems of pasture plant introduction, *Aust. J. Bot.* **2**, 1–21.

Hartley, W (1958a) Studies on the origin, evolution, and distribution of the Gramineae. II. The tribe Andropogoneae, *Aust. J. Bot.* **6**, 115–28.

Hartley, W (1958b) Studies on the origin, evolution and distribution of the Gramineae. II. The tribe Paniceae, *Aust. J. Bot.* **6**, 343–57.

Hartley, W. (1964) The distribution of grasses. In C. Barnard (ed.), *Grasses and Grasslands*, Macmillan: London, Ch. 3.

Hauman, L. (1925) Etude phytogéographique de la Pactogonie, *Bull. Soc. Bot. Belg.* **58**, 105–79.

Heyligers, P. C. (1965) Vegetation and ecology of the Port Moresby–Kairuku area. In *Lands of the Port Moresby. Kairuku area, Papua-New Guinea, Land Res. Ser.* **14**, CSIRO: Australia.

Holmes, C. H. (1946) Grasslands and their afforestation in Ceylon, *Indian For.* **72**, 6–11.

James, P. E. (1941) *Latin America*, cited by G. M. Roseveare (1948) *The Grasslands of Latin America*, Imp. Bur. Past. and Fld Crops, Bull. 36, William Lewis: Cardiff.

Johnston, A. and **I. Hussain** (1963) Grass cover types of West Pakistan, *Pakistan J. For.* **13**, 239–47.

Keay, R. W. J. (1959) *Vegetation Map of Africa South of the Tropic of Cancer*, Oxford Univ. Press.

Koeppe, C. E. & de Long G. C. (1958) Weather and Climate, McGraw Hill Book Company

Komarek, E. V. (1968) Lightning and lightning fires as ecological forces, *Proc. Tall Timbers Fire Ecol. Conf.* **8**, 169–97.

Komarek, E. V. (1972) Lightning and fire ecology in Africa, *Proc. Tall Timbers Fire Ecol. Conf.*, **11**, 473–511.

Monticelli, J. V. (1938) Anotaciones fitogeograficas de La Pampa Central, *Lilloa* **3**, 251–382.

Moore, C. W. E. (1964) Distribution of grasslands. In C. Barnard (ed.), *Grasses and Grasslands*, Macmillan: London, Ch. 11.

Olindo, P. M. (1972) Fire and conservation of the habitat in Kenya, *Proc. Tall Timbers Fire Ecol. Conf.* **11**, 243–57.

Parodi, L. R. (1930) Ensayo fitogeografico sobre el partido de Pergamino, *Rev. Fac. Agron. (Buenos Aires)* **7**, 65–271.

Phillips, J. (1959) *Agriculture and Ecology in Africa*, Faber and Faber: London.

Pittier, H. (1926) *Manual de las plantas usuales de Venezuela*, Min. Agric.: Caracas.

Pittier, H. (1942) *La Mesa de Guanipa. Ensayo de Fitogeographia*, Min. Agric.: Caracas, pp. 57.

Pratt, D. J., P. J. Greenway and **M. D. Gwynne** (1966) A classification of East African rangeland, with an appendix on terminology, *J. Appl. Ecol.* **3**, 369–82.

Rattray, J. M. (1960) *The Grass Cover of Africa*, FAO Agric. Studies, No. 49, Rome.

Roseveare, G. M. (1948) *The Grasslands of Latin America*, Imp. Bur. Past. and Fld Crops, Bull. 36, William Lewis: Cardiff.

Stewart, O. C. (1956) Fire as the great force employed by man. In H. L. Thomas (ed.), *Man's Role in Changing the Face of the Earth*, Univ. Chicago Press, pp. 115–33.

Van Steenis, C. G. Y. J. (1958) *Vegetation Map of Malaysia*, UNESCO: Paris, Humid Tropics Proj.

Weaver, J. E. (1954) *North American Prairie*, Johnson: Lincoln, Nebraska.

West, O. (1972) Fire, man, and wildlife as interacting factors limiting the development of vegetation in Rhodesia, *Proc. Tall Timbers Fire Ecol. Conf.* **11**, 121–45.

Whyte, R. O. (1964) *The Grassland and Fodder Resources of India*, Indian Council of Agr. Res. Sci, Monograph No. 22.

Whyte, R. O. (1968) *Grasslands of the Monsoon*, Faber and Faber: London.

Whyte, R. O., G. Nilsson-Leissner and **H. C. Trumble** (1953) *Legumes in Agriculture*, FAO Agric. Studies No. 21, Rome.

Wood, J. G. and **R. J. Williams** (1960) Vegetation. In *The Australian Environment*, CSIRO: Melbourne, pp. 67–84.

Chapter 7

Plant introduction, evaluation and utilization

Plant introduction can be described as the movement of genera, species, varieties, selections, ecotypes, etc. as plant parts, such as seeds, rhizomes, stolons, crown splits, rooted cuttings, etc., into regions where they have not grown previously. The introduction, evaluation and utilization of new materials can be divided into four stages:
1. Collection – the search and gathering of plants and plant parts.
2. Quarantine and maintenance – the examination of materials for sanitation, propagation and holding of new germplasm.
3. Characterization – the assessment of attributes which might be useful.
4. Utilization – the movement of the material into a breeding programme or into commerce.

This chapter deals with the procedures, techniques and schemes used by scientists and practising animal husbandrymen in locating new germplasm of grasses and legumes, assessing their value and determining their potential as a pasture, forage or fodder crop.

Inaccurate introduction records often results in a selection being moved from place to place as a 'new find'. A standardized and systematized system was developed in Australia for the introduction and assessment of pasture and fodder species (Williams *et al.*, 1976). Similar techniques have been utilized at the southeastern plant introduction station in the USA. This approach is needed where animal husbandry is an important aspect of the national economy and where tropical grasses and legumes form the basic feedstuffs for livestock.

Movement of plant species

Historically man has always carried crop plants with him to new areas and brought new ones into his environment. Of primary concern to primitive man was the provision of adequate food supplies for human consumption. With the domestication of animals he also had to provide feedstuff for livestock and thus maintained and carried forage grasses and legumes on migrations (Flannery, 1969). Many species became diffused along ancient and modern trade routes by land and sea, from one market centre to another and from one herder to another. They moved with invading armies and general migrations of people, in the ballast of ships and railroads, and in the packing material or fodder of caravans. These forms of dispersion are still going on, and at greater speed, today. In recent years there has been intentional explorations and importations of new species, types and selections by crop introduction agencies and by private entrepreneurs.

Medicago sativa (lucerne or alfalfa) provides an excellent example of the opportune and intentional dispersal of a forage and fodder crop (Hanson, 1966; Whyte, 1958). A native of the temperate regions of western Asia-media (the southeastern Caucasian range, northwestern Iran, and steppes of Turkistan), it is probably the oldest of recorded forages. It followed the path of civilization from east to west, reaching Greece during the Persian invasions of 492–490 BC. Three hundred years later seeds were taken to Italy and North Africa, carried by the Moors to Spain and taken by the Spaniards to Mexico in the sixteenth century. The species moved to Peru, Chile, Argentina and Uruguay about the same time, but was unknown in North America until the nineteenth century when it entered by two routes: California during the Gold Rush of 1851–54 as 'Chilean clover' and Colorado from Mexico via the Spanish mission.

Sources of new plants

The world's major food and feed plants were widely dispersed during the period of colonial expansion. Plant explorations were organized by many countries during the nineteenth century, mainly those in the temperate zones.

Centres of diversity

Through numerous plant-collecting expeditions during 1920–34, Vavilov (1949–50) assembled plant materials from all over the earth and established a plant-breeding centre at the Institute of Plant Industry in Leningrad. His studies showed that variation in crop plants is not evenly distributed over the earth and that the bulk of diversity is confined to relatively few restricted geographical areas or centres where the greatest number of genetic types (varieties) are found (Zohary, 1970). Crops listed for each centre included cultivated legumes, some of which are used as forages, but not pasture and forage grasses. In many cases centres of diversity for certain crops were found far from the area designated as the centre of origin. This caused Vavilov to distinguish between primary centres (where domestication originated) and secondary centres (diversity after domestication). It is now recognized that places of origin are more diffused and variation is more widespread than indicated by earlier studies. In view of recent findings of blurred boundaries surrounding places of origin, influence of introgressive hybridization, and increased knowledge of polyploid complexes, it is more appropriate to use the term 'centres of diversity' rather than centres of ôrigin.

Little is known about the centres of diversity of pasture and forage grasses and legumes. Regions of distribution of several grasses have been determined, showing areas which have the greatest diversity of species. Hartley and Williams (1956) listed three general regions of origin of grasses:

1. Tropical and subtropical Africa – species such as *Andropogon, Brachiaria, Cenchrus, Chloris, Cynodon, Digitaria, Eragrostis, Hyparrhenia, Melinis, Panicum, Pennisetum, Setaria, Sorghum, Themeda* and *Urochloa*.
2. Tropical Central and South America – *Axonopus, Eriochloa, Euchlaena, Ixophorus, Paspalum* and *Tripsacum*.
3. Eurasian region – temperate zone species.

Based on various reports the general regions of origin of legumes are as follows (browse plants not included):

1. Tropical America – *Calopogonium, Centrosema, Desmodium, Leucaena, Phaseolus, Macroptilium, Rhynchosia, Stylosanthes* and *Teramnus*.
2. Tropical Africa – *Dolichos* (Lablab), *Glycine* and *Indigofera*.
3. Asia – *Cajanus, Clitoria, Dolichos* (Lablab), *Glycine, Phaseolus, Pueraria* and *Stizolobium*.
4. Eurasian – temperate zone species.

Grazinglands and pastures

Semi-natural grazinglands and old, sown pastures provide sources for collecting pasture and forage species. Persistence and resistance to the trampling of animals are important characteristics of grazing plants, and those which survive are frequently low yielding. In some instances, ecotypes and selections developed in isolated areas have been given local names, e.g. within *Panicum maximum* – 'Coloniao' in Brazil, 'pajarito' in Colombia, 'Sabi' in Rhodesia, 'Solai' in Kenya.

Scientific institutions

Introduction gardens should not be overlooked as important sources of germplasm. In fact, movement of materials among countries comes largely from collections established at agricultural institutions. Plant breeders usually maintain seed stocks of introduced materials and most are willing to share their own creations and selections.

Climatic homologues

Regions with homologous climates and similar vegetation associations have been identified so that plant collectors can narrow their search for specific plant types (Hartley, 1954, 1963; Henry, 1970; Holdridge, 1967; Whyte *et al.*, 1959). Their classification is based on climatic data such as rainfall, temperature, rainfall–evaporation ratio, length of dry season, humidity, frost, etc. Some include information about soils, topography, latitude and plant cover in general. Many of the systems embrace regions which are too large for practical use and none show latitudinal analogues or seasonal analogues between different latitudes.

Agrostological regions

By surveying floristic lists available in world literature and arranging the presence of grass species into class ranges, Hartley (1954) prepared an agrostological index. Ten class groups (0–9) were listed and each served as an index which showed the percentages of the five major grass tribes. For example, Class 5 comprised the following rankings: Andropogoneae – 18.9–23.5 and 8.9–11.0 per cent for the eastern and western hemispheres, respectively; Paniceae – 21.3–26.5 and 29.3–36.5 for the two hemispheres; Eragrosteae – 12.9–16.0 for both hemispheres; Agrosteae – 12.1–15.0, and Festuceae – 24.9–31.0

Certain localities had identical class groupings, e.g. British Honduras and the Ivory Coast, Ceylon and Formosa. Several had like indices for some tribes but unlike for others. A number of plant explorations from Australia have been made using information from similar agrostological regions when supplemented with data of environmental conditions. For example, a close coincidence of the natural distribution of *Cenchrus ciliaris* was found between Central Australia and that in parts of Africa and Asia (Hartley, 1963). Subsequent explorations led to the introduction of valuable materials from these regions. The search for genetic material may be facilitated by using information of previous introductions to develop a network of genetic resource date (Burt and Williams, 1979; Williams *et al*, 1980). Robinson et al (1980) also pointed out that geographical relationships of species such as *Stylosanthes* can be delineated with isozyme patterns of existing accessions.

Procurement of plant materials

Plant exploration

Plant-collecting expeditions are the oldest and most effective means of procuring plant materials. Centres of diversity provide the greatest possibility of locating the widest array of genetic variability, but plant explorers should also visit the relevant scientific institutions in the region, as well as survey plant introduction gardens and herbaria. The plant exploration should be carefully planned, giving attention to the establishment of objectives, knowledge of the geography and climate of the region to be explored, study of the ecology and agriculture of the collecting areas, a survey of scientific and collecting facilities, techniques of collecting samples and specimens, identification of collections and a report on the project (Bennett, 1970; Hymowitz, 1971; Whyte, 1958; Winters, 1966).

Exchange of plant materials

The assembled collections of seeds and plants for evaluation by governments and public institutions are usually exchanged freely. The ease of movement depends on the quarantine regulations of the country into which the introduction is made. Extreme regulations in some countries hinder rapid movement and evaluation of new germplasm. It is not uncommon that cattlemen are responsible for the movement of plant materials within a country and from one country to another. A good example is the introduction of *Digitaria decumbens* (Pangola grass) into Colombia by a rancher visiting in Florida. Sometimes this is done irresponsibly. The development of tissue and embryo culture should facilitate movement of valuable genetic materials across country borders and perhaps avoid certain quarantine problems.

Purchase

Seeds and plants may be purchased from individuals and commercial companies. This is more likely to occur with selected cultivars and varieties released for commerce and where there are laws for the protection of breeders who develop new and recognized types (called breeders' rights).

Donations

A considerable number of introductions come from individuals travelling in a foreign country who collect materials of personal interest and/or see an item and think it might have value 'back home'.

Evaluation

Assessment and characterization of new pasture plants begin when the first notes are made by the collector and continue until the materials are discarded or utilized. Prior to the release into commerce an entry will have been grown out, observed and studied many times by many people.

Quarantine

New materials, about which nothing is known, are received by a government agency, such as a plant introduction station, and grown in single, short rows up to 3 years for the following (Kahn, 1966):
1. A quarantine check to ensure that plants are free of diseases and pests and not likely to become serious weeds.
2. Build up seed supplies or vegetative material for more extensive tests.
3. Preliminary evaluation and descriptions on growth habit, flowering and seed-setting, variation, pests and diseases persistence, etc. A decision is made at this time to eliminate the entry or to maintain planting materials for further testing.

Maintenance

In countries that collect and evaluate tropical grasses and legumes, the plant introduction centre has the responsibility for increasing and maintaining seed supplies and vegetative materials. An effort is made to accumulate several hundred grams of seed for storage under controlled conditions of temperature and humidity. Accurate records are highly important and are usually coded and computerized. Descriptive information is published and made available to interested parties.

Many tropical countries have not established plant introduction centres, so that maintaining planting materials becomes the responsibility of the pasture agronomists. Frequently, detailed records are not kept so that introductions may be lost or duplicated. Seed storage facilities may be inadequate, making it necessary to regrow the selection every 2 or 3 years in order to maintain viable stocks.

Early testing

The performance potential of pasture or forage crops is usually determined from trials in an introduction garden with plots of sufficient size to superimpose different treatments. Entries are established in single rows but more desirably in larger plots, e.g. 3 × 10 m, if ample seed supplies are available.

The entries should be arranged so that accessions of one species or those of a similar growth habit are in adjacent rows to facilitate note-taking. Within each group of species a standard cultivar is included as a check for making comparisons. Sufficient distance must be left between rows or plots (at least 1.0 m) and between ranges (2 or 3 m) to prevent mixtures of creeping and twining types. Herbicides may be used to maintain such borders, but more frequently they are hand-weeded and cultivated. For non-creeping entries of perennial species the establishment of sod-forming grasses provide convenient borders that are easily maintained.

There is always the question of what management practices to use for early testing, and weeding is usually the first problem. It is generally agreed that for preliminary evaluation, weeds should be removed from plots but left in the second stage, noting the botanical composition to obtain a record of competitive ability. The seasonal pattern of uninterrupted growth provides valuable information and allows observations on flowering and seed-setting. If plots are large enough, they can be divided for cutting treatments to determine regrowth potential. Application of fertilizers will be guided by the inherent fertility of the soil and the conditions under which the selections are to be grown when released for commercial use. Irrigation may be necessary if there is doubt of obtaining ample seeds for storage and distribution, otherwise, supplemental water does not permit valid assessment, especially if drought tolerance is important. Appropriate *Rhizobium* species is an absolute essential for testing legumes.

Notes which should be taken include seedling vigour, seasonal growth pattern, time of flowering, flower colour, seed set, plant height, leafiness, tillering, density of ground cover, diseases and pests, weed competition, drought and frost tolerance, lodging of tall-growing types, regeneration of growth during the second season, etc. Preliminary data on forage yield and regrowth potential are needed. Many of these characteristics can be rated by the use of a numerical scale to facilitate recording. Notes are recorded in a bound field book and data sheets could be arranged in a form for direct computerization. It is important to record the date that notes are taken. Field plans with plot locations identified on paper are valuable and permanent markers set in the field assure that treatments can be correctly identified.

The most critical, and perhaps most valuable, method of assessing grazing type is to place the new introduction into a pasture consisting of species with which it must compete, and keep records on competition and persistence. Then more detailed agronomic studies such as method of establishment, seeding rates, mixtures, yields, etc. should be carried out.

Forage production and quality

Promising accessions are moved into trials of larger plots with sufficient replication and proper designs for statistical comparisons. These are usually plots of about 3 × 6–10 m that may be used for different agronomic studies under cutting. Such trials comprise varietal (selections, strains lines) tests for determining herbage yields, chemical analyses, botanical compositions and possibly nutritional studies.

Based on comparative performance, selections are made for further assessment, such as dates of establishment, methods of densities of sowing, responses to applied fertilizers, heights and frequencies of cutting, grass–legume combinations, regional testing and preliminary animal evaluation. Such small-plot studies permit the rapid evaluation of a large number of introductions, lines, selections, cultivars and varieties at a minimal cost when compared to animal trials. Data obtained are useful in choosing the limited number of species and cultivars which are finally evaluated in grazing trials.

Regional testing

After early testing in the introduction garden, and during the time of advanced trials, the materials having potential as pasture and forage types are chosen for adaptation trials located at strategic sites within a country or region. Depending on facilities and personnel, they can be simple demonstrations with observations made or more elaborate and detailed trials to complement the advanced testing.

Animal evaluation

Only those accessions which have proved outstanding in the preceding trials should be chosen for detailed animal studies under grazing and feeding. These experiments should be carried out in paddocks of sufficient size to carry several animals each, properly replicated, conducted over several seasons and preferably done at several locations. At this stage there should be close cooperation of the animal scientist, pasture agronomist and economist.

Utilization

The aim of pasture grass and legume collection, introduction and evaluation is the ultimate release of superior cultivars for commercial production and use as pasture, hay or silage. From the thousands of introductions only a few reach the commercial market.

Some are discarded but many are kept for research purposes and as sources of germplasm for plant-breeding and improvement programme.

Domestication of chance introductions

It frequently happens that unintentional introductions occur when seeds are inadvertently scattered along the roadside, near a railway siding, close to a dockside, airport or discarded in packing refuse. The seeds germinate, plants develop, produce flowers and more seeds. Under favourable conditions seeds scatter and plants extend over considerable distances. It happens that they sometimes become a part of the local weed population, but occasionally a grass or legume comprises an important component of the vegetation grazed by livestock. Such was the case of *Stylosanthes humilis* (Townsville stylo) in northeastern Queensland, Australia (Humphreys, 1967). This legume was found widely spread around Townsville during the 1920s, being relished by cattle. It is an annual, producing a fair amount of herbage and a heavy seed crop during the 3–4 month rainy season. The dry herbage is utilized as foggage (standing hay) during the dry season and seeds are eaten as well. It is thought that seeds of this legume came originally in ship refuse from Brazil in the early twentieth century and conditions near the port of Townsville were suitable for the selection and development of an ecotypic population. This one legume aided considerably in the transformation of northeast Queensland and Northern Territory from a region where one animal wandered over about 20 ha of poor quality native grass to one that supports one animal on slightly more than 1.0 ha. Townsville stylo seeds are now produced commercially and several new selections have been made among variants occurring in the diverse population.

Collections are sometimes introduced only to remain almost unknown in a nursery for many years until their potential as a pasture crop is discovered. *Digitaria decumbens* (Pangola grass) was brought from the Transvaal region of South Africa to Florida in 1936 where it rested in the introduction nursery for about 20 years (Oakes, 1965). A visitor noted its growth characteristics, took stolons to a nearby experimental station and distributed materials for local grazing. In the new environment others recognized its potential as a pasture crop and within a few years vegetative material (the grass being a sterile hybrid) was spread over the global tropics.

Direct multiplication of introductions

The vast majority of the recognized cultivars of tropical pasture and forage grasses and legumes are direct multiplications of introductions. They may have undergone some mass selection, either natural from repeated seed harvests and sowings or imposed by the plant breeder who made selections from surviving populations and bulked seeds from chosen genotypes. A number of named cultivars were developed by this method in Australia and in Kenya. Seeds of *Desmodium intortum* and *D. uncinatum* were collected in South America, introduced into Australia, evaluated in Queensland for several years. The seeds were multiplied and released as commercial cultivars Greenleaf and Silverleaf respectively (Hutton, 1970). A similar situation occurred for Verano (*Stylosanthes hamata*) (Rural Research, 1973). The Masaba variety of *Chloris gayana* originated from a seed sample collected near Endebess, Kenya, and grown at the Kitale Research Station (Bogdan, 1965). Unselected seed were distributed to farmers, some of whom grew seeds for commercial production. The Pokot variety of *C. gayana* was developed by bulking seeds from five outstanding plants selected for their vigour, persistence and rapid development of stolons. The original seed came from a collection made in the West Pokot district of Kenya. Makueni guinea grass (*Panicum maximum*) arose as a local seed collection from Makueni in the Machakos district of Kenya, taken to Kitale in 1962 and is now produced commercially. Cynodon IB.8, a selection of *Cynodon nlemfuensis* var. *nlemfuensis* collected from the Lake Manyara area of Tanzania, was tested at the University of Ibadan in Nigeria and released for general distribution (Chheda, 1968). It is a highly self-sterile tetraploid which produces only stolons, is easy to establish vegetatively, is drought tolerant and superior to the existing *Cynodon* types in West Africa.

Germplasm source for breeding purposes

The vast majority of introductions are utilized as a source of germplasm for recombinations of types in pasture and forage crop-breeding programmes. They may not be used immediately but held in reserve, with seeds being stored in cold storage, sometimes called a germplasm bank, until the plant breeder has need for materials possessing specific attributes. Siratro was developed in Australia by crosses between selections of *Phaseolus atropurpureus* (now called *Macroptilium atropurpureum*) which came from Mexico (Hutton, 1970). Nandi Setaria Mark 2 originated in Kenya from a single-plant mass recurrent selection for 3 years, during which time superior plants were recombined in a polycross arrangement to form the base for subsequent selection (Bogdan, 1965). The germplasm source came from an original Nandi variety which was the product of direct multiplication of unselected seeds collected in the Nandi district of Kenya.

A cooperative approach

Introduction, evaluation, and utilization of pasture and forage crops requires the efforts of many people, especially the plant breeder, pasture agronomist, soil fertility specialist and animal nutritionist. Table 7.1 summarizes an integrated approach for the characterization of herbage species, varieties and selections which might have potential as pasture or fodder.

Table 7.1 *Scheme of pasture and forage crop characterization (adapted from Mott and Moore, 1969)*

1. Introduction	Observation Selection	Breeder lines
2. Small plot clipping	Varietal tests Chemical analyses *In vitro* digestibility	Regional tests Advanced selections
3. Agronomic Management	Sowing densities Fertilizer studies Grass–legume mixtures Cutting treatments	Cultivars
4. Animal response	*In vitro* digestibility *In vivo* digestibility Stocking rates	Product per animal Product per hectare Economic returns
5. Feeding systems	Pasture sequences Hay and silage Supplementary feeds	Animal output Economic returns

References

Bennett, E. (1970) Tactics of plant exporation. In O. H. Frankel and E. Bennett (eds) *Genetic Resources in Plants – Their Exploration and Conservation*, IBP Handbook No. 11, Blackwell: London, pp. 157–79.

Bogdan, A. V. (1965) Cultivated varieties of tropical and subtropical herbage plants in Kenya, *E. Afr. Agr. For. J.* **30**, 330–58.

Burt, B. L. and **W. T. Williams** (1979). Strategy of evaluation of a collection of tropical herbaceous legumes from Brazil and Venezuela. III. Use of ordination techniques in evaluation, *Agro-Ecosystems* **5**, 135–46.

Chheda, H. R. (1968) Cynodon IB. 8, an improved variety of Cynodon suitable for pastures in Southern Nigeria, *Ann. Confr. Agric. Soc. Niger.*, Ibadan (mimeo).

Flannery, K. V. (1969) Origins and ecological effects of early domestication in Iran and the Near East. In P. J. Ucho and G. W. Dimbleby (eds), *The Domestication and Exploitation of Plants and Animals*, Gerald Duckworth: London, pp. 72–100.

Hanson, A. A. (1966) Introduction of grasses and legumes for forage, *Proc. Int. Symp. Plant Introd., Escuelas Agric. Panamericana*, Honduras, pp. 109–18.

Hartley, W. (1954) The agrostological index: A phytogeographical approach to the problems of pasture plant introduction, *Aust. J. Bot.* **2**, 1–21.

Hartley, W. (1963) The phytogeographical basis of plant introduction, *Genetica Agraria* **17**, 135–60.

Hartley, W. and **R. J. Williams** (1956) Centres of distribution of cultivated pasture species and their significance for plant introduction, *Proc. 7th Intl. Grassld Cong.*, pp. 190–201.

Henry, J. M. (1970) The development of agrobioclimatic techniques. In O. H. Frankel and E. Bennett (eds), *Genetic Resources in Plants – Their Exploration and Conservation*, IBP Handbook No. 11, Blackwell: London, pp. 205–20.

Holdridge, L. R. (1967) *Life Zone Ecology*, Tropical Science Center: San Jose, Costa Rica.

Humphreys, L. R. (1967) Townsville lucerne: history and prospect, *J. Aust. Inst. Agric. Sci.* **33**, 3–13.

Hutton, E. M. (1970) Tropical pastures, *Adv. in Agron.* **22**, 2–73.

Hymowitz, T. (1971) Collection and evaluation of tropical and subtropical Brazilian forage legumes, *Trop. Agric. (Trin.)* **48**, 309–15.

Kahn, R. P. (1966) Plant quarantine aspects of plant introduction. *Proc. Int. Symp. Plant. Introd., Escuelas Agric. Panamericana*, Honduras, pp. 55–60.

Mott, G. O. and **J. E. Moore** (1969) Forage evaluation techniques in perspective, *Proc. Nat. Conf. Forage Quality Eval. Util*, Lincoln, Nebraska, (Mimeo.) pp. L 1–10.

Oakes, A. J. (1965) Digitaria collection from South Africa, *Trop. Agric. (Trin.)* **42**, 323–31.

Robinson, P. J., R. L. Burt and **W. T. Williams** (1980) Network analysis of genetic resources data. II. The use of isozyme data in elucidating geographical relationships, *Agro-Ecosystems* **6**, 111–18.

Rural Research (1973) *Stylosanthes hamata*, CSIRO Quarterly No. 82, pp. 7–10.

Vavilov, N. D. (1949–50) Phytogeographic basis of plant breeding, *Chron. Bot.* **13**, 16–54.

Whyte, R. O. (1958) *Plant Exploration, Collection and Introduction*, FAO Agric. Studies, No. 41, Rome.

Whyte, R. O., T. R. C. Moir and **J. P. Cooper** (1959) *Grasses in Agriculture*, FAO Agric. Studies, No. 32, Rome.

Williams, R. J., R. L. Burt and **R. W. Strickland** (1976) Plant introductions. In *Tropical Pasture Research, Principle and Methods*, Commonw. Bur. Past. Fld Crops, Bull. 51, Ch. 5.

Williams, R. J., R. L. Burt, B. A. Pengelly and **P. J. Robinson** (1980) Network analysis of genetic resources data. I. Geographical relationships, *Agro-Ecosystems* **6**, 99–109.

Winters, H. F. (1966) Mechanics of plant introduction, *Proc. Int. Symp. Plant Introd., Escuelas Agric., Panamericana*, Honduras, pp. 49–53.

Zohary, D. (1970) Centers of diversity and centers of origin. In O. H. Frankel and E. Bennett (eds) *Genetic Resources in Plants – Their Exploration and Conservation*, IBP. Handbook No. 11, Blackwell: London, pp. 33–42.

Chapter 8

Grassland improvement: establishment and renovation of the sward

Native or naturalized pasture and forage species are usually susceptible to high grazing pressures and respond poorly to increased soil
fertility. Since carrying capacity and yearly animal production depend
largely on the feeding value of mature forage during the dry season,
manipulation of existing grasslands to produce higher animal returns
is frequently limited. Where there is a drastic shortage and distribution of rainfall and where soils are extremely infertile; droughty and
highly erodible, there is a need for maintaining the stability of grazinglands. Where moisture conditions are adequate; and economical
returns are favourable, the replacement of existing species and use
of fertilizers should be given consideration. This operation calls for
some means of sod renovation or land tillage, fertilization, oversowing and establishment of sown pasture species.

Selection of species or cultivars

In selecting a new grass or legume; or in replacing an older one with
an improved cultivar of the same species; the following points should
be considered:
1. Adaptation – to the general region and to local conditions.
2. Intended use – continuous or rotational grazing, hay, silage, green
   chop, ration grazing, soiling.
3. Species or cultivar characteristics – ease of establishment; palatability, length of vegetative stage of growth, response to applied
   fertilizers, persistency, tolerance to drought, grazing, cutting and
   burning, herbage yield and regrowth potential, seeding habit, ease
   of eradication.
4. Availability of seed or planting material.
5. Value of the land – especially if the new grazingland is to be
   intensively used.
6. Topography of land – mechanizable or steep.
7. Type and quality of animals to be grazed or fed.
8. Managerial skill of the cattleman.

Source of seed

Seed or transplanting materials refer to true seeds, or vegetative materials, plant parts such as stems, stolons, rhizomes and crown pieces (called crown splits and sometimes referred to as propagules or sprigs).

Lack of viable seed places a serious constraint on the development and expansion of sown pastures and improvement of natural grasslands in many parts of the tropics. Seeds of tropical grasses and legumes are not readily available on a commercial basis, except in Australia and parts of East Africa.

Seeds are hand-harvested in localized areas but usually are not cleaned and processed. They often contain high amounts of waste, such as pieces of stems, leaves, spikelet branches, empty florets, immature seeds and extraneous material. Quality and viability are extremely low and germination may not exceed 1 per cent. Some farmers or cattlemen establish a seed source nursery and may share with neighbours or operate a small local business for distribution. In some countries the Ministry or Department of Agriculture and Extension Service maintain nurseries for distribution of seeds to farmers and ranchers.

Improvement of natural and semi-natural grazinglands

Bush thinning and control

Under rangeland conditions the thinning of bushes and shrubs helps to increase grass cover. Adequate grazing management practices must also be used to maintain the grasses and palatable herbs and to avoid encroachment of undesirable species.

Grazingland management

Periodic rests from grazing during critical growth periods permit the accumulation of plant nutrient reserves and aid seed formation. A system of deferred-rotational or seasonal grazing will allow time for seed development and formation and facilitate natural reseeding of deteriorated grazinglands. Such areas should be protected or lightly grazed during the time of seedling establishment of volunteer stands.

Oversowing grazinglands and rehabilitation of denuded areas

The improvement of natural grasslands usually requires some means of bush eradication and overseeding with or without tillage. In the arid and semi-arid regions, bush-clearing and land tillage may not give immediate economical returns and should be a part of a planned, long-range improvement programme.

Successful establishment of grasses and legumes sown into natural and semi-natural grazinglands lies in the choice of appropriate species, application of suitable techniques of land treatment, use of fertilizers, especially phosphate and appropriate methods of sowing (Bogdan and Pratt, 1967a; Bogdan, 1977; Borget, 1969; Kornelius *et al.*, 1978; Jones, 1979; Gomide *et al*, 1979; Cook, 1980). The simplest approach is direct overseeding into natural or native grasslands, but usually stands are poor.

Studies in Nigeria showed that *Stylosanthes guianensis* could be established by feeding seeds to cattle and grazing restricted areas (Foster, 1961). This is an extravagant use of seeds with an uneven distribution, but once the legume is established the cattle facilitate its spread in subsequent years.

Seeds of stylo have been broadcast by hand after burning, after intensive grazing and after light soil scarification. In Zambia, sowing into hand-hoed strips or into shallow grooves scratched by sticks gave a density of 24 000 plants/ha within 10 months (Rensburg, 1965). Oversowing after burning and light discing without covering the seed gave similar stands in the Ivory Coast (Letenneur, 1971), Nigeria (Haggar *et al.*, 1971), and Uganda (Stobbs, 1969). In the Congo (Risopoulos, 1966) stylo became established without burning, even though cattle were allowed to graze immediately after sowing (Table 8.1). In general, light discing and a subsequent passage of a bush cutter improved the legume population. The number of plants per m² were low (about 3.5), however, even at the highest level of sowing. Since this is a bunch-type legume, a desirable stand should possess from 12 to 15 plants/m²

Stylosanthes humilis and *S. hamata* have been oversown with or without soil disturbance into Australian rangelands where rainfall varies between 750 to 900 mm (Gillard, 1977; Humphreys, 1967; Woods, 1969; McKeague *et al.*, 1978). Cultivation will improve

Table 8.1 *Oversowing natural grazingland with Stylosanthes guianensis, number of plants/1 000 m² (Risopoulos, 1966)*

Herbage treatment	Mechanical treatment		
	None	Light discing	Passage with bush-cutter
1. Grazed, no burning, immediate grazing	1 565	1 745	2 680
2. Burning, immediate grazing	1 605	2 540	1 475
3. Burning, rest 1 year before grazing	1 235	2 505	3 475

Note. Sowing density: 2.5 kg/ha; counts made 3 years after sowing.

establishment, but seed prices and ease of access to land determine the economics of this practice. Seedlings and plants do not tolerate heavy shade so that destruction of trees and native grass provides higher initial stands and more dense populations than sowing into the regularly grazed rangeland. Sowing into strips of lightly disked *Heteropogon contortus* growing among sparse timber has met with considerable success (Graham, 1968).

Weather conditions during establishment are of primary importance and the condition of the seedbed may be of less consequence (Norman, 1960). In arid and semi-arid regions, germination and growth of seedlings are dependent on subsequent rains rather than stored moisture, whether sown into cultivated, harrowed or raked seedbeds. Near Darwin, Australia, annual sorghum (*Sorghum intrans*) predominates in open woodlands, and hampers the establishment of legumes oversown into the native grazinglands (Stocker and Sturtz, 1966). The grass is highly susceptible to fire damage, however, and can be controlled by burning the plant residue shortly after the rains begin. Use of fire allows Townsville stylo to become well established when sown soon after burning or later in the wet season (Table 8.2). Any sorghum seedlings which emerge are eliminated by grazing without damage to the legume.

Table 8.2 *Establishment of Stylosanthes humilis and control of Sorghum intrans near Darwin, Australia (data from Stocker and Sturtz, 1966)*

Treatment*	Plant density[†]		DM yield[‡]
	S. humilis	Sorghum	Sorghum (kg/ha)
1. Unburned	95	5 205	355
2. Burned	458	218	33
3. Burned; 100 kg superphosphate	518	198	32
4. Burned; 2,4,5-T	538	190	26
5. Burned; 2,4,5-T; 100 kg superphosphate	800	233	34

* Burned shortly after wet season started; 2,4,5-T at 0.4 per cent a.i. ester applied 2 months before burning.
† No. of plants/0.01 ha; data taken in April after treatment and sowing in December.
‡ 1st year DM yields of Townsville stylo up to 690 kg/ha on all treatments.

Six perennial and two annual legumes were broadcast and drilled, with and without single superphosphate, into a native grassland pasture of savanna woodland in north Queensland (Downes, 1967). The experimental area comprised scattered trees of *Eucalyptus* spp., *Heteropogon contortus* (bunch spear-grass), *Themeda australis* (kangaroo grass) and naturalized legumes, such as *Glycine tomentosa*, *Rhynchosia minima* and *Indigofera* spp. The area was burned in August, during the dry season, and stocked until January, during a part of the dry and rainy periods. Sowings were made in late November after light rains in October and November. In general, the perennial legumes were less vigorous during the year of establishment as compared to annual legumes, but their density increased in the second year. Seed placement in the soil was superior in terms of seedling population and vigour, the drilled treatment showing an overall advantage of 40 and 10 per cent for these characteristics, respectively, over the broadcast treatment. Application of fertilizer gave an increase of about 10 per cent in plant density and 18 per cent in seedling vigour for data taken at the end of the rainy season (4 months after sowing). No data were obtained on long-term persistency of the legumes or competition with native grasses.

At Kitalie, Kenya (about 1 890 m elevation) pelleted seeds of *Desmodium intortum, D. uncinatum, Trifolium semipilosum* and *Stylosanthes guianensis* were oversown on a closely grazed natural grassland dominated by *Hyparrhenia* spp. (Keya *et al.*, 1971). Seedling survival 3 months after sowing was 13.6, 13.4, 11.6 and 8.7 per cent, respectively, based on the number of pure germinating seeds. Fourteen months after sowing, the *Desmodiums* constituted more than 50 per cent of the botanical cover, *Trifolium* 40 per cent and *Stylosanthes* 27 per cent. The *Desmodium* spp. are likely to contribute less than 25 per cent of the vegetative cover, however, under intensively managed grazing conditions without proper fertilizer applications and judicious managerial skill.

In Queensland, Australia, Coaldrake and Russell (1969) broadcast four legumes into ash after burning Brigalow (*Acacia harpophylla*). *Macroptilium atropurpureum* cv. Siratro and *M. lathyroides* established well but did not survive beyond the year of seeding. Lucerne gave a poor stand but persisted into the third year. *Glycine javanica* (= *G. wightii*) establishment was extremely sparse and few plants survived. After 3 years of prolonged drought, *Cenchrus ciliaris* dominated the entire experimental area.

Legumes used for oversowing natural and naturalized grazinglands in other studies include *Glycine wightii* and stylo in Rhodesia (Smith, 1963), stylo into *Urochloa* (Falvey, 1979), *Centrosema pubescens, Dolichos formosus, Alysicarpus rugosus* and *Desmodium intortum* in Zambia (Rensburg, 1965), *Glycine*, centro, stylo, *D. intortum, Lotononis bainesii*, and *Macroptilium atropurpureum* in Hawaii (Motooka *et al.*, 1967), and stylo into *Imperata* in Papua New Guinea (Chadhokar, 1977; Gibson, 1980). Established stands have varied from poor (extremely sparse plant density) to good (25 to 30% botanical composition). The higher plant densities were usually associated with more intensive land tillage practices, which provided favourable seedbeds and with fertilization, especially phosphate. Even with fair establishment, a serious problem with most legumes has been persistency beyond the year of seeding, particularly in arid and semi-arid regions.

Grasses have been used for oversowing native grasslands and reseeding denuded areas caused by overgrazing, wind and water erosion. In Kenya species which have shown potential for improvement of grazinglands in low rainfall areas include *Cenchrus ciliaris, C. setigerus, Chloris gayana, C. roxburghiana, Cynodon dactylon, Enteropogon macrostachyus*, a perennial species of *Dactyloctenium, Eragrostis* spp. and *Latipes senegalensis*, all of local origin (Bogdan and Pratt, 1967a; Pratt, 1964). On the dry alluvial soils of the Rift Valley in Kenya, 'cultivating' with rigid tines attached to a tractor tool bar, spaced at 30 or 60 cm, and set to penetrate the soil to a 15 cm depth, gave an increased number of plants of sown species during the year of establishment as compared to other treatments (Table 8.3). Placement of cut branches on the soil surface also favoured seedling development. By the end of the third season fair amounts of *Cenchrus ciliaris* and *Eragrostis superaba* remained, with scattered plants of *Latipes*. The others had been eliminated with little noticeable change in native grass flora.

Scratch ploughing suffices under conditions of more favourable rainfall where furrows serve as a coarse seedbed. The shallow furrows made along the contour catch many of the seeds which are washed along the surface and provide protective sites for germination and seedling growth (Pratt, 1963; Schedler, 1969). Drilling grass seeds into closely grazed or cut swards may improve establishment (Owen and Brzostonski, 1967), but the more intensive mechanical operations are indicated for less favourable sites (Pratt and Knight, 1964; Rensburg, 1965). More dense stands occur when seedings are made

Table 8.3 *Plant densities of sown grass on denuded land of the Rift Valley, Kenya, having scattered growth of Acacia (Pratt, 1964)**

Soil treatment[†]	Sept. 1958		Oct. 1959		Sept. 1960	
	Branches	None	Branches	None	Branches	None
1. Close tine cultivation	73	62	111	45	12	7
2. Spaced tine cultivation	86	67	76	63	17	17
3. Strip cultivation	69	36	26	12	7	2
4. Scratch ploughing	51	35	30	10	10	1

* No. of plants of sown grasses per 100 ft; treatment and oversowing in May, 1958; cut branches of shrubs laid on soil of appropriate treatments and no branches on others (none).

† (1) Rigid tines spaced at 30 cm on tractor tool bar, set to penetrate soil to 15 cm depth; (2) tines spaced at 60 cm; (3) tines at 60 cm, 4.75 m strip cultivated and 14.2 m strip not cultivated; (4) scratched tines, 10 cm in depth, made by hand jembes at about 6 m intervals.

after tillage, with care taken not to cover seeds too deeply. Under low rainfall conditions, livestock should be excluded from the area for several weeks to allow time for plants to establish, and in the more arid regions grazing should be deferred for the first season after sowing.

Studies in the semi-arid region of Kenya during a season of only 475–500 mm rainfall illustrate the danger of overseeding as well as underseeding. With high seeding rates of a grass mixture, the increased competition at the seedling stage resulted in dispropor-tionately greater mortality. Five months after sowing, 3.83 plants were established per square yard (approximately 4/m²) at the opti-mum rate of 10 lb/acre. This represented a recovery rate of 9.3 per cent of the total seed sown, but wide variation existed among species. The recovery of *Eragrostis* reached 61.8 per cent, which is excep-tionally high and indicated the high adaptability of this species to the specific location as well as its agressiveness in respect to the other species. In general, a much lower recovery can be expected and under some conditions may not exceed 5 per cent. An optimum established plant density is considered to be 1 plant per square foot (about 10 plants/m²). Knowing the number of seeds per pound (or kg) of a given species, the pure germinating seed content, and accept-ing a 5 per cent recovery, one can calculate the required seeding rate. For example, *Cenchrus ciliaris* cv. Rongai has about 220 000 seeds per pound. To obtain a 5 per cent recovery, 20 seeds would be needed per square foot, which totals 871 000 seeds per acre. This number is equivalent to about 4 lb of pure germinating seed.

In the more humid areas of the tropics, grasses can be introduced into natural and naturalized grasslands with less tillage. For example, in the eastern plains of Colombia ('La Libertad' Agricultural Exper-iment Station, 450 m elevation, 3 500 mm annual rainfall) studies of oversowing after close grazing, mowing or burning showed that spe-cies such as *Melinis minutiflora, Hyparrhenia rufa* and *Panicum max-imum* dominated the natural vegetation within one rainy season (ICA, 1957–72). Light scarification of the sod by disking improved plant density. In addition, transplanting vegetative pieces of *Digitaria decumbens, Brachiaria* spp. and *Axonopus micay* gave good results. At another site near Popayan, located at about 1 750 m and with more than 1 500 mm annual precipitation, *D. decumbens* regenerated growth and predominated when transplanted into a weedy sward of native *A. micay* which had been clipped near the ground (Crowder *et al.*, 1962). Within one year the *Digitaria* constituted 85 per cent of the sward whether transplanted directly or placed in shallow fur-rows marked in the sod.

In the Guinea savanna of Nigeria *Andropogon gayanus* is fre-quently oversown into cleared areas by broadcasting seeds across the direction of prevailing winds and without soil disturbance (Rains,

1963). With favourable soil moisture seedlings develop to thicken the sward.

Strip planting

The planting or sowing of forage and fodder crops in narrow belts or strips in active grazinglands has been suggested as a means of improving the pasture and providing supplemental feed for grazing animals. In Zambia, consideration was given to establishing belts of fodder species such as *Leucaena leucocephala*, *Cajanus cajan*, *Dolichos lablab* (syn. *Lablab purpureus*), *Mucuna pueriens* and *Desmodium intortum* (Rensburg, 1965). *Cajanus cajan* and *D. lablab* sown in rows along the contour provided extra browsing and protein supplement for cattle grazing natural grasslands during the dry season in Brazil (Schaffhausen, 1963). In Queensland, strip planting of Townsville stylo into native grazinglands dominated by *Heteropogon contortus* increased the herbage available for grazing, as well as the stocking capacity, and facilitated the spread of the legume into non-seeded areas by seeds passing through the grazing cattle (Graham, 1968). Sowing a grass mixture into cultivated strips 4.75 m in width, and leaving a 14.2 m belt not cultivated or sown (Table 8.3), reduced the cost of cultivation and served as a catchment for seeds washed from the cultivated area (Pratt, 1964). Certain species of grass with light-weight seeds which are wind blown, such as *Dichanthium aristatum* and *Hyparrhenia rufa*, will invade adjacent areas when plants are allowed to develop seed heads.

Establishment of sown pastures and forage crops

Sown pastures are improved grazinglands comprising introduced grasses and/or legumes for temporary or permanent grazing. They may be established by (1) improvement of native or naturalized grasslands by some form or sod disturbance, fertilization and introduction of new species by sowing or transplanting, (2) renovation of a previously developed pasture, (3) development of previously cultivated land and (4) development of forested land.

Supplemental forages are grasses and legumes suitable for cutting and feeding as fresh material, conserving as hay or silage or using for temporary grazing. When grazed they are sometimes called supplemental pastures.

Land clearing, preparation and fitting

Large forested areas can be cleared by contract logging operations or, if there is no interest in lumbering, large trees can be pulled down by use of a heavy anchor chain weighing 50–60 tonnes, drawn between two tractor crawlers. A heavy steel ball attached to the chain holds the loop near ground level and helps in breaking or uprooting large trees. Smaller trees and shrubs may break, bend or be unharmed. A bulldozer with attached blade, usually having heavy chisels fixed to the bottom of the blade, is used for stumping, uprooting small trees and bushes, and wind-rowing to allow drying and burning. Soil dragged into the wind-rows, and ashes from burning, are spread over the land by bulldozer.

In parts of the tropics a complete land-clearing operation, including stumping, can be arranged in exchange for the timber to be used as firewood or for charcoal. Trees are felled, cut and split, bushes slashed, and stumps uprooted by hand, leaving the land ready for tillage.

Forested areas may also be cleared by using a large mechanical cutter, which fells trees up to 20 cm in diameter, picks them up, chops all parts into small pieces and spreads them over the soil as a mulch. Such equipment is expensive and would only be economical for large-scale operations or custom service.

The Holt Bush Breaker, which is a heavy roller with blades, smashes small trees and shrubs when drawn over bushed woodlands by heavy tractor crawlers (Pratt, 1966). The smaller woody types are bent over but tend to become upright within a short time, so that additional cutting is often needed. Studies have shown that the effectiveness of this equipment depends on the model of the machine, type of vegetative cover and soil characteristics. On certain soil types covered with sparse vegetation, the blades loosen the soil and leave a coarse seedbed without additional cultivation. On heavy soils, however, there may be compaction so that seedlings would have difficulty in becoming established.

Trees and brush can be killed by chemical treatment using 2,4,5-T or 2,4-D, mixtures of these, and formulations of Picloram, fenuron and arsenic (Little and Ivens, 1965). Chemicals are generally used as aqueous sprays but some come as granules or pellets for spreading over the soil surface. The basal trunk of trees and shrubs may be cut and the surface brushed or sprayed with a solution of the chemical, allowing runoff on to the soil so as to affect the budding zone of roots. On extensive areas with thick growth the brush-killers may be applied by plane, helicopter or motorized, low-volume spray equipment. It is usually necessary to repeat the treatment as a single application does not always completely kill the woody plants.

Time of sowing and transplanting

A well-prepared seedbed is the foundation and basis of successful seedling and sward establishment. Sown grasses and legumes require a finely granulated soil surface, which is firm and free of weeds. On loose, friable soils, use of a heavy piece of equipment to roll over the soil, known as a cultipacker, leaves a firmed seedbed for the sowing operation. Heavy soils should not be rolled as this contributes to the formation of a surface crust and a tine-type cultivation is preferred on hard-capped loams and soils subject to erosion. Prior to land preparation and during the fitting operations, attention must also be given to water control.

Sowing shortly after the rains begin takes advantage of soil nitrogen made available by mineralization, but coincides with the period of severe rainstorms and strong weed competition. With heavy downpours excessive runoff of water occurs and either washes seeds out of position or covers them too deeply. In addition, the first rains are frequently unreliable and short periods of water stress may affect seedling survival. In regions having a single rainfall peak and a short rainy season, early sowing is important to take advantage of the soil moisture and short growing period. Where bimodal rains occur, it is advisable to wait until later in the season or to sow at the beginning of the second wet period as storms are generally less severe. As the soil temperature increases moisture stress is accentuated, and damage to seeds of some species occurs at 50 °C or higher, especially if germination has begun.

Method of sowing

Grasses and legumes are sown alone or in mixtures by one of the following methods:

1. Broadcasting

(a) Seeds spread by hand, taking care to assure uniform distribution. Dividing the seeds into two equal parts and sowing in one direction, then cross-directionally, is an aid to equal dispersal. Wind is a factor to be considered, especially with light-weight seeds that may be blown over some distance, e.g. *Melinis minutiflora* and *Hyparrhenia rufa*. Mixing the seeds with materials such as sawdust, sand, fertilizer, etc. helps to distribute the seeds more evenly. (b) Seeds distributed by use of a 'Cyclone Seeder', a small mechanical device, consisting of a heavy canvas seed bag attached to a frame having a slotted and movable opening, a fluted dispersal pan turned by a hand crank, and straps which can be placed around the neck, allowing one to walk freely as the seeds are spread in a fan-like fashion. Attention must be given to movement across the field so that strips are not missed and overlapping does not occur. (c) Mechanical equipment – motorized or tractor-drawn implements built on the principle of the cyclone seeder; ordinary cereal grain drills with seed hoses hanging free. (d) Aerial – specially equipped airplanes, particularly for inaccessible hill grazinglands and extensive rangelands.

Broadcasting can give satisfactory stands when conditions are favourable for rapid seed germination and seedling growth. It has application for coarse seedbeds in parts of the tropics where preparation of a smooth surface may lead to excessive washing, for sowing during the rainy season when mechanical equipment cannot be moved on to the land and for areas too steep for cultivation. In accessible areas, seeds can be covered by dragging a brush (a cut tree trunk) across the soil surface, using a tine or disc harrow (set for discs to run almost straight), or a roller. With small seeds covering by rainfall will probably be ample, especially on coarse seedbeds.

2. Drilling

Equipment has been designed for spacing of rows 25–30 cm, proper placement and even distribution of both seeds and fertilizers. This ensures greater success in seedling establishment than the use of broadcasting methods. Models of many cereal grain drills have seeding attachments which fit on to the grain box. They operate from the gear mechanism to deliver seeds through the regular hoses behind the discs, but there is a tendency for deep coverage. Special seed drills are equipped to handle waste seeds and those with appendages, e.g. *Cenchrus ciliaris*.

3. Rows

Sowing in rows facilitates weeding and cultivation, especially during the early stage of establishment. Maize and cotton planters can be adapted for this purpose, or cereal grain drills may be used by blocking off the appropriate number of openings. Rows are sometimes marked by use of tines or chisels spaced on the tractor tool bar, with seeds or vegetative pieces placed by hand into the shallow furrows. With grass–legume mixtures, one may be sown in rows and the other broadcast or they may be placed in alternate rows, depending on the species, seedling vigour and habit of growth.

Row widths vary according to the habit of growth and the intended use of the pasture or forage crop, as well the rainfall regime and available soil moisture. Studies in Colombia (rainfall above 750 mm annually) showed that grasses grown for silage and greenchop or soiling should be placed in rows separated by 0.75–1.0 m (ICA, 1957–72). For establishment of species used for pasture, such as *Panicum maximum, Hyparrhenia rufa, Cenchrus ciliaris* and *Dichanthium aristatum*, spacings of 30–60 cm were more appropriate for good ground cover and optimum production of herbage. The spreading type legumes, e.g. *Centrosema pubescens, Pueraria phaseoloides, Macroptilium atropurpureum, Desmodium intortum* and *D. uncinatum* established well when sown in 1.0 m width rows. Lucerne gave optimum forage yields when drilled in 25–30 cm rows. In Kenya (630–890 mm rainfall or less), experiments with *Cenchrus ciliaris* showed that sowing in 3.0 ft rows (approx. 90 cm) have higher dry matter yields than sowing in 1.0 and 2.0 ft rows or broadcasting seeds (Sands *et al.*, 1970). In Nigeria (800–1 000 mm annual rainfall), growing *Andropogon gayanus* in 50 cm rows gave higher herbage yields than other widths (Rains, 1963).

4. Cultipacker seeder

This pulverizer-packer has been used for shallow seedings on the lighter soils. It consists of rollers arranged in tandem and a gear-operated seed box set above and between the two tiers of rollers. Seeds fall immediately behind the first roller and mostly between the small ridges which it forms. The ridges are split by the second roller, lightly covering the seed and compacting the soil. This equipment should not be used on most heavy soils as it may cause crusting of the surface.

5. Companion crops

The sowing of grasses into crops such as maize, sorghum and millet (sometimes called nurse crops) has shown varying degrees of success. *Andropogon gayanus* became well established when sown with maize and soybean in northern Nigeria, but use of *Stizolobium* spp. and *Pennisetum pedicellatum* were too competitive and gave poor stands of the grass (Haggar, 1961). Two cuttings of the companion crop at 40 and 80 days after sowing increased the plant population of *Andropogon* but decreased total herbage production, as compared to a single cutting at 80 days. Undersowing *Chloris gayana* into *Pennisetum americanum* planted in 1.0 m rows and 0.75 m hill spacings gave adequate stands by the time of millet harvest (Rains, 1963). The pasture grass was sown broadcast and in rows, and both methods resulted in good cover. In western Kenya (1 275–1 600 m elevation), broadcasting *C. gayana* into maize after the first weeding (maize about 15 cm) gave 55 per cent ground cover by the time of harvest but caused some decrease in maize grain yield (Golden, 1967). Grass establishment approached 60 per cent when sown at the time maize reached 30 cm height and had little effect on grain yield. Grazing for 7 days after maize harvest and then passing a gyromower left a mulch on the soil surface which favoured moisture retention. After applying 100 kg/ha of sulphate of ammonia, the grass was ready for grazing in 1 month. Waiting until the maize grew to about 1.0 m before sowing gave a 32 per cent grass cover due to competition for light and moisture, nor was the pasture ready for grazing by maize harvest. Similar results were obtained in Colombia when seeds of *Panicum maximum*, *Hyparrhenia rufa* and *C. gayana* were sown, and stolons of *Digitaria decumbens* were transplanted into maize at different dates from planting (ICA, 1957–72) and in Malawi with *C. gayana* and *Desmodium uncinatum* (Thomas and Bennett, 1975).

6. Aerial seeding

In regions of rough terrain and dense vegetation, and where extensive areas are to be sown, aerial seedings have been effective for the establishment of sown pastures. This was demonstrated in Hawaii (2 500 mm annual rainfall) where the vegetation of a steep-sided valley of about 17 ha was treated by aerial application of herbicide, followed by burning, then seeding and fertilizing from the air (Motooka *et al.*, 1967). A complex mixture of seven legumes and two grasses was used, with the seeds being pelleted to ensure even distribution of unequal seed sizes. Within 8 weeks from sowing *Panicum maximum* was heading and after 5 months many of the legumes had begun to flower. Stands were not uniform due to micro-environmental effects. Of particular interest was the ready establishment on slopes receiving the morning sunlight, and sparse development or nil on those exposed to the afternoon sun.

There is no assurance that any one method will result in better establishment than another. So many factors determine whether a seed germinates, the seedling grows, the plant develops and then survives. These include the adaptability of the species, the viability of the seed itself, the extent of land preparation and tillage, depth of placement in the soil, the relationship of seed contact with soil particles, compactness of the soil surface, availability of soil moisture and intensity of rainfall, water runoff and soil erosion, soil temperature, seed treatment, soil- and seed-borne nutrients, damage from insects and pests, grazing or cutting management (Anderson, 1968; Bogdan and Pratt, 1967b; Crowder, 1960; Jones and Jones, 1971; Smith, 1970).

Sowing rate (density)

Commercially grown and processed seeds have a high percentage of pure germinating seeds, especially those grown under the regulations of seed certification schemes. Seeds harvested locally may contain high amounts of waste and have low germination, sometimes less than 1 per cent.

Seeding recommendations range from 10 kg/ha to more than 40 kg/ha for grasses, and from 3 to 15 kg/ha for legumes (Borgét, 1969; Chaverra *et al.*, 1962; Gontijo *et al.*, 1969; Jones and Jones, 1971; Rey and Matta, 1966; Risopoulos, 1966). Generally, 5–10 kg of grass and 2–6 kg of legume seeds of high quality suffice to obtain desirable stands. This has been illustrated from sowings of *Panicum maximum* var. *trichoglume* and *Glycine javanica* cv. Tinaroo (=G.

wightii) made in northern Queensland (Tow, 1967). Increased seeding rates resulted in higher plant densities, with *Panicum* populations exceeding *Glycine*. The numbers of plants per unit area of both species declined over time, with the lower rates showing a greater percentage survival. The legume had no consistent effect on grass survival, probably due to the less aggressive characteristic of the legume seedlings. *Glycine* survival was affected by both its own density and that of the grass, the greatest survival occurring on plots of less dense grass-seeding and at higher densities of legume-seeding. In pure stand, the lowest rate of 3 kg/ha of *Glycine* would have been sufficient for good establishment since there was an average of more than one plant per square foot.

Sowing densities vary with the methods of seeding. In western Kenya, the optimum rate for *Cenchrus ciliaris* was 4.4 kg/ha for broadcast, 2.2 for 30 cm rows, 1.1 for 60 cm rows and 0.55 for 90 cm rows (Sands *et al.*, 1970). Several studies in Colombia showed that sowing 5.0 kg/ha of viable lucerne seeds in 25 cm rows gave forage yields equal to heavier rates of 10, 15 and 25 kg (ICA, 1957–72). Broadcasting 10 kg/ha compared favourably with the higher quantities of seed. Higher seeding rates were required for the larger-seeded legumes, e.g. 15–20 kg/ha of *Lablab purpureus* and *Cajanus cajan*.

Depth of sowing

This is usually related to seed size, seedling emergence and survival of small-seeded species. The optimum depth of most grasses and small-seeded legumes lies between 1.0 and 3.0 cm. In Nigeria, seedling emergence of *Andropogon gayanus* declined markedly with sowings made deeper than 2 cm (Bowden, 1963). The same results were obtained with *Chloris gayana* and *Panicum maximum* in Rhodesia (Smith, 1967). At 5 and 7.5 cm, 41 and 12 per cent of *Panicum* seedlings emerged, but 2 per cent and nil of *Chloris*, respectively.

The highest emergence of *Stylosanthes guianensis* and *S. humilis* in Australia occurred from placement between 0.6 and 1.3 cm, with none from a 5 cm depth (Stonard, 1969). The larger-seeded legumes such as *Centrosema pubescens*, *Leucaena leucocephala*, *Clitoria ternatea* and most *Crotalaria* spp. may be placed at 2.5–5.0 cm below the soil surface, but smaller-seeded types such as *Pueraria phaseoloides*, *Glycine wightii*, *Calopogonium mucunoides*, and most *Desmodium* spp., no more than 2.5 cm (Table 8.4). Those with much smaller seeds, such as *D. intortum*, *Lotononis bainesii* and some of the *Indigofera* spp., should be covered no more than 0.5 cm or left on the soil surface of a coarse seedbed to be covered by splashing raindrops.

Moisture in the top 1.0 cm soil layer fluctuates more than at

Table 8.4 *Emergence in percentage of legumes sown at various depths, derived savanna of Nigeria (Adegbola, 1964)*

Species	Surface	2.5 cm	5.0 cm	7.5 cm
Centrosema pubescens	56	71	80	80
Calopogonium mucunoides	26	32	34	11
Pueraria phaseoloides	14	59	30	25

2.5 cm and lower depths so that germination of seeds and seedling establishment under stress conditions may be erratic and patchy with the more shallow placements. Some compromise must then be made for sowing at a depth of greater moisture fluctuation, but at which seedlings can emerge.

Seed treatment

1. Scarification

Seeds of most tropical legumes are highly impermeable to water imbibition, because of hard seed coats, so that some form of scarification is needed to improve germination. The degree of hardness varies among species, seed lots, and to some extent stage of maturity at harvest.

Commercial harvesting and processing ruptures the seed coats of many species, but hand-harvested seeds require treatment. Several methods of scarification are as follows:

(a) Mechanical

Hand rubbing with sandpaper or carborundum paper may suffice but a number of seeds are so hard that the germination remains low. Pricking the seed coat with a pin or sharp-pointed knife is effective but arduous and time-consuming, especially with small seeds. Seed-scarifying equipment is available as small laboratory and large commercial models. These consist of gear-operated baffles operating within a cylinder or drum which is lined with an abrasive. In using these types, care must be taken not to damage the seeds by cracking or removal of the seed coat due to over-scarification. Species which can be mechanically scarified include *Centrosema pubescens, Clitoria ternatea, Glycine wightii, Pueraria phaseoloides, Calopogonium mucunoides, Macroptilium atropurpureum, M. lathyroides*, certain *Desmodium* and *Indigofera* spp.

(b) Acid treatment

Placing seeds in concentrated sulphuric acid for 3–5 min, followed by thorough washing and drying is an effective means of scarifying *Stylosanthes guianensis* and the above species, except for *Clitoria* and centro. Acid-resistant plastic bags are useful in enclosing seeds for submerging in the acid and subsequent washing. Extreme caution is required so as not to splash acid on the body and clothing. Several grasses, such as *Paspalum* and *Brachiaria* spp., have hard seeds and respond to acid treatment.

(c) Hot water

Soaking seeds of *Leucaena leucocephala* in water heated to about 80 °C for about 5 min, rapid cooling and drying improves germination. Another method is to bring water to about 90 °C and remove from the source of heat; place the seeds into the hot water and allow to remain until the water temperature drops to 40 °C, then remove, cool and dry. This is especially useful for *Stylosanthes humilis*. Heat treatment is also effective for stylos such as *S. hamata*, *S. scabra* and *S. viscosa* (Gilbert and Shaw, 1979).

2. Inoculation

Seed inoculation with the appropriate *Rhizobium* ensures that bacteria necessary for nodulation are introduced at the time of sowing. Herbage legumes grown in many parts of the tropics, however, produce nodules without inoculation because the 'cowpea-type' rhizobia is naturally present where native or naturalized legumes are growing or were recently grown. A number of tropical legumes are promiscuous in their acceptance of rhizobia and will normally nodulate from the native cowpea *Rhizobium*, e.g. *Calopogonium muconoides*, *Cajanus cajan*, *Lablab purpureus*, *Macroptilium atropurpureum*, *Pueraria phaseoloides*, *Stylosanthes guianensis* (some selections are specific, however), *S. humilis*, *Teramnus uncinatus* (Norris, 1972). Others are specific and must be inoculated with selected strains of *Rhizobium* for effective nodulation, e.g. *Centrosema pubescens*, *Leucaena leucocephala*, *Lotononis bainesii*, *Trifolium semipilosum*. A *Desmodium*-type *Rhizobium* is needed for the species of this genus. Non-specific bacteria infect the roots of some legumes so that the presence of nodules is no assurance that the relationship is symbiotic nor that the rhizobia are effective nitrogen fixers.

It is important that the correct *Rhizobium* is used for inoculation of tropical legumes. The 'cowpea-type' rhizobia are tolerant to acid soil conditions and produce alkaline end-products (Norris, 1965). This contrasts with the temperate zone legumes which grow in near-neutral or alkaline soils. The rhizobia associated with this group produce acid end-products. Thus, the *Rhizobium* used for temperate legumes is not appropriate for tropical legumes. An exception exists for *Leucaena leucocephala* in that it hosts a specific *Rhizobium* which has the alkaline-producing habit (Norris, 1972; Norris and Date, 1976). Inoculants for tropical legumes are commercially available from most seedsmen. They can be airmailed anywhere in the world and kept satisfactorily in a refrigerator for many months. Despite the fact that legumes may nodulate without inoculating the seeds, it is always advisable to add the appropriate commercial inoculant which ensures effective symbiosis and proper growth of the legumes (Ferrari *et al.*, 1967; Norris, 1972; Rotar *et al.*, 1967).

Most inoculum contains a sticker and is ready for use. If not, a water slurry can be prepared by simply mixing the inoculum with water and thoroughly wetting the seeds with the mixture. Syrup, such as honey or a commercial sticker (methyl cellulose), can be mixed with water to aid adhesion of the inoculum to the seeds.

3. Pelleting

Seed pelleting is the coating of grasses and legumes with a layer of fine powder held on to the seed by a sticker. This is done to assure even flow of seeds from the drill or aerial applicator, aid inoculation of legumes under favourable soil conditions, introduce inoculum into soil already inhabited with other types of rhizobia, benefit sowing into dry soil and conditions of high soil temperatures, allow the mixing of seeds with superphosphate prior to sowing without injury to the seeds or to the *Rhizobium*, overcome the effects of soil acidity on effective nodulation and seedling development and repel insects, birds or rodents (Brockwell, 1962; Diatloff, 1971; Norris, 1967; Norris and Date, 1976; Plucknett, 1971; Russell *et al.*, 1967).

Temperate zone legumes generally require liming of the soil for proper inoculation, symbiosis and N fixation by rhizobia, and optimum growth of plants. When seed pelleting became popularized in the temperate zone, lime was used as a coating agent and this tradition was carried over into the tropics. It has been shown, however, that tropical legumes are highly efficient in their uptake and utilization of nutrients and make reasonable growth at levels of soil Ca which are acute for temperate legumes (Norris, 1972). Liming is generally not needed for the growth of tropical legumes, except on very acid soils (e.g. pH of 4.5 or lower), nor is it essential for growth of *Rhizobium*. Thus, it is not the appropriate coating for pelleting seeds of tropical legumes, except for *Leucaena leucocephala*.

Phosphorus is usually the limiting factor in the establishment of tropical legumes. Many tropical soils contain low amounts of this element so that phosphate application is needed for rapid seedling development and plant growth. Rock phosphate which has been ground to pass a 300-mesh sieve should be used for pelleting tropical legume seeds (Date and Cornish, 1968; Diatloff, 1971; Norris, 1972). Its consistency may vary, depending on the source, and cause some physical problems in the pelleting process due to caking of the dust and sloughing from the seed (Plucknett, 1971). In Hawaii, calcium silicate (TVA slag) proved suitable as a coating material (Motooka *et al.*, 1967). This material supplies a small quantity of Ca which is useful in highly acid soils, but does not ·cause a highly alkaline reaction.

Two adhesives have been commonly used in seed pelleting: 4 per cent methyl cellulose and 40 per cent gum arabic. The former is more economical in that 4 kg is as effective as 40 kg of gum arabic, but the latter is more widely available in most tropical countries.

The pelleting process consists of (a) preparing the adhesive by mixing with hot water and then cooling, (b) adding the proper quantity of inoculum to the adhesive solution and thoroughly mixing, (c) tumbling the seed and adhesive–inoculant solution in a barrel, concrete mixer or bucket, (d) adding the pelleting material with continued tumbling until seeds are evenly coated and separated and (e) drying the pelleted seeds (Diatloff, 1971; Plucknett, 1971). Various seeds require different quantities of adhesive and pelleting material because of size variation. Seeds of different sizes should be pelleted separately. Many seedsmen offer a commercial pelleting service for a small cost per unit weight of seed.

4. Use of fungicides and pesticides

Legume seedlings are highly susceptible to fungal diseases and many may be killed shortly after emergence, commonly called 'damping-off'. Grasses also succumb to seedling diseases. The seeds and seedlings of both may be damaged by insects, and ants sometimes carry seeds into their beds. A number of fungicides and pesticides for seed treatment are available from seedsmen and may go under different trade names. Advice should be sought before using them for treating legume and grass seeds. Commercial seeds frequently have been treated at the time of processing and this should be indicated on attached labels or printed on the seed container.

Vegetative establishment

A number of tropical grasses are vegetatively propagated by use of stems, stolons and crown splits. Some species do not develop viable seeds and must be reproduced asexually, e.g. *Digitaria decumbens, Cynodon nlemfuensis* var. *nlemfuensis*, and *Pennisetum typhoides* × *P. purpureum* hybrids. Others produce small quantities of seeds so that establishment is more easily and effectively accomplished by using vegetative materials, e.g. *Cynodon dactylon* cv. Coastal and cv. Coast-cross 1, *Brachiaria mutica, Eriochloa polystacha, Axonopus scoparius*, and *Pennisetum purpureum*. Some are transplanted due to a lack of commercial seeds or difficulty of local harvest in certain regions, e.g. *Panicum maximum, Hyparrhenia rufa, Brachiaria decumbens, Paspalum notatum, Pueraria phaseoloides, Stylosanthes guianensis* and *Medicago sativa*.

Several methods of establishment are:

1. Burn or heavily graze the native or naturalized grazingland and set vegetative pieces in holes opened at random by hand using a hoe, spade or sharp-pointed object.

2. On steep slopes open a furrow along the contour and lay the vegetative material along the side of the furrow so that the tips extend above the soil when covered with a hoe or spade.

3. On arable land and level sites prepare the land to form a friable seedbed, mark furrows at 0.50 to 1.0 m apart and place the propagules in the furrow, covering by hand or a cultivator. With most grasses the stems, stolon pieces and crown splits are set from 0.25 to 1.0 m within the row, but with elephant grass the entire stem can be layered in the furrow and lightly covered.

4. Use of a grass transplanter which attaches to the drawbar of a tractor, has a strong chisel or plough for opening a furrow, seats from which operators place the propagules into the furrow and a press wheel for firming the soil. A tree or sweet potato transplanter also works well.

5. Broadcasting vegetative material over the soil surface and partially covering with a disc. Regeneration of growth may be poor and subsequent ground cover may be slow and sparse with this method.

A method used to establish guinea grass in parts of South America is to fell the forest by hand labour, wind-row the brush and trunks by bulldozer and burn the debris after drying. Maize is then planted (frequently by hand) without fertilizer, thus taking advantage of the forest fallow. Crowns of the grass are collected by hand and separated into splits, leaving the roots attached since stems of this species do not regenerate growth. Four men can collect in a single day sufficient vegetative material for 1 ha and six men can transplant 1 ha in a day. The crown splits are spaced 30–50 cm apart between the maize rows when the plants are 0.80–1.0 m tall. The grass regenerates growth and then develops seeds at about the time of the maize harvest. After the grass seeds drop, the grass is grazed rather heavily, then rested during the time of seed germination and seedling establishment at the beginning of the rainy period. Returns from the maize crop cover the cost of establishing the grass.

Transplanting crown pieces is a common method of establishing lucerne at the higher elevations in the Andean zone of South America. A farmer obtains a small quantity of seeds to establish a nursery, then transplants the seedlings in 1.0 m rows at about 0.50 m within the row. After one or two growing seasons the plants may be dug up and divided into crown splits for additional transplanting. This method is used on smallholdings where daily cutting and feeding to dairy cattle is practised.

Distribution of material for vegetative propagation is seldom a commercial enterprise. On a local basis, however, a governmental agency or a livestock farmer will sell vegetative pieces of grasses, usually in small quantities, for the establishment of a source nursery, or by the wagonload (truckload) for larger transplantings. Stoloniferous grasses such as Pangola, pará and kikuyu can be vegetatively propagated for commercial distribution (Nicholls and Plucknett, 1971). Stems and stolons cut into sections with one or two nodes readily develop roots when placed in a proper medium, such as sphaghum moss with added plant nutrients. The propagating medium can be put in small containers, such as the 'Jiffy' peat pots used by vegetable and flower growers. Grass propagules set into these receptacles resume growth and are ready for transplanting within 3–4 weeks. They are easily handled for shipping, short-term storage and direct transplanting without removal from the pots. Since the plants are well rooted they continue growth when set into the field. From 500 to 1 000 small pots would be sufficient for 1 ha.

With stoloniferous or creeping grasses, more rapid ground cover occurs with close spacing in rows, as compared to wider spacing and broadcasting vegetative material over a well-prepared soil surface. Transplanting stems or stolons of *Digitaria decumbens* on 0.25 × 0.25 m or 0.50 × 0.50 m centres gave dry forage yields of about 2.0 tonnes/ha within 2 months from establishment in Colombia, as compared to less than 1.0 tonne for transplanting on 1.0 × 1.0 m centres (Varela and Crowder, 1960). Spreading equivalent quantities of material over the soil surface and discing caused a 25–40 per cent reduction in hay produced at the first harvest. After three harvests were taken at 6-week intervals, no differences in ground cover or yields of forage were noted among the different methods. Stolons renewed growth more quickly than stem pieces.

Spacing stem pieces of *Pennisetum purpureum* and *Axonopus scoparius* at 0.50 × 0.50 m gave higher forage yields in Colombia than at 2.0 × 2.0 m (Lotero *et al.*, 1967, 1969). More dense stands occurred with stem cuttings having one or two nodes exposed, rather than laying entire stems in furrows.

Grass-legume mixtures

Grasses are usually the predominant species in sown pastures and grazinglands because of their greater agressiveness, stronger competition for nutrients and higher resistance to intensive grazing, trampling of animals and burning as compared to legumes. None the less, there are advantages for striving to maintain legume–grass mixtures in the sward available for grazing.

A perennial grass companion species is a desirable feature of a leguminous pasture for the following reasons:
1. To increase total herbage production. Not all grass–legume com-

binations, however, will exceed certain legumes in total forage yield and N-fertilized grass may outyield a pure legume stand.

2. To ensure stability of production. The growth patterns of both grass and legume species varies, and their rhythmic development may be different because of availability of nutrients and moisture. Grasses usually begin growth more quickly after early rains following the dry season and thus increase earlier grazing.

3. To increase the energy value of the pasture. The grasses are generally lower in crude protein content than legumes and provide the bulk of the energy ration.

4. To suppress the invasion of weeds. Under proper grazing management the perennial grasses maintain more dense sward and help to reduce the ingress of annual grasses and broad-leaved weeds.

Legume species are added to grass pastures in order to:

1. Increase the amount of crude protein available for the grazing animal. This is particularly important when the grasses begin to mature and during the dry season when they are dry and highly lignified.

2. Extend the grazing period into the dry season. Many legumes remain green throughout much of the dry period and if the plants become brown, or even die, their nutritive value is maintained at a much higher level than that of grasses. In many areas of the tropics it is possible to maintain a year-round feed supply of pasturage with the correct combinations of legumes and grasses and with judicious management of grazinglands and grazing animals.

3. Provide N for the companion grasses. The N fixed by legumes is not usually directly transferable to the associated grass in large quantities but becomes available after the nodules are sloughed from the roots. It is not uncommon that the legume will provide 100–200 kg/ha or more of N on an annual basis.

In compounding mixtures, information is needed on the compatibility, i.e. associative ability, of the grasses and legumes when grown in combinations. Of particular importance is the early seedling vigour, habit of growth, peak herbage production, time of flowering and seed development. With seeds of equal viability the ratio of legumes to grass should be about 3 : 2. Under good rainfall and planting conditions a 12–15 kg/ha seed mixture may be comprised of 7–9 kg of standard quality of legume and 4.5–6.0 kg of grass seeds. Most grasses grow more rapidly in the seedling and establishmental stages of development, so that the higher percentage of legume is important to encourage and favour their establishment in the sward. Seed size will modify the quantities within the mixture, of course; for example, stylo seeds are considerably smaller than those of centro so that the amounts should be adjusted on the basis of seed numbers per kilogram and the ultimately desired botanical composition.

The number of different species comprising a pasture mixture varies from a single grass and legume (simple) to 6 or 8 legumes and 3 or 4 grasses (complex). If considerable information is available about species adaptability, growth habits and responses to management treatments, the more simplified mixtures could be used. In many instances, however, such data are not available so that the more complex combinations are sown with the assumption that the different species will be sorted out, depending on climatic, soil and management conditions. Under most conditions the more complex mixtures tend to revert to simpler combinations. For the development of long-term improved pastures it is advisable to include at least two compatible grasses and two legumes. This allows for differences in sowing conditions, variation in germination and seedling growth due to localized soil and moisture situations, provides for a more dense sod and sward which helps prevent the ingress of weeds and gives a greater opportunity for diverse pasture and grazing management. To avoid delay in the stand establishment and to provide for more rapid ground cover and earlier grazing, mixtures should contain a grass and legume which germinate rapidly and produce early, even if these species die out after the first or second season. For example, *Melinis minutiflora* and *Macroptilium lathyroides* or *Calopogonium mucunoides* might be included in the mixture as early producers. One should not be discouraged if some of the more desirable legume species are less prevalent in the first season since some are slow in becoming established, such as *Centrosema pubescens, Glycine wightii, Macroptilium atropurpureum* and *Pueraria phaseoloides.*

Many studies of grass–legume combinations have been made and recommended mixtures are frequently listed for localized conditions (Anderson, 1968; Bogdan and Pratt, 1967b; Grof, 1981; Lazier, 1980; Rensburg, 1969; Suttie, 1965). The specific combinations are related to climatic conditions, soil fertility and applications of fertilizers, amount of rainfall and availability of soil moisture, elevation and intended use. Most have been examined under small-plot clippings trials and for short durations of time and are not always indicative of the performance to be expected under grazing conditions or over long periods of time.

A question of time and method of sowing grass–legume mixtures frequently arises. Most are sown simultaneously, even though separate operations may be performed due to differences in seed sizes or seed appendages, and to the need for transplanting vegetative materials. Since grasses are generally more aggressive than legumes, it has been suggested that the latter be established first, followed by sowing of grasses a short time later. This type of operation, however, meets with difficulty because of timing, weed control and labour demands. In an effort to minimize the grass and legume competition the two have been sown in alternate rows, grasses in rows with legumes broadcast, legumes in rows and grasses broadcast and alternating strips of the two. No single method can be recommended as a general guide and local pasture agronomists and agricultural extension officers should be consulted regarding mixtures, fertilizers and sowing methods.

Fertilization for establishment

A principal limitation to the successful establishment of legumes in the tropics is the widespread low phosphorus content of the soil. Many soils are heavily leached and highly acidic, being low in Ca as well as P. Most grasses grow over a wide range of soil pH, however, and tropical legumes tolerate the acidic condition provided phosphorus is available, the Ca : Mg ratio is favourable, and the aluminium (Al) content is not sufficiently high so as to restrict root development. On some soils sulphur (S), potassium and molybdenum (Mo) may be deficient. Small pot experiments provide a guide for determining soil nutrient deficiencies, along with foliar analysis and the use of indicated critical values which species and cultivars have for the principal nutrients (Andrew and Robins, 1969a, 1969b; Mohamed Saleem, 1972).

For the development of a legume or legume–grass pasture it is imperative that phosphate be applied before or at the time of sowing. The quantity needed will vary with the legume species, but, as a general rule, from 125 to 250 kg/ha of superphosphate will be ample for the initial establishment of the legume, with higher quantities needed on soils having a high P-fixing capacity. The use of single superphosphate will also take care of S needs, since it contains about 10 per cent of this element. Molybdenized superphosphate has been widely used in Australia and would probably be of benefit in other regions of the tropics. Phosphorus is important since it increases N fixation of the legume plant, improves herbage yields by as much as 2.5 per cent and raises the N level of the forage by 3.3 per cent or greater (Norris, 1967; Shaw *et al.*, 1966). Magnesium is needed in the legume nodule for proper symbiosis (Anderson, 1956) and S is essential for protein formation (Jones and Robinson, 1970).

Weed control

A well-prepared seedbed is the first step toward control or suppression of weeds and boosts the growth of the sown pasture species. This consists of deep ploughing of the land and thorough harrowing. Wherever feasible, a second light harrowing after about 3 weeks will destroy many weed seedlings which have developed with favourable soil moisture. *Imperata cylindrica* is a widespread and prevalent weed in many regions of the tropics following shifting cultivation and where fire is regularly used to burn the accumulated and dead plant material. A dry season fallow, followed by additional tillage at the beginning of the rainy season, use of fertilizers and the seeding of an adapted legume–grass mixture helps to reduce the population of this grass as well as other weeds.

Adequate fertilization at the time of seedbed preparation or the renovation practice is the second step for weed control and is needed for early growth of the desirable pasture legumes and grasses. Many of the weedy species thrive under low soil fertility conditions and are less competitive when the soil fertility level is improved. Growing a cash or food crop which is well fertilized and cultivated prior to pasture establishment is another means of reducing the weed population during the early growth period of the sown pasture mixture.

The method of sowing may favour the early development of the pasture species. Seeding in rows with proper placement of fertilizer below or to the side of the seeds gives an advantage to the young legume and grass seedlings.

Selective herbicides can be applied as pre- or post-emergence sprays, but usually their use is not economical unless the pasture is to be intensively utilized. Even then, an ecological approach will usually be as effective for the establishment of a relatively weed-free pasture sward as chemical treatment.

Despite these precautionary measures, some weedy plants usually emerge from the soil at about the same time, or even before, the seeded legumes and grasses. If their numbers are excessive the area can be mowed when the plants reach a height of 20–30 cm. Clipping the young pasture species hinders their growth and development but rids the area of most broad-leaved weedy types and also tends to suppress annual grasses. Light grazing might be helpful, but care must be taken not to damage the pasture species due to the trampling of animals or uprooting of the young seedlings. Hand-slashing can be employed, especially in the areas of heavy weed infestation and depending on the availability of labour.

First season management

Protection from grazing after seedling emergence, or regeneration of growth in the case of vegetatively propagated species and until plants attain well-formed root systems is necessary to ensure good stands of sown grasses and legumes. This suggests a time interval of 6–8 weeks after sowing or transplanting, or until plants reach a height of 40–60 cm. Light grazing could then be practised. With vegetatively propagated pasture types, such as *Digitaria decumbens* or *Cynodon* spp., grazing can be more intense than with the seeded types. Heavy grazing should be avoided, however, when legumes are included, since they usually grow more slowly than the grasses. Grazing should also be less intense in the more arid than in the humid regions since plant growth will be less rapid. In fact, it has been suggested that grazing or cutting of grasses, such as *Andropogon gayanus*, be deferred until the end of the first growing season so as not to affect performance adversely in subsequent years (Rains, 1963).

The grasses used as cutting crops, such as *Pennisetum purpureum*, *Axonopus scoparius* and *Tripsacum laxum*, should not be harvested until the plants are about 2.0 m high. A cutting level of about 20 cm above ground level is best to assure good establishment and vigorous growth after harvest.

Renovation

Sown pastures frequently deteriorate under conditions of improper management, such as inadequate maintenance fertilizer, overgrazing and prolonged drought. Deterioration may result from a reduced stand of legumes and grasses, predominance of grass or invasion of undesirable species. When this occurs, a decision must be made to attempt some means of renovation or to destroy the old sod and reseed the pasture. Several procedures can be employed to remove any excess herbage, partially destroy the sod and scarify the soil, fertilize, overseed, and cover the seed. Accumulated herbage may be heavily grazed, burned or mowed with removal of the plant residue. A simple scheme would then be to apply fertilizer and overseed without any attempt to disturb the remaining sod. This process is likely to be disappointing in terms of seedling establishment and improved sward density. A greater degree of success would be obtained with some means of partially destroying the sod by scarifying and lightly tilling the soil. This can probably be accomplished most effectively by using a disc harrow set so as to cut through the old sod and into the surface soil layer. The weight of the harrow and angle of the discs should be varied, depending on the density of the remaining sod. With some grasses, such as *Pennisetum clandestinum*, several discings may be needed and, in fact, the old stolons and rhizomes raked into piles for burning. With bunch grasses, such as *Panicum maximum* and *Hyparrhenia rufa*, care must be exercised not to uproot the entire crown, but with heavy stands the discs must cut away enough of the grasses so as to leave soil exposed as a seedbed for the oversown legumes.

Fertilizers may be applied before or after the sod scarification process. It should be emphasized that the addition of plant nutrients, particularly phosphate, may determine whether the oversown legumes become established or fail. Seeds can be sown by one of the methods previously discussed.

After disturbing the old sod and reseeding, it is likely that the old grasses will regenerate growth and develop more rapidly than the sown legumes and grasses. Several light grazings should be made to remove the herbage of the old grasses and reduce the competition for the newly emerged seedlings. Overgrazing, of course, must be avoided.

The renewal of grasslands by use of conventional techniques is expensive, time-consuming and may not always provide adequate stands of sown legumes and grasses. The use of a sod- or chisel-seeder has been successfully employed for the renovation of old grazinglands in the temperate zones and should prove beneficial in the tropics (Northwood and McCartney, 1969). Various implements have been designed for direct sod-seeding into old pastures without prior tillage operations. The basic design follows the principles of a regular grain drill with the exception of a heavy chisel or tine for cutting a slit into the undisturbed sod and into which the seed and fertilizer are distributed. Prior to the direct seeding operation, the pasture sward must be heavily grazed, mowed or treated with a contact, non-residual herbicide such as paraquat orglyphosate so as to reduce competition for the germinating seedlings of the sown species. The use of a contact herbicide has an advantage in that growth of weeds and grasses is suppressed and the dead material provides a mulch during the germination and early seedling growth of the seeded legumes and grasses. It does not kill the old, established pasture species which regenerate growth within a few weeks. The cost of using a contact herbicide may exceed that of burning, heavily grazing or mowing, but not of ploughing and cultivating. In fact, the entire renovation procedure must be considered in terms of the risks and economic returns involved in renewing the grazingland as compared to complete destruction, land-fitting, fertilizing and reseeding. Where mixed agriculture is practised, growing a cash or food crop after several years of grazing, then resowing a legume–grass mixture may be the most desirable means of rehabilitating the pastureland.

References

Adegbola, A. A. (1964) Forage crops research and development in Nigeria, *Niger. Agric. J.* **1**, 34–9.

Anderson, A. J. (1956) Molybdenum as a fertilizer, *Adv. Agron.* **8**, 163–202.

Anderson, G. D. (1968) Promising pasture plants for Northern Tanzania. VI. Practical suggestions for pasture improvement, *E. Afr. Agric. For. J.* **34**, 106–16.

Andrew, C. S. and **M. F. Robins** (1969a) The effect of phosphorus on the growth and chemical composition of some tropical pasture legumes. I. Growth and critical responses of phosphorus, *Aust. J. Agric. Res.* **20**, 665–74.

Andrew, C. S. and **M. F. Robins** (1969b) The effect of potassium on the growth and chemical composition of some tropical and temperate pasture legumes. I. Growth and critical percentages of potassium, *Aust. J. Agric. Res.* **20**, 999–1007.

Bogdan, A. V. (1977) *Tropical Pasture and Fodder Plants*, Longman: London.

Bogdan, A. V. and **D. J. Pratt** (1967a) *Reseeding Denuded Pastoral Land in Kenya*, Min. Agric. and An. Husb.: Nairobi.

Bogdan, A. V. and **D. J. Pratt** (1967b) Observations on some grass–legume mixtures under grazing, *E. Afr. Agric. For. J.* **36**, 35–8.

Borgét, M. (1969) Résultats et tendances présentés des recherches fourragères de l'IRAT, *Agron. Trop. (IRAT)* **24**, 103–38 (Fr. & Eng.).

Bowden, B. N. (1963) Studies on Andropogon gayanus Kunth. I. The use of A. gayanus in agriculture, *Emp. J. Exptl Agric.* **31**, 267–73.

Brockwell, J. (1962) Studies on seed pelleting as an aid to legume seed inoculation. I. Coating materials, adhesives, and methods of inoculation, *Aust. J. Agric. Res.* **13**, 638–49.

Chadhokar, P. A. (1977) Establishment of stylo (Stylosanthes guianensis) in kunai (Imperata cylindrica) pastures and its effect on dry matter yield and animal production in the Markham Valley, Papua New Guinea, *Trop. Grassld* **11**, 263–71.

Chaverra, H., A. Wieczorek, A. Bastidas and **L. V. Crowder** (1962) *Mezelas de los Pastos con Cereales*, Bol. Divulg. No. 10, Div. Invest. Agropecuria (DIA), Min. Agric., Colombia.

Coaldrake, J. E. and **M. N. Russell** (1969) Establishment and persistence of some legumes and grasses after ash seeding on newly burned Brigalow land, *Trop. Grassld* **3**, 49–55.

Cook, S. J. (1980) Establishing pasture species in existing swards. A review, *Trop. Grassld* **14**, 181–7.

Crowder, L. V. (1960) *Establecimiento y mantenimiento de Pastos en Colombia*, Bol. Divulg. No. 9, Div. Invest. Agropecuaria (DIA), Min. Agric., Colombia.

Crowder, L. V., A. Michelin and **J. Vanegas** (1962) Establecimiento de pangola en potreros viejos del pasto micay, *Agric. Trop. (Colombia)* **18**, 277–85.

Date, R. A. and **P. S. Cornish** (1968) A comparison of lime and rock phosphate for pelleting temperate and tropical legume seed, *J. Aust. Inst. Agric. Sci.* **34**, 172–4.

Diatloff, A. (1971) Pelleting tropical legume seed, *Queensld Agr. J.* **97**, 363–6.

Downes, A. (1967) Establishment of legumes in pastures of savanna woodland in North Queensland, *Queensld J. Agric. Anim. Sci.* **24**, 23–9.

Falvey, L. (1979) Establishment of two Stylosanthes species in a Urochloa mozambicensis dominant sward in the Daly River Basin, *J. Aust. Inst. Agric. Sci.* **45**, 69–71.

Ferrari, E., S. M. Souto and **J. Dobereiner** (1967) Efecto da temperatura do solo Na nodulaçao e no desenvolvimento da soja perene, *Pesquisa Agropec. Bras.* **2**, 461–8.

Foster, W. (1961) Note on the establishment of a legume in rangeland in Northern Nigeria, *Emp. J. Exptl. Agric.* **29**, 319–22.

Gibson, T. (1980) Pasture research in alang-alang (Imperata cylindrica) (L) Beauv. areas in Northern Thailand, *Biotrop* **5**, 193–230.

Gilbert, M. A. and **K. A. Shaw** (1979) The effect of heat treatment on the hardseedness of Stylosanthes scabra, S. hamata cv. Verano and S. viscosa CP I 34904, *Trop. Grassld* **13**, 171.

Gillard, P. (1977) Establishment of Townsville stylo in native perennial grass pastures, *Aust. J. Exptl Agric. Anim. Husb.* **17**, 784–8.

Golden, J. R. (1967) Undersowing as a means of establishing the ley in Western Kenya, *E. Afr. Agric. For. J.* **32**, 274–81.

Gomide, J. A., J. A. Obeid and **J. M. Oliveira** (1979) Introduçao de leguminosas tropicais em pastegems de gramineas, *Rev. Brasileira Zootecnica* **8**, 593–609.

Gontijo, R. M., F. A. Gomide, D. J. Sykes and **E. H. Vilela** (1969) Estudo sôbre establecimento de gramineas forrageiras, *Rev. Ceres* **16**, 107–20.

Graham, T. G. (1968) Strip planting of Townsville lucerne in spear grass, *Queensld Agric. J.* **94**, 544–50.

Grof, B. (1981) The performance of Andropogon gayanus–legume associations in Colombia, *J. Agric. Sci. (Camb.)*, **96**, 233–8.

Haggar, R. J. (1961) Use of companion crops in grassland establishment in Nigeria, *Exptl Agric.* **5**, 47–52.

Haggar, R. J., P. N. de Leeuw and **E. Agishi** (1971) The production and management of Stylosanthes gracilis at Shika, Nigeria. II. In savanna grassland, *J. Agric. Sci. (Camb.)*, **77**, 437–44.

Humphreys, L. R. (1967) Townsville lucerne: History and prospect, *J. Aust. Inst. Agric. Sci.* **33**, 3–13.

ICA Informe Anuales (1957–72) *Programa de Pastos y Forrajes, Instituto Colombiano Agropecuaria (ICA)*, Min. Agric., Colombia (Mimeo Series).

Jones, C. A. (1979) The potential of Andropogon gayanus in the Oxisol and Utisol savannas of tropical America, *Herb. Abst.* **49**, 1–8.

Jones, R. K. and **J. J. Robinson** (1970) The sulphur nutrition of Townsville lucerne (Stylosanthes humilis), *Proc. Intl. 11th Grassld Cong.*, pp. 377–80.

Jones, R. K. and **R. M. Jones** (1971) Agronomic factors in pasture and forage crop production in tropical Australia, *Trop. Grassld* **5**, 229–44.

Keya, N. C. O., F. J. Olsen and **R. Holliday** (1971) Oversowing improved pasture legumes in natural grassland of medium altitudes of Western Kenya, *E. Afr. Agric. For. J.* **37**, 148–55.

Kornelius, E., M. G. Saveressig and **W. J. Goldert** (1978) Pastures establishment and management in the cerrado of Brazil. In P. A. Sanchez and L. E. Tergas (eds), *Pasture Production in Acid Soils of the Tropics, Seminar Proc.*, CIAT: Colombia, pp. 147–66.

Lazier, J. R. (1980) The performance of four persistent forage legumes with Brachiaria mutica in Belize under grazing using observational techniques and clipping, *Trop. Agric. (Trin.)*, **57**, 353–61.

Letenneur, L. (1971) Le Stylosanthes gracilis: Synthèse des Travaux Effectués en Cote d'Ivoire et ses Possibilités d'Extension. *Seminar For. Crops Res. Africa*, Univ. Ibadan, Nigeria.

Little, E. C. S. and **G. W. Ivens** (1965) The control of bush by herbicides in tropical and subtropical grassland, *Herb. Abst.* **35**, 1–12.

Lotero, J., J. Bernal and **G. Herrera** (1967) Distancia de siembra y aplicación de nitrogeno en pasto elefante, *Rev. Inst. Colombiano Agropec.* **2**, 123–33.

Lotero, J., J. Bernal and **G. Herrera** (1969) Distancia de siembra y dosis de nitrogeno en pasto Imperial, *Rev. Inst. Colombiano Agropec.* **4**, 147–57.

McKeague, P. J., C. P. Miller and **P. Anning** (1978) Verano – a new stylo for the dry tropics, *Queensld Agric. J.* **104**, 31–5.

Mohamed Saleem, A. M. (1972) Productivity and chemical composition of Cynodon IB 8 as influenced by level of fertilization, soil pH and height of cutting, Ph. D. Thesis, Univ. Ibadan, Nigeria.

Motooka, P. S., D. L. Plucknett, D. E. Saiki and **D. R. Younge** (1967) *Pasture Establishment in Tropical Brushlands by Aerial Herbicide and Seeding Treatments of Kauai*, Hawaii Agric. Exptl Sta., Tech. Prog. Rep., No. 165.

Nicholls, D. F. and **D. L. Plucknett** (1971) 'Packet' planting techniques for tropical pastures, *Hawaii Farm Sci.* **20**, 10–1.

Norman, M. J. T. (1960) The establishment of pasture species on arable land at Katherine, N. T., *Tech. Paper Div. Land Reg. Survey No. 8*, CSIRO: Canberra.

Northwood, P. J. and **J. C. McCartney** (1969) Pasture establishment and renovation by direct seeding, *E. Afr. Agric. For. J.* **35**, 185–9.

Norris, D. O. (1965) Acid production by Rhizobium – a unifying concept, *Plant and Soil* **22**, 143–66.

Norris, D. O. (1967) The intelligent use of inoculation and lime pelleting for tropical legumes, *Trop. Grassld* **1**, 107–21.

Norris, D. O. (1972) Leguminous plants in tropical pastures, *Trop. Grassld* **3**, 159–70.
Norris, D. O. and **R. A. Date** (1976) Legume bacteriology. In N. H. Shaw and W. W. Bryan (eds), *Tropical Pasture Research, Principles and Methods*, Bull. 51, Commonw. Agric. Bur. Ch. 7, pp. 134–74.
Owen, M. A. and **H. W. Brzostowski** (1967) Grass establishment under semi-arid conditions in central Tanganyika. *Trop. Agric. (Trin.)* **44**, 275–91.
Plucknett, D. L. (1971) *Use of Pelleted Seed in Crop and Pasture Establishment*, Univ. Hawaii Cir. Coop. Ext. Ser., No. 446.
Pratt, D. J. (1963) Reseeding denuded land in Baringo district, Kenya, *E. Afr. Agric. For. J.* **29**, 78–91.
Pratt, D. J. (1964) Reseeding denuded land in Baringo district, Kenya. II. Techniques for dry alluvial sites, *E. Afr. Agric. For. J.* **29**, 243–60.
Pratt, D. J. (1966) Bush control studies in the drier areas of Kenya. II. An evaluation of the Holt IXa 'Bush Breaker' in *Tarchonanthus* acacia thicket, *J. Appl. Ecol.* **3**, 97–115.
Pratt, D. J. and **J. Knight** (1964) Reseeding denuded land in Baringo district, Kenya. III. Techniques for capped red loam soils, *E. Afr. Agric. For. J.* **30**, 117–25.
Rains, A. B. (1963) *Grassland Research in Northern Nigeria*. Samaru Misc. Paper No. 1, Zaria.
Rensburg, H. J. van (1965) Development of planted fallows of short duration to replace long-term bush fallow, *Afr. Soil.* **10**, 381–4.
Rensburg, H. J. van (1969) Legume/grass pastures in Zambia, *Farming Zambia* **4**, Suppl. 1–4.
Rey, G. E. and **P. J. Matta** (1966) *Los Pastos Cultivados en Costa Rica*. Bol. Tecn. No. 51, Min. Agric. y Ganaderia: San Jose.
Risopoulos, S. A. (1966) Management and use of grasslands, *FAO Past, For. Crops Studies No. 1 (MR/32496)*, Democratic Republic of the Congo.
Rotar, P. P., U. Urata and **A. Bromdep** (1967) *Effectiveness of Nodulation on Growth and Nitrogen Content of Legumes Grown on Several Hawaiian Soils, With and Without Proper Rhizobium Strains*, Hawaii Agric. Exptl Sta. Tech. Prog. Rep., No. 158.
Russell, M. J., J. E. Coaldrake, and **A. M. Sanders** (1967) Comparative effectiveness of some insecticides, repellants, and seed-pelleting in the prevention of and removal of pasture seeds, *Trop. Grassld* **1**, 153–66.
Sands, E. B., D. B. Thomas, J. Knight and **D. J. Pratt** (1970) Preliminary selection of pasture plants for the semi-arid regions of Kenya, *E. Afr. Agric. For. J.* **36**, 49–57.
Schaaffhausen, R. von (1963) Metodos economicos para forma e melhorar pastagens com as leguminosas Dolichos lablab, Guandu e Soja perene, *Rev. Soc. Rural Bras.*, **43**, 16–7.
Schedler, P. W. (1969) Scratch-ploughing. A simple technique for the reclamation eroded pasture land in semi-arid regions, *World Crops* **21**, 274–6.
Shaw, N. H., C. T. Gates and **J. R. Wilson** (1966) Growth and chemical composition of Townsville lucerne (Stylosanthes humilis). I. Dry matter yield and nitrogen content in response to phosphate, *Aust. J. Exptl Agric. Anim. Husb.* **6**, 150–6.
Smith, C. A. (1963) Oversowing pasture legumes into the Hyparrhenia grasslands of northern Rhodesia, *Nature (Lond.)* **200**, 811–12.
Smith, C. J. (1967) Sowing dryland pastures, *Rhod. Agric. J.* **64**, 69–70.
Smith, G. A. (1970) The effects of five different methods of establishing Stylosanthes humilis and S. guyanensis in non-arid American tropics, *Proc. Soil and Crop Sci. Soc. Florida* **30**, 68–74.
Stobbs, T. H. (1969) Animal production from Hyparrhenia grassland oversown with Stylosanthes gracilis, *E. Afr. Agric. For. J.* **35**, 128–34.
Stocker, G. C. and **D. J. Sturtz** (1966) The use of fire to establish Townsville lucerne in the Northern Territory, *Aust. J. Exptl Agric. Anim. Husb.* **6**, 277–9.
Stonard, P. (1969) Effect of sowing depth on seedling emergence of three species of Stylosanthes, *Queensld J. Agr. An. Sci.* **26**, 55–60.

Suttie, J. M. (1965) The establishment of pastures at medium elevations, *Kenya Farmer* **109**, 16–7.

Thomas, D. and **A. J. Bennett** (1975) Establishing a mixed pasture under maize in Malawi. I. Time of sowing, *Exptl Agric.* **11**, 257–63.

Tow, P. G. (1967) Sowing rate, survival and productivity of greenpanic -Glycine mixtures, *Queensld J. Agric. An. Sci.* **24**, 141–8.

Varela, J. and **L. V. Crowder** (1960) Metodos de establecimento del pasto pangola *Agric. Trop.* (*Colombia*) **16**, 400–9.

Woods, L. E. (1969) A survey of Townsville stylo (Townsville lucerne) established in the Northern Territory up to 1969, *Trop. Grassld* **3**, 91–8.

Chapter 9

Agronomic and other management practices

Pasture agronomy is an essential aspect of livestock farming. A basic knowledge of pasture plant–soil–climate interrelationships facilitates the application of agronomic management practises to assure optimal production of nutritious herbage and at the same time maintain soil fertility, desirable plant populations and botanical associations. Forage yield data from cutting trials provide information on the potential production of herbage that is available for hay, silage, green-chop feeding, soiling, zero-grazing and foggage. Knowing the potential herbage yields, and deducting a waste factor for trampling and rejected stems (usually 20–25%), one can estimate stocking rates.

Potential yields of grasses and legumes

The annual dry matter yields from grasses vary from less than 0.5 tonnes/ha on the arid *Aristida* native grasslands of the sub-Sahara to more than 100 tonnes/ha on heavily fertilized *Pennisetum purpureum* plantings in the humid, lowland tropics. Herbage production of legumes is also variable, but the accumulated mass of forage does not approach that which can be obtained with grasses. Annual dry matter yields selected as being among the highest recorded from experimental plots are shown in Table 9.1. It will be noted that the highest potential for production of grasses occurs in the humid tropics where water is not a limiting factor. In these areas the daily solar energy input is lower (400–500 cal/cm^2 per day) than in the subtropics (sometimes exceeding 700 cal/cm^2 per day during the summer), but there is comparatively little seasonal variation (Black, 1956). Thus, plant growth continues at a rapid rate throughout the year when soil moisture and nutrients are supplemented. In the temperate and subtropical regions, growth is reduced during periods of low temperatures, but the extended day of more than 18 h in the summer months of the high latitudes allows a great increase in dry matter accumulation. For this reason the total annual

Table 9.1 *Annual dry matter yields of selected grasses and legumes*

Climate and location	Species	Dry matter (t/ha)	Applied N (kg/ha)
Temperate humid			
Aberystwyth, UK (Cooper, 1969)	*Lolium perenne*	25.2	706
TeAwa, New Zealand (Suckling, 1960)	*L. perenne*	26.6	NL*
	Dactylis glomerata	22.0	NL
	Trifolium pratense	26.5	–
Subtropical			
Texas, USA (Fisher and Caldwell, 1959)	*Cynodon dactylon*	31.8	1 422
Queensland, Australia (Shaw *et al.*, 1965; Henzell, 1970)	*Paspalum plicatulum*	31.9	451
	Pennisetum clandestinum	30.0	1 120
California, USA (Loomis and Williams, 1963)	*Sorghum vulgare*	39.7	?
	Medicago sativa	32.4	–
Tropical			
Naitaima, Colombia (Annual Report, 1962–63)	*Pennisetum purpureum*	130.0	1 320
Puerto Rico (Vicente-Chandler *et al.*, 1959)	*P. purpureum*	83.4	2 244
El Salvador (Watkins and Lewy-Van Severen, 1951)	*P. purpureum*	85.4	?
Naitaima, Colombia (Crowder, 1967)	*Hyparrhenia rufa*	56.7	400
Puerto Rico (Vicente-Chandler *et al.*, 1959)	*Panicum maximum*	54.0	1 795
Trinidad (Adeniyi and Wilson, 1960)	*Digitaria decumbens*	49.3	269
Nigeria (Ademosun and Chheda, 1974)	*Cynodon nlemfuensis*	25.7	84
Mauritius (Anslow, 1957)	*Leucaena leucocephala*	34.6	–
Ivory Coast (Cadot, 1965)	*Stylosanthes guianensis*	17.4	–
Hawaii (Younge *et al.*, 1964)	*Desmodium intortum*	9.9	–

* NL = Non-limiting (fertilizer added as nutrients removed).

yields of certain species are equal to, or exceed, those of some grasses and legumes in the tropics. For example, *Trifolium pratense* in New Zealand gives dry matter yields greater than those of *Stylosanthes guianensis* in the Ivory Coast and *Desmodium intortum* in Hawaii. Irrigated lucerne in California however, produces less tonnage of hay than in Colombia.

The high dry matter yield of 130 tonnes/ha of *Pennisetum purpureum* in Colombia (Annual Report, 1962–63) came from the first-year harvest with cuttings made at 9-week intervals, 1 320 kg/ha of N applied every 6 weeks and supplemental irrigation during periods of drought stress. The figure of 85.4 tonnes reported from El Salvador (Watkins and Lewy-van Severen, 1951) was also taken during the first year after establishment. The 86.0 tonnes/ha yield from Puerto Rico, however, represents a 3-year average (Vicente-Chandler *et al.*, 1959).

The potential yields given in Table 9.1 greatly exceed those commonly obtained under experimental conditions. It is not unusual, however, to encounter reports of more than 40 tonnes/ha of dry matter accumulated during the growing season in the wet tropics (Adeniyi and Wilson, 1960; Little *et al.*, 1959; Nordfeldt *et al.*, 1951; Vicente-Chandler *et al.*, 1964; Wollner and Castillo, 1968). In the subtropics, production is often limited by seasonal water stress and low temperatures so that annual dry matter yields of the more productive species rarely exceeds 25–30 tonnes/ha (Bryan and Sharpe, 1965; Doss *et al.*, 1966; Prine and Burton, 1956; Shaw *et al.*, 1965). The total harvested from many grasses, however, does not surpass 10–15 tonnes. It is even less likely that the livestock farmer will approach the high yields quoted above. A number of limiting factors impose restrictions on plant growth and development, and potential yields are mainly determined by (1) species, (2) available light energy and its conversion by the plant for production of dry matter, (3) seasonal high and low temperatures, (4) inputs of water, (5) nutrients obtained from the soil and applied fertilizers and (6) management systems for maximizing cutting and grazing intensities to coincide with the seasonal feed requirements of livestock (Cooper, 1970; Holmes, 1968; Mitchell, 1963).

Fertilization and liming

Poor nutritional status of grasses frequently restricts animal carrying capacity, but higher productivity of tropical grasses can be expected with proper exploitation and management. In general, output of grasslands and grazinglands can be increased by application of fertilizers. The magnitude of increase is influenced chiefly by the climatic and edaphic factors and their interaction with the species composition of the grazingland. There may be occasions when appreciable and expected response to fertilizer application may not occur due to loss of nutrients by volatilization, leaching, fixation by the soil, transformation into other compounds and depletion of essential elements. Sizable amounts of nutrients are needed for maximum herbage yields and often exceed that which most soils can supply, especially under intensive management systems.

Some soils are found to be low in available plant nutrients so that improved pasture plants cannot be established until deficiencies are remedied. On soils with higher fertility it is possible to develop improved pastures without applied fertilizers, but even then nutrient deficiencies eventually limit growth of plants.

The nutrient status of the soil can be assessed by taking soil samples and making chemical analyses. This information is often supplemented with pot culture, using test plants that are sensitive to nutrient deficiencies and that respond to applied chemical elements (Andrew and Fergus, 1976). In addition, field experiments, plant analysis and the appearance of the foliage provide clues as to the nutrient status of the soil. Grasses and legumes differ in their nutrient requirements, and definition of these needs is important is seedling establishment, plant growth and maintenance of a desirable botanical combination.

Factors influencing use of fertilizers

When bringing newly cleared land into pasture production or when renovating depleted grazingland the choice whether or not to apply fertilizers may be easy, but under marginal conditions it becomes more difficult and critical. Many pastures species respond dramatically to applied fertilizer and it might be assumed that its use would be almost axiomatic. Checking actual usage on pastureland in the tropics and subtropics, however, indicates that this practice has not been widely adopted. In most tropical countries the amount of fertilizer applied to pastures is too small to be included as a statistic. In northern Australia, it was estimated that less than 1 per cent of the 105 million ha considered suitable for pasture improvement had been treated with fertilizers by the mid-1960s (Davies and Eyles, 1965). In Queensland, the most developed area of Australia in terms of tropical pasture improvement, very small amounts of fertilizer were used before the 1960s. In 1955–56, pastures received approximately 7 per cent of the total superphosphate used in this state. With the introduction of suitable legumes and their associated *Rhizobium* spp., this figure climbed to nearly 40 per cent in the early 1970s (Williams and Andrew, 1970).

Factors to be considered in the use of fertilizers for grazinglands are:

1. Economic return.

Under conditions of intensified beef and dairy production system, some supplemental plant nutrients are required for optimal output and these usually come from the bag. With less intensive livestock systems it may be more economical to increase the area of grazing-land or to rotate pastures and fertilize cash crops so as to capitalize on residual soil nutrients, or to reduce the number of cattle being grazed. Other points to consider include the type and quality of cattle, price of animal product, cost of fertilizer credit and returns on investments, the level of technology and its appropriateness and the managerial skill of the operator or overseer.

2. Species present and botanical composition

Many native or naturalized species do not respond to applied fertilizer in terms of increased herbage production. There may be, however, a substantial increase in the protein or mineral content of the forage. Application of fertilizer to natural and semi-natural grazingland is unlikely to provide an economic return, since they are located in drier regions where responses are too marginal or too small to be profitable. In areas of limited rainfall, above 650 mm but concentrated in a short rainy season of 3–4 months, the use of commercial fertilizers, especially phosphorus, can mean the difference between establishing and maintaining an improved pasture or attempting to manage natural grassland.

Species differ in their fertility requirements, some grasses and legumes being able to persist under conditions of low soil fertility, e.g. *Melinis minutiflora*, *Cynodon dactylon* and *Stylosanthes humilis*. Grasses generally respond to the application of N and legumes to P. If legumes are to be maintained in significant proportion, it is important to apply phosphate at regular intervals. Significant leaching of nutrients occurs in the upper soil layers of the humid tropics, causing deficiencies among shallow-rooted species. The deeper-rooted types, however, can utilize nutrients that are leached downward, thus recycling them in the plant–soil–animal relationship.

3. Nutrients in the soil and their availability

The quantity of nutrients available in the soil for plant use varies with soil type, available moisture, temperature, microbiological activity and previous use of the land. Some nutrients are readily available in the soil solution; others are released by mineralization; while some are fixed within the soil fraction and may or may not be released over time.

4. Quantity of nutrients removed

The extraction of nutrients from the soil is roughly proportional to the yields of herbage, but is also a function of the chemical composition of the herbage. Thus, a grass yielding 20 tonnes/ha annually of dry matter and containing 1.5 per cent of N would remove 300 kg of this element. Additional N would be found in the roots and stubble, but most of this is eventually returned to the soil. That contained in the herbage would be completely lost under cutting conditions, unless soiling is practised. The chemical constituents of grasses are highly variable, depending on the fertility status of the soil, applied fertilizers, plant species, rate of plant growth, interval of cutting and competition among species for soil nutrients. Vicente-Chandler *et al.* (1974) found that well-fertilized grasses when cut and taken off the land removed annually the equivalent of 2 tonnes of a 16-6-25 N-P-K fertilizer (Table 9.2). Not even a highly fertile soil could provide these quantities of nutrients, making high rates of topdressing necessary to sustain high levels of forage production.

Legumes also remove large quantities of soil nutrients, whether grown alone or in mixture where the herbage is cut for hay, silage or greenchop. A 10 tonnes/ha hay crop of lucerne would extract approximately 250 kg N, 25 kg P, 200 kg K, 60 kg Co, 25 kg Mg and 30 kg S or more. Other legumes have about the same chemical composition as lucerne, but percentages of individual components vary widely with sampling periods.

The amount of nutrients removed by animals grazing on pastures is relatively small, as shown by the data given in Table 9.3. The figures were compiled in the temperate zone of New Zealand (Hall *et al.*, 1965) and are not directly applicable to the potentially longer growing season in the tropics. They illustrate, however, that fertilizer requirements of grazed pastures are less than for grasses being cut and removed from the land. Much of the nutrient component of the forage is returned under grazing, with about 80 per cent of the N, P and K ingested by the animal being excreted in the faeces and urine. Fattening cattle produce about 8 tonnes of faeces annually, which contains about 70 kg of N, 18 kg of P_2O_5 and 60 kg of potash. This represents a sizable quantity of plant nutrients under an intensive livestock production system of 4–6 animals per hectare.

Table 9.2 *Herbage yields and nutrients removed annually in grasses harvested by cutting, Puerto Rico (Vicente-Chandler et al., 1974)**

Species	DM (t/ha)	N (kg/ha)	P (kg/ha)	K (kg/ha)	Ca (kg/ha)	Mg (kg/ha)
Brachiaria ruziziensis	32.9	333	54	442	150	77
Cynodon nlemfuensis	27.8	380	64	460	148	53
Pennisetum purpureum	27.7	332	70	554	105	69
Brachiaria mutica	26.4	337	47	421	126	87
Digitaria decumbens	26.1	329	52	393	119	74
Panicum maximum	25.3	317	48	399	163	109
Melinis minutifolia	14.5	227	35	228	62	48
Average	25.8	322	53	415	125	74

* Grasses cut every 60 days; topdressed with 440 kg/ha of N, 71.5 kg of P, 440 kg of K annually in six equal doses; soil limed to obtain a pH of 6.0.

Considering these high returns of nutrients, and looking at the data of low fertility losses shown in Table 9.3, one might surmise that pastures under grazing require little fertilizer. It must be remembered, however, that grazing cattle are not highly effective in maintaining soil fertility because of the irregular distribution of faeces and urine, which tends to be concentrated in limited areas (Peterson *et al.*, 1956a; Lotero *et al.*, 1966). These spots are over-fertilized, causing excessive leaching of N and K and rank herbage growth that is usually avoided by the cattle. The small green areas are sometimes called 'manure spots' or 'dung spots'. Other areas do not benefit from the excreta, so that herbage growth is poor and overall yields decline.

5. Loss of nutrients from the soil

The red clay, acid soils which are high in iron oxides fix large amounts of applied P in forms not readily available to plants. Rapid mineralization occurs when rains begin after the dry season, resulting in a flush of available nutrients, especially N. Heavy rainfall causes high leaching rates, especially of N and K in permeable soils. In some instances more than 50 per cent of N and 20–40 per cent of K may not be recovered in the herbage. They may not be completely lost in the drainage water, but instead remain in lower soil profiles and can be recycled by deep-rooted pasture species.

Nutrient recycling in grazinglands

Nutrients essential for plant growth pass through a cycle from the soil, or from the air in some instances, to the plant and are returned directly in the plant residue or through the grazing animal to the soil. The quantity and availability of nutrients in the cycle determine to a great extent the herbage production of the grazingland and the

Table 9.3 *Loss of fertility, expressed as fertilizer ingredients (kg/hectare) by disposal of animal produce and carcasses (Hall et al., 1965)*

Animal	Ammonium sulphate	Superphosphate	Potash	Calcium carbonate
1. Milking cows; 2,728 litres milk and 125 kg B.F. per hectare	72.7	27.3	13.6	9.6
2. Beef cattle, one head 454.5 per hectare*	52.7	35.0	2.5	18.2
3. Sheep, six (68.2 kg. each) per hectare*	47.7	21.8	21.8	9.6
4. Fat lambs, six (34.1 kg each) per hectare*	25.9	10.9	10.9	4.5

* Animals raised and fattened on the farm; for those purchased and brought in for fattening the losses are negligible.

yield of animal output. The nutrient pool of the soil is composed of an organic fraction, dead plant material, organisms, excreta, etc. Grazing and browsing animals furnish an important link in the recycling of nutrients between the plant and soil. In the past, ungulates that roamed over the North American plains, the grasslands and savannas of Africa and the pampas of Argentina probably contributed to the formation of fertile soil (Mott, 1970). In contrast, soils of the tropical savannas of South America and Australia are less fertile and productivity of grasses is inferior. Ungulates were sparse or absent in the latter areas. Livestock and wild ungulates play a less important role in soil formation today but are important in the recycling mechanism.

Sources of nutrients include (1) parent materials from which the soil was derived, (2) return of plant residues to the soil, (3) return of excreta by grazing animals, (4) application of commercial fertilizers, and (5) nutrients from the atmosphere.

Cycling of nitrogen in grazed pastures

A simplified scheme (Fig. 9.1) for the cycling of N in grazed grass–legume pastures was described by Henzell (1970). Nitrogen made available to plants from the soil is derived from applied fertilizers, precipitation and non-symbiotic fixation. Soil N is also derived from mineralization of the organic complex (humus). Not all of the soil nitrogen is available for plant use as some will be lost by volatilization, leaching and biological denitrification owing to microbial reduction of nitrites and nitrates.

Legumes may supply substantial quantities of N to the soil organic complex through fixation by the symbiotic relationship with rhizobia or absorbed by the plant from that available in the soil. When legume plants are eaten by animals a portion of the N is returned to the soil organic complex by excretion of urine and dung. Nitrogen is also returned directly to the soil as plant residue, including decomposed nodules.

Grasses and other herbaceous plants add to the soil organic complex. Grasses and forbs may be eaten by livestock with a portion of the N being returned to the soil by excretion along with that returned in the plant residue.

In this system animals play an important role in the recycling of N. The system depicts an intensive management scheme whereby animals are confined to the pasture. Nitrogen is added by plant material other than that grown on the pastureland and none is removed as hay, silage or seeds. Some N is removed in animal products such as meat, milk or wool.

Most grazinglands in the tropics, however, are not intensively managed, particularly those in drier environments. Under such conditions of extensive management, available N is largely derived from

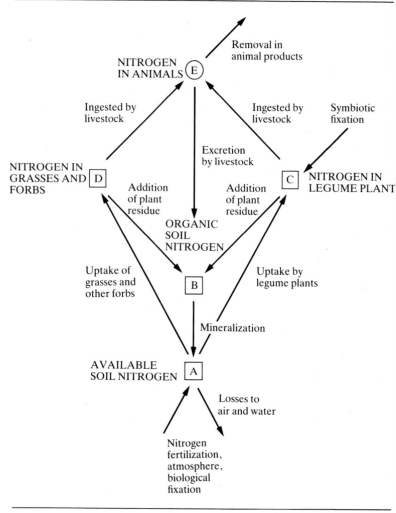

Fig. 9.1   Cycling of nitrogen in grazed pastures (redrawn from Henzell, 1970).

the soil reserves with small input from atmospheric accessions or biological fixation. In this low-fertility system an almost closed N cycle exists with little opportunity for increased output.

Nitrogen fertilization

Dry matter yield

Most tropical soils are deficient in nitrogen and heavy applications are required to produce high yields of grass with high protein content. The favourable response by forage grasses to applied N in terms of increased yield and chemical composition is well known and documented. Data on dry matter yields are usually obtained from trials in which the herbage is cut. The practical objectives of carrying out N fertilizer trials under cutting conditions are (1) to find the rates of applied N that optimize dry matter production at a level of favourable economic return and (2) to reduce the number of treatments that would need to be studied under animal grazing conditions. The spectacular effects of N applied to responsive grasses is depicted in Fig. 9.2, which shows curves fitted to data from several

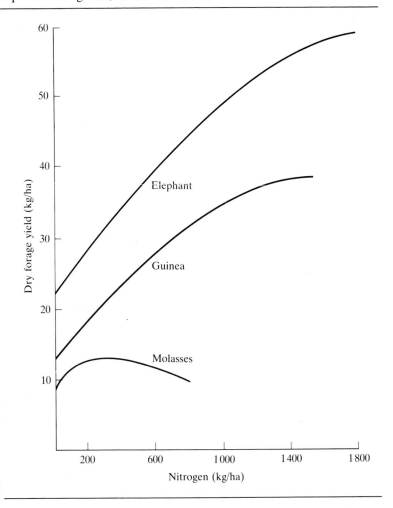

Fig. 9.2 Response of grasses to applied nitrogen, forage cut and removed Crowder, 1977).

sources (Crowder, 1977). Dry matter production increases almost linearly (Fig. 9.3) with successive increments of N fertilizer (Crespo *et al.*, 1975; Crowder, 1974; Grof and Harding, 1970; Hendy, 1972; Henzell, 1970; Oyenuga and Hill, 1966; Salette, 1970; Vicente-Chandler *et al.*, 1974; Wollner and Castillo, 1968). Most N fertilizer studies have been carried out with pure grass stands, located on soils previously cropped and fertilized, and with the herbage removed. Grass species respond differently with yields of such grasses as elephant, guinea, Pangola and *Cynodon*, increasing linearly with annual rates of 600–800 kg/ha of applied N, the curve being less pronounced beyond 1 000 to 1 200 kg/ha. Some grasses such as molasses (*Melinis minutiflora*) do not benefit greatly with application of N.

Differences in response to applied N in terms of dry matter production are due to factors such as species, stages of growth, amount and time of N applied, soil moisture and climatic conditions. The dramatic yield increases of dry matter have been recorded under humid or irrigated conditions. There is little or no response to applied N in the arid and semi-arid regions found in many parts of the tropics and subtropics.

Crude protein in herbage

In evaluating N fertilizer trials, dry matter yield is not the only criterion to be considered. Crude protein content of the herbage holds

Fig. 9.3 Improved cultivars of grasses such as *Digitaria decumbens* respond markedly to applied nitrogen. The effect is less dramatic under grazing conditions.

importance. Nitrogen content of herbage is determined by means of chemical analysis, and crude protein calculated by multiplying this value by 6.25. Crude protein yield per unit area of land is often reported along with the yield of dry matter. This is simply a function of dry matter yield and crude protein percentage. Crude protein yields thus increase as dry matter yields increase.

Many researchers have shown that N fertilization increases the crude protein content at any stage of plant growth. Data obtained in Puerto Rico by Vicente-Chandler *et al.* (1974) will be used to illustrate the effects of applied N (Fig. 9.4). Several grass species were included in these studies, with harvests being made at 60-day intervals and all herbage removed. Nitrogen in the form of ammonium sulphate was applied in six equal topdressings over the year. A sharp increase in crude protein content occurred in all grasses as increments of applied N increased. Species differed at each level, but tended to be less divergent at the highest levels. Taking an average of species, crude protein content increased from 6.2 per cent with-

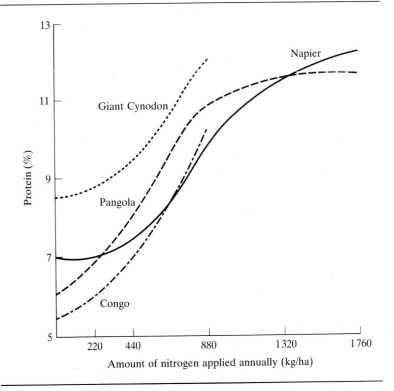

Fig. 9.4 Effect of N fertilization on crude protein content of grasses; herbage cut and removed every 60 days (Vicente-Chandler *et al.*, 1974).

out N to 10.2 and 12.3 per cent with 880 and 1 760 kg/ha, respectively, of N applied annually. Feeding trials carried out with herbage of these grasses showed that crude protein was readily digested by ruminants regardless of the N treatment. Only traces of nitrates which can be toxic to cattle were found in the forage, even at the highest level of applied N. Other studies have also shown that crude protein content of grasses rises sharply with increasing levels of nitrogen (Burton and Jackson, 1962; Chheda and Akinola, 1971a; Crespo, 1974; Crowder *et al.*, 1963; French, 1957; Michelin and Crowder, 1959; Olsen, 1975; Whitney and Green, 1969). With applied N the grass herbage frequently contains 15 per cent or more of crude protein on a dry weight basis during periods of rapid growth, but may drop to 5 per cent or less in the dry season. It appears that increased N content of fertilized grass comes in part from luxury consumption.

Seasonal effect on nitrogen content

The nitrogen content of grasses changes during the season of growth. Established grasses take up N rapidly as growth resumes when rains begin after the dry season and soil moisture is readily available. Equally, their N content drops rapidly as the soil N supply diminishes. A flush of N occurs because of mineralization of soil organic matter becoming wet after being dry, but it then steadily declines as the season progresses (Birch, 1960). This seasonal decline was illustrated by Henzell and Oxenham (1964) who studied *Chloris gayana, Setaria sphacelata* (syn. *S. anceps*) and *Paspalum plicatulum* in Queensland, Australia (Fig. 9.5). Nitrogen content declined as plants became older, the decrease being sharpest in leaf blades. This decline was more distinct than noted in flowering stems. As leaf blades matured, their tissue accumulated greater amounts of cell wall constituents causing a porportional decrease in N content. In contrast, the tissue of flowering stems comprised cell wall materials so that stems always had less N than leaves.

Time of nitrogen application

The data of Henzell and Oxenham (1964) provided evidence that a constant N supply is needed to regulate the N content of warm-season grasses at all stages of plant growth. Most studies, however, have been carried out under a cutting regime where an attempt was made to maintain grasses in a vegetative stage of growth (Crowder, 1977; Henzell, 1971; Quinn *et al.*, 1961; Rodel, 1969; Vicente-Chandler *et al.*, 1974; Velasquez *et al.*, 1975; Whitehead, 1970; Whitney, 1974; Wilkinson and Langdale, 1974). Under these conditions, fertilizer N was divided into split applications in an effort to sustain growth, provide even distribution of herbage and maintain a fairly constant N content of herbage. The value of split application

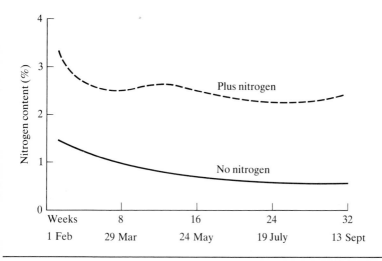

Fig. 9.5 Nitrogen content of above-ground parts, average of three grasses over the growing season (redrawn from Henzell and Oxenham, 1964).

was found to depend on the source of N, quantity of N used, number and time of topdressing, frequency of cutting, species characteristics and climatic conditions. In areas of heavy rainfall, more frequent applications reduced losses of N due to leaching and volatilization. In areas of lighter rainfall, a single application resulted in total dry matter yields comparable to split applications. Heavy applications can cause burning of new emerged tillers and fast growth after application to the detriment of yields later in the season. The number of applications did not markedly affect annual dry matter yields or protein content of the herbage, both of which increased sharply with additional increments of applied N. Applying N in two annual applications gave yields of dry matter and protein content that were higher immediately following fertilization and lower in subsequent harvests, as compared to four and six annual applications. With grass species that remained vegetative, such as *Pennisetum clandestinum*, selections of *P. purpureum*, and *Cynodon nlemfuensis*, a rather high level of crude protein was maintained (3–4%N) with frequent topdressings of N. In contrast, some grass species, such as *Cenchrus ciliaris*, *Chloris gayana* and selections of *Panicum maximum*, produced flowering stems regardless of N topdressing. A greater response occurred, and more forage was produced per kilogram of N, as the length of the cutting interval was extended. More N was needed, however, to sustain high levels of protein content in the herbage as plants became older. Grasses responded more strongly to applied N during the season of rapid plant growth, i.e. favourable moisture and temperature, than during the drier seasons when days were cooler and shorter. Protein content, however, was higher during periods of slower growth due to reduced dry matter yield but continued uptake of nitrogen.

Forms of nitrogenous fertilizer

Nitrogen exists in several fertilizer forms such as ammonium sulphate, ammonium nitrate, sodium nitrate, urea and anhydrous ammonia. Ammonium sulphate has been the most commonly used form in the tropics, but recently urea has become more popular, largely because of the lower cost per unit of N. Several studies have shown that all forms markedly increase dry matter yields of grasses with little difference in total production (Henzell, 1971; Figarella *et al.*, 1972; Lotero *et al.*, 1968; Vicente-Chandler and Figarella, 1962; Villamizar and Lotero, 1967). This is illustrated by data obtained in Puerto Rico (Figarella *et al.*, 1972) with five forms of N fertilizer applied to established Pangola grass (Table 9.4). These data showed that response to urea was slightly less than to the other forms, as has been observed by other researchers. Studies in Georgia, which lies just north of the subtropics in the United States, with Coastal bermudagrass showed that 1 kg of N from urea was about 80 per cent as effective as 1 kg of N from ammonium nitrate or nitrate of soda in terms of herbage yield (Burton and DeVane, 1952; Burton and Jackson, 1962). In Queensland, Australia, the mean yields of Rhodes grass from plots fertilized annually with 224 and 448 kg/ha of urea over a 7-year period were 87 per cent of that of sodium nitrate, 89 per cent of ammonium sulphate and 94 per cent of ammonium nitrate–limestone (Henzell, 1971). The physical properties of

Table 9.4 *Yields of Pangola grass as influenced by forms of N fertilizer in Puerto Rico, average of 3 years (Figarella et al., 1972)*[*]

N source	N rate (kg/ha)	Dry matter[†] (t/ha)	CP forage (%)	N recovery (%)	Soil pH[‡]
Nil	0	5.37	6.4	–	5.5
Urea	628	19.89	7.7	40.6	4.8
Ammonium sulphate	632	20.46	8.6	55.3	4.5
Ammonium nitrate	617	20.11	8.1	46.4	4.9
Urea + CaCO$_3$	584	20.34	7.3	41.8	5.2
Ammonium nitrate–lime	632	20.91	7.9	44.4	5.5
LSD[05]	–	N.S.	0.6	8.8	–

[*] All plots fertilized abundantly with phosphorus and potassium; nitrogen applied in six equal applications annually.
[†] Dry matter determined by cutting at intervals of 45 days with herbage removed from plots; average of 3-year period; no significant differences among N treatments.
[‡] Soil samples taken from all plots at end of experiment and tested for pH.

urea, as well as those of ammonium nitrate, are inferior to ammonium sulphate, being more hygroscopic. In addition, there is evidence that urea is less effective than ammonium sulphate in stimulating growth of grass on alkaline soils.

With continued application of most forms of nitrogen fertilizer there is a general decline of herbage yields. This situation is caused by various unfavourable soil conditions such as decreased soil pH, possible phosphorus fixation brought about by increased soil acidity, increased exchangeable aluminum and decreased exchangeable calcium and magnesium. Ammonium sulphate has a rapid and deleterious effect on soil pH and leaching of bases, especially calcium and potassium. In Colombia, topdressing Pangola grass with ammonium sulphate over a 5-year period caused a notorious decline in pH (Fig. 9.6). At the beginning of the experiment the pH averaged 5.8 but dropped measurably with increasing increments of ammonium sulphate. At the end of the experiment the pH in the 0–20 cm soil layer had dropped to 3.5 with a total application of 5 000 kg/ha of N. Dry matter yields declined significantly as compared to those obtained with topdressings of nitrate of soda, averaging 6.4 and 7.8

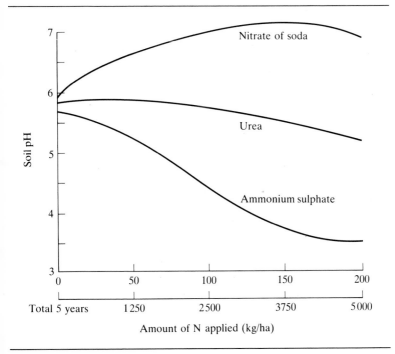

Fig. 9.6 Effect for forms and amounts of nitrogen on soil pH in Colombia (Villamizar and Lotero, 1967).

tonnes/ha per cutting (five cuts/year) with ammonium sulphate and nitrate of soda, respectively. The change in soil pH did not perceptibly alter the plant population of Pangola grass.

From a detailed study of soil changes resulting from annual application of 1 320 kg/ha of N in the ammonium sulphate form Abruña *et al.* (1958) in Puerto Rico noted that total exchangeable base content was markedly affected. After 3 years and 3 960 kg/ha of N the content dropped from 21.9 to 4.0 milli-equivalents/100 g in the 0–60 cm soil layer. The decrease is equivalent to about 19.8 tonne/ha of calcium carbonate. In Georgia, the application of ammonium nitrate to a Coastal bermudagrass sod distinctly reduced the exchangeable K content of the lower horizons of a red–yellow podzolic soil (Burton and De Vane, 1952). It is likely that this occurred because of greater uptake of potassium by the grass and by more rapid leaching.

Grass species and soil types influence to a great extent the effect of applied N on herbage yields and alteration of pH. Application of 400 kg/ha of urea over a 2-year period to *Axonopus scoparius*, a bunch grass, growing on an Oxisol in Colombia caused the soil pH to decrease about one unit (Crowder *et al.*, 1963) After the fourth harvest the dry matter, yields began to decline sharply with rates of N above 50 kg/ha. With continued cutting and removal of the forage, a common practice since this species is used for daily fresh feeding, plant growth declined, the plant population decreased and herbage production fell dramatically (Table 9.5). A slight change in the acidity of these soils causes a release of iron and aluminum into the soil solution. Roots of *A. scoparius* are extremely sensitive to the presence of both elements, as noted by additional field and glasshouse trials. Visual examination of roots showed deformities in terms of restricted extension of the root cap, brittleness and dis-

Table 9.5 *Influence of nitrogen fertilization on forage yields of Axonopus scoparius* growing on an Oxisol, Medellin, Colombia (Crowder et al., 1963)*

Nitrogen[†] (kg/ha)	Dry matter per harvest (t/ha)					Soil ph[‡]
	1st	3rd	5th	6th	10th	$\bar{x}$
Nil	3.00	3.25	3.70	2.83	5.48	5.8
50	2.25	2.40	3.86	2.81	2.89	5.3
100	2.72	3.44	1.00	0.94	0.66	5.1
150	2.48	2.78	1.64	1.15	0.60	4.7
200	2.78	2.92	1.10	0.76	0.29	4.5

* Grass established by vegetative propagation; stem pieces set at 0.50 m apart in rows separated by 1.0 m; three cuttings made and herbage discarded.
[†] Nitrogen applied as urea after each harvest; P_2O_5 at 100 kg/ha and K_2O at 50 kg/ha applied at transplanting and after the fifth harvest; approximately five harvests/year.
[‡] pH in the 0–20 cm soil layer after the tenth harvest.

tended portions of the roots and discoloration from white or bright yellow to a reddish-orange. Studies with this species on an alluvial soil at the same experimental station showed that comparable amounts of N-boosted forage yields were achieved due to increased basal tillering and extended root development.

In general, all forms of N fertilizers are highly soluble and readily available after application. With anhydrous ammonia, a slight lag in growth of grasses occurs since only those roots adjacent to the bands made by the applicator feet have access to this element in abundance. In contrast, all roots in the upper soil layer have access to broadcast forms of N when soil moisture is plentiful. There is relatively little difference in the crude protein content of grasses, or in the chemical composition, when the various forms of N fertilizer are used (Burton and DeVane, 1952; Burton and Jackson, 1962; Crowder *et al.*, 1963; Villamizar and Lotero, 1967; Vicente-Chandler *et al.*, 1974), nor in the palatability of herbage (Burton *et al.*, 1958). In the humid tropics, the residual N effect is of short duration and the benefits in terms of increased plant growth disappear within a few months. A major portion of the applied N is absorbed by the plants or lost by leaching and volatilization within the first 5 months after topdressing.

All forms of N fertilizer discussed are effective providers of this nutrient to grasses used for pasture and forage. The cheapest source in terms of cost per kilogram of N should thus be used, taking into consideration the cost of application and of the $CaCO_3$ required to neutralize residual acidity. Theoretically, 538 kg of $CaCO_3$ is needed for each 100 kg of N applied as ammonium sulphate (21% N) to neutralize the hydrogen ions liberated by the decomposition of this fertilizer in the soil. Urea (46% N) requires 180 kg $CaCO_3$ per 100 kg of N applied. In contrast, nitrate of soda (16% N) causes an alkaline reaction in the soil, equivalent to the addition of 180 kg of $CaCO_3$ per 100 kg of N applied (Villamizar and Lotero, 1967).

In recent years, fertilizers that have controlled the release of N have been examined (Allen and Mays, 1974). Application of sulphur-coated urea (SCU) to *Dactylis glomerata* (Beaton *et al.*, 1967) and to *Festuca arundinacea* (Mays and Terman, 1969a) in one dosage did not improve total productivity but maintained uniform seasonal distribution of forage and crude protein as compared to more variable distribution by ammonium sources. A single application of SCU gave comparable results with Coastal bermudagrass. Yields were similar with use of SCU as the same amount of N applied in four topdressings of ammonium nitrate (Mays and Terman, 1969b). The effect of SCU applied at an annual rate of 360 kg/ha to *Cynodon nlemfuensis* growing on a Utisol was investigated in Puerto Rico (Vicente-Chandler *et al.*, 1974). Cuttings were made at 60-day intervals over a 1-year period. A single application of SCU produced almost 20 per cent more total dry matter than 300 kg/ha of urea, and was almost equal to six equal topdressings of the latter.

Slow-release and conventional nitrogenous fertilizers were applied to an established Pangola grass pasture in subtropical Australia (Lowe and Cudmore, 1978). A single application of the slow-release forms totalled 672 kg/ha, while comparable amounts of the conventional forms were split into several yearly applications. Over a 4-year period, slow-release urea formaldehyde gave the highest forage yields and the residual effect persisted into the fourth season. Nitrogen recovery of this form equalled that of the more frequently applied ammonium nitrate and urea.

Advantages cited for SCU include (1) season-long effective N in a single application, (2) more even distribution of herbage and avoidance of excess early production with a high N content, (3) application of an appreciable quantity of N without foliage burn, (4) little volatilization and reduced potential leaching losses and (5) increased soil sulphur level. In selecting a N fertilizer to boost grass herbage yields, however, the cattleman must consider not only effective utilization of N but also availability of the product and the economics of its use.

Nitrogen recovery

The efficiency of N fertilization of grasses is usually estimated in terms of kilograms of dry matter produced per kilogram of N applied. This depends to a great extent on the amount of N used and frequency of cutting, i.e. stage of growth at harvest time. For the first 400 kg/ha applied on an annual basis, from 50 to 65 kg of dry matter per kilogram of N will be obtained with a cutting interval of about 45 days and adequate soil moisture to assure continued regrowth. Above this level of applied N the efficiency will sharply decline. In drier areas it may be reduced by one-half. The efficiency of N utilization increases as the underground plant parts and soil organic matter increase. The first increments of applied N essentially increase dry matter yields and further increments raise the N content of the grass.

Another way of measuring N efficiency is to determine the amount recovered in harvested forage. This value is obtained by subtracting the amount of N found in herbage produced by the unfertilized control from the total recovered annually when N fertilizer is supplied, and dividing by the amount of N applied. The percentage of N recovered depends principally on the amount of N applied, time of application, soil moisture, source of N, production of forage, N content of the herbage and varietal or species differences. Recovery figures may exceed 85 per cent at low rates of applied N and with frequent cutting but diminish rapidly with higher levels.

In Puerto Rico, return of applied N varied from 40.6 to 55.3 per cent with different forms of N (Table 9.4) (p. 162). On Pangola grass in Colombia, recoveries were 57 per cent for ammonium sulphate, 55 for nitrate of soda and 46 for urea (Villamizar and Lotero, 1967). With Coastal bermudagrass in Georgia, Burton and Jackson (1962) found an average recovery (using 110, 240 and 480 kg/ha of N) of 81.3 per cent for ammonium nitrate, 80.2 for ammonium sulphate, 78.4 for anhydrous ammonia and 60.2 for urea.

Henzell (1971) working with *Chloris gayana* in Queensland, noted 'marginal' recoveries of 60.4 per cent for sodium nitrate, 56.5 for ammonium nitrate–limestone, 52.5 for ammonium sulphate and 50.4 for urea. Marginal N recovery was calculated by subtracting the nitrogen yield of a 224 kg/year treatment from that of the corresponding 448 kg/year treatment and expressing the difference as a percentage of 224. This was done to avoid the confusing effects of legume–nitrogen fixation on the nil plots due to an invasion of an indigenous legume. In studies such as these a native or naturalized legume will often appear when phosphate and potash are added to the soil.

In Australia, the N recovery was lower in dry than in wet seasons (Henzell, 1971), and in Georgia (Prine and Burton, 1956) 88 per cent of applied N was recovered in a wet year but only 47 per cent in a dry year. Chheda and Akinola (1971a) in Nigeria reported 93 per cent recovery from *Cynodon nlemfuensis*, as compared to about 45 per cent with two other selections of *Cynodon* when fertilized with 165 kg/ha of N per year and cut at 6-week intervals. Different grass species in Puerto Rico, when topdressed annually with 220, 480 and 960 kg/ha of N, gave the following average recoveries: *Digitaria decumbens* 57.8 per cent, *Pennisetum purpureum* 53.6, *Brachiaria ruziziensis* 53.1, *B. mutica* 49.9, *Panicum maximum* 49.3 and *Cynodon nlemfuensis* 40.9 (Vicente-Chandler *et al.*, 1974). Such wide fluctuations in percentage recoveries have been observed in almost all experiments with nitrogenous fertilizers. Most studies have been carried out with improved cultivars or selections. Native or naturalized grasses are much less responsive to applied N and recovery rates are decidedly lower. Norman (1962) in Australia found that recovery values of a mixture of *Sorghum plumosum*, *Themeda australis* and *Chrysopogon fallax* fluctuated between 5.6 and 7.1 per cent.

Even though recovery values of over 90 per cent have been reported, those more commonly obtained range between 30 and 65 per cent. Although it is commonly found that fertilizer N is lost from soil–plant systems (Allison, 1966), it is difficult to account for the portion not recovered. Blue (1970), working with *Paspalum notatum* in Florida, found that 29 per cent of N applied in the

112 kg/year treatment and 21 per cent in the 224 kg treatment could not be accounted for in a field experiment. A lysimeter experiment at the same location showed at 15 per cent loss from labelled ammonium nitrate (168 kg/ha of N) applied to Coastal bermudagrass (Brown and Volk, 1966). An analysis of N removal in the herbage of *Chloris gayana* and that found in the stubble and roots to a 30 cm depth showed that 15 per cent or more of applied N was not recovered (Henzell, 1971, Australia). A complementary study using a gas lysimeter suggested that leaching of N from the soil may have occurred (Henzell *et al.*, 1968). Other ways whereby N can be lost include runoff in surface water, chemical and biological immobilization, volatilization and denitrification.

Effect of nitrogen fertilizer on grass–legume mixtures

The effect of fertilizer nitrogen on the proportion of desirable pasture species is as important as boosting the yield of forage. Continued use of high N fertilizer causes a rapid decline in the legume component of subtropical and tropical grass–legume combinations, as noted by several researchers. The *Desmodium intortum* population in established mixtures of *Pennisetum clandestinum* and *Digitaria decumbens* in Hawaii was reduced to less than 10 per cent during the first year with application of 410 kg/ha of N as ammonium sulphate (Whitney, 1970). The N was split into equal amounts topdressed at 5-week intervals. In contrast, the legume comprised 50 per cent of both pasture combinations without N fertilization.

Use of 360 kg/ha per year in four split applications over the growing season to a number of grass–legume mixtures in Queensland eliminated the legumes within a period of 12 months from the beginning of fertilization (Jones *et al.*, 1969). Lower amounts of N at 75 and 240 kg/ha also reduced the legume fraction in a grass–Siratro (*Macroptilium atropurpureum*) pasture (Jones, 1967). Grasses in association with other legumes at the same location showed N deficiency at the beginning of the growing season and legume regrowth was delayed. It was anticipated that application of N prior to legume regrowth at the end of the season would boost total production without eliminating the legume (Jones, 1970). Two rates of N, 110 and 360 kg/ha, were topdressed as split dosages on well-established pastures of Siratro + *Chloris gayana* and Siratro + *Setaria anceps*, both of which had been adequately fertilized with other plant nutrients. Both pastures responded in a similar manner and data for the latter mixture is given in Table 9.6. Nitrogen depressed legume yield in every year except with the lower amount of fertilizer in the first year. A gradual decline in the legume occurred until it was eliminated at the 330 kg level. The deleterious effect of applied N on legume composition of the sward appeared to be cumulative. A

Table 9.6 *Herbage yields of Siratro (Macroptilium atropurpureum) in mixture with Setaria anceps and fertilized with urea, Queensland, Australia (Jones, 1970)**

Year	(kg/ha) Legume dry matter [†]		
	0 N	110 kg N	360 kg N
1962–63	795	790	190
1963–64	2 410	1 015	95
1964–65	605	260	10
1965–66	1 660	700	0

* Urea split and broadcast over pastures before legume growth in spring and after cessation of growth in fall.
† Forage yields obtained by sampling prior to grazing four times per year; hand separations made for legume and grass; yields significantly different at 0.01 level.

similar decline of lucerne and *Glycine wightii* in mixture with *Panicum maximum* var. *trichoglume* was noted by Kleinschmidt (1967) in another location of southeastern Queensland. In Rhodesia (Zimbabwe), the legumes *Lotononis bainesii* and *Trifolium semipilosum* persisted satisfactorily when grown in mixtures with six rhizomatous or stoloniferous grasses and amply fertilized with phosphate and potash, but not including N (Clatworthy, 1970). In most combinations the legumes were eliminated within 3 years with annual top-dressing of 224 kg/ha of N as nitrolime made in three split applications. A lower rate of N, 112 kg/ha, affected the legumes to a less extent but their contribution to the total herbage was significantly reduced.

Competition for nutrients, water and light are major factors influencing botanical composition in plant associations. Tropical and subtropical grasses have a more rapid growth rate and are more aggressive than legumes. They also extract more N from the soil in grass–legume mixtures (Henzell *et al.*, 1968), and are more persistent under frequent defoliation. Applied N reduces nodulation of the legume which further interferes with its competitive capability. Grasses and legumes generally have different cycles of growth in terms of vegetative and reproductive phases. Their management in mixtures is, therefore, difficult, and becomes more complicated with applied N.

Response of native pasture to nitrogen

Species adapted to native, unproductive grazinglands have the capacity to survive and dominate the plant community under conditions of limited nutrient supply. In these situations phosphorus is more likely to be a limiting factor for increased herbage productivity than N (Bishop, 1977; Garza *et al.*, 1971; Harker, 1962; Lotero *et al.*, 1965; Norman, 1962; Olsen and Santos, 1975; Poultney, 1959;

Tapia, 1971). Appreciable increases in yields of dry matter occur only when both nutrients are supplied, and response to applied N may be negligible in the absence of phosphate.

In Australia (Norman, 1962) found that three grass species in a mixed stand responded differently to nitrogen treatment. Botanical separations showed that yields of *Sorghum plumosum*, when harvested after flowering and seeding, more than doubled when top-dressed with 176 kg/ha of N at the beginning of the rainy season, from 450 to 900 kg of dry matter over a 7-month growth period. *Chrysopogon fallax* showed a trend towards response to N, yielding about one-half as much dry matter as *S. plumosum*. In contrast, the yields of *Themeda australis* sharply declined, from 1320 to 550 kg/ha of dry matter harvested from the control and N treatments, respectively. Crude protein content of the grasses increased slightly with applied N, from 2.1 to 3.1 per cent, but in the absence of P. In the presence of P, N content did not increase with applied N, even though herbage yields were higher (Table 9.7). This suggested that the additional N taken up by the plants was fully elaborated in the photosynthetic process, leaving the N content of the forage at the same base level as the control. Over the 3-year period of the experiment, the N content of grasses increased from 0.23 per cent in the first season to 0.52 per cent in the third season. It is likely that with further annual topdressings of N the carbon–nitrogen ratio of the soil, plus accumulation of additional organic matter, would have progressively declined. Thus, a higher proportion of the applied N would have been available for plant assimilation.

Table 9.7 *Dry matter yields* of native pasture in response to nitrogen and phosphate fertilizer[†] (Norman, 1962)*

Nitrogen[‡]	Phosphate treatment[‡]			
	0	P_1	P_2	$\bar{x}$
0	2 155	2 255	2 500	2 300
N_1	2 365	3 080	3 445	2 960
N_2	2 235	3 905	4 420	3 520
$\bar{x}$	2 250	3 080	3 460	

* Yields are in kg/ha with one cutting made at maturity. S.E. for N and P treatments = ±103 kg/ha, for the interactions ± 180 kg.

[†] Data from Katherine, Northern Territory, Australia with the pasture composed largely of *Sorghum plumosum*, *Themeda australis* and *Chrysopogon fallax*; average of all grasses over 3 years.

[‡] N_1 = 495 kg/ha of ammonium sulphate (21% N) applied as split dressing over a 3-year period.

N_2 = 1 090 kg/ha of ammonium sulphate split over 3 years.

P_1 = 165 kg/ha of superphosphate (21% P_2O_5) first year and 82.5 kg/ha second and third years.

P_2 = 330 kg/ha of superphosphate first year and 165 kg/ha second and third years.

In the above study, harvests were made at the time of plant maturity, which accounts in part for the low N content of herbage. More frequent cutting would have shown a higher N content of the grasses during the peak of vegetative growth. It must be remembered, however, that in the Northern Territory of Australia, availability of feed during the prolonged dry season determines animal stocking rate. Herbage must be accumulated *in situ*, therefore, for dry-season grazing.

Nitrogen fertilization to native pastures is not an efficient means of improving feed production. More intensive utilization of the pasture during the wet season when the grasses are more nutritive is not practical, because of species vulnerability due to overgrazing. When this occurs, carbohydrates are not transferred to the roots as reserve for next season's growth. An alternative for increasing feed supply in the dry season might be oversowing of native pastures with a legume such as Townsville stylo as discussed in Chapter 8.

Nitrogen fixation by organisms associated with tropical and subtropical grasses

Nitrogen fixation by microorganisms associated with grasses is more likely to occur in tropical rather than in temperate environments (Döbereiner, 1966, 1968; Neyra and Döbereiner, 1977; Weier, 1980). This biological phenomenon holds tremendous potential for increased production of quality herbage and addition of nitrogen to the soil reservoir through recycling by grazing animals. Species such as *Andropogon, Brachiaria, Cynodon, Digitaria, Hyparrhenia, Melinis, Panicum, Paspalum, Pennisetum, Saccharum* and *Zea* have the capability of associative fixation (Day *et al.*, 1975; DePolli *et al.*, 1977). They all possess a carbon pathway (C_4 dicarboxylic acid) in photosynthesis that is more efficient than the C_3 pathway in temperate grasses. This results in higher net photosynthesis, greater energy conversion and increased growth potential as compared to temperate grass species. Wide variation occurs among grasses in N fixed by the associative organisms, but estimates of about 1.5 g/ha of N per day have been made for Pangola and bahia grasses (Day *et al.*, 1975).

Nitrogen fixation is mediated by the enzyme nitrogenase which substitutes photosynthetically derived energy and requires 15 molecules of ATP for each molecule of N_2 reduced. Nitrogenase activity in the rhizosphere and inside roots of field-grown tropical grasses is measured by the acetylene reduction method (Döbereiner *et al.*, 1972; Döbereiner and Day, 1973; Smith *et al.*, 1976).

The bahia grass – *Azotobacter paspali* association, described in Brazil, has been the most intensely studied (Döbereiner, 1966; Döbereiner and Campbelo, 1971; Döbereiner and Day, 1973). Activity of this organism occurred largely in the rhizosphere of grass roots, showing greatest activity during the growing season. Accumulation of N persisted into the dry season and enhanced growth and N content of herbage. Ecotypes of bahia and elephant grass differed significantly in nitrogenase activity on their roots; native forms of the grasses were more efficient nitrogen fixers. This suggested that commercial grass cultivars have been selected for response to higher mineral N levels. Such variability among grass types poses a challenge to plant breeders to promote this characteristic through breeding and selection.

The fate of fixed nitrogen was ascertained by studies of bahia grass growing in vermiculite with nitrogen-free solution (Döbereiner and Day, 1975). At no time did the grass plants display signs of N deficiency. After 2 months, the quantity of nitrogen had increased several-fold as plants developed. Plants contained a high percentage of the nitrogen, but a significant quantity remained in the vermiculite, indicating that a portion would be made available to associated vegetation.

In Brazil, seasonal variation in nitrogenase activity was observed in elephant and Pangola grasses grown in the field, with and without applied N. Highest rates of N fixation occurred during the rainy season and when warm temperatures prevailed (Döbereiner and Day, 1975; Day *et al.*, 1975). Additions of 20 kg/ha of fertilizer N every 2 weeks did not interfere with nitrogenase activity, even after application of 160 kg/ha. Further studies with bahia grass growing in pots showed a decline in organism activity with the use of an ammonium form of nitrogen but not with a nitrate form.

The discovery of N fixation in a broader spectrum of tropical grasses projects far-reaching implications in the development of forage–livestock feeding systems (Döbereiner and Day, 1975, 1976). The organism *Spirillum lipoferum* (reclassified as *Azospirillum lipoferum* cited by Hubbell, 1978) was found concentrated in the cortex cells of Pangola grass. This organism appears to be associated with the C_4 grasses which have malate as one of the primary photosynthetic products. *Spirillum lipoferum* utilizes malate as an energy source and in association with grasses may be dependent upon them for this product as a substrate. Similar strains of *Spirillum* were isolated from guinea grass and other nitrogen-fixing grass species. In Florida, a *Spirillum* was found inside grass roots, and surface-disinfected sections showed that nitrogenase activity was associated with the roots and not in the rhizosphere soil (Smith *et al.*, 1976). These findings suggested a symbiotic relationship as well as a specific *Spirillum* relationship with grass species. Forty different genotypes representing five tropical genera were inoculated with *Spirillum* and compared with their uninoculated counterparts. Inoculated 'Transvala' digitgrass (*Digitaria decumbens*) and guinea grass (*Panicum maximum*) yielded 163 and 150 per cent of their uninoculated controls, respectively. The guinea grass also gave higher contents of crude protein than the uninoculated grass. A field study of bulrush millet (*Pennisetum americanum*) and guinea grass compared four rates of N fertilizer as ammonium nitrate (nil, 20, 40 and 80 kg N/ha applied after seedling emergence and nil, 10, 20 and 40 kg after the first harvest) superimposed over plots inoculated with *Spirillum* and uninoculated checks. Dry matter yield increases due to *Spirillum* inoculation occurred in both grasses when totals of 40 and 80 kg of N/ha were applied. It was calculated that N fixation by the bacteria added an additional 40 kg N/ha. These data indicated that an initial N fertilization was required to incite a response to inoculation in the low N residual, sandy soil where the trials were carried out. Applied N may not be needed to stimulate early plant growth in soils with a higher N residual. The forage yield data showed a linear response to applied N fertilizer on inoculated plots but a curvilinear response on inoculated plots. These differences substantiated an inoculation response. At the higher rates of applied N, responses were reduced due to repression of the nitrogenase enzyme activity. These results demonstrated a symbiotic relationship of bacteria in grasses and a system that can increase production of dry matter or reduce N fertilizer requirements of certain grasses.

Phosphorus fertilization

Highly weathered soils of the tropics are generally deficient in phosphorus and the short-term efficiency of P fertilizers is usually low until the P-fixation capacity is satisfied and a residual P supply accumulates. Early research with phosphate fertilizers in the tropics was located on soils with a high fixation capacity (Russel *et al.*, 1974). This led to the belief that P fixation is a distinguishing characteristic of tropical soils. The acidic, red latosols (so-called lateritic soils) are noted for high P fixation (deGeus, 1969). P response curves for these soils are sigmoidal, rather than parabolic (Campbell and Keay, 1965). The amount of P required to satisfy the fixation capacity of such soils is rather large, especially with pasture crops where fertilizers are broadcast rather than concentrated in a row as with field crops. In Hawaii, application of rates up to 320 kg/ha of P applied as superphosphate to a Pangola grass–*Desmodium intortum* pasture gave increased forage yields and beef cattle liveweight gains (Younge and Plucknett, 1966). Annual dry matter yields of the grass–legume combination were 3.2, 14.2 and 19.2 tonnes/ha and liveweight gains 250, 1 100 and 1 500 kg/ha with nil, 330 and 1 320 kg/ha of P, respectively. After 12 years without additional P application these massive amounts continued to maintain forage yields. In the fourth to sixth years yields for these treatments averaged close to 2.5, 12.0 and 17.0 tonnes/ha of dry matter and in the tenth to eleventh years 1.5, 5.0 and 10.0 tonnes/ha, respectively (Fox *et al.*, 1971). An initially large investment is required for application of such high amounts of phosphate but might be economical under intensive pasture utilization. Banding of fertilizer, especially phosphate, at sowing time reduces the initial amount needed and creates a favourable condition for seedling growth (Fenster and Leon, 1978). Roots tend to concentrate in the band, however, unless additional fertilizer is mixed into the soil prior to sowing.

In contrast to such spectacular results, many soils in tropical and subtropical areas respond to modest applications of P with pronounced increases in forage production (Cloutier, 1971; Crowder *et al.*, 1963; Crowder and Chheda, 1977; Stephens and Donald, 1958; Trumble, 1952; Vicente-Chandler *et al.*, 1974). On these soils the residual P is of considerable value. When pastures are grazed a large portion of the soil P taken up by grasses and legumes is returned in the excrement of grazing livestock or in plant residue. Thus, long-term P requirements of such soils will be lower than initial applications.

Forms of phosphorus fertilizers

The most common forms of phosphate fertilizer used on tropical pastures are the ordinary (single) and triple superphosphates. They are particularly effective in soils with low to moderate fixation capacities. The single superphosphate is preferred in some areas since it contains sulphur. The less soluble forms, such as rock phosphates, may be more effective and economical in acid soils that fix large quantities of P. Rock phosphates are more reactive in acid soils and cost one-third to one-fifth that of superphosphate in terms of elemental P. Higher recovery of applied P can be expected over a short term from both superphosphates than from rock phosphates (Michelin *et al.*, 1974) but the latter have an advantage in permanent pastures because of residual effects. A mixture of both might be preferred, considering the costs of elemental P and repeated top-dressings required for the more soluble forms.

A study carried out with Pangola pastures in Florida (Hodges *et al.*, 1966) illustrates the effect of different phosphate fertilizers on soil P content, forage yields and P content, and animal performance (Table 9.8). Soil at the experimental site was classified as a fine sand with about 3.0 per cent organic matter and a pH of 4.9. Total annual rainfall averaged 1 400 mm over the period of study. The experimental herds and replacements consisted of grade Brahmin cows, being placed on trials at 2 or 3 years of age. The same herd remained on each phosphorus treatment throughout the study, receiving no feed other than grazing. Forage production during 1955–58 averaged 50 per cent higher on the P-treated pastures than on the nil-P check. This margin dropped to 23 per cent in the following 5-

Table 9.8 *Effect of phosphate fertilizers on forage yields and P content of Pangola grass, P soil level and animal performance in Florida (Hodges et al., 1966)*

Treatment	Total P_2O_5* (kg/ha)	Soil P, 0–10 cm 1956 (ppm)	1961 (ppm)	Air dry forage/yr[†] 1955–58 (t/ha)	1959–63 (t/ha)	Herbage P 1958 (%)	1963 (%)	Livewt gains/ha[‡] 1955–58 (kg)	1959–63 (kg)
1. No phosphate	0	47	49	9.3	10.7	0.10	0.06	135	130
2. Superphosphate	240	61	56	12.1	11.8	0.20	0.07	253	193
3. Superphosphate + lime	240	83	66	14.8	13.4	0.25	0.09	264	242
4. Rock phosphate	835	494	385	14.9	15.0	0.35	0.25	234	200
5. Triple superphosphate	215	59	62	14.0	10.9	0.23	0.08	256	190
6. Basic slag	195	139	112	14.3	14.9	0.23	0.14	295	225

* Superphosphate at 55 kg P_2O_5/ha from 1947–54 reduced to 25 kg 1955–58; 1.0 t/ha rock phosphate in 1947, 1950, 1953; 550 kg/ha basic slag 1947–54, 330 kg 1955–58; treatment 3 received 1 100 kg calcic lime/ha in 1947, 1950 and 1953; all treatments 1.0 t/ha dolomitic lime in 1955 and 1 100 kg calcic lime in 1959, except treatment 2; minor elements in 1947 and 1953; all treatments split dressing of 110 kg/ha of N as ammonium nitrate and single dressing 55 kg K_2O annually.
† Forage yields from quadrats moved after each harvest.
‡ Paddocks of 6.0 ha divided into quarters and rotationally grazed.

year period when residual P was tested. It should be noted that the yields from rock phosphate and basic slag treatments did not decline because of the higher residual phosphorus of these two less soluble forms. Soil P of the nil treatment remained the same throughout the experiment, with herbage from these paddocks containing lower than the nutritional minimum of 0.15 per cent P recommended by the National Research Council (1963). With annual topdressing of superphosphate, herbage P remained at an adequate level for beef cattle growth but dropped sharply when fertilization was discontinued. The maintenance of sufficient herbage P in pastures receiving rock phosphate was consistent with the residual levels measured in the soil of this treatment. Two cows on the nil-P treatment developed P-deficiency symptoms while calves were being suckled but both returned to normal when the calves were weaned. Data on liveweight gain was a combination of calf production and net again or loss in weight of cows. A greater weight increase occurred on all phosphate treatments during the base period of applied fertilizers, i.e. 1955–58, and in subsequent years after phosphate amendment was discontinued, 1959–63. Pastures receiving phosphate produced about 90 per cent more liveweight gain than the check during 1955–58 and kept a 60 per cent margin during the residual P period. In comparing animal gains with forage yields, it appeared that cattle production declined appreciably during the residual period while forage yields were not consistently affected. Weaned calf percentages showed no consistent differences among cows on the phosphate treatments nor the nil-P treatment. Calving percentage on the superphosphate treatment plus lime, however, was consistently higher than the rock phosphate treatment.

In Campinas, Brazil, a trial with *Glycine wightii* compared several sources of phosphate, namely superphosphate, Olinda and Araxá rock phosphates, thermaphosphate and bone meal, alone at 120 kg/ha of P_2O_5 and in combination with 4.0 tonnes/ha of limestone (Neme, 1966). The experiment was located on a Latosol, Terra Roxa, having 4.2 per cent organic matter, 0.13 me P per 100 g of soil, 4.10 me Ca and a pH of 5.4. The phosphate fertilizers and limestone were topdressed about 3 months after establishment of Glycine. An unusually dry period delayed harvests for about 8 months, after which eight cuttings were made over a 3-year period. The nil-P treatment averaged 1.9 tonnes/ha of dry matter per harvest. Thermophosphate boosted yields 25 per cent, Olinda rock phosphate 16 per cent and the other three sources about 10 per cent. Addition of limestone gave a 5 per cent advantage of superphosphate and the rock phosphates.

On a dark red Latosol near Brasilia, two highly soluble rock phosphates, hyperphosphate from North Africa and North Carolina, provided similar early growth of *Stylosanthes humilis* and *Brachiaria mutica* (grown alone) as ordinary superphosphate when applied at 86, 345 and 1 380 kg/ha of P_2O_5 (Yost *et al.*, 1976). The low solubility Araxá rock phosphate resulted in about one-half the production as superphosphate, but when thermally altered as themosphosphate the performance was equal to superphosphate. A comparison of yields 1 year after application showed that P availability of Araxá rock phosphate increase with time.

Biosyser is a slow release phosphate fertilizer made from a mixture of ground rock phosphate and sulphur, inoculated with sulphur-oxidising *Thiobacillus thio-oxidans* (Swaby, 1975). The pattern of phosphorus and sulphur release provides long-term benefits for tropical legumes, with accumulated herbage yields from biosyser application exceeding that of single superphosphate (Jones and Field, 1976; Partridge, 1980). Greater initial response occurs with single supersphosphate but repeated applications are needed for sustained production. From information available up to the present date, it seems that the effect of pelleted biosyser will persist for about three growing seasons.

Response of grasses to phosphorus fertilizers

The P requirements of grasses, unlike that of N, depends more on soil properties than on the grass species. Phosphorus is likely to be a limiting factor when grasses are established on land not previously cultivated and fertilized with phosphate. Even then, there may be need for additional P fertilizer on some types. Jordan *et al.* (1966) suggested addition of P to Coastal bermudagrass when soil values (with 0.05 M HCl + 0.025 M H_2SO_4 extraction) were less than 25 ppm of phosphorus. Woodhouse (1969) also emphasized the maintenance of 20–25 ppm of soluble P in the soil for increased production of this grass. Dry matter yields increased linearly as the soluble P in the soil was raised to 10 ppm, then became curvilinear (Fig. 9.7). Phosphorus is less mobile in soils than N and K and remains fairly close to the area in which it is placed. Once the fixation capacity is satisfied by incorporation of phosphate fertilizers into the soil, a single annual or biennial application of P fertilizer can be top-dressed to maintain the nutrient requirement of grasses. This indicates that important feeder roots are located near the soil surface.

The amount of P removed by grasses in Puerto Rico, when forage was cut and removed, ranged from 35 to 70 kg/ha (Table 9.2 p. 153). The optimum levels of P fertilizer in an Oxisol and a Utisol were about 70 kg/ha of P annually for *Pennisetum purpureum* and *Cynodon nlemfuensis* and 50–60 kg for other grasses. Elephant grass responded strongly to 70 kg/ha of P on two soils with little previous P fertilization (Fig. 9.8). The P content of herbage increased with rates up to 140 kg annually, having about 0.17 per cent when cut at a 60-day interval. The P content of herbage may have been in-

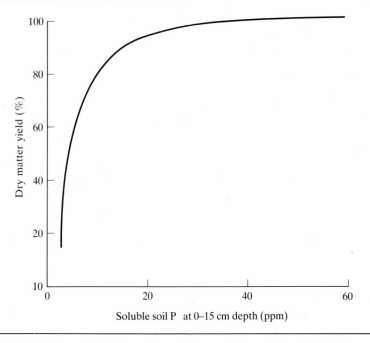

Fig. 9.7 Relative yield of Coastal bermudagrass as influenced by soluble P (adapted from Wilkinson and Langdale, 1974; taken from Jordan *et al.*, 1966, and Wood-house, 1969).

fluenced by the high level of N applied (660 kg/ha in six equal applications), which tended to reduce the accumulation of this nutrient relative to lower amounts of N. Elephant, guinea and Pangola grasses did not respond in forage production or P content to phosphate applications during 4 years on another soil previously in sugar-cane and well fertilized. This soil released about 47 kg/ha of P annually, as compared to 24 kg released by the other two soils.

Application of 25, 50 and 75 kg/ha of P as triple superphosphate to *Cynodon nlemfuensis* in Nigeria (Saleem, 1972) increased dry matter production by 4, 17 and 20 per cent, respectively, over a 3-year period without complementary N. With annual topdressings of 165 and 248 kg/ha of N, the same levels of P gave increases in herbage production ranging up to 35 per cent. A part of the benefit derived from applied P was contributed to the increased numbers of tillers as compared to the nil-P treatment. Low recovery of P in early harvests indicated rapid fixation in the soil probably as aluminum, iron and calcium phosphates. Total soil P increased over time, and hence P was later released and made available to the grass. Phosphorus recovery over the 3-year study varied between 85 and 69 per

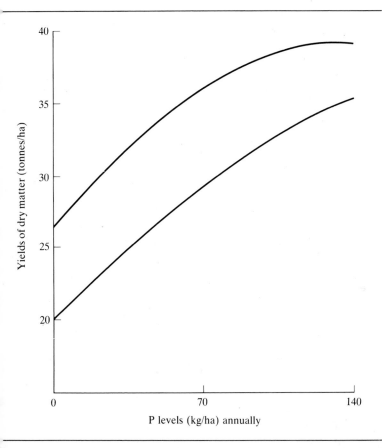

Fig. 9.8   Yields of elephant grass growing on two soils in Puerto Rico when fertilized with triple superphosphate; P content of herbage in parentheses (Vicente-Chandler *et al.*, 1974).

cent with 50 and 75 kg/ha of applied nutrient, respectively. After 2 years of cutting and removing grass, the soil test values showed more than adequate amounts of P available in treatments receiving the two rates initially. Under grazing, of course, much of the P would have been recycled. Thus, residual P would support further production of grass or subsequent cropping. Variable responses of grasses in terms of forage yields have been noted by researchers, depending on previous fertilizer regimes and soil management practices (Boswingle, 1961; Crowder *et al.*, 1963; Kein *et al.*, 1976; Velasquez *et al.*, 1975). In most of tropical and subtropical Australia, application of superphosphate to correct P deficiency is based on soil analysis with 125–250 kg annual topdressing for maintenance of grass–legume pastures (Jones and Jones, 1971). In general, applied

P increases the content of this nutrient in herbage but does not alter the crude protein content.

Response of legumes to phosphorus fertilizers

The dominating importance of P in the establishment, productivity and persistence of legumes has been recognized in the tropics and subtropics. Whether legumes are established and grow successfully depends directly on availability of P, either residual in the soil or applied as fertilizer. A résumé of soil fertility studies in the humid tropics of Queensland, Australia, showed that P fertilizers significantly increased plant growth in 52 experiments with legume-based pastures (Teitzel and Bruce, 1970). There were only three trials in which a response was not measured. Soil analyses indicated a low P content wherever a response occurred.

Application of phosphate greatly enhanced seedling establishment and early growth of several legumes in the acid soils of the eastern plains in Colombia (Spain, 1975) and in Uganda (Olsen and Moe, 1971). Of equal importance was the improved legume nodulation in the latter study. Phosphorus is essential in nodule development and N fixation. The effect of P on nodule dry weight was more vividly described by Gates (1970) at the Cunningham Laboratory in Queensland (Fig. 9.9). Phosphorus application stimulated nodulation at successive harvests in a series of studies with *Stylosanthes humilis* cv. Commercial carried out in the field, in the glasshouse and under controlled environment. An increase in the pinkness of the nodule was observed with added increments of P, indicating greater activity of the associated *Rhizobium*. Phosphorus content of the nodules was increased by P treatment. There was a positive interaction with harvest, suggesting that this effect began with incipient nodulation. Thus, under conditions of P deficiency it might be expected that demands for this nutrient during nodule formation would be considerable and thus seriously compete with seedling growth. The dramatic effects of P supply on nodulation was also observed in seedlings of *Leucaena leucocephala*, *Lotononis bainesi* and *Macroptilium lathyroides*.

That legumes respond positively and significantly to applied P was portrayed by data obtained by Bruce (1974) with *Stylosanthes guianensis* cv. Schofield in north Queensland. Three experiments at two sites were superimposed on established commercial pastures of the stylo (Fig. 9.10). The soil had a previous history of superphosphate application at the time of pasture establishment. Companion grasses, *Melinis minutiflora* and traces of *Panicum maximum*, occurred with the stylo in experiments one and two. For each experiment the established pasture was cut to about 30 cm above ground level, plots of 8 × 10 m or 3 × 5 m marked and fertilizers broadcast by hand. In addition to application of single superphosphate (9.6% P)

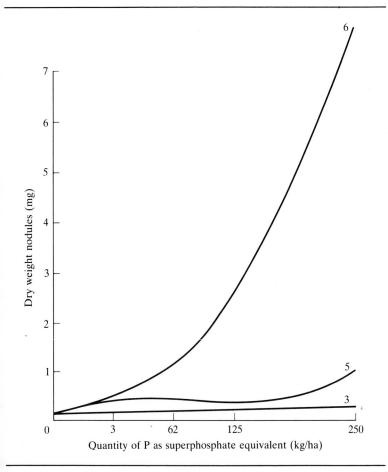

Fig. 9.9   Effect of P on nodule formation of *Stylosanthes humilis* at harvests 3 (17 days), 5 (23 days) and 6 (26 days) (adapted from Gates, 1970).

muriate of potash (125 kg/ha), copper sulphate (11 kg) and zinc sulphate (11 kg) were included. Yields were measured 4–5 months after fertilization, just when the stylo began to flower. A second harvest was made in experiment one during year two. Dry matter yields increased rather sharply with amounts up to 375 kg/ha of superphosphate (36 kg P) in experiments 2 and 3 and the second harvest of experiment one. Increments of 500 and 625 kg of superphosphate gave additional yields during the first year of experiment one. Stylo comprised 75 and 53 per cent of the total herbage for harvest in years one and two of experiment one, 93 and 100 per cent of experiments two and three. Phosphorus in stem tips of stylo rose from 0.10 per cent in the check plot to 0.25 per cent with 65 kg/ha of P.

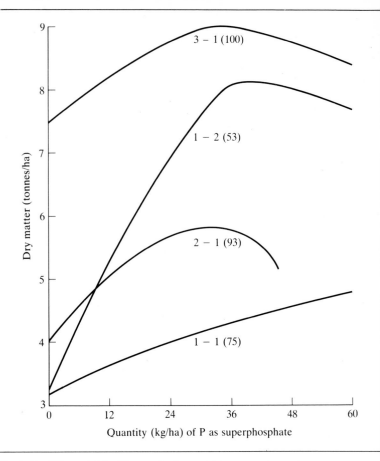

Fig. 9.10 Effect of P fertilizer on herbage yields of stylo in north Queensland; experiment 1, two harvests, experiments 2 and 3 one harvest; figures in parentheses are per cent legume (adapted from Bruce, 1974).

That in the plant tops (cut at 5 cm above ground level) measured 0.05 and 0.18 per cent, respectively. Phosphorus content of the herbage varied over the season. In the stem tips it reached a maximum just before flowering and then declined rapidly. As flowering proceeded P was mobilized from the leaves and transported to the inflorescences. Phosphorus content of the tops attained a maximum in the early vegetational stage of growth then declined gradually as the plants developed and matured.

Other studies have substantiated the beneficial effects of phosphate application on productivity of tropical and subtropical legumes (Andrew and Jones, 1978; Fisher and Campbell, 1972; Grof, 1966; Jones, 1968, 1974; Piñedo and Santhirasegarum, 1973;

Robertson *et al.*, 1976). The real value of applied P to legumes is derived in output of animal product. In an evaluation of fertilizer requirements of tropical-based legume pastures, Andrew and Bruce (1975) pointed out the need for crop-logging or monitoring of the soil–plant–animal relationships. The full effect of nutrients, such as P, can only be achieved over a period of time to allow full expression on dry matter production of the legumes, amount of N produced, use of N by the grass and animal, balance of legume to grass and overall animal production.

The efficiency of superphosphate application tends to decline over time unless a maintenance supply is provided on an annual basis. Several studies by the Division of Tropical Agronomy in northeast Queensland showed that relative small amounts of P applied annually to legume pastures maintained animal liveweight gain over several years (Table 9.9). Annual topdressing of 10–25 kg/ha of P kept the soil nutrient supply at a level for optimal plant growth and boosted animal production several-fold. To derive sustained benefits the cattleman needs to be knowledgeable about the nutrient status of the soil, nutrient status of pasture plants, productivity of the pasture and animal output.

Response of native pastures to phosphorus fertilizers

Improvement of native or naturalized pastures by direct fertilization has been disappointing in terms of increased herbage yields, animal

Table 9.9 *Animal production from four sites in northeast Queensland in relation to fertilizer usage and available soil phosphorus (Andrew and Bruce, 1975)*

Location and P treatment (kg/ha per yr)	Duration (yrs)	Animal LWG (kg/ha per yr)	Soil P* (ppm)
1. Rodds Bay	9		
Control		63	5
Stylosanthes humilis + 10 P		136	10
Stylosanthes humilis + 20 P		123	40
2. Narayen	5		
Control		34	8–10
Macroptilium atropurpureum + Cenchrus ciliaris + 25 P		135	
3. Townsville	5		
Control		23	6
Timber removed + Stylosanthes humilis + 10 P		173	
4. Beerwah[†]	8		
Control		80	3
Sown pasture + 10 P		176	19
Sown pasture + 20 P		268	20

* N/100 H_2SO_4 extraction; (1) eighth year, (2–3) initial, (4) seventh year.
[†] A mixture of several improved grasses and legumes.

production and economical returns. Most such grazinglands are composed of unproductive grasses, but sometimes with a low percentage of native or naturalized legumes. The response to applied fertilizers has largely been to a combination of N and P (Tables 9.7 p. 170 and 9.10). Herbage yields can sometimes be increased several-fold, but total production is decidedly inferior to that obtained from sown pastures that are fertilized.

Response of unproductive grazinglands to applied P probably depends on the species that have been able to persist under the inherent nutrient status of the soil. In extremely infertile soils grasses that are unresponsive to applied P predominate. With greater inherent soil fertility more responsive species are likely to be found. In Colombia, Lotero *et al.* (1965) examined various combinations of N, P, K and Ca topdressed over plots of a naturalized pasture composed of *Paspalum notatum, Axonopus compressus, Sporobolus* spp. and traces of *Desmodium* spp. A pronounced response to applied P occurred in terms of increased herbage yields and higher percentage of legume. The data given in Table 9.10 are averages of 19 harvests over a 5-year period when N was topdressed after every cutting and P applied initially and after every fifth cutting. At the end of the experiment *P. notatum* was the predominant grass on plots that had received N and P.

An increase in naturally occurring legumes, such as *Tephrosia, Desmodium, Rhynchosia, Dolichos* and *Stylosanthes*, was observed with P application to natural pastures on sandy soils of the Tanzanian coast (Schmidt and Watkins, 1968). Phosphate alone resulted in total herbage yields of 2 880 kg/ha per year as compared to 4 230 kg where all nutrients were supplied. After 18 months there was no change in the grass cover, which remained predominantly *Digitaria mombasana* and *Heteropogon contortus*. The increased

Table 9.10 *Fertilization of a naturalized pasture located at 1 800 m in Colombia (Lotero et al., 1965)*[*]

Limestone and N		Dry matter yields (kg/ha)[†]		
		P_0	P_{150}	P_{300}
Ca_0	N_0	220	450	500
	N_{50}	600	1 415	1 735
Ca_4	N_0	495	505	870
	N_{50}	1 295	1 555	1 755

[*] Pasture initially consisted of *Paspalum notatum, Axonopus compressus, Sporobolus* spp. and *Desmodium* spp.; Ca = limestone at nil and 4.0 t/ha topdressed 2 months prior to other treatments; N = nil and 50 kg/ha applied as urea after every cutting; P = nil, 150 and 300 kg/ha applied as triple superphosphate initially and after every fifth cutting.

[†] Forage cut and removed from plots; yields significantly different for treatments at cutting dates and totals.

legume component was considered of doubtful value, since most native types were not relished by cattle and contributed little to the total production of forage. This indicated, however, that improved-type legumes could be established with correction of soil nutrients. It also appeared likely that superphosphate might encourage higher producing grasses since *Panicum maximum* increased on plots receiving this nutrient. In a subsequent study with application of 100 kg/ha of triple superphosphate (44% P_2O_5) Hendy (1975), however, found no increase in animal production and no improvement in botanical composition of naturalized pastures. In northwest Queensland, native grazingland consisting of *Dichanthium fecundum, Eulalia fulva, Astrebla* spp., *Iseilema* spp. and the native legume *Rhynchosia minima* did not respond to applied monosodium phosphate at rates up to 100 kg/ha of P (Bishop, 1977). Application of 254 kg/ha of N increased yields of forage by 56 per cent in the first year and by 22 per cent in the second year. No interaction of N and P occurred. Phosphorus content of herbage increased from 0.8 per cent in the control to 0.18, 0.22 and 0.24 per cent in treatments receiving 25, 50 and 100 kg/ha of P. This influence of P persisted into the third year. Norman (1962) also noted an increase in herbage P with increasing levels of applied phosphate to native pasture in the Northern Territory. The change, however, only represented an increase from 0.023 to 0.072 per cent P; thus, the general level of P was extremely low. Tothill (1974), in southeast Queensland, found that application of 48 kg/ha of molybdenized P and 138 kg of N did not alter the composition of *Heteropogon contortus* in a natural pasture, but *Themeda australis* declined rapidly and was replaced by less desirable grass species.

These data suggest that native grazinglands in general have evolved under a survival mechanism of low nutrient requirement. Grasses and legumes do not respond to added P, or only slightly, in terms of forage productivity, but grasses sometimes respond to added N. It would thus appear that such grassland vegetation is best suited to an extensive, open-range grazing system without use of fertilizer. Low phosphate content of herbage can be more economically corrected by mineral supplementation to animals.

Critical percentage of phosphorus in herbage of grasses and legumes

Critical concentration of a plant nutrient is described as that content above which forage yield response to further fertilization of that nutrient is doubtful. Andrew and Robins (1969a, 1971) and Johansen *et al.* (1980) determined the critical concentrations of P in the tops of several grasses and legumes (Table 9.11) by growing plants of cultivars or selections of each species in pots, adding different increments of P, sampling just prior to flowering and analysing the herbage for P content. Results of several species were verified in the

Table 9.11 *Critical percentages of P (dry matter basis) in the tops of grass and legume species*

Grasses	% P	Legumes	% P
Cenchrus ciliaris	0.26	Centrosema pubescens	0.16
Chloris gayana	0.23	Desmodium intortum	0.22
Cynodon dactylon*	0.14	D. uncinatum	0.23
C. nlemfuensis[†]	0.22	Glycine. wightii	0.23
Digitaria decumbens	0.16	Lotononis bainesii	0.17
Melinis minutiflora	0.18	Macroptilium atropurpureum	0.24
Paspalum dilatatum	0.25	M. lathyroides	0.20
Pennisetum clandestinum	0.22	Medicago sativa	0.24
Panicum maximum	0.19	Stylosanthes guianensis[‡]	0.16
Sorghum almum	0.20	S. humilis	0.17
Setaria anceps	0.22	Vigna luteola	0.25

* Data from Jordan *et al.* (1966);
[†] Saleem (1972).
[‡] Bruce (1974); all others from Andrew and Robins (1969a, 1971).

field. The critical concentration was assumed to lie at the point where the curve began to change decidedly from almost linear to curvilinear in terms of relative yield and P content of herbage (Fig. 9.11). The P concentration in dry matter of plant tops for all species increased with added amounts of this nutrient. In the legumes study of Andrew and Robins (1969a) all species had approximately equal percentages of P in their tops (0.30%) with the highest nutrient treatment, except for *Lotononis bainesii*, *Medicago sativa* and *Vigna luteola*. These three species attained levels of 0.35 per cent or greater. The highest P treatment of grasses gave concentrations of about 0.25 per cent for *Digitaria decumbens* and *Sorghum almum*, 0.46 per cent for *Setaria anceps* and ranges between these values for the others (Andrew and Robins, 1971). In each instance, dry matter yields increased as the percentage of P in the herbage increased, and the critical concentrations were related to near maximal production. Some legume species such as *Stylosanthes guianensis*, *S. humilis* and *L. bainesii* produced maximum herbage at relatively low applications of P and at lower plant P concentrations than others, such as *Macroptilium atropurpureum* and *Medicago sativa*. The superiority of *S. humilis* to mobilize P and perform well when growing on soils with low available phosphate was previously demonstrated (Andrew, 1966; Humphreys, 1967; Moody and Edwards, 1978). The lower P-requiring legumes originated in regions with soils low in P content, and those having higher P critical levels originated in regions with soils high in available P. A comparable situation exists among the grasses. *Cynodon dactylon* and *Melinis minutiflora* are found growing naturally on acid soils with low P availability, whereas *Setaria anceps*, *Chloris gayana* and *Cenchrus ciliaris* occur on soils having higher contents of P.

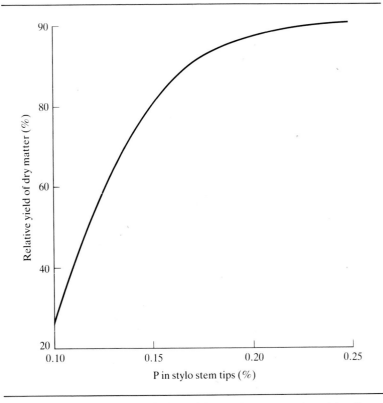

Fig. 9.11   Relationship between herbage yields and P content at time of first flowering (Bruce, 1974).

Potassium fertilization

The influence of K on productivity of pasture species or cultivars varies with the requirement of this nutrient for plant growth, ability of plants to extract the nutrient from the soil, P fertilization and in the case of grasses N fertilization. Most soils supply rather large and widely varying quantities of K during the first 2 or 3 years after pasture establishment. By the third or fourth year soil K may decline, especially when the herbage is removed and the nutrient supply not replenished. This is more likely to occur with grasses than with legumes. Since about 80 per cent of K consumed by animals is returned to the soil via excreta, the need for maintenance application will be considerably lower under grazing than under cutting conditions. On soils inherently deficient in K, applications of this nutrient will be needed for establishment and maintenance of grasses and legumes.

In Puerto Rico, Vicente-Chandler *et al.* (1974) found that high-yielding grasses, when cut and forage removed, utilized about 440 kg/ha of K on an annual basis. Some of the K uptake probably represented luxury consumption so that actual requirements were lower. They observed significant increases in dry matter yields of such grasses as elephant, guinea, Pangola and giant *Cynodon* with K rates up to 220 and 400 kg/ha. On some soils, 880 and 1 320 kg enhanced herbage yields of elephant and giant *Cynodon*. The effect of applied K on giant *Cynodon* is shown by data in Table 9.12. The grass responded strongly to 1 320 kg of applied K on unlimed and to 440 kg on limed soil. Apparently, reduced leaching of K from the limed soil decreased the need for K fertilization. It should be recalled that these dramatic results were obtained when herbage was cut and removed from the experimental area. Intensive management of Coastal bermudagrass also appeared to require from 250 to 400 kg/ha of applied N to maintain adequate exchangeable K in the soil (40–50 ppm) and produce maximal dry matter yields (Jordan *et al.*, 1966; Woodhouse, 1968). On soils having higher inherent levels of K, and with less intensive management of grasses, the response to K will be less pronounced (Crowder *et al.*, 1963; Saleem, 1972).

It was noted by Clement and Hopper (1969) that grasses absorb more K than is required for maximum forage production with a plentiful supply of this nutrient in the soil. This luxury consumption is a feature common to other crops, but becomes of greater importance in grasses due to periodic removal of immature herbage. Thus, redistribution of excess K for further utilization within the plant does not occur. Luxury consumption can be reduced to some extent by making split applications of the K fertilizer. The amount of K uptake by a grass appears to be determined by the total forage yield, concentration of this element in the soil and application of other nutrients, especially N. Application of 880 kg/ha of N per year to *Pennisetum purpureum*, *Panicum maximum* and *Brachiaria mutica* in Puerto Rico resulted in the removal of four times the amount of

Table 9.12 *Influence of applied K and limestone on dry matter yields of Cynodon nlemfuensis in Puerto Rico (Vicente-Chandler et al., 1974)*[*]

K supplied annually (kg/ha)	Yields of grass (t/ha)	
	Unlimed	Limed[†]
0	5.39	9.74
400	17.61	24.61
1 200	29.15	29.68

[*] K supplied as potassium sulphate annually; 660 kg/ha per year of N as ammonium sulphate topdressed in six equal amounts; 110 kg of P added yearly; cuttings made at intervals of 60 days and herbage removed.
[†] Soil limed to 70 per cent base saturation.

applied K (Vicente-Chandler *et al.*, 1959). Removal of K from soil by Coastal bermudagrass exceeded the applied rates of 55 and 110 kg/ha, so that up to 220 kg/ha was required to make up the annual loss in an experiment lasting 11 years (Woodhouse, 1968). Saleem (1972) reported that *Cynodon nlemfuensis* removed more K than the 220 kg/ha of this element applied as muriate of potash. Insignificant reductions in soil-exchangeable and non-exchangeable K in the upper 25 cm soil layer suggested K utilization by roots at deeper depths. Coastal bermudagrass also extracted K from deep layers of sandy soil (Jackson *et al.*, 1959; Woodhouse, 1968). The recycling of K from lower depths favours utilization of this element previously leached from the above regions.

A K content of less than 1.0–1.5 per cent in 60-day-old grass in an intensive management system suggested a deficiency of this nutrient to Vicente-Chandler *et al.* (1974) in Puerto Rico. A higher content in the herbage indicated luxury consumption. Smith (1972) studied K uptake by cultivars of *Setaria anceps* when plants were grown in pots filled with a latosol-type soil and fertilized with different increments of potassium sulphate. Plant tops were cut at 3 cm above soil level 20 days after sowing and K treatments applied. A uniform harvest was then made after 32 days when plant tillers had six to seven expanded leaves. At harvest, all cultivars growing in pots with less than 40 kg/ha of K equivalent displayed visual K deficiency symptoms. Critical plant K levels were estimated (90% of maximum relative yield) as being 1.6 per cent for Narok cultivar, 1.8 for Nandi and 1.3 for Kazungula. It was pointed out, however, that these values may not be particularly useful in assessment of their status, since sodium (Na) can substitute for K in some plant biological processes. In addition, factors such as seasonal conditions, stage of plant growth, plant part, supply of other nutrients and competition among species when grown in associations may affect K levels in grasses.

The response of tropical legumes to applied K fertilizers has been variable (Bryan and Evans, 1973; Jones and Freitas, 1970; Teitzel, 1969). Andrew and Robins (1969b) found that several species of legumes growing in pots responded in growth up to an equivalent of 220 kg/ha of applied K as potassium chloride. The optimum treatment for most species (*Centrosema pubescens, Glycine wightii, Lotononis bainesii, Macroptilium lathyroides, M. atropurpureum, Medicago sativa, Stylosanthes humilis*) appeared to be about 110 kg/ha. Dry matter yields of *Desmodium intortum* and *D. uncinatum* were depressed with application 220 kg/ha or more of KCl. Addition of 27.5 kg/ha had little effect on K concentration in the herbage, even though yields increased. Thereafter K percentage rose with increasing amounts of this nutrient. Ranges were measured from 0.30 per cent K in the herbage of some species in the nil treatment to more than 3.0 per cent with addition of 300 kg/ha of KCl. It was estimated that the critical percentages ranged from about 0.6 in *S. humilis* to 1.2 per cent in *M. sativa*.

Legumes may be at a competitive disadvantage for soil K when grown in association with certain grasses, since the latter appear to have a greater absorptive capacity for this nutrient. Lucerne is particularly sensitive to low levels of soil K, and quickly displays symptoms of a deficiency when combined with cool-season grasses at the higher elevations and with warm-season grasses at the lower elevations of the tropics (personal observations).

Effects of lime on herbage production

Limestone application to pastures in the tropics should be considered as a source of nutrient Ca rather than a soil amendment. Even the most acid soils contain sufficient Ca to meet the requirements of grasses for many years and most soils contain ample Ca for tropical and subtropical legumes. If lime application is needed for maximum productivity of legumes it is in modest amounts as illustrated by studies on an Oxisol in the eastern plains of Colombia by Spain *et al.* (1975). Their data indicated that only 150 kg/ha of lime as $CaCO_3$ was needed to give nearly maximum herbage yields of *Centrosema pubescens, Desmodium intortum, Pueraria phaseoloides* and *Stylosanthes guianensis*. The first response, as depicted by Fig. 9.12, was probably to Ca; however, a benefit from Mg cannot be ruled out. A depressing effect occurred with 1 000 and 2 000 kg/

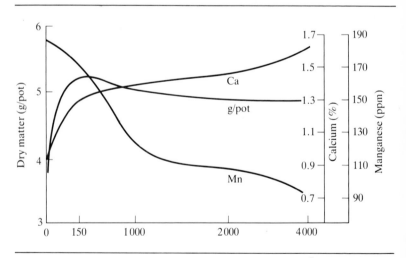

Fig. 9.12 Effect of applied lime on dry matter yields, Ca and Mn concentration, average of four tropical legumes in Colombia (Spain *et al.*, 1975).

ha of $CaCO_3$ equivalent with a slight benefit from 4 000 kg. In most trials, lime at 1 tonne/ha has been the first increment and it is likely that this exceeds the most beneficial range. For example, Olsen and Moe (1971) obtained no response in dry matter production of *D. intortum* and *S. guianensis* with 2.5 tonnes or more of applied lime.

Tropical pasture legumes are more efficient than their temperate counterparts in extracting Ca from the soil. At a Ca level of 0.1 m.e. per 100 g of soil (3% of total exchange capacity) in a soil of 5.5 pH, *Centrosema pubescens, Desmodium uncinatum, Indigofera spicata, Macroptilium lathyroides* and *Stylosanthes guianensis* did not show Ca deficiency and were well nodulated. *Medicago sativa, M. tribuloides, Trifolium repens* and *T. fragiferum*, in contrast, exhibited severe Ca deficiency symptoms and were poorly nodulated in Queensland (Andrew and Norris, 1961). The tropical forms showed little response to applied lime but the temperate forms showed a pronounced increase in herbage yields. Some tropical legumes such as *Leucaena leucocephala* (Wu, 1964) and to some extent *Glycine wightii* (Diatloff and Luck, 1972; Andrew and Hutton, 1974) do respond to lime application on acid soils.

High-yielding grasses under intensive cutting management in Puerto Rico responded strongly to applications of limestone (Vicente-Chandler *et al.*, 1974). Forage yields did not perceptibly change in the first year but almost doubled in the third and fourth years with addition of an initial 8.8 tonnes/ha of lime. The response to liming on the acidic soils was attributed to a marked reduction in exchangeable Al rather than an accompanying increase in soil Ca. Yields continued to rise beyond the point of an abundance of Ca for plant growth. Elephant grass roots on unlimed plots displayed arrested, coralloid-type growth, a typical symptom of severe Al toxicity. Applying lime increased the Ca content of all grasses. Elephant and Pangola contained about 0.40 per cent and guinea about 0.60 per cent of this nutrient. Liming decreased the manganese (Mn) content of soil and herbage but did not alter P or Mg contents. Spain *et al.* (1975) also noted a decline in manganese percentage of legumes and grasses as levels of lime increased in comparison to the nil treatment, but did not observe a change in dry matter production of *Brachiaria mutica, Hyparrhenia rufa* and *Melinis minutiflora* growing on a soil with ph 4.3.

In the practice of pasture fertilization, sufficient Ca as a plant nutrient for tropical grasses and legumes may be supplied in the form of phosphate fertilizers. Simple superphosphate contains about 20 per cent Ca, triple superphosphate about 15 per cent and basic slag 45–60 per cent.

Calcium can be transported from lower soil depths to upper layers by roots. In Nigeria, Saleem *et al.* (1975) grew 'Cynodon

IB.8' (*C. nlemfuensis*) in a sandy loam soil having a pH of 4.2. A layer 50 cm deep, contained in open-bottomed boxes 1.8 × 2.0 m, rested on a clean undisturbed surface of the same soil series but having a pH of 5.3. Lime at rates of 0.4, 0.8, 1.3 and 2.5 tonnes/ha were incorporated into replicated boxes for comparison with a nil treatment. Rooted cuttings of the grass were transplanted into the boxes, allowed to become established, a uniform cutting made and the stubble topdressed with 168 kg/ha of N as urea and 112 kg of K_2O as muriate of potash. This fertilizer was repeated after every third harvest taken at intervals of 5–6 weeks. Cuttings were made at 10 cm above ground level and herbage sampled for dry matter and chemical analyses. Irrigation provided at 3 cm weekly during the dry season maintained vigorous growth of the grass. Soil samples collected from the 0–23 cm layer within the boxes after every harvest were used to determine available P, Ca and pH. At the end of the experiment a pit dug around the edge of each box allowed the removal of soil samples from below at two levels of the undisturbed soil, i.e. 0–23 and 23–46 cm, for determination of pH. Forage yields were comparatively uniform during the first five harvests, being about 2.5 tonnes/ha per cutting. They diminished about 30 per cent in the sixth and another 15 per cent in the eighth, despite repeated N topdressing, then levelled off in the ninth and tenth cuttings. The decline may have been due to the removal of soil nutrients, namely P, as it was found that '*Cynodon* IB.8' requires about 0.22 per cent P in herbage for optimum production. Also, repeated cutting at 10 cm above ground level, with removal of herbage and lack of a rest period, probably did not allow accumulation of carbohydrate reserves and other plant nutrients. The optimum cutting height of this grass should be about 18 cm (Chheda and Saleem, 1972).

Liming increased availability of P and Ca in the experimental soil and the percentage of these in the herbage for the first five harvests. Both declined in subsequent harvests, probably due to initial utilization by the grass, removal by herbage and fixation in the soil. A steady rise in pH from 4.2 occurred in the 0–23 cm experimental soil layer of the nil lime, 0.4 and 0.8 tonne/ha treatments. By the end of 11 months the pH had approached the original value of 6.2 at about the sixth harvest where 0.8 and 1.3 tonnes/ha of lime were added. pH values of the nil and 0.4 tonne/ha treatments were 5.4 and 5.8, respectively. A decline of pH in the undisturbed soil occurred below the boxes of these treatments (Table 9.13). A gain of Ca in the soil within the boxes suggested translocation of Ca ions from the lower soil horizons of the undisturbed soil through the deeper-feeding grass roots. Simpson (1961) also noted an increase in pH in soils covered with *Chloris gayana* and postulated that a transfer of cations to the surface soil had occurred through roots of the grass

Table 9.13 *Effect of lime on pH of undisturbed soil below each treatment and total Ca in experimental soil within boxes containing 'Cynodon IB.8' (Saleem et al., 1975)**

Limestone (t/ha)	pH of soil below boxes[†]		Total Ca within boxes	
	0–23 (cm)	23–46 (cm)	Beginning (ppm)	End (ppm)
nil	4.6	5.2	1 350	1 420
0.4	4.8	5.4	1 460	1 500
0.8	4.9	5.9	1 580	1 650
1.3	5.6	5.9	1 730	1 660
2.5	6.0	6.1	1 980	1 680

* Experimental soil had an initial pH of 6.2 after forest clearing but this had dropped to 4.2 at the time of beginning the experiment due to continued application of ammonium sulphate; a 50 cm layer was enclosed in open-bottomed boxes and rested on undisturbed soil of the same type but not heavily cropped.
† Initial pH of undisturbed soil 5.3 in 0 to 23 and 5.8 in 23–46 cm layers, respectively.

and deposit of plant residue. The mobilization of P and K from lower soil layers by deeply penetrating roots and their accumulation in the topsoil after decomposition of plant organic matter has been observed in *Pennisetum purpureum* and other grasses (Vine, 1968).

'Cynodon IB.8' growing on soils subjected to several years' cropping produced 10–12 tonnes/ha of forage (Chheda, 1974). The herbage contained an average of 1.22 per cent N, 0.25 per cent P, 1.74 per cent K, 0.37 per cent Ca and 0.34 per cent Mg, thus mobilizing over 120 kg N, 26 kg P, 170 kg K, 36 kg Ca and 28 kg Mg, per hectare annually. Another 30 per cent probably occurred in roots, stolons and stubble. In the present experiment approximately 20 tonnes/ha of aerial dry matter were produced under irrigation over the period of 1 year. About 14 per cent of 'Cynodon IB.8' roots occur in the soil layers below 30 cm with some reaching depths of 1.75 m (Mackenzie and Chheda, 1970). Using P_{32}, Nye and Foster (1961) found that a grass fallow derived 30 per cent of its P from soil layers below 25 cm in which 19 per cent of roots existed. Thus, a conservative estimate of 20 per cent nutrient uptake by 'Cynodon IB.8', and using 10 tonnes/ha dry matter production, the following quantities of nutrients would be 'pumped-up' from soil layers below 30 cm: N, 24, P, 5.0; K, 35, Ca, 7.3 and Mg 5.5 kg/ha. These figures compare favourably with those obtained by Nye and Greenland (1960) in a forest fallow. This upward translocation of nutrients can be an important factor counterbalancing losses from topsoil by leaching.

Sulphur fertilization

S-deficient soils generally exhibit one of the following properties: high in allophane or oxides, low in organic matter and often sandy

(Sanchez, 1976). Soils may be deficient in S where annual burning occurs, since about 75 per cent of this nutrient is volatilized by fire. Field response to applied S was demonstrated by Anderson and Spencer (1950) in subterranean clover growing in Australia. Deficient plants had leaves that were pale green to yellow and stunted, even though inoculated with an effective strain of *Rhizobium*. The nodules were fewer in number and smaller than on healthy plants. S-deficiency symptoms resembled those of N deficiency, but addition of N fertilizer did not improve plant colour or growth. The effect of S deficiency was on the host legume in which protein synthesis was modified. Since the plant demand for N was decreased nodulation was indirectly affected.

Stylosanthes humilis plants growing in a S-deficient soil near Townsville, Queensland, displayed yellow colouring while those receiving 4.5 kg/ha of S as anhydrous $NaSO_4$ grew normally (Jones and Robinson, 1970). Maximum dry matter yields occurred with this rate of S, being about 4.4 tonnes/ha as compared to 3.8 tonnes produced from the nil-S treatment. Increasing S to 12 kg/ha gave higher N contents of herbage. Apparently, fixation of N by bacteria in nodules is not impaired as much as synthesis of protein. It is likely that small quantities of S satisfy the requirements of the legume and enable the rhizobia to fix sufficient N for the plant to attain maximum herbage yield permitted by environmental factors. With additional S the bacteria are stimulated to fix more N, thus increasing the N content and crude protein yield of the plant but not dry matter yield. A striking feature of a complementary glasshouse study with nil and 34 kg/ha of S was the redistribution pattern of S within the plant. In the S-treated mature plants 60 per cent of the total S absorbed had been translocated to the seed heads. In contrast, S-deficient plants absorbed about one-ninth the amount of S and translocated only 25 per cent to the seed heads, more than 50 per cent remaining in the roots. A later study showed that mature leaves contained 0.10 per cent S while seeds contained 0.36 per cent (Robinson and Jones, 1972). A tentative critical level of 0.10 per cent S concentration at full flowering was proposed. Jones and Quagliato (1973) observed improved plant growth of *S. guianensis* immediately after establishment in soils of the Campo Cerrado of Brazil and of *Centrosema pubescens* and *Glycine wightii* several months after establishment.

In São Paulo State of Brazil, certain pastures of *Paspalum notatum* in a grazing trial received 250 kg/ha of N in split applications. Other received an equal amount of N plus 200 kg/ha of P_2O_5 from ordinary superphosphate applied when the trial was established (McClung and Quinn, 1959). An initial response to N occurred on all pastures, but after about 6 months the N-treated paddocks began to deteriorate in terms of herbage production. At the end of 1 year their condition resembled that of the unfertilized pastures and in another 6 months, after additional N treatment (total of 450 kg/ha), plants displayed extreme chlorosis and some had died. Plants on the P-treated pastures grew normally with a continued response to applied N. Small plots (0.9 × 6.0 m) within the N-treated pastures were then topdressed with 40 kg/ha of S as $CaSO_4$, $NaSO_4$ and elemental S. Improvement in grass colour was observed within 1 week after the sulphates were applied and within 2 weeks plants returned to a normal green colour. Dry matter yields 26 days after sulphate application doubled and in 91 days tripled those of the nil S treatment. The effect of elemental S was delayed but by 91 days equalled that of the sulphates. Thus, with intensification of grass-based pastures in the tropics and subtropics monitoring of S is important and is of even greater consequence in developing grass–legume pastures.

Micronutrients and tropical pasture species

As fertilizer inputs increase for intensive pasture production, deficiencies of micronutrients are more likely to occur (Bruce, 1978). Use of higher-analysis fertilizers will tend to accentuate these deficiencies. Furthermore, deficiencies are more apt to appear as greater numbers of different types of legumes and grasses having higher nutrient requirements are introduced.

A classic example of a micronutrient deficiency is the widespread need for molybdenum (Mo) in Australia. Mo is essential in the formation of nitrate reductase and plays an important role in the symbiotic fixation of N (Anderson, 1956). Its concentration in the legume nodule is about six times that of the whole plant (Jensen and Betty, 1943). Mo deficiency resembles N starvation and plant symptoms are likely to occur when legumes grow on soils low in available N. The micronutrient becomes less available with increasing soil acidity. Liming often releases sufficient Mo to meet plant requirements (Williams and Andrew, 1970). In fact, some reports of pasture response to lime were due to increased availability of Mo. For example, Watson (1960) reported that liming a soil in Malaysia to raise the pH from 5.0 to 6.0 resulted in increased herbage yields of centro and kudzu. The same effect was also obtained by applying sodium molybdate. Mo responses have been obtained with all of the tropical and subtropical legumes used in northern Australia, both in establishment and maintenance (Mannetje *et al.*, 1963; Swain, 1959; Truong *et al.*, 1967; Williams and Andrew, 1970). A large percentage of the phosphate applied to pastures in Australia has been molybdenized so as to apply about 1 kg/ha of Mo to legume-based pastures. On certain soils reapplication of this micronutrient on a regular basis is needed to maintain optimal herbage production.

There is evidence that some species require higher amounts or more frequent applications than others, e.g. *Glycine wightii* (Luck and Douglas, 1966).

Vicente-Chandler *et al.* (1974) reported that high-yielding grasses can take up about 75 kg/ha of Mg and losses of this nutrient can occur by leaching. Heavy applications of K tend to suppress the absorption of Mg, thus increasing its need by grasses. They found a slight response by Pangola grass to applied Mg in terms of forage production on some soils after 4 years of intensive cropping. The herbage contained about 0.23 per cent Mg. Elephant grass did not respond and contained 0.25 per cent Mg in 60-day-old grass. This was assumed to be a critical concentration for growth and development. Mg requirements of most grasses will generally be met through normal liming practices on soils where limestone is needed.

Boron (B) deficiency was noted in lucerne growing on certain soils in Colombia (Crowder *et al.*, 1959) but was corrected with application of 25–30 kg/ha of borax. Andrew and Bryan (1955) found B to be marginally deficient in the coastal sand regions of Queensland. Cobalt deficiency was recognized by Mannetje *et al.* (1976) in Malaysia and led to low animal production.

Aluminium (Al) toxicity is likely to occur in tropical soils with a pH of 4.5 or less. The effect on the plant is largely through root damage with restriction of terminal elongation and branching, along with diminished nodulation. Legume species show a striking difference in tolerance of Al in the soil solution. Andrew and Vanden Berg (1973) found that *Stylosanthes humilis* and *Desmodium intortum* were relatively insensitive to levels as high as 2 ppm of Al in the soil solution. *Glycine wightii* was highly sensitive to low levels and lucerne plants began to die at 1 ppm of Al.

Manganese toxicity is sometimes compounded with Al toxicity since both ions are progressively released with increasing soil acidity. Some species of tropical legumes are sensitive to high levels of Mn, thus limiting their suitability to certain soils. Andrew and Hegarty (1969) examined Mn toxicity associated with yields of several legumes in Queensland. *Centrosema pubescens* and *Stylosanthes humilis* were relatively insensitive, showing about a 5 per cent reduction in yield with 1 600 and 1 140 ppm, respectively. Increasing Mn to 40 ppm decreased production by 30 per cent in centro and 50 per cent in Townsville stylo. Lucerne, Glycine and Siratro were more sensitive, exhibiting a toxicity threshold of 400–800 ppm and a 5 per cent yield decline. Production of herbage dropped to 15 per cent or less of this threshold with 40 ppm of Mn. Silverleaf desmodium and Phasey bean fell into an intermediate category. The detrimental effect of Mn can be offset by liming (Fig. 9.12).

Cutting management

The interval between harvests of grasses and legumes profoundly affects herbage production, nutritive value, regrowth potential, botanical composition and species survival. In general, an extended period between cuttings has the following effects:

1. An increase in the percentage content of dry matter, crude fibre, lignin and cell wall. With increasing plant maturity older leaves contain increased cell wall constituents and thus reduced intercellular space as well as condensed cellular inclusions. In addition, midribs and leaf sheaths attain a greater percentage of fibre and lignin, older leaves senesce and lose water, stems elongate and are less succulent.

2. An increase, then a decrease, or fluctuation, in total dry matter production and nitrogen-free extract. An initial increase in yield occurs due to greater accumulation of total photosynthate, but as leaves mature and senesce leaching from them occurs along with translocation of photosynthate from actively photosynthesizing leaves to the senesced leaves.

3. A decrease in leaf : stem ratio, percentages of crude protein (CP), mineral constituents (P, K, Ca, Mg), and soluble carbohydrates. With stem elongation the percentage of leaf declines. This brings about a relative decrease in cellular inclusions since cell walls of older leaves and of stems thicken. A part of the decrease in CP and minerals is brought about by a dilution effect because of increased dry matter yield.

4. An increase, and then a decrease in the amount of N uptake by the plant and N recovery. More frequent cutting stimulates plant development and sustains biological processes, thus a greater demand for N. As plants mature, these activities decline. Also, a dilution effect of N occurs with increased forage production.

5. A rapid decline in animal intake and digestibility. More mature herbage is less nutritive and thus less appealing to the grazing animal than juvenile and nutritious material. Digestibility diminishes as cell wall structure increases so that passage through the animal is slowed down.

Increased dry matter yields with extended cutting intervals are consequences of additional tiller and leaf formation, leaf elongation and stem development (Akinola *et al.*, 1971a; Dovrat *et al.*, 1971; Fisher, 1973; Michelin *et al.*, 1968; Robertson *et al.*, 1976). The period of maximum forage production varies with different grass species (Haggar, 1970; Nourrissat, 1965). More nutritious herbage is obtained with reduced cutting intervals so that an optimal harvest scheme must be chosen to obtain a balance between quality and increased herbage yield. Cutting too frequently reduces total forage yields, depletes carbohydrate reserves, causes a decline in root development, favours weed invasion and adversely affects regrowth potential (Borgét, 1969; Chheda and Akinola, 1971b; Akinola *et al.*, 1971b; Jones, 1973; Perez and Lucas, 1974; Rains, 1963). At Ibadan, Nigeria, '*Cynodon* IB.8' developed more roots and achieved deeper soil penetration under a 6-week than a 4-week cutting scheme (Mackenzie and Chheda, 1970).

Cutting near ground level increases total and seasonal forage production over a short period as compared to more elevated cutting heights, but plants are adversely affected in the same way as too-frequent harvesting. Over time plants become weakened, stands thin out, weeds invade and spots of bare ground between plants lead to soil erosion. A cutting height of 15–18 cm above ground level in Nigeria was needed to maintain stands of *Andropogon gayanus* (Rains, 1963) and '*Cynodon* IB.8' (Chheda, 1974) but a *Cynodon*–centro mixture performed equally with 5, 10 and 15 cm cutting heights (Moore, 1965). Grass–legume combinations are difficult to maintain under cutting management, however, because their growth habits differ and patterns of plant growth and development do not coincide.

Height and frequency of cutting studies provide leads to agronomic practices that give some insight relative to grazing intensity. They cannot substitute for animal experimentation. It is likely that pasture agronomists have spent more time and effort than necessary in carrying out cutting management experiments. This is a more economical means, however, of evaluating grasses and legumes alone and in combinations on an extensive scale before more exhaustive animal studies.

Irrigation

Herbage and fodder production and distribution in the tropics and subtropics closely follow the seasonal rainfall pattern. Use of supplemental water for sustained plant growth during periods of water stress during the rainy season and throughout the dry season has been suggested as a means of improving animal output. From 2 to 3 cm of water as precipitation or as irrigation is needed weekly for continuous forage production. Limited studies indicate that an uninterrupted supply of herbage can be maintained with added water, especially when N fertilizers are topdressed (Barnes, 1970; Crowder *et al.*, 1960; Kinch and Pipperton, 1962; Rivera-Brenes *et al.*, 1961; Vasquez, 1965; Osman, 1979). Fritz (1971) found that *Chloris gayana* yielded up to 40 tonnes/ha per year under irrigation in Réunion as compared to 25 tonnes under natural rainfall conditions of about 1 100 mm. At Tarna (Niger) *Pennisetum purpureum* with 2 132 mm per year of water (rainfall plus irrigation) and 600 kg/ha of N yielded 340 tonnes/ha of fresh material when cut seven times (Borgét, 1971). *Macroptilium atropurpureum*, receiving the same amount of water, produced 100 tonnes/ha per year of fresh herbage. Production of both was about double that expected under normal rainfall conditions. Few comparative studies of animal output with and without irrigation are available. Irrigated Pangola grass pasture that was topdressed at frequent intervals with N and rotationally grazed provided 2 760 kg/ha of liveweight gain at the Parada Experiment Station in northeast Queensland (Norman, 1974). This is more than three times that expected without irrigation. It is unlikely, however, that the herbage yields and animal output of the livestock farmer will approach those of experimental conditions. Under intensive management, however, high levels of output can be expected. For example, a dairyman near Medellin, Colombia (1 500 m elevation), grew elephant grass and practised soiling, made cuts every 30–40 days and harvested 40–50 tonnes/ha of fresh material each time (Crowder, 1977). After every second cutting he passed a horse-drawn cultivator between the rows of grass to incorporate manure and stable refuse returned to the land and to open channels for irrigation. He did not recognize that the operation also enhanced aeration of the soil. Washings from the milking stalls were collected in a pit and returned to the grass planting. Supplemental water was provided by gravity flow during periods of moisture stress. Annual dry matter yields were calculated at about 76 tonnes/ha, an amount sufficient to maintain 12 milking-grade Holsteins that were supplemented with 1 kg of concentrate per 4 kg of milk. There is a need for studies in the subtropics and tropics to provide a quantification of efficient water use and economic returns.

Recycling nutrients from dung and urine spots

Studies in the temperate zone indicate that mature cattle produce about 25 kg of faeces and 10 kg of urine per animal daily. On the average, fresh cattle faeces contains 0.38 per cent N, 0.18 per cent P_2O_5, 0.22 per cent K_2O and fresh urine 1.10 per cent N, 0.01 per cent P_2O_5 and 1.15 per cent K_2O (Salter and Sollenberger, 1939; Doak, 1952). Cattle produce an average of 12 defecations and 8 urinations per animal daily. Each defecation covers about 0.093 m^2 (1.0 ft^2) and each urination 0.279 m^2 (3 ft^2) (Peterson *et al.*, 1956a). Frame (1970) showed that dung spots have the equivalent of over 112 kg/ha of P and 224 kg of K. Urine spots contain about 336 kg/

ha of N and 560 kg of K. Even though a portion of the deposited N and K may be lost by volatilization, leaching and other means, 50 per cent or more is recycled in the production of forage.

Deposit of the excreta by grazing animals results in a distribution over the pasture unlike that obtained by uniform mechanical distribution of fertilizers (Peterson *et al.*, 1956a). Loss of nutrients occurs from the time of deposition of the excreta through volatilization, leaching, surface runoff of water, absorption by plants, etc. At some point in time the supply of a given nutrient being added to the soil by dung or urine will be balanced by that lost from the root zone. Lotero *et al.* (1966) found that this equilibrium condition for N and K of urine spots was reached within 130–200 days during the spring and summer in North Carolina. They suggested that this rapid depletion rate limited the proportion of the pasture being influenced by urine spots to about 15 per cent of the total area. In essence, this has an effect of doubling or tripling herbage production in a limited area of the pasture and leaving a high proportion unaffected. Thus, urine deposit is not an effective means of maintaining the fertility status of the overall pasture. In a system of intensive pasture management with high stocking rates, however, the excretal return of such elements as P and K, which are not too rapidly lost from the root zone, should provide a substantial contribution to the fertility of the pasture. Several factors affect the influence that nutrients applied in the excreta will have on the fertility status of the soil. Peterson *et al.* (1956b) identified the principal factors to be (1) stocking rate which determines excretal density, (2) loss rate of a given element from the root zone and (3) effective area per excretion.

Nutrient loss from dung spots proceeds at a slower rate than from urine spots, but up to 80 per cent of the N may dissipate from dung lying on the soil surface in a warm climate (Gillard, 1967). Persistence of droppings can lead to fouling, since animals avoid grazing in the immediate vicinity of the faeces. Some attention has been given to mechanical scattering of droppings in order to reduce the concentration effect and to distribute the nutrients over a wider soil area. To be effective, however, this operation would need to be performed at rather frequent intervals.

In Asia and Africa coprophagous insects evolved along with cattle, so that dung spots do not accumulate in the pasture or on the soil surface. When dung-feeding insects attack a dropping, they bury most of the dung within 48 h. The remainder is scattered over the soil, so that it is exposed to the air. The buried portion is formed into balls in which the female beetle lays an egg. There a larva develops, pupates and the adult beetle emerges to dig its way to the soil surface in search of more dung. Insects that feed on cattle dung are not found as part of the natural fauna in South America and Australia. In the latter, indigenous beetles feed on the pellet-like droppings of native marsupials but they cannot cope with the large, wet droppings of introduced domestic cattle. Since 1967 African and Asian beetles have been introduced into Australia with three beneficial effects (1) they free pastures from dung accumulation, (2) they fertilize the soil and (3) they help control pests (Bornemissza, 1960; Gillard, 1967; Rural Research in CSIRO, 1972). Several species became well established, but *Onthophagus gazella* spread more rapidly and has been more active. Within 2 years it had colonized 400 km of northern Queensland around Townsville and penetrated 80 km inland, closing the 80 km gaps between release sites. Without the beetles, cattle droppings in Australia often last several months, or even a year, and cover considerable pasture area. With rapid multiplication of the beetles practically all droppings are destroyed in the colonized area. The insects act as effective agents in dispersing nutrients by burying dung and spreading it over the soil surface, thus improving soil fertility (Bornemissza and Williams, 1970).

The buffalo fly (*Haematoblia exigua*) was introduced into northern Australia from Timor in the 1820s when buffaloes were brought into the continent. The small, blood-sucking flies irritate cattle and interfere with resting and feeding. The cattle rub themselves to relieve the irritation, thus causing extensive sores over the shoulders, along the neck and underside and around the eyes. The fly breeds exclusively in buffalo and cattle dung. Introducing the dung beetle has been effective in reducing the number of larvae that survive in a dung spot. They cannot prevent the fly from laying eggs in the fresh droppings but burial of the dung effects some control. Larvae in the dung scattered over the soil are exposed to the sun and air so they are soon desiccated and die (Bryan, 1973). Disposal of dung by beetles is more rapid during the wet season and the buffalo fly nuisance appears to have abated in this period. Beetle activity slows down with the onset of the cool dry season so that dung disintegration diminishes. Fortunately, fly activity is also reduced during this part of the year. It is likely that other species of dung beetle that are more active under cooler conditions can be introduced.

About 1 year after releasing *Onthophagus gazella* around the Townsville area a thriving population of dung beetles was found on an island about 7 km from the coast. Movement was even more impressive in Hawaii where the beetles made the 50 km crossing between the islands of Maui and Hawaii. This provides evidence that these beneficial insects have considerable capacity for dispersal once they have colonized an area.

Weed control and mowing

The invasion of sown pastures and grazinglands by herbaceous and grassy weeds, shrubs, small trees and poisonous plants may pose a difficult problem for cattlemen and dairymen. If left unheeded, the undesirable plants frequently overrun the area so that the value of the land for grazing is completely destroyed. Pastures should be established with as few weeds possible and some means of keeping weedy plants under control should be employed at all times. Invasion by weeds and brush is always a sign of poor pasture management and usually results from inadequate fertilization and overgrazing the desirable species. Many of the weedy plants are highly adaptable to low soil fertility and readily invade and rapidly flourish under conditions of overgrazing. The most effective, efficient and cheapest means of keeping weeds out of established pastures is to follow a regular programme of fertilization and judicious grazing management practices. This promotes the growth and development of desirable vegetation in its competition with the undesirable types. Even with strict pasture and grazing management, however, some of the more aggressive weedy plants gain headway, making it necessary to impose some form of control and eradication.

It would be impossible to identify all of the weedy species in this discussion or even to describe a general group having a common occurrence. The pasture of agronomist and livestock manager should consult weed and pasture publications and weed specialists for details of local conditions. A few weeds will be discussed, however, because they are sufficiently widespread to deserve special mention and to illustrate some of the common difficulties encountered.

Imperata cylindrica (commonly known as swordgrass but also called lalang, alang-alang and cogon in parts of Asia, blady grass in Australia, spear-grass in Africa) is a strongly rhizomatous perennial, tolerant of burning and highly persistent on low fertility soils. It is prevalent under systems of shifting agriculture, and is an indicator plant of reduced soil fertility, showing the need for return to forest or bush fallow. In Southeast Asia the var. *major* is strongly invasive with tall-growing plants reaching a height of 2.0 m or more. A lower-growing type *I. cylindrica* var. *africana* occurs throughout Africa but is equally aggressive as the Asia species (Kasasian, 1971). In South America the var. *condensata* and in India the var. *latifolia* are less serious pests but invade fields of cash and food crops and mismanaged grazinglands. They are all easily recognized by the characteristic silky white panicles which are spike-like and stand out as brilliant plumes. The roots of *Imperata* penetrate deeply into the soil and their production plus the rhizomes may be double that of the above-ground plant parts, making it difficult to eradicate. Whenever present, the grass becomes a difficult problem during the establishment and development of newly sown pastures. Deep turning and rotary cultivation repeated five or six times before sowing the improved pasture grasses and legumes suppress the regeneration of growth from rhizomes. Dalapon at 10–15 kg/ha and glyphosphate at 20 kg/ha give control for a time but repeated treatment during the year is needed to exhaust the regenerative capacity of the rhizomes (Ivens, 1971; 1980). The dosage of dalapon can be reduced by following it with paraquat after 3–4 weeks. A period of 6–8 weeks is needed for degradation of the dalapon before sowing the pasture species. Once the desired grasses and legumes become established and then are adequately fertilized and properly grazed the *Imperata* can be kept under control and should disappear. It is relatively intolerant of shading so that maintaining a dense sward helps to suppress the weedy grass. In the early vegetative stage of growth the young leaves are eaten by livestock but with maturity they become stiff and upright with rough, sharp edges and a long, tapering needle-like point. Oversowing with a legume, such as stylo, improves pasture quality and animal performance (Chadhokar, 1977).

Sporobolus pyramidalis (cat's-tail grass) is a densely tufted perennial with leaf blades having a long, fine point and producing tough, lignified culms, sometimes reaching 1.5 m in height. The grass occurs naturally along roadways and trails and is common in many African grazinglands. Seeds are abundant so that volunteer plants readily invade disturbed areas and sown pastures which are overgrazed or deteriorated because of low soil fertility. The plants are highly drought tolerant and rapidly crowd out the more desirable grazing species. In the early growing season cattle and sheep graze the young leaves but they soon become wiry and are avoided. A large number of woody flowering stems develop and are not accepted by livestock. Thus, many more seeds are produced and young seedlings quickly colonize bare, compact soil. The grass may be found as the dominant species or growing in association with weedy-type *Eragrostis* and *Aristida* spp. There are other *Sporobolus* species such as *S. capensis* and *S. smutsii* that can become troublesome weeds. Once they are firmly established the value of the pasture declines rapidly, so that rehabilitation consists of land-turning, fertilization and reseeding. Local patches may be dug out and reseeded with desirable pasture species. Invasion and spread into sown pastures can be controlled by management so as to maintain a vigorously growing cover of desirable grazing species.

A number of the *Paspalum* species are considered desirable pasture species, occurring naturally in tropical America and used in other areas of the tropics and subtropics. Several may become weedy pests, however, when grazinglands are mistreated. *Paspalum milligrana, P. urvillei* (Vasey grass) and *P. virgatum* are tufted perennials which grow prolifically when moisture is sufficient. Young growth of the latter two species is palatable and nutritious, but when plants are old, with coarse stalks, they are avoided by cattle. Even in the young stage of growth, *P. milligrana* is not acceptable. The grass clumps of the three species enlarge with increased tillering and new seedlings invade overgrazed areas and bare spots between the desirable pasture plants. All three are difficult to eradicate by grubbing alone. They can be controlled by spraying their base and leaves with a solution of 4 kg/ha of dalapon or 30 kg of Chlorea in 200 litres of water (Vicente-Chandler *et al.*, 1974). The latter is a mixture of sodium chlorate (40%), sodium metaborate (51%) and monuron (2.5%). Use of powdered Chlorea is also effective by dropping 25–50 gm on to each clump of grass. The treatment may need to be repeated 2 or 3 months later to finish off surviving tussocks. The pasture should not be grazed for 3 weeks after application of the herbicide.

Paspalum fasciculatum (bamboo grass) is a trailing weedy grass having large stems that readily regenerate growth when cut into sections. The creeping stems spread rapidly over a wide area with ascending stems reaching a height of 1.5 m or more. The dense foliage quickly suppresses development of all other species, leaving a dominant mass of coarse, rank and non-palatable herbage. Even the young tender leaves of the grasses are not usually eaten by cattle. It is more commonly found in low-lying wetter areas, making eradication and control difficult because of rapid regeneration of growth from the cane-like stems. Close cutting and removal of the herbage at the start of the dry season, followed by deep turning and repeated diskings, reduce the regenerative capacity of the stems. Even so, they are highly tolerant of desiccation. Close cutting and repeated applications of dalapon, TCA or sodium chlorate may be effective, but time must elapse before the land can be sown with pasture species. Growing a clean cultivated crop also suppresses the grass, after which sown pastures can be established.

Paspalum conjugatum (sourgrass) is common in most places throughout the tropics and is highly tolerant of shade. It spreads by seeds and vegetatively by stolons. The creeping stems form a dense sward and suppress the growth of other species. Cattle reject the herbage at all stages of growth and many seed heads develop. The seeds are washed by surface runoff into other sites where seedlings quickly establish and are strongly competitive with other grasses and legumes. This grass is fairly resistant to dalapon but can be controlled by use of amitrole followed by paraquat (Kasasian, 1971). It can also be controlled by turning the land at the beginning of the dry season, harrowing and fallowing, fertilizing and fitting for the establishment of improved pasture grasses and legumes with the coming of the rainy season. This *Paspalum* species is more easily kept under control by good pasture management than those discussed above. It thrives under conditions of low soil fertility, whereas they flourish on more fertile soils.

Cymbopogon species are common constituents of the more open grazinglands of East Africa. They are coarse, tussocked, perennial grass with rough and fibrous leaves growing to a height of 1.5 m with rank flowering culms reaching 2.0 m or more. *Cymbopogon afronardus* occurs in the Nyaza district of Kenya and the Ankole district of Uganda where its presence can reduce the value of grazingland to the point of being useless for cattle. Leaves of the grass possess a distinctive odour when crushed, coming from bitter aromatic oils. Cattle avoid the *Cymbopogon* when other grazing is available, so that a heavy mass of vegetation accumulates to a depth of 1.5–2.0 m. The semi-prostrate tillers and leaves of a single plant may spread over a circle of 2 m and smother all other vegetation. Following burning in the dry season, the tufts of *C. afronardus* quickly tiller and the leaves grow rapidly, producing a heavy foliage which suppresses the development of other grasses. Furthermore, many seeds have formed during the dry season and the vigorous seedlings of this grass readily colonize open spaces and bare ground. Cattle can be forced to eat the young leaves, so that mob-stocking to maintain the sward at no more than 15 cm height suppresses the *Cymbopogon* and allows an increase of more desirable grasses such as *Brachiaria decumbens* and *Panicum maximum* (Harrington and Thornton, 1969). Application of N also favours the frequency of occurrence and dominance of the more acceptable species. Hoeing out the *C. afronardus* is also an effective means of improving pastures infested with this species but strict control of burning and stocking rate must be practised to limit the spread of the *Cymbopogon*. Systems of management have not been developed to eradicate existing stands.

These are but a few of the weedy grasses that are present in native and naturalized grazingland and invade improved pastures due to mismanagement, primarily the lack of a regular maintenance fertilizer programme and overgrazing. Some are actually considered suitable for grazing and are used for pasturage under marginal conditions of livestock-farming.

In like manner, a large number of broad-leaved weeds present troublesome problems to the livestock husbandryman. Once established, they are difficult to eradicate, and the plants are unpalatable to livestock. Most have strong tap-roots with considerable nutrient reserve so that mowing has little effect on their survival. They are drought tolerant and highly competitive during periods of water stress. Even under unfavourable conditions seed-setting occurs, adding another dimension to their aggressiveness.

A common herbaceous weed of grasslands, pastures and waste lands is the group of pantropic *Sida* species. They are low-growing, usually twiggy, green-stemmed perennial herbs with simple, alternate leaves, toothed around the margin, having yellow flowers and well-developed tap-roots. *Sida alba, S. cordifolia* and *S. rhombifolia* occur throughout the tropics and other species are found locally. The tough, wiry stems are not eaten by livestock and over a period of time this weed becomes the dominating species. Mowing only gives temporary control, because of the deep, woody tap-roots. Young seedlings are killed by spraying with 2,4-D or MCPA, but older plants are not affected by these herbicides. Leaves of the *Sida* species are frequently yellow due to the mottling or mosaic effect of a virus.

Several of the *Triumfetta* species occur throughout the tropics and sometimes become severe invaders of grazinglands. *Triumfetta rhomboidea* is widespread around the globe and is common in East Africa, along with *T. flavescens* (Ivens, 1971). *Triumfetta lappula* occurs in pastures, along roadsides and in waste places of the tropical Americas (Cardenas *et al.*, 1972). Plants have a strong tap-root, erect stems, branched and pubescent leaves growing as much as 1.5 m, axillary clusters of small, yellow flowers, and small prickly fruits. Pre-emergence treatment with 2,4-D usually restricts seedling emergence, and foliar application of established plants is an effective control measure. Most species can be kept in check by occasional and controlled burning of the grassland.

Several members of the nightshade family (Solanaceae) are widely scattered over the tropics, generally inhabiting waste places but frequently invading pastures where they develop into dense herbaceous thickets that are avoided by cattle. *Solanum panduriforme* (Sodom Apple) is a shrubby perennial occurring in East Africa and is noted as being an important problem in several areas (Ivens, 1971). Mowing encourages its growth and spread from an extensive rhizome system. Plants have prickly stems and produce small, yellow, leathery fruits that resemble tomatoes. In Tanzania they are reportedly eaten in times of famine, but in Rhodesia (Zimbabwe) are considered poisonous when unripe. *Solanum nigrum* is a smaller-growing, non-spiny, cosmopolitan annual herb with clusters of black or orange berries. It is primarily a weed of arable and waste land,

but grows among a wide range of crops. The unripe fruits are poisonous and cattle have been affected by grazing the foliage. Both are
susceptible to 2,4-D and 2,4,5-T. Some regrowth of the *S. panduriforme* takes place after spraying but two applications give a high
degree of eradication.

Mimosa pudica (sensitive plant) is a diffusely spreading perennial
herb that may act as an annual in regions with 6 months or more of
drought. It flourishes in many tropical lowland areas and in medium
elevations, occurring in open fields, waste places along roadways,
becoming a serious weed in pastures. The plants are not generally
accepted by cattle since they possess prickly thorns and bristly hairs.
During long dry periods the plants turn brown and above-ground
parts may die, but the crowns are generally persistent and regenerate growth when rains occur. Seed pods are freely formed, being
dispersed to give rise to volunteer plants. *M. invisa* is a vigorous,
spreading annual or perennial with spiny stems, but less aggressive
than *M. pudica*. Effective control measures for these weedy types
have not been developed. Applicators of herbicides such as Tordon
suppress their growth but will also damage the associated grass.

The woody and semi-woody plants constitute a special and distinct group of woody or noxious plants. A large number of the truly
woody forms belong to the members of the native or naturalized
flora of the grazing region or of peripheral areas that are less suitable for grazing (King, 1966). Species of *Acacia* present a common
problem in grasslands throughout the tropics and subtropics. The
flat-topped shrubby trees of this genus are characteristic features of
the wooded savannas of Africa, particularly East Africa. Under intensive grazing their numbers increase because of reduced competition from other components of the grassland community. Dense
thickets develop and are difficult to clear. The trees sprout rapidly
from the base when cut and the thorny shoots near the ground interfere with grazing to a greater extent than the original trunk. Young
shoots are susceptible to fire but older trees are highly resistant to
burning. Repeated cutting keeps *Acacia* regrowth under control but
rarely gives a complete kill of the stump. Chemical control depends
on the species and stage of plant growth (Ivens, 1971). After clearing an area, application of 2,4,5-T, or one of the 2,4-D/2,4,5-T
mixtures marketed as bush-killers, in combination with diesel oil, to
the bark around the base of the trunk provides partial control. Repeated treatments are usually needed for effective kill. In Australia
the leguminous tree *Acacia harpophylla* (brigalow) predominates
over a wide area and the landscape is sometimes referred to as 'brigalow' lands. The tree is difficult to erradicate in development of
improved pastures because of the sprouts ('suckers') that develop

freely on roots more than 5 mm in diameter. The sprouts can be controlled by spraying with 2,4,5-T and shallow ploughing. Competition from sown pastures retards their development along with intermittent heavy stocking with sheep (Coaldrake, 1970).

The genus *Prosopsis* (mesquite) is a leguminous woody shrub or tree that provides considerable forage for browsing in the arid and semi-arid regions of the Americas and elsewhere. Species have also been used to stabilize windswept sand and as wood for fuel. *Prosopsis juliflora* and several related species are especially aggressive in invading abused rangeland. The leaves and pods are relished by cattle and many herds have survived during periods of severe drought because of the forage provided. The seeds are indigestible and are readily spread in the droppings, so the dense and near inpenetrable thickets develop. By shading and competition for the scarce soil moisture, growth of the valuable range grasses is restricted and frequently they are eliminated. The *Prosopsis* is controlled by spraying 2,4-D and 2,4,5-T in an emulsion of oil. Since vast areas are usually involved, aircraft applications of low gallonage are commonly made (King, 1966). Control of mesquite in Texas (Fisher and Quinn, 1959) increased gain in weight of yearling steers and the liveweight gain per hectare by an average of 18 per cent as compared to the check treatment.

Tabernaemontana fuchsiaefolia ('Leiteiro') is one of Brazil's most serious pasture pests with up to 2 500 trees per hectare. It is characterized by an extensive lateral root system capable of producing new sprouts. Plants are difficult to control effectively or economically by hand or mechanical methods. Studies by Quinn *et al.* (1956) showed that application of Karmex as a wettable powder or granules by hand equipment, or by airplane, killed mature trees as well as seed sprouts. This product has a slow, cumulative toxic action, the effect spreading laterally for more than 6 m through the large underground lateral roots. These researchers pointed out the critical factor of application time to obtain effective control. Treatments applied after a flush of new leaves following the advent of rains gave more positive results than during the dry, cool months. It was further noted than an adequate grass cover or seeds ready to germinate is important for regeneration of the grass sward and to provide competition against subsequent weed invasion.

Lantana camara is a spreading shrub with prickly stems, growing to more than 3 m. The leaves are aromatic and pubescent. Individual flowers are small but several are borne in a showy cluster, having a number of colour forms. The most troublesome type has pinkish purple flower heads with yellow centres. It is principally an inhabitant of waste land and abandoned fields but readily invades overgrazed pastures. There are several garden varieties used as ornamentals. *Lantana canescens* and *L. trifolia* are of similar growth habit but the latter is smaller without prickles. All produce large quantities of seed that are widely distributed by birds. Dense and thorny thickets develop where *Lantana* becomes established and little grass can survive. The leaves are poisonous to livestock. When a thicket is cleared a copious regrowth of suckers appear and, even though roots are dug out, numerous seedlings emerge. Herbicide treatment has been somewhat variable with overall spraying of the foliage causing dieback, though seldom preventing regrowth. Application of 2,4-D or 2,4,5-T in diesel oil to the basal bark generally kills the above-ground parts but repeated treatment is needed for control. Biological control methods have also been successful against *L. camara*. In Hawaii and Fiji dramatic results were achieved with the introduction of a lace-bug (*Teleonemia scrupulosa*) which eats the leaves, and a seed-destroying fly (*Ophiomyia lantanae*) (Ivens, 1971).

Opuntia species (cactus, prickly pear) originated in tropical America but are widely distributed over the tropics and subtropics, having been introduced as a hedge plant. Once the hedge is established, it spreads rapidly from the cut or broken fragments of stems that fall to the ground and readily take root. Within 3–4 years considerable areas of dense thicket develop and large expanses of grazingland can be covered. In fact, millions of hectares have become infested since the latter part of the nineteenth century. Dense prickly pear may comprise from 1 200 to 2 000 tonnes/ha of fresh plant material (King, 1966). *Opuntia* species are particularly susceptible to control by biological methods. Introduction of the moth *Cactoblastis cactorum* into Australia reduced vast stands to decayed pulp in the late 1920s and early 1930s. The nocturnal moths are free-flying and deposit their eggs near the growing point of the cactus. The larvae are heavy feeders on the internal tissue of the thickened leaves, causing a breakdown and opening the way for invasion of pathogenic organisms that bring about final decay. The cochineal insect *Dactylopius tomentosus* has also been effective for biological control in some areas, but the insect does not spread rapidly so that periodic distribution will be needed to maintain control (Ivens, 1971). A spineless cactus has been developed and distributed to provide a source of livestock feed in arid and semi-arid regions (Semple, 1970). Efforts are being made to select types resistant to attacks of these insects.

Diseases and insects

Prevalence and severity of grass and legume pests have received little attention in the tropics and subtropics, even though the vegetation of pastures provides the media for development of a wide spectrum of pathogenic fungi, bacteria, viruses and nematodes. When viewed over a broad geographical region, diseases of epidemic proportions have seldom occurred or were not recognized in tropical pastures. Periodic outbreaks of a disease of infestation of an insect on a regional basis, however, occurs and causes pronounced local devastation.

A stunting disease of *Digitaria decumbens* caused by a virus and transmitted by *Sogoto furcifera* was found in Surinam in 1958 (Dirven and vanHoof, 1960). It was later identified in Guyana (Biessar, 1966) and reported in Peru by Revilla (1967) who gave other hosts as *Panicum purpurascens* (syn. *Brachiaria mutica*) and *Paspalum conjugatum*. This disease was viewed with great alarm since Pangola is a single genotype and widespread in the tropics. A so-called 'A-24 Taiwan' strain (actually *Digitaria pentzii*) and other species of *Digitaria* were not affected by the virus. Another virus, *Digitaria striata*, was described in Australia (Greber, 1979). Neither virus became widespread nor appeared in epidemic proportions. Liu (1969) found a pathogen of rust on Pangola growing in Puerto Rico and identified it as a variant of *Puccinia oahuensis*. Other *Digitarias, Panicum maximum, B. mutica* and *Axonopus compressus* appear to have resistance to the rust-causing organism.

The first leaves of certain elephant grass strains often display small grey spots incited by *Piricularia grisea* but the disease causes little or no apparent damage in terms of herbage yields. The causal agent of another leaf spot *Helminthosporium sacchari* also attacks elephant grass, causing severe lesions on the leaves and stems. In some instances leaves turn brown as if burned. Several selections are resistant, one of which is Merker.

A bacterial disease caused by *Xanthomonas axonoperus* drastically reduces plant growth and occasions death of *Axonopus scoparius* and *A. micay* in Colombia (Castaño *et al.*, 1964). The pathogen is mainly transmitted by cutting knives but can be carried by grazing animals and man. Studies confirmed that sexual and apomitic types existed in *A. scoparius*, an important cutting grass in the medium elevations of Colombia and Ecuador. Resistant plants were found among progenies of sexual types. One promising selection was multiplied and released commercially by vegetatively propagated stem cuttings.

A common leaf spot caused by *Cercospera fusimaculans* occurs on certain selections of *Panicum maximum* and appears to interfere with plant processes, but no data are available to substantiate this general observation. A number of grass selections, including those that have evolved under natural conditions appear to have tolerance or resistance. Guinea grass florets are also subjected to invasion of *Claviceps maximemsis*, the causal agent of ergot. This organism impedes seed development under certain conditions but the mass of deformed tissue does not contain the alkaloid commonly associated with this fungal growth. Several other grasses such as *Melinis minutiflora* and *Brachiaria mutica* may be infected with a species of the same organism. The ergot organism *C. paspali* is so prevalent on *Paspalum dilatatum* in southeastern USA and in Australia as to inhibit production of viable seed, and if eaten in large quantities may cause abortion.

Theis (1953) described several diseases on grasses in Puerto Rico: *Claviceps maximemis*, *Phyllosticta panici* and *Cerospora fusmaculans* on *Panicum maximum*; *Helminthosporium sacchari* and *Piricularia grisea* on *Pennisetum purpureum; Uromyces leptodermus* (causal agent of a rust) on *Brachiaria mutica*. Simmonds (1966), Bradley and Rogers (1978) and Lenne *et al.* (1980) described other pasture diseases and pests in Australia.

Virus or virus-like diseases are prevalent among tropical pasture legumes, for example the native or naturalized *Rhynchosia* spp. commonly display a leaf mosaic. The virus-like disease 'legume little leaf', caused by a mycoplasma, frequently occurs on *Desmodium intortum, D. uncinatum, Lotononis bainesii* and *Stylosanthes humilis*, sometimes destroying plants (Hutton, 1970). *Centrosema pubescens, Glycine wightii, Leucaena leucocephala* and *Macroptilium atropurpureum* show field resistance to little leaf. The fungal organism *Rhizoctonia* spp. causes severe leaf stem and root damage to *M. atropurpureum* cv. Siratro (Jones *et al.*, 1969; Sonoda *et al.*, 1971; Sonoda, 1976; Hutton and Beal, 1977). In fact, this is a primary limitation to the adaptability and productivity of Siratro in the humid tropics receiving 1 200 mm or more of precipitation. Entire plant populations of *Stylosanthes* spp. can be completely eliminated by *Colletorichum* spp. Devastating epidemics have been recorded in Florida (Kretschmer *et al.*, 1974) and in Australia (Irwin and Cameron, 1978). The disease is more likely to develop late in the growing season when it markedly reduces seed yields but may cause death of plants over a wide area within the field. *Botrytis cinera* attacks the inflorescences of stylos, causing flora abortion and reduced seed yields (O'Brien and Pont, 1977).

Damage caused by insects inhabiting or invading a pasture may not always be easily recognized. Aphids, particularly the sugar-cane yellow aphid (*Sipha flava*), attack leaves and stems of Pangola grass and *Hemarthria* species and may reduce herbage yields 30 per cent or more. Plants infested by aphids may turn yellow, eventually reddish and sometimes die. As one walks through infested Pangola pastures the aphids adhere to clothing and/or exudate from the insects covering shoes and legs of trousers. Attacks are particularly harmful during dry periods. The insect can be controlled by spraying with Diazinon AG 500 or Malathion 85 in 400 litres/ha of water or by low volume spraying with these amounts in 40 litres/ha of water (Vicente-Chandler *et al.*, 1974). Grasses with resistance to the aphid have been encountered (Oakes, 1978). Rhodes grass scale (*Antonia graminis*) attacks stems of Pangola at soil level, causing severe damage and death of plants in localized areas. A chinch bug, *Blissus leucopterus*, feeds on all grasses but may be especially harmful to Pangola and guinea grasses. Damage usually occurs at ground level and affected areas appear in a circular pattern. The insect can be controlled with biweekly application of Diazinon AG 500 or Sevin 50. The fall armyworm (*Laphygma fugiperda*) and the grassworm (*Mocis repanda*) can defoliate large areas of both grasses and legumes in a short period of time. Once the plants are well established the effects are less severe since they are larger and growing vigorously. Small seedlings, however, can be completely destroyed. Several species of leaf hoppers are sometimes observed in large numbers in pastures and one, *Orosius argentatus*, transmits the 'little leaf' mycoplasma (Hutton and Beal, 1977). Spittle bugs are small sucking insects that excrete a white froth or spittle around themselves during the larval stage. They cause a withering of the infested stems and blasting of the seed heads. The red-banded spittle bug that causes dieback in *Cynodon dactylon* can be controlled by removing the previous season's growth, close grazing or frequent mowing. A bean fly, *Amnenus quadrituberculatus*, sometimes causes heavy damage to legumes such as *Glycine*, *Desmodium* and Siratro by larval root-feeding. Late-season heavy grazing reduced incidence of oviposition sites and cultivation in early autumn exposed young larvae and destroyed eggs in Australia (Braithwaite and Rand, 1970).

The root-knot nematodes (*Meloidogyne* spp.) are a serious pest of most vegetable and field crops because of wide host range and rapid production. Of the grass species tested, *Cynodon dactylon* has consistently shown resistance or tolerance to the root-knot nematodes and has been used in rotations as a sod crop preceding susceptible crops (McGlohon *et al.*, 1961). Studies in Nigeria showed that *Cynodon nlemfuensis* (*Cynodon* IB.8) effectively controlled the nematodes (Adeniji and Chheda, 1971). After growth in soil heavily populated with *Meloidogyne* spp. the grass roots were free of galls and no nematodes were observed within the root tissues. Glasshouse studies with the grass grown in soil inoculated with egg masses and second stage larvae of the nematode indicated that the soil had become devoid of larvae after 3 months. In the field rapid reduction of the root-knot nematode population in the soil occurred in

3–6 months, with economic control of the larvae within 12 months. After 18 months the grass was turned under and land prepared for tomatoes susceptible to the nematode. No plants showed signs of root-galling. Observations in Puerto Rico indicated that Pangola grass has a suppressing effect on the nematode population of the soil. Bananas following this grass grew more vigorously and were more productive than those following a clean cultivated crop (personal observation). In tests with five species of *Meloidogyne* in Georgia, Minton *et al.* (1967) noted that *Desmodium intortum* was one of the most resistant but *D. uncinatum* was susceptible. In Queensland, Colbran (1964) also recorded that *M. javanica* was parasitic on *D. uncinatum*.

Pests and diseases in pastures may go unnoticed or the severity of their damage may not be recognized by the casual observer. The pasture agronomist should suspect presence of a pest or disease if herbage production is lower than expected and no other satisfactory explanation for the discrepancy is in evidence. First the problem must be identified then its importance assessed and a decision made whether to attempt control measures. Identification may be simple if disease symptoms or insects are obvious. Sometimes, however, these are not clearly recognized or may be complicated with involvement of several organisms. Assistance by a specialist may be needed to establish specifically the primary cause of the problem and to assess the extent of damage. In devising field-control measures, consideration should be given to changes in management practices that could modify or reduce the magnitude of a pest or disease. In some instances use of a pesticide may be required, but in accordance with prescribed and precautionary measures and perhaps in consultation with a specialist.

References

Abruña, F., R. W. Pearson and **C. Elkins** (1958) Quantitative evaluation of soil reaction and base status changes resulting from field applications of residually acid-forming nitrogen fertilizers, *Soil Sci. Soc. Amer. Proc.* **22**, 539–42.

Ademosun, A. A. and **H. R. Chheda** (1974) Regional evaluation of Cynodon genotypes. II. Ile-Ife area, *Niger. Agric. J.* **11**, 25–30.

Adeniji, M. O. and **H. R. Chheda** (1971) Influence of six varieties of Cynodon on four Meloidogyne spp, *J. Nematology* **3**, 251–4.

Adeniyi, S. A. and **P. N. Wilson** (1960) Studies on Pangola at ICTA, Trinidad. I. Effects of fertilizer application at time of establishment, and cutting interval, on the yield of ungrazed Pangola grass, *Trop. Agric. (Trin.)* **37**, 271–82.

Akinola, J. O., H. R. Chheda and **J. A. MacKenzie** (1971a) Effects of cutting frequency and levels of applied nitrogen on productivity, chemical composition, growth components and regrowth potential of three Cynodon strains. 2. Growth components: tillering and leaf area, *Niger. Agric. J.* **8**, 63–76.

Akinola, J. O., J. A. MacKenzie and **H. R. Chheda** (1971b) Effects of cutting frequency and levels of applied nitrogen on the productivity and chemical composition, growth components and regrowth potential of three Cynodon strains. 3. Regrowth potential, *W. Afr. J. Biol. Appl. Chem.* **14**, 7–12.

Allen, S. E. and **D. A. Mays** (1974) Coated and other slow-release fertilizers for forages. In D. A. Mays (ed.), *Forage Fertilization*, ASA, CSSA, SSSA: Madison, Wisconsin, Ch. 26.

Allison, F. E. (1966) The fate of nitrogen applied to soils, *Adv. Agron.* **18**, 219–58.

Anderson, A. J. (1956) Molybdenum as a fertilizer, *Adv. Agron.* **8**, 163–202.

Anderson, A. J. and **D. Spencer** (1950) Sulphur in nitrogen metabolism of legumes and non-legumes, *Aust. J. Sci. Res.* (*B*) **3**, 431–49.

Andrew, C. S. (1966) A kinetic study of phosphate absorption by excised roots of Stylosanthes humilis, Phaseolus lathyroides, Desmodium uncinatum, Medicago sativa and Hordeum vulgare, *Aust. J. Agric. Res.* **17**, 611–24.

Andrew, C. S. and **R. C. Bruce** (1975) Evaluation of fertilizer requirements of tropical legume-based pastures, *Trop. Grassld* **9**, 133–9.

Andrew, C. S. and **W. W. Bryan** (1955) Pasture studies on the coastal lowlands of subtropical Queensland. I. Introduction and initial plant nutrient studies, *Aust. J. Agric. Res.* **6**, 265–90.

Andrew, C. S. and **I. F. Fergus** (1976) Plant nutrition and soil fertility. In N. H. Shaw and W. W. Bryan (eds), *Tropical Pasture Research*. Commonw. Bur. Past. Fld Crops, Bull. 51, Hurley: England, Ch. 6.

Andrew, C. S. and **M. P. Hegarty** (1969) Comparative responses to manganese excess of eight tropical and four temperate pasture legume species, *Aust. J. Agric. Res.* **20**, 687–96.

Andrew, C. S. and **E. M. Hutton** (1974) Effect of pH calcium on the growth of tropical pasture legumes, *Proc. 12th Intl. Grassld Cong.*, pp. 23–8.

Andrew, C. S. and **R. K. Jones** (1978) The phosphorus nutrition of tropical legumes. In *Mineral Nutrition of Legumes in Tropical and Sub-Tropical Soils, Proc.* Brisbane; Australia, pp. 259–311.

Andrew, C. S. and **D. O. Norris** (1961) Comparative responses to calcium of five tropical and four temperate pasture legume species, *Aust. J. Agric. Res.* **12**, 40–55.

Andrew, C. S. and **M. F. Robins** (1969a) The effect of phosphorus on the growth and chemical composition of some pasture legumes. I. Growth and critical percentage of phosphorus, *Aust. J. Agric. Res.* **20**, 665–74.

Andrew, C. S. and **M. F. Robins** (1969b) The effect of potassium on the growth and chemical composition of some tropical and temperate pasture legumes. I. Growth and critical percentages of potassium, *Aust. J. Agric. Res.* **20**, 999–1007.

Andrew, C. S. and **M. F. Robins** (1971) The effect of phosphorus on the growth chemical composition and critical phosphorus percentages of some tropical grasses, *Aust. J. Agric. Res.* **22**, 693–706.

Andrew, C. S. and **P. J. Vanden Berg** (1973) The influence of aluminum on phosphate-sorption by whole plants and excised roots of some pasture legumes, *Aust. J. Agric. Res.* **24**, 341–51.

Annual Report (1962–63) *The Rockefeller Foundation Program Agricultural Science*: New York, p. 99.

Anslow, R. C. (1957) Investigation into the potential productivity of "acacia" (Leucaena glauca) in Mauritius, *Review Agric. Sucr. Ile Maurice* **38**, 99–102.

Barnes, D. L. (1970) Irrigated pastures, *Rhod. Agric. J.* **67**, 154–9.

Beaton, J. W., W. A. Hubbard and **R. C. Speer** (1967) Coated urea, thiourea, urea-formalderhyde, hexamine, oxamide and oxidized nitrogen-enriched coals as slowly available sources of nitrogen for orchardgrass, *Agron. J.* **59**, 127–33.

Biessar, S. (1966) *The Stunting Virus Diseases of Pangola Grass and the Reaction of Digitaria spp. to the Disease in British Guiana*, FAO Plant Prot. Bull. 14, pp. 60–2.

Birch, H. F. (1960) Nitrification in soils after different periods of dryness, *Plant and Soil* **12**, 81–96.

Bishop, H. G. (1977) The response of nitrogen and phosphorus fertilizer of native pasture on the balbirni land system in North-west Queensland, *Trop. Grassld* **11**, 257–62.

Black, J. N. (1956) The distribution of solar radiation over the earth's surface, *Arch. Met. Geophys. Bioklim. Ser.* **B7**, 165–89.

Blue, W. G. (1970) Fertilizer uptake by Pensacola bahiagrass (Paspalum notatum) from Leon fine sand, a spodosol, *Proc. 11th Int. Grassld Cong.*, pp. 389–92.

Bornemissza, G. F. (1960) Could dung eating insects improve our pastures, *J. Aust. Inst. Agric. Sci.* **26**, 54–6.

Bornemissza, G. F. and **C. H. Williams** (1970) An effect of dung beetle activity of plant yield, *Pedobiologia* **10**, 1–7.

Borgét, M. (1969) Résultats et tendaces présentés des recherches fourragères de l'IRAT, *Agron. Trop. (Paris)* **24**, 103–55.

Borgét, M. (1971) Recherches fourragéres menées par l'IRAT en Afrique tropicale et à Madagascar, *Seminar For. Crops West Africa*, Ibadan, Nigeria (mimeo).

Boswingle, E. (1961) Residual effects of phosphorus fertilizers in Kenya, *Emp. J. Exptl Agric.* **24**, 136–42.

Bouton, J. H. and **D. A. Zuberer** (1979) Response of Panicum maximum Jacq. to inoculation with Azospirillum brasilenae, *Plant and Soil* **52**, 585–90.

Bradley, R. H. and **D. J. Rogers** (1978) Pests of Pangola grass in north Queensland pastures, *Queensld Agric. J.* **104**, 320–4.

Braithwaite, B. M. and **J. R. Rand** (1970) The pest status of Amnemus spp. in tropical legume pastures in north coastal New South Wales, *Proc. 11th Int. Grassld Cong.*, pp. 676–81.

Brown, M. A. and **G. M. Volk** (1966) Evaluation of ureaform fertilizer using nitrogen-15 labelled materials in sandy soils, *Soil Sci. Soc. Am. Proc.* **30**, 278–81.

Bruce, R. C. (1974) Growth response, critical percentage of phosphorus, and seasonal variation of phosphorus percentage in Stylosanthes guyanensis cv. Schofield topdressed with superphosphate, *Trop. Grassld* **8**, 137–44.

Bruce, R. C. (1978) A review of the trace element nutrition of tropical pasture legumes in Northern Australia, *Trop. Grassld* **12**, 170–83.

Bryan, R. P. (1973) The effect of dung beetle activity on the numbers of gastrointestinal helminth larvae recovered from pasture samples, *Aust. J. Agric. Res.* **24**, 161–8.

Bryan, W. W. and **R. R. Evans** (1973) Effect of soils, fertilizers and stocking rates on pastures and beef production on the Wallum of southest Queensland. 1. Botanical composition and chemical effects on plants and soils, *Aust. J. Exptl Agric. Anim. Husb.* **13**, 516–29.

Bryan, W. W. and **J. P. Sharpe** (1965) The effect of urea and cutting treatments on the production of Pangola grass in southeastern Queensland, *Aust. J. Exptl Agric. Anim. Husb.* **5**, 433–41.

Burton, G. W. and **E. H. De Vane** (1952) Effect of rate and method of applying different sources of nitrogen upon the yield and chemical composition of bermudagrass, *Agron. J.* **44**, 128–32.

Burton, G. W., J. E. Jackson and **B. L. Southwell** (1958) Does nitrogen source effect the palatability of Coastal bermudagrass? *Agron. J.* **50**, 172–3.

Burton, G. W. and **J. E. Jackson** (1962) Effect of rate and frequency of applying six nitrogen sources on Coastal bermudagrass, *Agron. J.* **54**, 40–3.

Cadot, R. (1965) Fodder crop experiments, *Afr. Soils* **10**, 461–70.

Campbell, N. A. and **J. Keay** (1965) Flexible techniques in describing a range of response curves of pasture species, *Proc. 9th Int. Grassld Cong.*, pp. 332–8.

Cardenas, J., C. E. Reyes and **J. D. Doll** (1972) *Tropical weeds*, Inst. Colombiano Agropec: Bogota.

Castâno, J. J., H. D. Thurston and **L. V. Crowder** (1964) Transmision de la gomosis en los pastos micay y Imperial, *Agric. Trop. (Colombia)* **20**, 379–87.

Chadhokar, P. A. (1977) Establishment of stylo (Stylosanthes guianensis) in Kunai (Imperata cylindrica) pastures and its effect on dry matter yield and animal production in the Markham Valley, Papua New Guinea, *Trop. Grassld* **11**, 263–72.

Chheda, H. R. (1974) Forage crop research at Ibadan. 1. Cynodon spp. In J. K. Loosli, V. A. Oyenuga and G. M. Babatunde (eds), *Animal Production in the Tropics*, Heinemann (Nigeria): Ibadan, pp. 79–94.

Chheda, H. R. and **J. O. Akinola** (1971a) Effects of cutting frequency and level of applied nitrogen on crude protein production recovery by three Cynodon strains, *W. Afr. J. Biol. Appl. Chem.* **14**, 31–8.

Chheda, H. R. and **J. O. Akinola** (1971b) Effects of cutting frequency and levels of applied nitrogen on the productivity and chemical composition, growth components and regrowth potential of three Cynodon strains. 1. Yield, chemical composition and weed composition, *Niger. Agric. J.* **8**, 44–62.

Chheda, H. R. and **M. A. M. Saleem** (1972) Effects of heights of cutting after grazing on yield, quality and utilization of 'Cynodon IB.8' pasture in Southern Nigeria, *Trop. Agric. (Trin.)* **50**, 113–19.

Clatworthy, J. N. (1970) Legumes and fertilizers as sources of pasture nitrogen. *Proc. 11th Int. Grassld Cong.*, pp. 408–11.

Clement, C. R. and **M. J. Hopper** (1969) The supply of potassium to high yielding cutgrass, *NAASQ* **72**, 101–9.

Cloutier, P. E. (1971) Agronomic factors in pasture and forage crops farm management in tropical Australia, *Trop. Grassld* 5, 245–54.

Coaldrake, J. E. (1970) The brigalow. In R. M. Moore (ed.), *Australian Grasslands*, Aust. Natl Univ. Press: Canberra, Ch. 9.

Colbran, R. C. (1964) Studies of plant and soil nematodes. 7. Queensland records of the order Tylenchida and the genera Trichodorus and Xiphinema, *Queensld J. Agric. Sci.* **21**, 77–123.

Cooper, J. P. (1969) Potential forage production. In L. Phillips and R. Hughes (eds), *Grass and Forage Breeding, Occasional Symp. No. 5*, The Brit. Grassld Soc., pp. 5–13.

Cooper, J. P. (1970) Potential production and energy conversion in temperate and tropical grasses, *Herb. Abst.* **40**, 1–15.

Crespo, G. (1974) Responses of six tropical pasture species to increasing levels of nitrogen fertilizer, *Cuban J. Agric. Sci.* **8**, 177–88.

Crespo, G., T. Rodriguez and **J. Perez** (1975) Potential response of guinea grass (Panicum maximum Jacq.) and Pangola (Digitaria decumbens Stent.) to nitrogen fertilization, *Cuban J. Agric. Sci.* **9**, 353–62.

Crowder, L. V. (1967) Grasslands of Colombia, *Herb. Abst.* **37**, 237–45.

Crowder, L. V. (1974) *Pasture and Forage Research in Tropical America*, Cornell Int. Agric. Bull. 28, Cornell Univ: Ithaca, N.Y.

Crowder, L. V. (1977) Potential of tropical zone cultivated forages. In *Potential of the World's Forages for Ruminant Animal Production*, Winrock Int. Livestock Trg Center, Arkansas, pp. 49–78.

Crowder, L. V. and **H. R. Chheda** (1977) Forage and fodder crops. In C.L.A. Leakey and J. B. Wills (eds), *Food Crops of the Lowland Tropics*, Oxford Univ. Press, Ch. 8.

Crowder, L. V., J. Lotero, A. Michelin, R. Ramirez, A. Wieczorec and **A. Bastidas** (1963) *Fertilización de Gramineas Tropicales y Subtropicales en Colombia*, D.I.A. Bol. Divulgación, No. 12. Min. Agric.: Colombia.

Crowder, L. V., A. Michelin, M. Martinez and **G. B. Baird** (1959) La produccion de alfalfa en Colombia. VII. Respuesta a la aplicación del borax en el Valle de Cauca, *Agric. Trop. (Colombia)* **15**, 387–95.

Crowder, L. V., R. L. Richardson and **A. McCormack** (1960) Producción de forraje de varias especies de gramineas adaptadas a las condiciones del clima calido de Colombia, *Agric. Trop. (Colombia)* **16**, 101–13.

Davies, J. G. and **A. G. Eyles** (1965) Expansion of Australian pasture production, *J. Aust. Inst. Agric. Sci.* **31**, 77–93.

Day, J. M., M. C. P. Neves and **J. Döbereiner** (1975) Nitrogenase activity on the roots of tropical forage grasses, *Soil Biol. Biochem.* **7**, 107–12.

deGeus, J. G. (1969) Some aspects of pasture development in the tropics, *Stikstof* No. **13**, 29–47.

DePolli, H., E. Matsuii, J. Döbereiner and **E. Salati** (1977) Confirmation of nitrogen fixation in two tropical grasses by 15N$_2$ incorporation, *Soil Biol. Biochem.* **9**, 119–23.

Diatloff, A. and **P. E. Luck** (1972) The effects of the interaction between seed inoculation, pelleting and fertilizer on growth and nodulation of Desmodium and Glycine on two soils in S.E. Queensland, *Trop. Grassld* **6**, 33–8.

Dirven, J. G. P. and **H. A. vanHoof** (1960) A destructive virus disease of Pangola grass, *Tydschrift Plantenziekten* **66**, 344–9.

Doak, B. W. (1952) Some chemical changes in the nitrogenous constituents of urine when voided in pasture, *J. Agric. Sci. (Camb.).* **42**, 162–71.

Döbereiner, J. (1966) Azotobacter paspali spp. n. uma bacteria fixadora na rizofera de Paspalum, *Pesq. Agropec. Bras.* **1**, 357–65.

Döbereiner, J. (1968) Non-symbiotic nitrogen fixation in tropical soils, *Pesq. Agropec. Bras.* **3**, 1–6.

Döbereiner, J. and **A. B. Campbelo** (1971) Non-symbiotic nitrogen fixing bacteria in tropical soils, *Pl. soil Spec. Vol.*, pp. 457–70.

Döbereiner, J. and **J. M. Day** (1973) Dinitrogen fixation in the rhizosphere of tropical grasses. In W. D. P. Stewart (ed.) *Nitrogen Fixation by Free-Living Microorganisms IBP Synthesis Series*, Cambridge University Press.

Döbereiner, J. and **J. M. Day** (1975) Potential significance of nitrogen fixation in rhizosphere associations of tropical grasses. In E. Bornenisza and A. Alvano (eds), *Soil Management in Tropical America* N. C. State Univ.: Raleigh, pp. 197–210.

Döbereiner, J. and **J. M. Day** (1976) Associative symbiosis and dinitrogen fixing sites. In N. E. Newton and C. J. Nyman (eds), *Proc. Intl. Symp. N. Fix.*, Wash. State Univ. Press, pp. 518–38.

Döbereiner, J., J. M. Day and **P. J. Dart** (1972) Nitrogenase activity and oxygen sensitivity of the Paspalum notatum-Azotobacter paspali association, *J. Gen. Microb.* **71**, 103–16.

Doss, B. D., D. A. Ashely, O. L. Bennett and **R. M. Patterson** (1966) Interaction of soil moisture, nitrogen and clipping frequency on yield and nitrogen content of Coastal bermuda grass, *Agron. J.* **58**, 510–12.

Dovrat, A., J. G. P. Dirven and **B. Deinum** (1971) The influence of defoliation and nitrogen on the regrowth of Rhodes grass (Chloris gayana Kunth). 1. Dry matter production and tillering, *Neth. J. Agric. Sci.* **19**, 74–101.

Fenster, W. E. and **L. A. Leon** (1978) Management of phosphorus fertilization in establishing and maintaining improved pastures on acid, infertile soils of tropical America. In P. A. Sanchez and L. E. Terges (eds), *Pasture Production in Acid Soils of the Tropics, Seminar Proc.*, CIAT: Colombia, pp. 109–22.

Figarella, J., F. Abruña and **J. Vicente-Chandler** (1972) Effect of five nitrogen sources applied at four rates to Pangola grass sod under humid tropical conditions, *J. Agric. Univ. P. R.* **56**, 410–16.

Fisher, C. E. and **L. Quinn** (1959) *Control of Mesquite on Grazinglands*, Texas Agric. Expt. Sta., Bull. 935.

Fisher, F. L. and **A. G. Caldwell** (1959) The effect of continued use of heavy rates of fertilizers on forage production and quality of Coastal bermudagrass, *Agron. J.* **51**, 99–102.

Fisher, M. J. (1973) Effect of times, height and frequency of defoliation on growth and development of Townsville stylo in pure ungrazed swards at Katherine, N. T., *Aust. J. Exptl Agric. Anim. Husb.* **13**, 389–97.

Fisher, M. J. and **N. A. Campbell** (1972) The initial responses to phosphorus fertilizer of Townsville stylo in pure ungrazed swards at Katherine, N. T., *Aust. J. Exptl Agric. Anim. Husb.* **12**, 488–92.

Fox, R. L., S. M. Hasan and **R. C. Jones** (1971) Phosphate and sulphate sorption by Latosols, *Proc. 1st Int. Symp. Soil Fert. Eval. (New Delhi)*, pp. 857–64.

Frame, J. (1970) Fundamentals of grassland management. Part 10: The grazing animal. *Scottish Agric.* **50**, 28–44.

French, M. H. (1957) Nutritional value of tropical grasses and fodders, *Herb. Abst.* **27**, 1–9.

Fritz. J. (1971) Recherches sur les productions fourragères effectuées par l'IRAT Réunion, 1963–1970, *Agron. Trop. (Paris)* **11**, 1248–69.

Garza, R., V. Perez, O. Chapa and **J. Monroy** (1971) Respuesta de gramas nativas a la fertilización de nitrógeno, fosfóro y potasio en el trópico humedo, *Tecn. Pecuaria (Mexico)* **18**, 54–61.

Gates, C. T. (1970) Physiological aspects of the rhizobial symbiosis in Stylosanthes humilis, Leucaena leucocephala and Phaseolus atropurpureus, *Proc. 11th Int. Grassld Cong.*, pp. 442–6.

Gillard, P. (1967) Coprophagous beetles in pasture ecosystems, *J. Aust. Inst. Agric. Sci.* **33**, 30–40.

Greber, P. S. (1979) Digitaria striata virus – a rhabdovirus of grasses transmitted by Sogotella kolophon (Kirk), *Aust. J. Agric. Res.* **30**, 43–51.

Grof. B. (1966) Establishment of legumes in the humid tropics of north eastern Australia, *Proc. 11th Int. Grassld Cong.*, pp. 1037–42.

Grof, B. and **W. A. T. Harding** (1970) Dry matter yields and animal production of guinea grass (Panicum maximum) on the humid tropical coast of north Queensland, *Trop. Grassld* **4**, 85–95.

Haggar (1970) Seasonal production of Andropogon gayanus. I. Seasonal changes in yield components and chemical composition, *J. Agric. Sci. (Camb.)* **74**, 487–94.

Hall, T. D., D. Meredith and **R. E. Altona** (1965) The role of fertilizers in pasture management. In D. Meredith (ed.), *The Grasses and Pastures of South Africa*, Central News Agency: Cape Town, Ch. 5.

Harker, K. W. (1962) A fertilizer trial on Paspalum notatum pasture. 1. The effects on yield, *E. Afr. Agric. For. J.* **27**, 201–3.

Harrington, G. N. and **D. D. Thornton** (1969) A comparison of controlled grazing and manual hoeing as a means of reducing the incidence of Cymbopogon afronardus Stapf in Ankole pastures, Uganda, *E. Afr. Agric. For. J.* **35**, 154–9.

Hendy, K. (1972) The response of pangola grass pasture near Darwin to the west season application of nitrogen, *Trop. Grassld* **6**, 25–32.

Hendy, K. (1975) Review of natural pastures and their management problems on the north coast of Tanzania, *E. Afr. Agric. For. J.* **41**, 52–7.

Henzell, E. F. (1970) Problems in comparing the nitrogen economies of legume-based and nitrogen-fertilized pasture systems, *Proc. 11th Int. Grassld Cong.*, pp. A112–19.

Henzell, E. F. (1971) Recovery of nitrogen from four fertilizers applied to Rhodes grass in small plots, *Aust. J. Exptl Agric. Anim. Husb.* **11**, 420–30.

Henzell, E. F., A. E. Martin, P. J. Ross and **K. P. Haydock** (1968) Isotopic studies on the uptake of nitrogen by pasture plants. IV. Uptake of nitrogen from labelled plant material by Rhodes grass and Siratro, *Aust. J. Agric. Res.* **19**, 65–77.

Henzell, E. F. and **D. J. Oxenham** (1964) Seasonal changes in the nitrogen content of warm-climate pasture grasses, *Aust. J. Exptl Agric. Anim. Husb.* **4**, 336–44.

Hodges, E. M., W. G. Kirk and **F. M. Peacock** (1966) Phosphate fertilizers on Pangola grass pastures, *Proc. 9th Int. Grassld Cong.*, pp. 915–18.

Holmes, W. (1968) The use of nitrogen in the management of pasture for cattle, *Herb. Abst.* **38**, 265–77.

Humphreys, L. R. (1967) Townsville lucerne: history and prospect, *J. Aust. Inst. Agric. Sci.* **33**, 3–13.

Hubbell, D. H. (1978) The potential of associative symbiosis between nitrogen fixing bacteria and forage grasses. In P. A. Sanchez and L. E. Tergas (eds), *Pasture Pro-*

duction in Acid Soils of the Tropics, Seminar Proc., CIAT: Colombia, pp. 139–44.

Hutton, E. M. (1970) Tropical pastures, *Adv. Agron.* **22**, 1–73.

Hutton, E. M. and **L. B. Beall** (1977) Breeding of Macroptilium atropurpureum, *Trop.* Grassld **11**, 15–31.

Irwin, J. A. G. and **D. F. Cameron** (1978) Two diseases of Stylosanthes spp. caused by Colletotrichum gloeosporioides in Australia and the pathogenic specialization of one of the causal organisms, *Aust. J. Agric. Res.* **39**, 305–17.

Ivens, G. W. (1971) *East African Weeds and Their Control*, Oxford Univ. Press: Nairobi.

Ivens, G. W. (1980) Imperata cylindrica (L.) Beauv in West African Agriculture, *Proc. Biotrop Workshop, Special Publ. 5*, pp. 149–56.

Jackson, J. E., M. E. Walker and **R. L. Carter** (1959) Nitrogen, phosphorus and potassium requirements of Coastal bermudagrass on a Tifton loamy sand, *Agron. J.* **51**, 129–31.

Jensen, H. L. and **R. C. Betty** (1943) Nitrogen fixation in leguminous plants. III. The importance of molybdenum in symbiotic nitrogen fixation, *Proc. Linn. Soc. NSW* **68**, 1–8.

Johansen, C., K. E. Merkley, and **G. R. Dolby** (1980) Critical phosphorus concentrations in parts of Macroptilium atropurpureum cv siratro and Desmodium intortum cv Greenleaf as affected by plant age, *Aust. J. Agric. Res.* **31**, 643–702.

Jones, M. B. and **L. M. M. de Freitas** (1970) Repostas de quatro leguminosas a fósforo, potássio, e calcario num Latossolo Vermulho-Amarelo de Campo Cerrado, *Pesq. Agropec. Bras.* **5**, 91–9.

Jones, M. B. and **J. L. Quagliato** (1973) Response of four tropical legumes and alfalfa to varying levels of sulphur, *Sulphur Inst. J.* **9**, 6–9.

Jones, R. J. (1967) Effects of close cutting and nitrogen fertilization on growth of siratro (Phaseolus atropurpureus) pasture at Samford, south-eastern Queensland, *Aust. J. Exptl Agric. Anim. Husb.* **7**, 157–61.

Jones, R. J. (1970) The effect of nitrogen fertilizer in spring and autumn on the production and botanical composition of two sub-tropical grass-legume mixtures, *Trop. Grassld* **4**, 97–108.

Jones, R. J. (1973) The effect of frequency and severity of cutting on yield and persistence of Desmodium intortum cv. Greenleaf in a sub-tropical environment, *Aust. J. Exptl Agric. Anim. Husb.* **13**, 171–7.

Jones, R. J. and **R. M. Jones** (1971) Agronomic factors in pasture and forage crops production in tropical Australia, *Trop. Grassld* **5**, 229–44.

Jones, R. K. (1968) Initial and residual effects of superphosphate on a Townsville lucerne pasture in northeast Queensland, *Aust. J. Exptl Agric. Anim. Husb.* **8**, 521–7.

Jones, R. K. (1974) Phosphorus response of a wide range of accessions from the genus Stylosanthes, *Aust. J. Agric. Res.* **26**, 847–62.

Jones, R. K. and **P. J. Robinson** (1970) The sulphur nutrition of Townsville lucerne (Stylosanthes humilis), *Proc. 11th Int. Grassld Cong.*, pp. 377–80.

Jones, R. K. and **J. B. F. Field** (1976) A comparison of biosuper and superphosphate on a sandy soil in the monsoonal tropics of northern Queensland, *Aust. J. Exptl Agric. Anim. Husb.* **16**, 99–102.

Jones, R. M., J. L. Alcorn and **M. C. Rees** (1969) Death of Siratro due to violent root rot, *Trop. Grassld* **3**, 137–9.

Jordan, C. W., C. W. Evans and **R. D. Rouse** (1966) Coastal bermudagrass response to application of P and K as related to P and K levels in the soil, *Soil Sci. Soc. Amer. Proc.* **30**, 477–9.

Kasasian, L. (1971) **Weed Control in the Tropics**, Leonard Hill: London.

Kein, L. T., O. C. Ruelke and **H. L. Bueland** (1976) Effects of fertilization on Pangola, Transvala digitgrass and Coastal bermudagrass, *Proc. Soil and Crop Soc. Fla.* **35**, 80–3.

King, L. J. (1966) *Weeds of the World, Biology and Control*, Interscience: New York.

Kinch, D. M. and **J. C. Pipperton** (1962) *Koa Haole, Production and Processing,*

Hawaii Agric. Expt. Sta., Bull. 129.

Kleinschmidt, F. H. (1967) The influence of nitrogen and water on pastures of green panic, lucerne and glycine at Lawes, south-eastern Queensland, *Aust. J. Exptl Agric. Anim. Husb.* **7**, 441–6.

Kretschmer, A. E., Jr., R. M. Sonoda and **J. B. Brolman** (1974) Morphologic, agronomic and disease susceptibility differences among Stylosanthes humilis accessions in south Florida, *Res. Center, Report No. RL, 1974*, Fort Pierce.

Lenne, J. M., J. W. Turner, and **D. F. Cameron** (1980) Resistance to disease and pests of tropical pasture plants, *Trop. Grassld* **14**, 146–52.

Little, S., J. Vicente-Chandler and **F. Abruña** (1959) Yield and protein content of irrigated Napier grass, guinea grass and Pangola grass as affected by nitrogen fertilization, *Agron. J.* **51**, 111–13.

Liu, J. L. (1969) Occurrence of rust of Pangola grass in Puerto Rico, *J. Agric. Univ. P. R.* **53**, 132–9.

Loomis, R. S. and **W. A. Williams** (1963) Maximum crop productivity: an estimate, *Crop Sci.* **3**, 67–72.

Lotero, J., G. Herrera and **L. V. Crowder** (1965) Respuesta de una pradera natural a la aplicación de fertilizantes, *Agric. Trop. (Colombia)* **21**, 229–32.

Lotero, J., A. Ramirez and **G. Hererra** (1968) Fuentes, dosis y metodos de aplicación de nitrógeno en pasto elefante, *Revista ICA (Colombia)* **3**, 113–21.

Lotero, J., W. W. Woodhouse, Jr. and **R. G. Peterson** (1966) Distribution and loss rate of N and K applied to the soil by grazing animals, *Proc. 9th Int. Grassld Cong.*, pp. 168–90.

Lowe, K. F. and **J. F. Cudmore** (1978) A comparison of slow-release and conventional tional nitrogenous fertilizers for an established Pangola grass pasture in a subtropical environment, *Aust. J. Agric. Exptl. Anim. Husb.* **18**, 415–20.

Luck, P. E. and **N. J. Douglas** (1966) Dairy pasture research and development in the near north coast centered on Cooroy, Queensland, *Proc. Trop. Grassld Soc. Aust.* **6**, 35–53.

MacKenzie, J. A. and **H. R. Chheda** (1970) Comparative root growth studies of 'Cynodon IB.8': an improved variety of Cynodon forage grass suitable for southern Nigeria and two other Cynodon varieties, *Niger. Agric. J.* **7**, 91–7.

Mannetje, L. 't., S. S. Ajit and **M. Murugaiah** (1976) Cobalt deficiency in cattle in Johore. Liveweight changes and responses to treatments, *MARDI Res. Bull.* **4**, 90–8.

Mannetje, L. 't, N. H. Shaw and **T. W. Elich** (1963) The residual effect of molybdenum fertilizer on improved pasture on a prairie-like soil in sub-tropical Queensland, *Aust. J. Exptl Agric. Anim. Husb.* **2**, 20–5.

Mays, D. A. and **G. L. Terman** (1969a) Sulphur-coated urea and uncoated soluble nitrogen fertilizers for fescue forage, *Agron. J.* **61**, 489–92.

Mays, D. A. and **G. L. Terman** *(1969b) Response of Coastal bermudagrass to nitrogen in sulphur-coated urea, urea and ammonium nitrate, Sulphur Inst. J.* **5**, 7–10.

McClung, A. C. and **L. R. Quinn** (1959) *Sulphur and Phosphorus Responses of Batatais Grass* (Paspalum notatum), IBEC Res. Inst., Bull. No. 18: New York.

McGlohon, N. E., J. N. Sasser and **R. T. Sherwood** (1961) *Investigations of Plant-parasitic Nematodes Associated with Forage Crops in North Carolina,* N.C. Agric. Expt. Sta., Tech. Bull. 148: Raleigh.

Michelin, A. and **L. V. Crowder** (1959) Influencia de los niveles y frequencia de aplicación de nitrógeno en la produccion del pasto pangola, *Agric. Trop. (Colombia)* **15**, 843–52.

Michelin, A., L. A. Leon and **A. Ramirez** (1974) Uso eficiente de fertilizantes en la producción de pastos en suelos acidos, *Suelos Ecuatoriales* **6**, 265–87.

Michelin, A., A. Ramirez, J. Lotero and **H. Chaverra** (1968) Frecuencia de corte y aplicacción de nitrógeno en Coastal bermuda, Pangola y pará en el Valle de Cauca,

Agric. Trop. (Colombia) **24**, 682–5.

Minton, N. A., I. Forbes and **H. D. Wells** (1967) Susceptibility of potential forage legumes to Meloidgyne species, *Plant Dis. Reptr* **51**, 1001–4.

Mitchell, K. J. (1963) Production potential of New Zealand pastureland, *Proc. N. Z. Inst. Agric. Sci.* **9**, 80–96.

Moody, P. W and **D. G. Edwards** (1978) The effect of plant on critical phosphorus concentration in Townsville stylo (Sylosanthes humilis H. B. K.), *Trop. Grassld* **12**, 80–9.

Moore, A. W. (1965) The influence of fertilization and cutting on a tropical grass-legume pasture, *Exptl Agric.* **1**, 193–200.

Mott, G. O. (1970) Nutrient recycling in pastures. In D. A. Mays (ed.), *Forage Fertilization*, ASA, CSSA, SSSA: Madison, Wisconsin, Ch. 15.

National Research Council (1963) *Nutrient Requirements of Domestic Animals. No. 4: Nutrient Requirements of Beef Cattle*, Nat. Acad. Sci.: Washington, D. C.

Neme, N. A. (1966) Adubos fosfatados e calcario na produçao de foragem de soja perenne (Glycine javanica L.) em Terra-Roxa, Mistruda (Latosol Roxo), *Proc. 9th Intl. Grassld Cong.*, pp. 677–81.

Neyra, C. A. and **J. Döbreiner** (1977) Nitrogen fixation in grasses, *Adv. Agron.* **29**, 1–38.

Nordfeldt, S., I. Issac, A. K. S. Tom and **L. A. Henke** (1951) *Studies of Napier grass. 1. Nutritive Value, 2. Optimum Feeding Level*, Hawaii Agric. Expt. Sta., Bull. 12.

Norman, M. J. T. (1962) Response of native pasture to nitrogen and phosphate fertilizer at Katherine, N. T., *Aust. J. Exptl Agric. Anim. Husb.* **2**, 27–34.

Norman, M. J. T. (1974) Beef production from tropical pastures. Part 1, *Aust. Meat. Res. Comm. Rev.* No. **16**, 1–23.

Nourrissat, P. (1965) Influence de l'époque de fauche et de la hauteur de coupe sur la production d'une prairie naturelle au Sénégal, *Sols Afr.* **10**, 365–77.

Nye, P. H. and **W. N. Foster** (1961) The relative uptake of phosphorus by crops and natural fallow from different parts of the root zone, *J. Agric. Sci. (Camb.)* **56**, 299–306.

Nye, P. H. and **D. J. Greenland** (1960) *The Soil Under Shifting Cultivation*, Commonw. Bur. Soils Tech. Commun. No. 51, Farnham Royal: England.

O'Brien, R. G. and **W. Pont** (1977) Diseases of Stylosanthes in Queensland, *Queensld Agric. J.* **103**, 126–8.

Oakes, A. J. (1978) Resistance in Hermarthria species to the yellow sugar-cane aphid, Sipha flava (Forbes), *Trop. Agric. (Trin.)* **55**, 377–81.

Olsen, F. J. (1975) Effects of nitrogen fertilizer on yield and protein content of Brachiaria mutica (Forsk.) Stapf, Cynodon dactylon (L.) Pers. and Setaria splendida Stapf in Uganda, *Trop. Agric. (Trin.)* **51**, 523–9.

Olsen, F. J. and **P. G. Moe** (1971) The effect of phosphate and lime on the establishment, productivity, nodulation and persistence of Desmodium intortum, Medicago sativa and Stylosanthes gracilis, *E. Afr. Agric. For. J.* **37**, 29–35.

Olsen, F. J. and **G. L. Santos** (1975) Effects of lime and fertilizers on natural pastures in Brazil, *Exptl Agric.* **11**, 173–6.

Osman, A. E. (1979) Productivity of irrigated tropical grasses under different clipping frequencies in the semi-desert region of the Sudan, *J. Range Mgmt.* **32**, 182–5.

Oyenuga, V. A. and **D. H. Hill** (1966) Influence of fertilizer applications on the yield, efficiency and ash constituents of meadow hay, *Niger, Agric. J.* **3**, 6–14.

Patridge, D. J. (1980) The efficiency of biosuper made from different forms of phosphate on legumes in the hill land in Fiji, *Trop. Grassld* **14**, 87–94.

Perez, I., F. and **E. Lucas** (1974) Cutting intervals and nitrogen fertilization in four cultivated pastures in Cuba, *Proc. 12th Intl. Grassld Cong.*, pp. 191–201.

Peterson, R. G., H. L. Lucas and **W. W. Woodhouse, Jr.** (1956a) The distribution of excreta by freely grazing cattle and its effect on pasture fertility: I. Excretal distribution, *Agron. J.* **48**, 440–4.

Peterson, R. G., W. W. Woodhouse Jr. and **H. L. Lucas** (1956b) The distribution of excreta by freely grazing cattle and its effect on pasture fertility: II. Effect of returned excreta on the residual concentration of some fertilizer elements, *Agron. J.* **48**, 444–9.

Piñedo, L. and **K. Santhirasegaram** (1973) Respuesta de algunas especies de pastos tropicales a la aplicación de fosfóro, Proc. IV Reunion Assoc. Latinoamer. Prod. *Animal*, CIAT: Colombia.

Poultney R. G. (1959) Preliminary investigations on the effect of fertilizers applied to natural grassland, *E. Afr. Agric. J.* **25**, 47–9.

Prine, G. M. and **G. W. Burton** (1956) The effect of nitrogen rate and clipping frequency upon the yield, protein content and certain morphological characteristics of Coastal bermudagrass (Cynodon dactylon(L.) Pers.), *Agron. J.* **48**, 296–301.

Quinn, L. R., G. O. Mott and **W. V. A. Bisshoff** (1961) *Fertilization of Guinea Grass Pastures and Beef Production with Zebu Steers*, IRI Res. Inst., Bull. No. 24: New York.

Quinn, L. R., K. L. Swiercznski, W. L. Schilmann and **F. H. Gullove** (1956) *Experimental Program on Brush Control in Braziah Pastures*, IBEC Res. Inst. Bull. No. 10: New York.

Rains, A. B. (1963) *Grassland Research in Northern Nigeria*, Samaru Misc. Publ. 1, Inst. Agric. Res., Ahmadu Bello Univ.: Zaria.

Revilla, V. A. (1967) *Una Virosis del Pasto Pangola en el Peru*, Estac. Agric., La Molina: Peru.

Rivera-Brenes, L., J. Torres and **J. Arroyo** (1961) Response of guinea, Pangola and Coastal bermuda grasses to different nitrogen fertilization levels under irrigation in the Lajas Valley of Puerto Rico, *J. Agric. Univ. P. R.* **45**, 123–46.

Robertson, A. D., L. R. Humphreys and **D. G. Edwards** (1976) Influence of cutting frequency and phosphorus supply on the production of Stylosanthes humilis and Arundinaria pusilla at Khon Kaen, north-east Thailand, *Trop. Grassld* **10**, 33–9.

Robinson, P. J. and **R. K. Jones** (1972) The effect of phosphorus and sulphur fertilization on the growth and distribution of dry matter, nitrogen, phosphorus and sulphur in Townsville stylo (Stylosanthes humilis), *Aust. J. Agric. Res.* **23**, 633–40.

Rodel, M. G. W. (1969) The effect of applying nitrogen in various ways on the herbage yields of giant Rhodes grass (Chloris gayana Kunth.), *Rhod. Agric. J.* **66**, 44–5.

Rural Research in CSIRO (1972) Dung beetles on the move, March, 1972, No. 75, pp. 2–6.

Russel, D. A., W. J. Free and **D. L. McCune** (1974) Potential for fertilizer use on tropical forages. In D. A. Mays (ed.), *Forage Fertilization*, ASA, CSSA, SSSA: Madison, Wisconsin, pp. 39–65.

Saleem, M. A. M. (1972) Productivity and chemical composition of Cynodon IB.8 as influenced by level of fertilization, soil pH and height of cutting, Ph.D. Thesis: Univ. Ibadan, Nigeria.

Saleem, M. A. M., H. R. Chheda and **L. V. Crowder** (1975) Effects of lime on herbage production and chemical composition of Cynodon IB.8 and on some chemical properties of the soil, *E. Afr. Agrir. For. J.* **40**, 217–26.

Salette, J. E. (1970) Nitrogen use and intensive management of grasses in the wet tropics, *Proc. 11th Int. Grassld Cong.*, pp. 404–7.

Salter, R. M. and **C. J. Sollenberger** (1939) *Farm Manure.* Ohio Agric. Expt. Sta., Bull. 605, Wooster: Ohio.

Sanchez, P. A. (1976) *Properties and Management of Soils in the Tropics*, Wiley: New York.

Schmidt, P. E and **R. Watkins** (1968) *Response of Natural Coastal Pastures to Fertilizers and Some Economic Considerations in Beef Production*, Annual Report of Livestock Breeding Sta., Tanga: Tanzania.

Semple, A. T. (1970) *Grassland Improvement*, Leonard Hill Press: London, Ch. 10.

Shaw, N. H., T. W. Elich, K. P. Haydcock and R. P. Waite (1965) A comparison of seventeen introductions of Paspalum species and naturalized P. dilatatum under cutting at Samford, southeastern Queensland, *Aust. J. Exptl Agric. Anim. Husb.* **5**, 423–32.

Simmonds, J. H. (1966) *Host Index of Plant Diseases in Queensland,* Dept. Prim. Ind.: Brisbane.

Simpson, J. R. (1961) The effects of several agricultural treatments on the nitrogen status of a red earth in Uganda, *E. Afr. Agric. For. J.* **36**, 158–63.

Smith, F. W. (1972) Potassium nutrition, ionic relations and oxalic acid accumulation in three cultivars of Setaria sphacelata, *Aust. J. Agric. Res.* **23**, 969–80.

Smith, R. L., J. H. Bouton, S. C. Schank, K. H. Quesenberry, M. E. Tyler, J. R. Milam, M. H. Gaskins and R. C. Littel (1976) Nitrogen fixation in grasses inoculated with Spirillum lipoferum, *Science* **193**, 1003–5.

Sonoda, R. M., A. E. Kretschmer and J. B. Brohmann (1971) Web blight of introduced forage legumes in Florida, *Trop. Grassld* **5**, 105–7.

Sonoda, R. M. (1976) Reaction of Macroptilium atropurpureum and related species to three diseases in Florida, *Trop. Grassld* **10**, 61–3.

Spain. J. M. (1975) Forage potential of allic soils of the humid lowland tropics of Latin America. In E. C. Doll and G. O. Mott (eds), *Tropical Forages in Livestock Production Systems*, ASA Spec. Publ. No. 24: Madison, Wisconsin, pp. 1–8.

Spain, J. M., C. A. Francis, R. H. Howeler and F. Calvo (1975) Differential species and varietal tolerance to soil acidity in tropical crops and pastures. In E. Bornemissza and A. Alvaro (eds), *Soil Management in Tropical America.* N.C. State Univ.: Raleigh, pp. 308–29.

Stephens, C. G. and C. M. Donald (1958) Australian soils and their responses to fertilizers, *Adv. Agron.* **10**, 167–256.

Suckling, F. E. J. (1960) Productivity of pasture species on hill country, *N. Z. J. Agric. Res.* **3**, 579–91.

Swaby, R. J. (1975) Biosuper-biological superphosphate. In K. D. McLechian (ed.), *Sulphur in Australian Agriculture*, Sydney Univ. Press.

Swain, F. G. (1959) Responses to molybdenum three years after previous application on red Basaltic soils of the far north coast of N.S.W., *J. Aust. Inst. Agric. Sci.* **25**, 51–4.

Tapia, M. E. (1971) *Pastos Naturales del Antiplano de Perú and Bolivia*, Publ. Misc. 85, IICA, San Jose: Costa Rica.

Teitzel, J. K. (1969) Responses to phosphorus, copper and potassium on a granite loam on the wet tropical coast of Queensland, *Trop. Grassld* **3**, 43–8.

Teitzel, J. K. and R. C. Bruce (1970) Soil fertility studies on the wet tropical coast of Queensland, *Proc. 11th Intl. Grassld Cong.*, pp. 1475–9.

Theis, T. (1953) *Some Diseases of Puerto Rico Forage Crops*, Fed. Expt. Sta. Bull. 51, Rio Piedras, P. R.

Tothill, J. C. (1974) The effects of grazing burning and fertilizing on the botanical composition of a natural pasture in the sub-tropics of south-east Queensland, *Proc. 12th Int. Grassld Cong.*, pp. 515–21.

Truong, N. V., C. S. Andrew and P. J. Sherman (1967) Responses by Siratro (Phaseolus atropurpureus) and white clover (Trifolium repens) to nutrients on solodic soils at Beaudesert, Queensland, *Aust. J. Exptl Agric. Anim. Husb.* **7**, 232–6.

Trumble, H. C. (1952) Grassland agronomy in Australia, *Adv. Agron.* **4**, 1–65.

Vasquez, R. (1965) Effects of irrigation and nitrogen levels on the yields of guinea grass, para grass and guinea-kudzu mixtures in the Lajas Valley, *J. Agric. Univ. P. R.* **49**, 389–412.

Velasquez, E. R., O. Larez and W. B. Bryan (1975) *Pasture and Livestock Investigations in the Humid Tropics: Orinoco Delta, Venezuela. II. Fertilizer Trials with Introduced Forage Grasses*, IRI Res. Inst., Bull. No. 43, New York.

Vicente-Chandler, J. and **J. Figarella** (1962) Effect of five nitrogen sources on yield and composition of napier grass, *J. Agric. Univ. P. R.* **46**, 102–6.

Vicente-Chandler, J., S. Silva and **J. Figarella** (1959) The effect of nitrogen fertilization and frequency of cutting on the yield of I. Napier grass, II. Guinea grass, and III. Para grass, *J. Agric. Univ. P. R.* **43**, 215–48.

Vicente-Chandler, J., R. Caro-Castros, R. W. Pearson, F. Abruña, J. Figarella and **S. Silva** (1964) *The Intensive Management of Tropical Forages in Puerto Rico*, Univ. P. R. Agric. Expt. Sta., Bull, 187.

Vicente-Chandler, J. F. Abruña, R. Caro-Castros, J. Figarella, S. Silva and **R. W. Pearson** (1974) *Intensive Grassland Management in the Humid Tropics of Puerto Rico*, Univ. P. R. Agric. Expt. Sta., Bull. 233.

Villamizar, F. and **J. Lotero** (1967) Respuesta del pasto pangola a diferentes fuentes y dosis de nitrógeno, *Revista ICA (Colombia)* **2**, 57–70.

Vine, H. (1968) Developments in the study of soils and shifting agriculture in tropical Africa. In R. P. Moss (ed.), *The Soil Resources of Tropical Africa*. Cambridge Univ. Press: London, Ch. 5.

Watkins, J. M. and **M. Lewy-Van Severen** (1951) Effect of frequency and height of cutting on the yield, stand and protein content of some forages in El Salvador, *Agron. J.* **43**, 291–6.

Watson, G. A. (1960) Interaction of lime and molybdate in the nutrition of Centrosema pubescens and Pueraria phaseoloides, *J. Rubber Res. Inst. Malaya* **16**, 126–38.

Weier, K. L. (1980) Nitrogen fixation association with grasses, *Trop. Grassld* **14**, 194–201.

Whitehead, D. C. (1970) *The Role of Nitrogen in Grassland Productivity*, Commonw. Bur. Past. Fld Crops Bull. 48, Hurley England.

Whitney, A. S. (1970) Effects of harvest interval, height of cut and nitrogen fertilization on the performance of Desmodium intortum mixtures in Hawaii, *Proc. 11th Int. Grassld Cong.*, pp. 632–6.

Whitney, A. S. (1974) Growth of kikuyu grass (Pennisetum clandestinum) under clipping. I. Effects of nitrogen fertilization, cutting interval and season on yield and forage characteristics, *Agron. J.* **66**, 281–7.

Whitney, A. S. and **R. E. Green** (1969) Pangola grass performance under different levels of nitrogen fertilization in Hawaii, *Agron. J.* **61**, 577–81.

Wilkinson, S. R. and **G. W. Langdale** (1974) Fertility needs of warm-season grasses. In D. A. Mays (ed.), *Forage Fertilization*, ASA, CSSA, SSSA: Madison, Wisconsin, Ch. 6.

Williams, C. H. and **C. S. Andrew** (1970) Mineral nutrition of pastures. In R. M. Moore (ed.), *Australian Grasslands*, Aust. Natl. Univ. Press: Canberra, Ch. 21.

Wollner, W. and **J. L. Castillo** (1968) The effect of different levels of N on the yield of Pangola (Digitaria decumbens Stent.), *Revta. Cub. Cienc. Agric.* **2**, 227–32.

Woodhouse, W. W. Jr. (1968) Long-term fertility requirements of Coastal bermudagrass. I. Potassium, *Agron. J.* **60**, 508–12.

Woodhouse, W. W. Jr. (1969) Long-term fertility requirements of Coastal bermudagrass. II. Nitrogen, phosphorus and lime, *Agron. J.* **61**, 251–6.

Wu, M. H. (1964) Effect of lime, molybdenum and inoculation of rhizobia on the growth of Leucaena glauca on acid soil, *J. Agric. Assoc. China* **37**, 57–60.

Yost, R. S., E. J. Kamprath, E. Lobato, G. C. Naderman and **W. V. Soares** (1976) Residual effects of phosphorus applications. In *Agronomic-Economic Research on Tropical Soils, Annual Report 1975*, N. C. State Univ.: Raleigh, pp. 28–32.

Younge, O. R., D. L. Plucknett and **P. P. Rotar** (1964) *Culture and Yield Performance of Desmodium intortum and D. canum in Hawaii*, Hawaii Agric. Expt. Sta., Bull. No. 59.

Younge, O. R. and **D. L. Plucknett** (1966) Beef production with heavy phosphorus fertilization in infertile wetlands of Hawaii, *Proc. 9th Intl. Grassld Cong.*, pp. 959–63.

Chapter 10

Grazing management and animal feeding

Grazing management consists of the wise and skilful manipulation of two basic biological systems: the pasture sward, i.e. the herbage available for grazing, and the grazing animal. In a livestock enterprise, the animal producer or manager attempts to regulate animal numbers so as to utilize a fluctuating supply of herbage effectively and efficiently. Factors affecting either system will ultimately influence output per animal (meat, milk, wool, reproduction) and pasture yield per hectare (herbage dry matter, botanical composition, feed units, animal product). In sown pasture husbandry the manager has more direct control of some factors, and can generally expect a higher return per unit of input and per hectare, than in an extensive range enterprise. The principal factors under direct control of the manager include (a) the choice of grasses and legumes for the pasture, (b) manipulation of agronomic practices in pasture maintenance, e.g. fertilization, weed control, (c) selection of livestock, type of animals as well as individuals, (d) choice of grazing system and (e) control of livestock and, indirectly, control of herbage quality and nutritive value by manipulation of livestock, e.g. intensity of grazing, and (f) use of supplementary feeding.

A vast majority of livestock in the tropics graze native and natural grasslands and relatively few are given access to improved pastures. Under most conditions, an excess of herbage generally prevails for a period of time in the rainy season, but an extreme shortage occurs in the dry season. Carrying capacity is too often based on the flush period, resulting in overstocking and overgrazing in the dry period. A rapid decline in herbage nutritive value of predominately grass-based grazinglands imposes an additional constraint, so that animals lose a high proportion of the liveweight gained in the rainy season.

Livestock-keeping in the tropics often represents a way of life as well as providing beef and milk and other amenities. It is only when modern agriculture and industralization emerge that increased demands for meat, milk and milk products, wool and other animal products arise. This forces changes in livestock-keeping and prompts consideration of intensified animal and grazingland management.

Two major contrasting lines of pasture improvement have developed in the tropical and subtropical areas (Osbourn, 1975). In the humid tropics of Central and South America, the Pacific Islands and Southeast Asia, vegetatively propagated grass and high levels of N fertilizer have been utilized. In the monsoonal areas use of legume-based pastures established from seeds and limited amounts of phosphate have been the major features of improved pastures, initially in Australia then extending into Africa, India and South America. The fossil fuel energy needed to fix and produce a unit of N fertilizer is about 5.5 times that for a unit of P as superphosphate. With the recent rise in prices of fuel and fertilizer and the anticipated shortage of fossil fuel it would seem more propitious to intensify research efforts and direct on-farm practices towards utilization of legumes in pasture and forage crop improvement.

Forage utilization methods

An efficient livestock production system requires that pasture herbage be converted to animal products. Plant growth takes place in an environment with a favourable combination of moisture, nutrients, temperature, light and space. In parts of the humid tropics this may occur throughout the year so that a continuous supply of herbage is available for grazing. In much of the tropics and subtropics, however, pasture production occurs in a defined rainy season, thus imposing adjustments in the grazing management system if herbage is to be distributed over a year. The lack of a 12-month feed supply is perhaps the most serious constraint to animal production in most of the tropics. In the temperate zone, and parts of the subtropics, a combination of grazing, hay and silage allows flexibility in planning a year-round animal feed programme, but forage conservation is not a common practice in the tropics. In some areas standover feed is accumulated for dry season utilization, e.g. use of Townsville stylo as 'standing hay', but is limited in scope. Most tropical grasses are unsuited for this practice since they become stemmy and lignified with the onset of dry weather. This occurs despite such management practices as late-season mowing, restricted grazing and N topdressing. Accumulation of legume forage has not been widely accepted as a means of providing dry season supplemental feed. Hay-making is usually difficult because of uncertain drying conditions. Thus, most livestock producers are confronted with a management system of undergrazing in the wet season and overgrazing in the dry season.

Continuous grazing

This is an extensive system of grazing in which livestock remain on the same pasture area for prolonged periods. It is the most common system used in the tropics and subtropics, generally being satisfactory for relatively low-yielding pastures. Under continuous grazing, the numbers of animals per unit area of land are usually kept constant. This causes either under- or overstocking as plant growth and seasonal conditions are not predictable. In much of the tropics the livestock producer or manager is apt to select a stocking rate based on herbage available in the wet season so that a feed shortage then exists in the dry season. Even with undergrazing during the flush period so as to accumulate a reserve for the sparse season, herbage quality declines so that livestock lose weight.

Spot-grazing is likely to occur when the pasture is undergrazed, causing a wide range in forage maturity as the season advances. Cattle repeatedly graze in 'patches', keeping the sward height at a lower level than in areas avoided. Nutritive value of herbage in these closely grazed sites exceeds that of the more mature forage. Continuous grazing usually favours production per head during the season of plant growth as compared to other systems because of more selective grazing. Selectivity in this case refers to the more nutritious species in a mixed sward and to specific parts of the plant. During the prolonged dry season this ceases to be a factor in terms of animal output.

Continuous grazing in humid areas permits the build-up of ticks and internal parasites of grazing cattle and possibly nematode infestation of the soil. Grazing young stock on the same pasture as older cattle causes heavy infestation of helminth parasites and retarded growth. These problems are considerably reduced under rotational grazing.

Rotation grazing

This system of grazing management is suited to intensive utilization of sown pastures, especially where irrigation is available. It is designed to obtain more uniform grazing of the pasture sward than continuous grazing. In most rotational systems, the pasture is cross-fenced into paddocks and the herd allowed access to one until the herbage is uniformly grazed to a given height, then the herd is moved into a second division, and so on. By the time animals return to the first paddock a uniform growth should have developed. This system eliminates to a great extent spot-grazing and admixture of mature and immature vegetation, as animals are forced to consume a major portion of the herbage. Forage yield is potentially higher from rotational than from continuous grazing.

The interval between grazings depends on pasture species, available moisture from rainfall or irrigation, applied fertilizer, livestock enterprise, etc. The scheduling of grazing and regrowth periods on a calendar basis and a fixed scheme of rotation by paddocks can seriously harm regrowth potential and plant persistency. Consideration should be given to phenological and morphological characteristics, to the physiological response of plant species or cultivars comprising the pasture sward and to the effects of grazing pressure on regrowth potential. Thus, the system should be designed so as to allow flexibility in rest and grazing periods. In an intensive system the rest period is used for fertilizing, mowing, and irrigating if these are practised.

Leader–follower groups

In this rotational grazing system animals are divided into two groups, high producers (milking cows and fattening cattle) and low producers (dry cows and reserves) (Blaser *et al.*, 1959; Evans and Hacker, 1973). The higher producers enter a paddock first to graze the more nutritious top growth or about one-half the grazeable herbage. This group remains on the paddock for 3–7 days. They are moved to a second paddock and followed by the second grazers to consume the residue left by the first grazers. In this way the higher-producing animals have access to the more nutritious and highly digestible portion of the herbage on offer.

Strip or ration grazing

This is an extreme form of rotation grazing. A small section of the pasture or grazing area is separated by an electric fence and the animals permitted to graze. As soon as the herbage is grazed to a given above-ground level the fence is moved and another strip made available. Often a second fence is moved behind to prevent grazing of regrowth and maintain more even distribution of excreta. The fences are usually moved daily. In some places a fence is not used but livestock are herded on to a given section. For example, lucerne is ration-grazed in the valley around Mexico City by close herding. The system provides a uniformly developed sward and allows near maximum utilization of herbage. It is applicable under conditions of highly productive pastures where high net returns are expected.

Deferred grazing

With sown pastures this system refers to setting aside certain pasture areas or paddocks for accumulation of hay *in situ* to be grazed at a later date. This is a common practice with Townsville stylo in northern Australia, but otherwise not widely practised. The feeding value of most grasses rapidly declines with maturity and drying out. The stand-over material generally provides no more, and perhaps less, than maintenance ration.

The practice also refers to a scheme of improving rangeland. By withholding cattle, plants attain maturity and develop greater root and crown reserves for subsequent regrowth. Furthermore, seeds form and are dispersed for stand improvement.

Fresh cut and daily feeding

This is sometimes called 'green-chop', 'soiling' or 'zero-grazing'. Forages such as elephant, guinea and Guatemala grasses are grown specifically for cutting as fresh herbage, transporting to penned livestock and feeding on a daily basis. In addition, sections of a pasture may be reserved for fresh cutting, or grasses growing along roadways and in out-of-the-way places may be cut at opportune times. These materials are cut by hand or by a tractor-drawn forage chopper. The difference in 'green-chop' and 'soiling' lies in handling of manure and feed residue. In a soiling practice they are returned directly to the land area where the forage is grown. With 'green-chop' they are not.

Hay and silage

These are not common forage conservation practices in most of the tropics and subtropics. Under intensive management practices, and especially for dairying, they may be utilized. They are discussed further in Chapter 12.

Continuous vs. rotational grazing

The livestock producer may sometimes question whether a pasture should be continuously or rotationally grazed. This is a difficult decision and no simple answer can be given. The livestock farmer holds interest in (1) liveweight gain, milk or wool production, (2) animal production per hectare and (3) an economic return. Animal output under grazing conditions is influenced by the amount of herbage consumed and its digestibility. Product per land area depends on animal performance and carrying capacity. The two are interdependent and impinge on the class of livestock, animal management, stocking rate, quantity and quality of forage, amount of herbage utilized and selective grazing (Blaser *et al.*, 1959). Economic return is largely dependent on the managerial skill of the livestock producer in manipulating and integrating these factors.

Controversy exists as to whether continuous or rotational grazing is superior in terms of animal and pasture performance. Records of rotational grazing date back for more than 350 years (Wheeler, 1962), and grazing management experiments have been in progress for over 60 years (Barnard and Frankel, 1964). Despite the many trials comparing the two systems, there is still marked disagreement about their relative merits. Research carried out in temperate zones indicates that grazing methods *per se* have little effect on animal production (Wheeler, 1962; Walton, 1981). The extensive and elaborate tests conducted by McMeekan and Walshe (1963) in New Zealand showed an 8 per cent advantage of milk production for rotational over continuous grazing. Blaser *et al.* (1969) also noted an advantage for milk and meat production when pastures were rotationally grazed. Benefits from the rotational system were attributed to the greater amounts of forage produced and conserved, thus permitting increased carrying capacities. Fewer experiments comparing methods of grazing have been carried out in the tropics. In general, the published data have shown comparable results to those of the temperate regions.

Thus, in deciding whether to use continuous or rotational grazing the livestock producer must critically analyse the local situation. Usually, with extensive grazing of native or naturalized grazing-lands, as well as sown pastures of low productivity, the continuous system would be advisable. This scheme is simpler, more easily managed and will be more efficient under these conditions. With intensive grazing of high-producing pastures, e.g. beef or milk production on N-fertilized grass, well-managed grass–legume pastures and where irrigation is accessible, the rotational system might be more appropriate. It should be remembered, however, that closer supervision and greater managerial skill will be needed for rotational grazing. Furthermore, full utilization of herbage by grazing may be necessary because of the greater difficulty of forage conservation in most of the tropics.

Comparison of grazing systems in sown pastures

On a sown pasture of centro and guinea grass in Queensland (3 225 mm annual rainfall), Grof and Harding (1970) compared continuous grazing and a two-paddock rotational system (2 weeks on and 2 weeks off). At a fixed stocking rate of 3.5 animals/ha, liveweight gains of beef cattle averaged 135 and 155 kg/year over a 2-year period for the two systems, respectively. Benefit of rotational grazing occurred in the wet season. In Uganda (1 389 mm rainfall) Stobbs (1969c) used East African Zebu steers to grass Siratro and guinea grass pastures over a period of 29 months. Treatments included continuous, three-paddock and six-paddock rotational grazing of 4.94 animals/ha. The continuous and three-paddock systems gave significantly higher gains per animal (99 and 95 kg/year) than the six-paddock system (84 kg), but did not differ from each other. Joblin (1963), also in Uganda, compared three-paddock rotations and strip-grazing of a *Hyparrhenia rufa* and *Chloris gayana* mixture, using 2.4 East African Shorthorn Zebu heifers ha/ between the ages of 1 and 3 years. After 8 months the difference in favour of paddock-grazing was 24 per cent and after 16 months (including a dry season) 14 per cent. Overall, the liveweight gains were 114 and 100 kg gain per animal per year. The difference was attributed to forced consumption of the whole plants under strip-grazing. A comparison of continuous and rotational systems was made in Rhodesia (Zimbabwee) (889 mm rainfall) with cows and calves grazing *Cynodon plectostachyus* fertilized with 335 kg/ha of N per year (Rodel, 1971). In a 4-month growing season 7.5 cows and 7.5 calves/ha were continuously grazed and the same among three paddocks (1 week on and 2 weeks off). Two other groups each of six cows and six calves were continuously grazed and strip-grazed. Rotational and strip-grazing did not improve liveweight gain of cows or calves, each making from 80 to about 100 kg gain over the grazing period.

In the Cauca Valley of Colombia (2 000 mm rainfall) no significant difference was noted when steers grazed N-fertilized Pangola grass (150 kg/ha per year) continuously or rotationally in a four-paddock system (Ayala, 1969). Daily gains ranged from 0.53 to 0.60 and 0.58 to 0.65 kg/head, respectively, over a 333-day period. In the eastern plains of Colombia (1 750 mm rainfall) steers grazing continuously on unfertilized molasses grass performed equally as well as those on rotational grazing. Under more intensive management and high N fertilization in Puerto Rico (2 000 mm rainfall), rotational grazing of five grasses (elephant, guinea, Pangola, pará and Giant Cynodon) gave a definite advantage over continuous grazing (Caro-Costas *et al.*, 1965). Gains of dairy heifers and milk production from lactating cows grazing small paddocks for 3–5 days, allowing re-growth for 21 days, exceeded that of continuous grazing. Another study in Puerto Rico compared rotational versus permanent grazing of kudzu–molasses grass pastures, using five 2-acre pastures over 3 consecutive years (Vicente-Chandler *et al.*, 1953). Annual liveweight gains averaged 486 and 450 kg/ha, respectively, for the two systems. It was found that five paddocks were required for efficient rotational grazing. The paddocks needed to be of such size for consumption of herbage in 5–7 days. If animals remained longer, early grazed plants renewed growth that was readily eaten. This was damaging since regeneration of growth utilized carbohydrate reserves, and if grazed before replenishment pasture recovery was delayed. When this occurred repeatedly yields diminished, stands declined and weeds invaded the pasture.

Effects of grazing system in native and naturalized grazinglands

On naturalized grazingland in northern Nigeria (800 mm rainfall), continuous grazing produced slightly higher liveweight gain (15.8 kg/ha per year) as compared to deferred and three-paddock rotational grazing (12.1 and 11.2 kg/ha, respectively). In the southern Guinea zone (1 200 mm rainfall), however, a rotational system was superior to continuous grazing (104 and 86 kg/ha, 2 years' total of 6 months' grazing per year, respectively) when stocked at one animal unit per hectare (Leeuw, 1971). The advantage was attributed to greater selectivity and rest periods that favoured recovery of desirable grasses. A comparison of continuous, rotational (2 weeks on and 6 weeks off), rotational with spelling and slashing of *Heteropogon contortus* plus *Bothrichloa bladhii* pastures in Queensland (756 mm rainfall) favoured the continuous system (Humphreys, 1977). Annual liveweight gains over 5 years averaged 85, 67 and 49 kg/head per year when steers were stocked at 0.41/ha. In Uganda (890 mm rainfall) a native pasture was cleared of *Cymbopogon afronardus*, leaving *Themeda triandra*, *Hyparrhenia filipendula*, *Brachiaria decumbens* and other minor species (Harrington and Pratchett, 1974). After 3 years of grazing by Ankole steers a continuous treatment gave average annual gains of 105 kg/ha and a rotational scheme of five paddocks (with the fifth reserved for dry-season grazing) gave 91 kg/ha. This occurred at a stocking rate of 0.8 steer/ha. Rotational grazing allowed the weedy grass *C. afronardus* to regenerate more rapidly than did continuous grazing. Over a 12-year period in Rhodesia (700–900 mm rainfall) Kennen (1969) found that continuous grazing of native grasslands consisting of *Hyparrhenia* and other species permitted greater liveweight gains on an individual animal as well as per hectare basis. When stocked at one steer per 4.5 ha those continuously grazed gained an average of 199 kg/head each 2 years, while those grazing an equivalent area in a four-paddock system gained 150 kg/head. Several other studies of native and naturalized grazinglands also confirmed that under conditions of low herbage production a continuous system gave increased gains per hectare or was equal to a rotational system (Norman, 1960 – Australia; Walker, 1968, Walker and Scott, 1968 – Tanzania; McKay, 1968 – Botswana; McKay, 1971 – Kenya; Thornton and Harrington, 1971 – Uganda).

Jackson (1972) examined the conception rate of cows on native grasslands of Rhodesia. Those grazed continuously conceived at a higher rate than those on short duration rotation (14 days grazing, 50 days rest), with variations ranging from 5 to more than 40 per cent among different herds. Continuous grazing enabled cattle to maintain the so-called sour grasses (those that tend to become coarse, stemmy and lignified) in a shorter and more vegetative stage of growth and therefore more nutritious. Resting allowed these species to reach the unpalatable stage of growth.

Experimentation in grazing management

Grazing experiments can be divided into two broad types:

1. Fixed number of livestock per unit area of land – the researcher decides a number for each pasture system and maintains the rate for the duration of the experiment. The fixed number can be employed in both continuous and rotational systems. In some trials a natural increase occurs because of calving and lambing. Set-stocking is a term sometimes found in the literature and refers to either a fixed number of animals or continuous grazing.
2. Variable number of animals per unit area of land – the stocking rate is varied by the investigator depending on availability of herbage in both continuous and rotational systems. A specialized form of variable stocking, known as 'put and take' grazing, has been utilized in some studies (Blaser *et al.*, 1956; Mott, 1960; Quinn *et al.*, 1965).

There has been much discussion and controversy regarding which category to use in evaluating animal and pasture productivity. Advantages and disadvantages have been reviewed in a number of publications. Perhaps the most valuable in comparative terms are those by Wheeler *et al.* (1973) and Wheeler (1962). For further information the reader is referred to Davies (1946); Jones and Sanderland (1974); Matches (1970); Mannetje *et al.* (1976); McMeekan and Walshe (1963); Mott (1960), Morley (1966); Morley and Spedding (1968); Murtagh (1975); Stobbs and Joblin (1966) and Willoughby (1970).

Definition of terms used in grazing studies (taken in part from Mott, 1960)

Stocking rate

The number of animals per unit area of land, e.g. head/ha or reciprocally ha/head. The number of animals, or animal days, per hectare is a direct measure of herbage output on the pasture, provided the stocking rate allows full utilization of the herbage on offer. If the investigator underestimates herbage production, or overestimates the amount that livestock will consume, then the pasture is not stocked at full capacity. The pasture is being undergrazed. If herbage production is overestimated, then livestock consume more than anticipated. Thus, the pasture is being overgrazed.

Grazing pressure

The number of animals per unit of available herbage. This represents an overall balance between the amount of herbage on offer and animal requirements. It is the rate of stocking at which animals fully utilize a predetermined amount of grazeable herbage. The rate of consumption should be such as to provide an adequate feed supply during the entire grazing season. The term grazing intensity is used interchangeably with grazing pressure.

Carrying capacity

The number of animals per unit area of land or the stocking rate to give optimal daily rate of gain per animal and optimal liveweight gain per unit area of land. The term grazing capacity is used interchangeably with carrying capacity.

Animal days per unit area of land (hectare)

The number of animals per unit area multiplied by the number of days that animals are on pasture. This is considered a measure of quantity, i.e. herbage yield of the pasture.

Output per animal (product per animal)

The animal performance expressed as rate of gain in meat-producing animals or milk production in lactating animals. The results are given on a daily basis or within a given period of time. This is usually considered a measure of herbage quality.

Produce per unit area of land (hectare)

A measure of liveweight gain or milk production, i.e. product per hectare = produce per animal per day × animal days per hectare.

Animal unit

A standardized term sometimes used to describe the number of animals on pasture. One animal unit is considered a 2-year-old beef cow, bull, steer or heifer. A yearling steer or heifer weighing 275–320 kg is equivalent to 0.75 animal unit; a weaner calf (8 or 9 months of age) weighing 160–200 kg, 0.50; a calf 4–8 months, 0.30; a ewe, ram or wether over 1 year, 0.20; a lamb up to 1 year, 0.15 units (Shultis, 1962). Instead of using animal units most investigators quote the weight of experimental animals.

Variable and fixed stocking rates

In carrying out grazing management studies the investigator selects stocking rates based on knowledge or predictable herbage available for grazing. The intent is to balance the number of animals per unit area of land so as to utilize effectively and efficiently a major proportion of the forage at a grazing pressure to assure continued production of desirable species.

Increasing the stocking rate beyond a given number of animals per unit area of land results in a decrease in output per animal but an increase in production per hectare. For example, Mears and Humphreys in Australia (1974) grazed kikuyu grass pastures with

Angus weaner cattle at a range of 2.2–16.6 head/ha. Liveweight gain per head over a 2-year period averaged 168, 128 and 73 kg/year at the low, medium and high stocking rates, respectively. Product per hectare with the heavier rates was related to rate of applied N. In Uganda, Stobbs (1969b) compared three fixed stocking rates and a variable rate using East African Zebu steers on *Hyparrhenia rufa* + *Stylosanthes guianensis* pastures. The cumulative liveweight gains for a 3-year period are shown in Fig. 10.1. Total gain at the low stocking rate of 1.65 steers/ha was 485 kg/ha, compared to 1 244 kg/ ha for the high rate of 4.94 steers/ha, even though the latter lost weight in the dry seasons. Steers at the low and medium rates gained about 0.286 kg/day and those at the high and variable rates 0.213 and 0.220 g/day, respectively. White Fulani steers grazing at 0.42 and 1.25 head/ha on cleared shrub savanna in northern Nigeria gave average daily gains from 0.18 to 0 kg during the dry season. At the Mokwa cattle ranch, Gudali bulls grazed semi-natural *Andropogon gayanus* grassland during the wet season at 1, 2 and 4 head/ha. The low rate gave daily and total liveweight gains of 0.49 and 84 kg/ ha, respectively (3-year average of 171 grazing days/year); 2 head/ ha, 0.28 and 96 kg; 4 head/ha was discontinued after 1 year because of overstocking (Leeuw, 1971). On steep slopes in Colombia, 18- to 24-month-old steers of a native race BOM (black and white colour, with black skin and ears) were stocked at 1 and 2 animals/ha on naturalized *H. rufa* pastures (Ramirez *et al.*, 1968). Daily gains varied through the year, being related to rainfall and growth of grass, but averaged 0.38 kg for the lighter stocking rate and 0.29 kg for the heavier rate. After 1 year the pastures grazed by two animals had deteriorated by overgrazing; bare patches developed due to the death of *H. rufa* plants and were invaded by broad-leaved weeds. Similar results of liveweight gains per animal and per hectare have been reported by other investigators (Bisset and Marlowe, 1974; Eng *et al.*, 1978; Harrington and Pratchett, 1974; Harker and McKay, 1962; Jones, 1974; Partridge, 1979; Riewe, 1961; Smith, 1966).

Generalized relationship between production and stocking rate

To predict whether a pasture enterprise will be profitable the animal producer needs to know (1) the rate at which beef gains increase or decrease with changes in herbage yield or stocking rate, (2) the costs involved in making these changes and (3) the price of beef produced (Hart, 1972). Costs of inputs are obtained from records and selling prices of beef can be estimated. Information about predicting beef yield is attainable from animal–pasture studies, usually of relatively short duration at experimental stations or localized cattle farms. These results are not readily extrapolated to an individual situation, however. Several models have been derived to show the

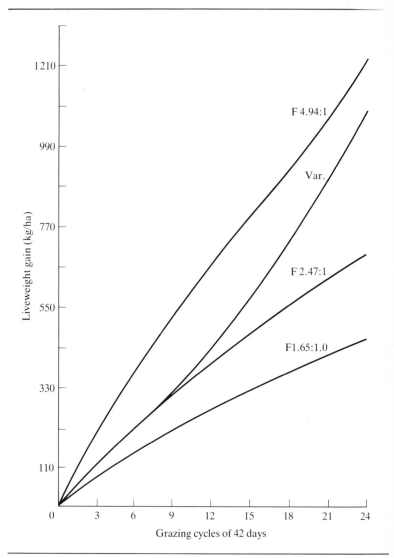

Fig. 10.1 Cumulative liveweight gain from *Hyparrhenia* + Stylo pastures at three fixed (F) stocking rates, 1.65, 2.47 and 4.94 steers/ha, and a variable (Var.) stocking rate in Uganda (redrawn from Stobbs, 1969b).

relative relationship between stocking rate and product per animal and product per unit area of land. The model proposed by Mott (1960) and modified by Heath *et al* (1973) has probably had the greatest influence on pasture and animal investigators in the tropics and subtropics. The relationships are given in relative terms of liveweight gain per animal and per hectare, based on an optimum grazing pressure (Fig. 10.2). The product per animal curve (Y) shows little decrease as stocking rate (X) increases from a lenient rate (undergrazed) to the optimum rate. There exists a rather narrow critical point at the optimum rate beyond which a marked decrease occurs. Product per hectare (Z) shows a very rapid rise as stocking rate increases from an undergrazed condition to the optimum rate. Maximum output takes place at a stocking rate slightly in excess of the optimum, then a sharp drop occurs. Some investigators consider that the nil productivity point extends beyond the relative stocking rate of 1.5. In contrast to the curvilinear model, a linear relation between liveweight gain per animal and stocking rate has been proposed by several researchers (Bennett *et al.*, 1970; Cowlishaw, 1969; Hart, 1972; Morley and Spedding, 1968).

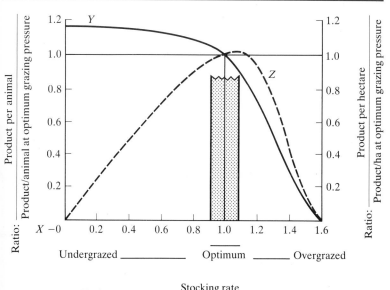

Y = gain/animal

Z = gain/unit area

X = grazing pressure

Figures are relative

Fig. 10.2 Relationship of product per animal and per hectare to stocking rate (redrawn from Heath, *et al.*, 1973). Figures are relative.

Jones and Sanderland (1974) analysed data collected from Setaria – Siratro pastures investigated by Jones (1974) in Queensland. They found that the linear relation between gain per animal and stocking rate more closely fit a linear model than the curvilinear model of Mott (1960). They examined other data from stocking-rate trials carried out over a wide range of environmental conditions, such as tropical and temperate legume–grass pastures and N-fertilized tropical and temperate grass pastures. Data originally used by Mott (1960) were also included in constructing the model shown in Fig. 10.3. The linear model predicts that gain per animal (y_a) at the optimum rate will be one-half that at the lower rate. Also, the negative value for liveweight gain per animal would occur at a stocking rate double the optimum rate and not 50 per cent greater as suggested by Mott (1960). They suggested that gain per hectare (y_h) would change gradually on both sides of the optimum rate and there would be no sharp declines. The authors conceded that the model represented different experiments and no single one covered the en-

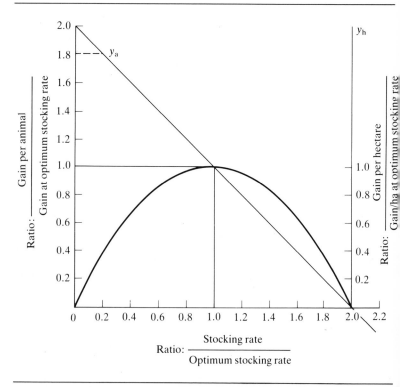

Fig. 10.3 Relationship between stocking rate and both gain per head and gain per hectare from various grazing experiments (redrawn from Jones and Sanderland, 1974).

tire range of stocking rates. For example, the stocking rates encountered were no lower than 0.18 of the relative optimum. Thus, this section was extrapolated by the dotted lines.

Fixed stocking rates

In this system a range of animals per hectare is chosen so as to bracket the optimum grazing pressure. Or, an equal number of animals is selected and unequal-sized paddocks used to obtain a range of stocking rates. The investigator must recognize that the quantity of forage will vary from time to time, depending on the season. Furthermore, selective grazing of certain areas within the pasture (patch grazing) will cause localized variation in stocking pressure. Experimental design, number of replications, field layout of pastures, use of supplements, replacement stock, etc. will vary with treatments. The experimenter should be familiar with these variables before embarking on planning grazing trials (Mannetje *et al.*, 1976).

Variable stocking rates

A 'put-and-take' system of grazing in which stocking rate is intentionally varied according to available herbage in the grazing season, so as to achieve a constant grazing pressure, was first described by Mott and Lucas (1952), then in more detail by Lucas (1962), Peterson and Lucas (1968) and Matches (1970). The method involves two classes of animals:
1. 'Testers' – those selected to remain on the experimental plot for the duration of the full grazing period and about which inferences are to be made, e.g. a measure of herbage quality by average daily animal performance.
2. 'Put-and-take' animals (also called 'regulators' or 'grazers') that are placed on paddocks when available herbage exceeds the daily requirements of testers and removed when quantity of herbage declines.

The put-and-take animals are kept on a supplementary pasture similar to that of the treatment under study. Experimental pastures must be frequently observed to determine whether to 'put' or 'take' animals. Preferably, the regulators are of the same type as testers but not necessarily. The production and weights of testers and regulators are measured at frequent intervals and the numbers of days on the experimental pastures are recorded for each grazing animal. Three methods have been proposed for computing animal output products for the put-and-take system:
1. Using data of all animals on a given treatment so that animal days/ha × average daily gain = product per hectare.
2. Obtaining data on production per head of testers and multiplying this by total number of grazing days, i.e. testers plus regulators.

3. Calculating the energy required for maintenance and production by all animals that grazed a given treatment, and expressing this as effective feed units in terms of the carrying capacity of 'tester grazing' days (Peterson and Lucas, 1968).

Comparison of fixed and variable stocking rates

Physical requirements to carry out trials using fixed or variable stocking rates are not widely different. Put-and-take experiments require slightly more land and animals than a range of fixed stocking rates, but only one stocking rate (as used in most studies) may have appeal to some investigators. The range of fixed stocking rates requires less skill and daily attention, which may be an advantage at smaller experimental stations for on-farm trials. Some investigators have experienced difficulty in deciding when to put-and-take animals due to the subjective judgement of available herbage, and the interpretation of data obtained from this system as compared to fixed stocking (Burns *et al.*, 1970; Stobbs, 1969a). It has been suggested that a wide range of stocking rates needs to be evaluated regardless of the system, and curves fitted for each treatment. Comparisons can then be made on whatever basis is relevant to a given question. In the put-and-take system this would add to the overall cost to a greater extent than the fixed system.

In the final analysis, both the fixed and variable methods have a place in carrying out grazing experiments. Wheeler *et al.* (1973) aptly noted that 'variable rates tend to be appropriate for the study of components of farm systems and fixed rates for the study of systems *per se*'. Grazing experiments are usually designed for long-term study, not readily modified, difficult to interpret and expensive to conduct. Thus, it is wise to spend considerable time in planning, selecting objectives, reviewing treatments and designing experiments before establishing pastures.

Factors affecting stocking rate

The optimum stocking rate, i.e. grazing intensity, will vary for each farm situation. Eventually, each animal producer will have to decide which system of pasture management and the appropirate stocking rate most suitable for his local situation. Several factors to be considered in making this decision include the following (taken in part from Humphreys, 1972).

Rate of forage growth

The herbage produced varies with species, available moisture, supply of soil nutrients, weed competition, topography of land, stocking rate, etc.

Seasonal variation in feed supply

In most parts of the tropics the herbage available for grazing fluctuates with the wet and dry periods. Also, rainfall is variable from year to year, making it difficult to predict production of forage. Feed sources other than from the pasture will alter the stocking rate in stress periods.

Fertilization scheme

The amount and time of N fertilizer applied to grass pastures determine to a great extent plant growth and thus affect carrying capacity. A regular maintenance programme, especially phosphate, is essential to stabilize legume-based pastures and to sustain animal productivity.

Nutritive value of herbage

Animals graze in a selective manner and thus improve their diet. They choose the more desirable species in a mixed sward and select the most nutritious parts of plants. It is not uncommon to find a higher proportion of a given species whithin ingested forage than would appear from the botanical composition of the pasture.

Botanical composition and ground cover

Grazing management of mixed pastures must be such as to maintain a favourable balance of species. If overgrazing occurs desirable species may disappear, leaving bare soil spots. Broad-leaved weeds and weedy grasses often fill in these gaps and may eventually predominate because animals avoid the undesirable plants.

Accessibility of forage

Animals may not have ready access to all parts of a pasture under conditions of extensive grazing. This might occur in areas far-removed from watering points.

Class of animals and nature of animal product

A breeding herd can be sustained at a higher stocking rate than feeder steers. Milking cows require higher-quality pastures than non-lactating cows. Sheep grown for wool are less vulnerable to stress than those grown for meat.

Type of experiment

A trial designed to evaluate persistence of legume or grass species would need to be stocked at a different rate for one to study N response of a single grass species. Stocking rates would likely differ when animals receive feed supplements or when irrigation is accessible in water-stress situations.

Beef production on tropical pastures

Properly managed cultivated grasses and legumes in tropical zones have the potential to improve forage quality and increase herbage yields several-fold over that of native and naturalized grazinglands. Cultivated types are defined as those displaying superiority in some trait as compared to common types. Their use in pastures would lead to dramatic improvement in production per animal and per unit area of land.

The most important factors in cattle-feeding are the quantity of energy-supplying constituents (total digestible nutrients, TDN) and digestible protein. TDN requirements for maintenance can be obtained by multiplying the weight of an animal by 0.008. A 400 kg animal requires about 0.4 kg of total protein daily, assuming 60 per cent digestibility. In addition to maintenance requirements, 3.5 kg of TDN and 0.5 kg of protein are needed per kilogram of liveweight gain. The 10 per cent protein content required in the feed of most cattle can be provided by herbage of improved pastures during the favourable season of plant growth. Cattle consume about 2.5 kg dry herbage (approximately 13.5 kg fresh weight) daily per 100 kg liveweight. Well-managed grass and grass–legume pastures can provide the TDN requirements for maintenance and liveweight gain of grazing cattle.

Beef production from fertilized grass

Cattle with genetic potential are capable of producing remarkable liveweight gains on high-quality pastures. In terms of maximizing beef production, grasses that respond to applied N have the greatest potential in regions with 1 200–1 500 mm annual rainfall and more where supplemental water is available for irrigation. A comparative estimate of beef production under different climatic conditions and management systems is given in Table 10.1. These figures represent a compilation of data from several experiments reported in the literature (Dirven, 1970). A dry season of 6 months was assumed for the monsoonal climate. In the humid tropics there would be sufficient rainfall for plant growth throughout the year. The data show that high liveweight gains per hectare can be expected with sown grasses and/or legumes and use of fertilizers. Yields of herbage, and therefore animal output, vary widely from region to region due to differences in total rainfall and distribution, soil fertility, botanical composition of the pasture, amount of fertilizer used, breed of cattle and stocking rate.

The potential liveweight gain based on theoretical expectation was exceeded at the Parada Research Station in north Queensland. Irrigated Pangola grass pasture that was topdressed at frequent in-

Table 10.1 *Estimates of liveweight gains from beef cattle in monsoon and humid tropical climates* * (*Dirven, 1970*)

Type of pasture and treatment	Climate	
	Monsoon (kg/ha)	Humid tropics (kg/ha)
Natural grazinglands		
Improved grazing management	20	90
Legumes oversown, fertilized	150	400
Cultivated pastures		
Grass–legume mixtures, fertilized	250	600
Nitrogen-fertilized grass	550	1 650

* Calculations based on data from 21 grazing experiments on various continents.

tervals with N and rotationally grazed with 12.4 steers/ha produced a mean of 2 760 kg/ha of liveweight gain over a 3-year period (Norman, 1974). In Florida, steers on St. Augustine grass reportedly gained 1.1 tonne of beef per acre in one year (approximately 2 200 kg/ha liveweight gain) when animals received a daily supplement of 450 g of cottonseed meal (Kidder, 1952). This output was not repeated in subsequent trials. Studies over a 10-year period, however, showed daily gains of 6.35 kg/ha in the summer months, but only 0.8 kg/ha in the winter period (Haines *et al.*, 1965).

Fig. 10.4 Cattle on native pasture in Tanzania. Their condition is a reflection of poor quality of herbage. Photograph courtesy of R. E. McDowell.

Fig. 10.5 Cattle on well-managed, highly nutritious Pangola grass in the Cauca Valley of Colombia. Photograph courtesy of R. E. McDowell.

In Puerto Rico beef production of intensively managed guinea, elephant and Pangola grass pastures was compared with that of elephant grass as fresh, chopped forage (Caro-Costas *et al.*, 1961). The pastures and elephant grass planting received a yearly total of 1 650 kg/ha of 14-14-10 fertilizer topdressed in six applications. Irrigation was supplied as needed to provide 380 mm of water weekly, including irrigation. Liveweight gain of elephant grass cut and fed was 1 500 kg/ha. Increasing the quantity of applied fertilizer gave higher liveweight gains per hectare when steers grazed hillslopes (Vicente-Chandler *et al.*, 1974). Use of 4 000 kg/ha of 15-5-10 fertilizer, split into four dressings per year, boosted liveweight gains per hectare by 67 per cent compared to 1 760 kg/ha of the same fertilizer (Table 10.2). Average daily gains per animal were not significantly altered by topdressed nutrients, a property commonly noted when yields of grass are increased by applied N.

In Queensland annual liveweight gains of steers on Pangola grass were (1) 1 106 kg/ha with 448 kg/ha of applied N and 5.6 animals/ha and (2) 699 kg with 168 kg of N and 4.3 animals/ha (Bryan and Evans, 1971). Fertilized Pangola pastures in the Caribbean gave liveweight gains of 1 180 kg/ha in the Virgin Islands (Oakes, 1960) and 1 288 kg/ha in Jamaica (Richards, 1965). Liveweight gains of 6-month-old weaner cattle grazing kikuyu grass in New South Wales varied from 380 to 1 056 kg/ha, according to N rates of 134–672 kg/ha in year one and 637–1 477 kg/ha in year two (Mears and Humphreys, 1974). In Brazil 200 kg/ha of N applied to guinea grass

Table 10.2 *Effect of nitrogenous fertilizer on annual production of intensively managed elephant grass pastures in Puerto Rico, 5-year average (Vicente-Chandler et al., 1974)*

Fertilizer (kg/ha)	Daily gain (kg/head)	Liveweight gain (kg/ha)	Carrying capactiy/ha* (no.)	TDN consumed[†] (kg/ha)
1 760	0.635	1 042	5.4	7 580
3 080	0.590	1 410	7.2	9 855
4 400	0.590	1 740	8.9	12 210
LSD[05]	np	270	1.0	–

* Calculated from bodyweight, days of grazing and liveweight gains.
[†] Based on one 275 kg animal making normal gains = 3.86 kg TDN daily requirement.

raised liveweight gains of Zebu steers to 704 kg/ha, compared to 301 kg for no N treatment (Quinn *et al.*, 1965).

In terms of liveweight gains per hectare per year, the response usually varies from 0.25 to 4.0 kg liveweight per kilogram of N for the first 100 kg/ha of annually applied N. The potential increases with active plant growth and production of high-quality forage. For example, Mears and Humphreys (1974) found that cattle growth was clearly related to availability of fresh herbage of N-fertilized kikuyu up to 600 kg dry matter per head. The response to increasing levels of applied N declines to about 1.5 and 1.0 kg liveweight gain per kilogram of N between 300 and 400 kg of N. Rates of daily gain average from 0.3 to 0.6 kg/head over the entire year. In Puerto Rico a number of beef breeds grazing solely fertilized grass pastures made daily gains of 0.68 kg, average of several years' data (Vicente-Chandler *et al.*, 1974). In Australia, Hereford and Shorthorn steers weighing about 250 kg/head and grazing various pastures gained 0.9–1.2 kg/head per day for short periods of time on young grass after early rains (Smith, 1970). This compares favourably with temperate-zone pastures. In the tropics, however, daily gains are seasonal and decline as plants age and become less nutritive.

These data emphasize that beef cattle gains under intensive management are directly related to applied N when conditions are favourable for plant growth. Continued high animal output from fertilized grass pastures also depends on the managerial skill of the cattleman and his ability to administer appropriate animal-pasture practices. Cattle numbers must be adjusted to the available herbage, even with irrigation.

Beef production from grass–legume pastures

Animal output from legume-based pastures is closely related to the percentage of legume in the mixture (see Fig. 16.5 p. 439). Total liveweight gain expected from such pastures, however, does not approach the potential of N-fertilized grass (Table 10.1 p. 243). Legumes cannot provide sufficient N for maximum growth of grass, so that dry matter yields of legume–grass mixtures are lower than N-fertilized high-yielding grasses. A comparison of beef production from legume–grass and N-fertilized grass pastures is shown in Table 10.3. An analysis of the table reveals that (1) relatively few direct comparisons of grass and grass–legume pastures have been made in the same experiment, (2) a range in liveweight gains occurs with N-fertilized grass and mixtures, reflecting divergent environmental effects, (3) liveweight gains from grass–legume pastures exceed grass alone at low levels of N fertilizer and (4) addition of legume in some instances is equivalent to more than 100 kg/ha of N fertilizer.

At the South Johnstone Research Station (2 500 mm rainfall) in Queensland, Grof and Harding (1970) examined the relative merits of a legume-based guinea grass pasture and pure guinea grass with and without applied N. Each pasture treatment consisted of two 1.6 ha paddocks on a rotational grazing scheme of 2 weeks on and 2

Table 10.3 *Comparison of beef production from N-fertilized grass and grass–legume pastures expressed in animal liveweight gain per hectare*[*]

Grass(es)	$N^†$ (kg/ha)	LWG (kg/ha)	Legume(s)	LWG (kg/ha)	Reference
1. Molasses[‡]	340	633	Kudzu	545	Vicente-Chandler et al. (1964)
2. Ruzy[‡]	165	650	Centro	511	Mellor et al., (1973a)
3. Pangola	168	699			
	448	1 106	Mixture[§]	507	Bryan and Evans (1971)
4. Guinea	165	585	Centro	450	Grof and Harding (1970)
5. Pangola	100	531	Centro	410	Aronovich et al. (1970)
6. Pará	100	260			
	200	310	Centro	305	Magadan et al. (1974)
7. Guinea + Rhodes + *Hyparrhenia rufa*	140	393	Centro and stylo	315	Stobbs (1969a)
8. Setaria	336	491	Siratro	256	Jones (1974)
			Desmodium	256	

[*] Taken in part from Stobbs (1975b).
[†] N applied to grass alone.
[‡] A factor of 1.1 used to convert from 1b/acre to kg/ha.
[§] Mixture of Phasey bean, greenleaf, Desmodium, Lotononis and white clover seeded; Phasey bean disappeared after establishment year, Lotononis and Desmodium on drier soil areas, white clover on wetter soil areas.

weeks off. The stocking rate was 4.25 steers/ha for each treatment. The cumulative liveweight gains for the three pastures are shown graphically in Fig. 10.6. In the first year gain was increased from 365 kg/ha per year on grass alone to 450 kg/ha by the inclusion of centro. A further increase to 585 kg/ha was obtained by applying 165 kg of N. At the end of 2 years the legume–grass pasture had yielded 35 per cent more liveweight gain than grass alone, and top-dressing a total of 330 kg/ha of N gave a 41 per cent yield over legume–grass. Working in the same area but off the experimental station, Mellor *et al* (1973b) recorded liveweight gains of 928 kg/ha from a recently established guinea grass–centro pasture. A gradual decline in productivity occurred over a 3-year period, however, and levelled off at about 560 kg/ha. The reduction appeared to be related to declining N status of the soil but may have also been an effect of Mo deficiency.

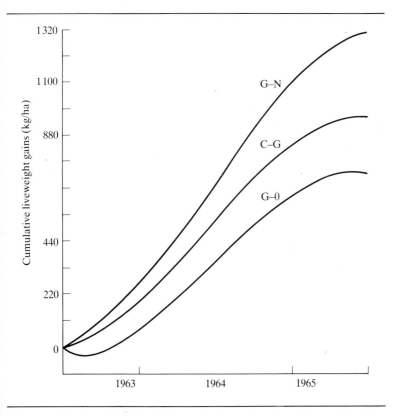

Fig. 10.6 Cumulative monthly liveweight gains on Centro–guinea grass (C–G), guinea grass alone with 165 kg/ha of N annually (G–N) and without N (G–0) (Grof and Harding, 1970).

At the Walkamin Research Station in Queensland, Miller and van der List (1977) examined beef production from pará grass (*Brachiaria mutica*) and *Setaria anceps* in mixture with several legumes under supplemental irrigation. Annual liveweight gains for the 3-year trial in Experiment 1 were 686 kg/ha for pará–Glycine, 522 for pará–stylo and 437 for pará–Siratro. The pastures were legume-dominant at the beginning of the trial. After 1 year grasses and legumes comprised about equal proportions. During the cool season of the second year most of the stylo died with some regeneration of seedlings. This pasture remained grass-dominant until the end of the experiment. Siratro became infected with *Rhizoctonia solani* in the wet season of each year and the percentage composition gradually declined. Glycine remained well balanced in mixture with pará throughout the trial. All pastures were rested 9 weeks in May–June of the second year and again in December–January of the third year because of lack of legume in the stylo and Siratro pastures. In Experiment 2 liveweight gains for 2 years averaged 714 and 670 kg/ha for pará–Glycine at two stocking rates, 5.1 and 3.4 steers/ha, respectively; 645 and 674 kg/ha, respectively, for Setaria–Glycine; 597 and 575 for Setaria–Siratro. When grazing began all pastures were grass-dominant (78–81%). After 9 months of grazing, the lightly stocked Setaria pastures contained 30–35 per cent legume and the pará pasture had about 60 per cent Glycine. The heavily stocked pastures contained about one-half these amounts of legume. After another year the Setaria pastures had only traces of legume, but Glycine comprised 25 and 40 per cent of the lightly and heavily stocked treatments, respectively, in the pará–legume pastures. Survival and productivity of legumes vary with management practices, type of soil and available moisture, but investigators and cattlemen have generally complained about maintenance of legumes.

Economics of beef production from N-fertilized grass and grass–legume pastures

An economic analysis of (1) Pangola grass alone with relatively low input of 168 kg/ha of N (annual liveweight gains of 699 kg/ha), (2) Pangola with 448 kg/ha of N (1 106 kg liveweight gain) and (3) Pangola with legumes (507 kg liveweight gain) showed that all three required large investments. The alternatives were high-cost and high-turnover systems with low internal rates of return (Firth *et al.*, 1974). All budgets were found to be sensitive to beef price and cost change. With an increase in produce price the high N budget was found to be far more sensitive to beef price than to cost of storing cattle or land input. Land value represented a large component in the cost system of the legume-based pasture budget. With relatively high beef prices, however, all three budgets appeared to hold attractive investment opportunity.

Nuthall and Whiteman (1972) used data from various legume-based and N-fertilized tropical pastures to review the economic evaluation of beef production. They fitted regression curves based on legume-based pasture productivity of 275, 385 and 495 kg liveweight gain per hectare and N levels of 55, 110, 220, 330 and 440 kg applied in four topdressings per year to grass alone. Using a beef price regime of $112 (Australian)/100 kg of beef, the 385 liveweight gain from a legume-based pasture showed a break-even point. Considering N at 17.6 cents/kg (current price at the time of study) the yield necessary to break even at 110 kg of N was 568 kg liveweight per hectare, and at 440 kg of N 1 113 kg liveweight gain. Their regression line showed that expected yields at these two levels of applied N would have been 396 and 1 015 kg liveweight gain per hectare. This suggested that only high rates of N fertilizer were competitive at the time of study.

Vicente-Chandler *et al.* (1974) constructed an economic model of beef production based on liveweight gains obtained from several highly fertilized grasses in Puerto Rico. They showed that an intensive grazing management system under conditions of year-round plant growth would give economic returns where a market exists for quality beef. Their calculations were based on a 60 ha farm unit, carrying 300 head weighing about 180 kg initially and sold 1 year later at about 385 kg. About once per decade an exceptionally dry year in this region of Puerto Rico would require supplementary feeding, increasing production costs considerably.

Economic analyses make use of data obtained from experimental studies carried out according to recommendations of the pasture agronomist and animal scientist and usually under their supervision. Often they find difficulty in translating results directly to the local conditions of an animal producer in terms of pasture improvement, animal input and expected returns from animal products.

Dairy production on tropical pastures

The nutrient requirements of lactating cows exceed that of beef cattle, and for high milk production dairy cattle need high-quality feedstuff. The amount of feed to produce 1 kg of liveweight gain per day is equal to that needed to produce 8–9 kg of milk per day. In addition to maintenance requirements, lactating cows require 0.3 kg TDN and 0.06 kg of protein per kilogram of milk containing 4.0 per cent fat. High-quality grasses can provide the feed required for a cow producing 10 litres of milk daily. A 550 kg cow at this level of production needs about 14 kg of dry matter with 7.0 kg of TDN, 1.1 kg protein, 32 g Ca and 26 g P (NAS, 1978).

Tropical forages for lactating cows

A compilation of milk production records from experiments in various parts of the tropics and subtropics showed daily yields varying from 6.5 to 15.0 kg/cow. The higher yields came from regions of higher latitudes and elevations (Stobbs, 1971a, 1975a). The main factor limiting milk production on well-fertilized tropical pastures appeared to be low intake of digestible energy. Jersey cows grazing fertilized pastures of Setaria and Rhodes grasses in Australia produced 68.2 per cent as much milk as those penned and fed lucerne hay (Hamilton *et al.*, 1970). Animals grazing 3-week regrowth of Setaria gave higher fat-corrected milk (FCM) yields (7.0 kg/cow per day) than those grazing 5-week-old Setaria (6.3 kg/cow per day). Transferring animals from a supplemented diet to sole grass pastures caused a notable decline in milk production (Fig. 10.7). Comparable results occurred in another experiment using identical twins (Dale and Holden, 1968). Cows grazing Glycine–kikuyu pastures produced about one-half the quantity of milk as those fed lucerne hay plus concentrate (about 10.5 kg of milk per cow per day and 21.5 kg, respectively, at the beginning of the experiment). When moved to the legume–grass pasture milk yields diminished by 0.4– 0.5 kg/week, reaching a plateau between 4.5 and 6.8 kg/cow per day. In both experiments animals gained weight over the trial period.

Milk production of cows grazing improved tropical pastures is considerably lower than those grazing temperate pastures. The data in Table 10.4 summarizes production obtained from Jersey cows on pastures in tropical Australia, on pastures in temperate zones, and fed concentrate rations (Stobbs and Thompson, 1975). Daily production varied from 9 to 12 kg/head at peak lactation when cows grazed young, actively growing, tropical forage plants. Friesian and other large dairy breeds are capable of higher production because of body size and higher feed intake. For example, Cowan *et al.* (1975) at the Kairi Research Station in Queensland reported average yields on unsupplemented guinea grass–Glycine pastures of 13.7 and 13.4 kg/ head (1.3 and 2.5 cows/ha and lactation periods of 278 and 259 days, respectively). In Puerto Rico, McDowell *et al.* (1977) obtained average yields of 11.64 kg/cow over a 5-year period (Holsteins, Brown Swiss and cross-breds) in a 270-day lactation. Animals were grazed at 2.5 head/ha and had access to a mixture of Pangola, giant Cynodon and pará grasses fertilized with 2.5 tonnes/ha of a 15-5-10

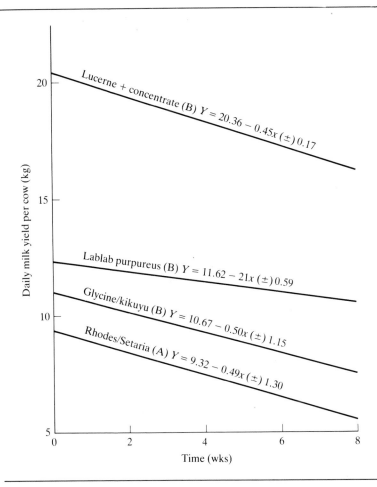

Fig. 10.7 Regression of milk yield per cow on time from placement on pastures: (A) Hamilton *et al.*, 1970; (B) = Dale and Holden, 1968.

formulation. Average daily production of individual animals exceeded 17 kg/head over the lactation period.

Fertilized Pangola grass pastures in Queensland have consistently produced about 10 per cent higher milk yields than Rhodes, Setaria and kikuyu (Stobbs and Thompson, 1975). Superiority of this species comes from the increased intake of digestible energy and is probably related to the high soluble carbohydrate content in Pangola herbage (Minson, 1967). High intake of *Lablab purpureus* kept daily milk production of Jersey cows grazing solid stands of this legume at about 10 kg/cow, as compared to average yields below

Table 10.4 *A summary of research findings of milk production per lactation from tropical and temperate zone pastures and concentrate rations* (Stobbs and Thompson, 1975)*

Diet[†]	Dry matter digestibility (%)	Milk production (kg/head)
1. *Tropical pasture*		
(a) Immature	60–65	1 800–2 000
(b) Semi-mature	50–55	1 000–1 400
2. *Temperate pasture*	70–80	3 300–3 800
3. *Concentrate ration*	80–85	4 400–4 900

* Data from Jersey cows only.
[†] Experiments in tropical Australia and different temperate regions.

7.0 kg/cow on Setaria and Rhodes grass (Fig. 10.7). Desmodium or Siratro alone gave daily yields of 7.7 kg/head as compared to Pangola grass which produced 9.0 kg (Stobbs, 1971b). Animals grazing the large-leaved legumes had difficulty harvesting sufficient herbage to satisfy their nutritional requirements. Milk yields from Jersey cows on *Trifolium semipilosum* in Queensland averaged 16 kg/cow per day over an extended time, probably due to increased intake and higher digestibility of the more dense herbage of this compact legume (Stobbs, 1975a). Using a leader–follower system of grazing 3-week regrowth of Rhodes grass and guinea grass, Stobbs (1978) noted that leader cows produced 8.0 kg of milk per day and followers 5.8 kg.

At Ibadan, Nigeria, the milk production of White Fulani cows did not vary significantly (about 5.5 kg/head per day) when grazed on different mixtures of giant Cynodon, elephant grass and guinea grass in mixtures with centro (Olaluku, 1972). Slight increases occurred when centro comprised more than 20 per cent of the herbage. Milk production declined as plants aged, but liveweight gains indicated that nutrient supplies were in excess of the genetic potential of the cows to produce milk. Despite prevalence of the tsetse fly in this area animals were not treated for the parasite, but maintained vitality when provided an adequate and nutritive diet. At the Shika Research Station in northern Nigeria, White Fulani cows on fertilized Pangola grass produced average daily milk yields of 5.8 and cross-bred Friesians 7.5 kg/head (Olayiwale, 1974).

Increased milk yields of dairy cows with the genetic potential for higher milk production can be obtained by pasture management practices to improve quality of ingested forage. One way of accomplishing this is by shortening the time spent on a given paddock in a rotational system of grazing. This was demonstrated in the Philippines, where reducing the grazing cycle of lactating cows on elephant, guinea and pará grass pastures to a 10-day rotation improved milk yields per cow and per hectare (Hardison, 1966). When rotationally grazed at intervals of 20–30 days the grasses contained sufficient crude protein for body maintenance and production of about 10 kg of milk daily, but the TDN intake would limit production to about 5 kg daily.

Milk production per hectare from tropical pastures

Studies of improved tropical pastures indicate that milk production per hectare approaches that of temperate pastures. Payne (1969) reported an average of 773 grazing days and up to 3 125 kg/ha of milk annually with guinea, elephant and pará grasses in the Philippines. He had suggested (1963) that improved pastures in the tropics should be capable of supporting five lactating cows per hectare, each producing 2 720 kg of milk annually, or a total of 13 600 kg/ha. In Queensland, Friesians grazing guinea grass–Glycine pastures at 2.5 head/ha yielded 8 220 kg of milk in a lactation of 259 days (Cowan *et al.*, 1975). Pastures were divided so that cows grazed a given paddock for 2 days and were off for 8 days before returning to the same paddock. Those grazed at 1.3 head/ha produced a total of 4 954 kg with an extended lactation of 278 days. At the lower stocking rate Glycine comprised 37.1 per cent of the herbage on offer (dry matter basis), but the legume component dropped to 13.9 per cent at the higher stocking rate. In Puerto Rico, Holsteins at 2.25 head/ha were given access to fertilized grass and no supplementation from time of weaning. They gave milk yields of 4 497, 8 132 and 8 573 kg/ha in their first, second and third lactations, respectively (Caro Costas and Vicente-Chandler, 1974). The increased yield was attributed to increased body size and larger intake of herbage. Cows grazing N-fertilized kikuyu grass in New South Wales produced up to 9 388 kg of milk annually when stocked at 4.94 animals/ha (Colman and Kaiser, 1974). A heavily N-fertilized (672 kg/ha) and irrigated Pangola grass pasture gave mean yields of 17 400 kg/ha of milk from Jersey cows and 22 400 kg/ha from Friesian cows without supplemental feed (Thurbon *et al.*, 1973).

Milk production per hectare increases with higher stocking rates, but pastures may be adversely affected in regard to legume component and survival of desirable grass species. McDowell *et al.* (1975) stated that a stocking rate of 2.0–2.5 lactating cows/ha permits selective grazing so that each individual will be able to obtain 2.3–2.4 per cent of her bodyweight. For cows on grazing alone the daily intake of crude protein and TDN would be sufficient for 11.4 and 10.3 kg of milk. Intake of about three times maintenance is needed for production at overall efficiency. With seasonal fluctuation in available forage, however, the ingested TDN is frequently no more than twice maintenance requirements. This suggests the need for supplementary feed to assure sustained milk yield.

Milk production on tropical pastures with supplementation

Feeding concentrates such as maize, molasses, urea and biuret to lactating cows on tropical pastures may increase milk production by 25 per cent or more. The additional milk yield, however, may not be economical unless animal performance is also considered. In general, low levels of supplementary feeding (1.3–2.3 kg/head daily of concentrates) boosts production 3–8 per cent. This gain usually does not justify the extra costs (Caro-Costas *et al.*, 1972; Dale and Holden, 1968; Dirven, 1965; Jeffery *et al.*, 1970; Joblin, 1966; Veitia, 1971). Several studies have shown, however, that supplementation at a ratio of 0.45 kg of concentrate to 1, 2 or 3 kg of milk will increase yields significantly. For example, Caro-Costas *et al.* (1972) fed Holstein cows grazing intensively managed grass pastures 0.45 kg of protein concentrate at the ratio of 1, 2, 3 and 4 kg of milk daily (Table 10.5). Total milk production and length of lactation increased with the first three treatments, but they did not differ significantly. Cows fed at these levels produced an average of 4 757 kg of milk during an average lactation of 279 days or 17.0 kg/head daily. At the lowest quantity of concentrate, the milk yield dropped sharply to 3 385 kg and the lactation period was reduced to 242 days. Slopes of lactation curves were not appreciably affected by the rates of 0.45 to 1, 2 and 3 kg, but the reduced rate of 0.45 kg of concentrate to 4 kg of milk caused a drop after the first month.

McDowell *et al.* (1977) used 282 Holstein cows that had completed one or more lactations to evaluate the following feeding regimes in Puerto Rico:
1. Pasture alone (mixture of Pangola, giant Cynodon and pará topdressed annually with 2.5 tonnes/ha of 15-5-10).

Table 10.5 *Effect of four levels of concentrate feeding on milk production of Holstein cows grazing heavily fertilized, intensively managed, tropical grass pastures in Puerto Rico** (Caro-Costas et al., 1972)

Concentrate 0.45 kg (kg milk)	Milk yield/ cow/lactation[†] (kg)	Lactation period (days)	Daily milk production (kg)
1.0	4 422	268	16.5
2.0	4 900	280	17.5
3.0	4 948	288	17.2
4.0	3 385	242	14.0

* Pastures fertilized with 2.5 t/ha of 15–5–10 formulation and stocked at 2.5 cows/ ha.
† Averages of 12 cows per treatment.

2. Pasture plus 0.45 kg molasses per 0.91 kg of milk in excess of 10 kg/day.
3. Pasture plus 0.45 kg maize per 0.91 kg of milk in excess of 10 kg/day.
4. Pasture plus 0.45 kg commercial concentrate (20% crude protein, 72.5% TDN) per 0.91 kg of milk irrespective of daily yield.
5. Pasture plus the commercial concentrate fed at 0.45 kg per 0.91 kg milk in excess of 10 kg/day.
6. Pasture plus 0.45 kg urea (4%) and molasses per 0.91 kg milk in excess of 10 kg/day.

A résumé of this study is given in Table 10.6 The five groups receiving supplement produced significantly higher milk and fat yields than those on pasture alone. Average milk yields of 3 142 kg/cow on fertilized grass was slightly less than recorded in Australia (3 289 kg/cow on guinea grass–Glycine and a comparable stocking rate of 2.5/ha, Cowan *et al.* 1975, and 3 256 kg on N-fertilized and irrigated Pangola at about four times the stocking rate, Thurbon *et al.*, 1973), but higher than normally expected. Cows grazing grass alone and supplemented with molasses or ground maize tended to have shorter lactations than those on other treatments. In terms of breeding efficiency, animals on medium level of supplements returned to oestrus sooner than cows on grazing alone or high-concentrate feeding. Level of feeding apparently affected reproduction of cows on grass alone. Those on high concentrate would have consumed less forage so that some factor other than reduced energy from consumption of grass influenced calving interval.

Cows on grass alone or fed low amounts of concentrates require ingestion of more herbage to meet their nutrient requirements than those fed more liberal amounts of concentrate. Data from Puerto Rico showed the following fresh forage intake to provide daily requirements of a 545 kg cow (Vicente-Chandler *et al.*, 1974):
1. Cows producing 20 kg/day and fed 0.45 kg concentrate per kilogram of milk, 22.7 kg fresh forage.
2. Cows producing 20 kg/day and fed 0.45 kg concentrate per 3.0 kg milk, 63.6 kg fresh forage.
3. Cows producing 10 kg/day and fed a 0.45 kg concentrate per kilogram milk, 29.5 kg fresh forage.
4. Cows producing 10 kg/day and fed 0.45 kg concentrate per 3.0 kg milk, 50 kg fresh forage.

Cows in one trial gave an average of 13.5 kg of milk daily over the lactation period when allowed access to fertilized grass. To produce this amount of milk and maintain body weight a 525 kg cow would have to consume 2 545 kg of TDN or about 4 240 kg of dry matter of 60 per cent digestibility over a 300-day lactation period (Caro-Costas and Vicente-Chandler, 1974). This is equivalent to about 14 kg of dry forage daily or 60 kg of fresh herbage containing

Table 10.6 *Effect of feeding systems on milk production of Holstein cows at Gurabo, Puerto Rico (McDowell et al., 1977)*

Feeding system*	Lactation (no.)	Milk/cow/ lactation (kg)	Lactation period (days)	Daily milk/cow (kg)	Supplement/ lactation (kg)	Calving interval (days)
1. Pasture alone	87	3 142	271	11.7	0	407
2. Pasture + molasses	73	3 670	284	12.9	528	392
3. Pasture + maize	48	3 672	281	13.1	594	392
4. Pasture + high concentrate feeding	79	4 482	291	15.4	2 062	409
5. Pasture + limited concentrate feeding	76	4 227	290	14.6	838	394
6. Pasture + urea and molasses	26	3 830	290	13.2	557	387

* See text for details of systems.

23 per cent dry matter. In the early stage of lactation cows would have to consume higher amounts.

Grass alone in the tropics, even though succulent and containing a relative high percentage of crude protein, may be insufficient as a feed to allow full expression of the milk production potential of lactating cows. In Australia, Flores *et al.* (1979) allowed 24 lactating Jerseys to graze 3-week herbage regrowth of *Chloris gayana* containing 18 per cent crude protein (dry weight basis). Some were fed *Leucaena* forage, some supplemented with formalin casein, and others unsupplemented. Those given *Leucaena* produced 10.3 kg of milk/head per day, those given formalin casein 10.1 kg, and those unsupplemented 9.6 kg.

Value of supplement

It is generally agreed that supplementary feeding of lactating cows grazing high-quality pastures may not be economical. Vicente-Chandler *et al.* (1974) prepared models for dairy herds of 150 milk cows based on data obtained in Puerto Rico to show that limited concentrate feeding would be profitable where rainfall is ample for near year-round growth of pasture plants. The increased milk yields obtained by McDowell *et al.* (1977) suggested the economic feasibility of using supplements at moderate levels of feeding (Table 10.6). Use of molasses, ground maize, limited concentrate feeding and urea plus molasses would be preferred to molasses or ground maize alone, since these may restrict the intake of protein needed by lactating cows.

In Venezuela, Combella *et al.* (1979) examined the effect of concentrate feeding on milk production and herbage intake of lactating Friesian heifers grazing *Cenchrus ciliaris*. Milk yields increased 0.27 kg per kilogram of concentrate but were not economical. For each kilogram of concentrate, herbage intake was decreased by 0.64 and 0.42 kg in the rainy and dry seasons respectively. Furthermore, the time spent grazing was reduced with supplementation.

Most studies of dairy production on improved pastures have been carried out in the subtropics, less adverse environments of the tropics and the higher elevations, largely with European breeds. Much of the milk in the tropics, however, comes from cross-bred and unimproved cows on native or naturalized grasslands. Although their genetic potential is restricted, milk yields of such animals can be increased to some extent by pasture management. The demand for milk and milk products will continue to increase as economies expand. Around urban centres this supply will undoubtedly come from intensively managed dairy operations highly dependent on concentrate feeding. Improved pastures under appropriate management practices could substantially reduce the need for high levels of concentrate feeding. Additional research is needed however, regarding the soil–plant–animal relationships as related to dairy production in the tropics.

Integration of sown (improved) pastures and native or naturalized grazinglands

The utilization of improved pastures in combination with native and naturalized grazinglands usually augments animal performance. Investigators often refer to schemes of integration but limited comparative data are available. Animal producers make little use of the practice, even though it has great potential for increased animal production. Integration can be broadly separated into two categories (Winks, 1975).

1. Integration of pastures

Separate land areas of native (naturalized) and sown pastures are grazed in combination. They may be rotationally grazed or utilized at distinct and strategic periods of the year or grazing seasons. A practice sometimes followed by animal producers is that of carrying the breeding herd on native pastures and then growing-out or fattening cattle on improved pastures.

2. Integration of species

An improved pasture species, usually a legume, is broadcast, sod-seeded, sown in strips or rows in a native pasture. This is a method of pasture improvement rather than pasture integration. The system of planting Leucaena in rows within a naturalized pasture of *Dichanthium caricosum* in Fiji so as to comprise 10 and 20 per cent of the land area represents integration of species (Patridge and Ranacou, 1974). Liveweight gains of steers on grass alone averaged about 70 kg/head of annual liveweight gain, on grass plus 10 per cent Leucaena 95 kg, and on grass plus 20 per cent Leucaena 180 kg. No adverse effects due to ingestion of Leucaena were noted as reported by Blount and Jones (1977), who found a chronic toxicity from mimosine and depressed liveweight gain when cattle grazed this legume.

Integration of pastures

In tropical and subtropical Australia, liveweight gains of beef cattle have shown increases in proportion to the time spent on Townsville stylo pastures compared to native grass pastures alone. This was amply illustrated by trials carried out by Norman (1970) where

Shorthorn steers grazed Townsville stylo and native pastures in (1) sequential grazing systems, i.e. during different seasons of the year and (2) complementary systems, i.e. Townsville stylo pastures 2 days and native pastures 5 days, Townsville stylo 4 days and native pastures 3 days. In both instances liveweight increased linearly with number of grazing days on Townsville stylo (Fig. 10.8).

At the Brian Pastures Research Station in Queensland, Scateni (1966) used native pastures for summer grazing and improved pastures of green panic (*Panicum maximum* var. *trichoglume*) and lucerne for cool-season grazing. Lucerne failed to persist under grazing and was substituted by N-fertilized green panic. Liveweight gain per head increased by 35 per cent in the integrated system compared to native pasture alone. Carrying capacity rose by 70 per cent with the combination. Later, Addison (1970) reversed the approach by grazing native pastures with or without supplement in the summer period and improved native pasture combinations in the cool season. Sown pastures included N-fertilized green panic and green

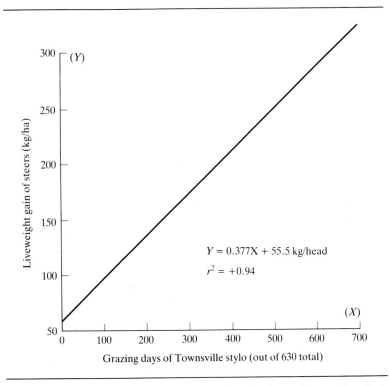

Fig. 10.8 Relation between liveweight gain and grazing time on Townsville stylo in sequential native pasture/Townsville stylo grazing systems (Norman, 1970).

panic mixed with lucerne or Siratro. Improved and native pastures were rotationally grazed (1 week on and 3 weeks off the improved). The integrated system gave a 76 per cent increase in liveweight gain over native pasture alone. Animals on native pasture and supplemented with 0.75 kg of cottonseed meal per head daily showed an average liveweight gain of 61.0 kg annually. At the Narayen Research Station, Mannetje *et al.* (1975) compared the performance of breeders grazing native grass, green panic–Siratro or a combination. During the cool, dry season overstanding herbage on sown pasture was grazed 25 per cent and native pasture 75 per cent of the time. Liveweight performance of bulls and weaning calves was about the same, but slightly higher for breeding cows on the combination at stocking rates of 4 ha/cow compared to 6 ha/cow on native pasture alone. There was a trend toward higher conception rate on the combination system.

Even though benefits in terms of animal performance have been obtained with integrated systems, there is limited information regarding desirable proportions of improved and unimproved pastures, management schemes, stocking rates and modifications needed for different classes of livestock. Most studies have been carried out with steers. Additional information is needed for breeding herds along with economical aspects of breeding and fattening stock.

Factors affecting grazing behaviour

The quantity and digestibility of feed ingested are major factors affecting animal productivity. Tropical pasture swards differ in yield of dry matter, vertical development and structure, density or compactness of the herbage, leaf : stem ratio, arrangement and accessibility of leaves, ease of prehension (disposition of shoots and leaves for ready biting) and removal of herbage from the plant. The sward structure is influenced by pasture species, fertilization and stocking rate. Growth habit of legumes range from the tall, stemmy *Desmodium tortuosum* to the prostrate rather open canopy of *D. intortum* and *Macroptilium atropurpureum* or dense small-leaved *Lotononis bainesii* and *Trifolium semipilosum*. Grasses vary from the upright, stemmy *Panicum maximum* and *Hyparrhenia rufa* to the stoloniferous leafy *Digitaria decumbens*. Fertilization of sown pastures increases growth rate and alters internode length and leafiness. The intensity of grazing modifies the amount of herbage on offer and canopy structure; for example, under lenient grazing a higher proportion of stems will be present.

Animal selectivity of herbage ingested

Herbage ingested by grazing animals may differ markedly in botanical and chemical composition from that on offer because of selective grazing. Grazing animals select different plant species and the more leafy parts of the plant (Theurer, 1970; Gardener, 1980; McLean *et al*, 1981). This is apparent from studies of clipped samples of herbage and that of samples recovered from oesophageal fistulas (Dradu and Harrington, 1972; Harrington and Pratchett, 1973; Marshall *et al.*, 1969). Brendon *et al.* (1967) found that oesophageal fistula samples collected from cattle grazing tropical pastures contained more crude protein and less crude fibre. They noted a 66 per cent increase in crude protein and an 8 per cent decrease in crude fibre of ingested herbage. It was suggested that animals selected specific plant parts while grazing. Analyses of plant components showed that the crude protein content of leaves was 55 per cent higher and the crude fibre 17 per cent lower than the whole plant value.

The botanical composition of fistula samples may be determined by one of the following methods (Harker *et al.*, 1964; Heady and Van Dyne, 1965; Sparks and Malechek, 1968; Van Dyne and Heady, 1965).
1. Manual separation – a sample of ingested material (5–10 g) is placed in a shallow dish containing water and forceps used to separate grasses, legumes and weeds.
2. Microscope point-hit – the sample is spread on to a tray and passed under a binocular microscope equipped with a cross hair. The plant fraction appearing under the cross hair is identified and recorded.
3. Microhistological technique – cuticle fragments of different species are identified under a compound or projection microscope.

In the process of grazing cattle first eat in a horizontal then in a vertical plane. They selectively graze in both directions. At first the animals consume the growing points which are the most juvenile and nutritious portion of the plant. Then they search for the younger leaves, followed by more mature ones. In this way the highest quality diet can be selected from the sward. At low stocking rates cattle can compensate for low quality of the entire sward. Removal of leaves reduces the potential for selection and as the sward is defoliated the animals are forced to consume more stemmy and less nutritious material.

Separation of the vertical components will show a notable difference in digestibility from the upper to the lower portion of a plant. The growing points (down to the first distinguishable node) are 60–70 per cent digestible. In a sward of guinea grass with 30 days regrowth the growing points comprise about 5 per cent of the total fresh weight assuming removal at 15 cm above ground level (Fig. 10.9). Mature leaves are almost as digestible as the growing points and make up about 25 per cent of the total. Older leaves, some of which are in a stage of senescence, constitute about 5 per cent of the plant fraction and are 50–55 per cent digestible. The upper-steam portion represents about 25 per cent of the plant weight and has a digestibility value of 45–55 per cent. The lower-stem portion forms about 40 per cent of the entire plant and is only 35–45 per cent digestible. Overall, the plant would approximate 50 per cent digestibility. If plants are allowed to grow for 60 days before grazing they have approximately 45 per cent digestibility. This is inadequate for high animal performance (McDowell, 1972).

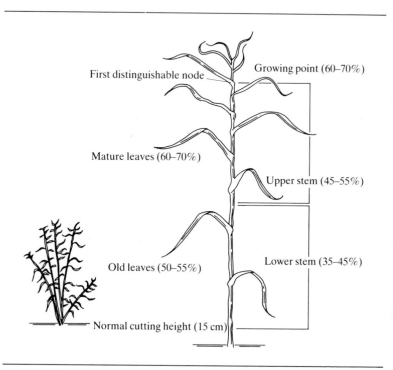

First distinguishable node

Growing point (60–70%)

Mature leaves (60–70%)

Upper stem (45–55%)

Old leaves (50–55%)

Lower stem (35–45%)

Normal cutting height (15 cm)

Fig. 10.9 Vertical components of bunch-type tropical grasses and their digestibilities (McDowell, 1972).

Grazing time

The length of time spent grazing and the rate of intake determine the amount of herbage ingested by the grazing animal. The grazing time reflects ease of prehension and removal of herbage from the plant. It is also influenced by quantity and quality of herbage as well as its accessibility. Cattle in the tropics and subtropics spend from 7 to 12 h/day grazing, 5.0 to 8.5 h ruminating and the remainder idling (lying, standing, drinking, walking). Those on tropical and subtropical pastures spend more time grazing than their counterparts on temperate pastures, even though large quantities of herbage are available. Stobbs (1974b) compared temperate and tropical grass species in Queensland and found a mean difference of about 20 per cent in favour of temperate types (Table 10.7). Hancock (1954) recorded grazing times of 6.0–6.5 h/day in New Zealand. Later Hancock and McMeekan (1955) and Brumby (1959) in New Zealand noted mean grazing times as low as 4.8 h but up to 9.0 h for lactating Jerseys with individual cows grazing 11.5 h/day. Animals grazed for longer periods under adverse pasture conditions.

Most information on grazing behaviour has been obtained by visual observation over periods of a few days. Several mechanical devices have been employed but the most successful has been a vibracorder attached to the neck of grazing animals and described by Stobbs (1970) and Stobbs and Cowper (1972).

By visual observation Payne *et al.* (1951) found that grade Friesian cows in Fiji spent from 6.5 to 10.5 h grazing, the time varying from month to month. In Kenya at 1 625 m elevation, lactating Jerseys grazed for 9.7 h, ruminated 6.6 (standing or lying) and idled 7.7 h/day when given access to elephant grass for 1.5 h after morning milking and then Rhodes grass for the remainder of the day (Goldson, 1963). Friesians in Uganda were pastured on Rhodes grass for about 16 h (Mugerwa *et al.*, 1973). Those in early lactation grazed for 8.1 h (47%). Dry cows that remained on pasture for 24 h utilized 8.6 h (35.6%) in grazing. In Queensland, Cowan (1975) reported that milking Friesian cows on guinea grass–Glycine pastures spent about 10 h grazing per day. Also in Queensland grazing time of dry and lactating Jerseys was measured automatically with vibracorders (Stobbs, 1970; Stobbs and Cowper, 1972; Stobbs and Hutton, 1974). They found that lactating cows grazed for longer periods than dry cows. In one trial lactating cows grazed immature grasses (*Chloris gayana, Digitaria decumbens, Setaria anceps, S. splendida, Pennisetum clandestinum*) an average of 8.5 h/day and legumes at the pre-flowering stage (*Desmodium intortum, Lablab purpureus, Macroptilium atropurpureum*) for 10.8 h. On more mature grasses and legumes the grazing time was extended (Table 10.7). The difference in grazing time was attributed to ease of prehension and removal of herbaceous material in grasses. Cows grazing pure legume swards had difficulty harvesting the large leaves and removing them from the more fibrous stems.

Table 10.7 *Grazing time, rate of biting and size of bite ingested by Jersey cows on pastures of various days regrowth and quality in southeastern Queensland (Stobbs, 1974b; Stobbs and Hutton, 1974)*

Pasture*	Grazing time (24 h)			Night grazing (%)	Rate of biting (24 h)			Bite size (O.M./bite)		
	Cow days	Total (h)	Range (h)		Cow days	Total (×100)	Range (×100)	Pastures sampled†	Mean (g)	Range (g)
1. Immature										
Temperate	56	7.7	6.8–8.7	31	16	288	250–364	4	0.43	0.31–0.71
Tropical	346	9.3	7.0–9.9	52	58	445	325–542	16	0.34	0.17–0.50
2. Mature										
Tropical grass	438	11.3	8.0–12.2	50	48	617	487–752	18	0.17	0.05–0.31
Tropical legume	209	12.0	9.8–13.1	49	–	–	–	–	–	–

* Temperate pastures comprised of *Avena sativa* and *Trifolium repens*; tropical grasses were *Chloris gayana*, *Digitaria decumbens*, *Setaria anceps*, *S. splendida* and *Pennisetum clandestinum*; tropical legumes were *Macroptilium atropurpureum* and *Desmodium intortum*; immature < 3-week regrowth, mature > 3-week regrowth.
† 3–4 animals grazing 3–6 pasture replicates.

In the temperate zones *Bos taurus* beef types graze between 6.5 and 8.5 h/day. Johnstone-Wallace and Kennedy (1944), working at Cornell University with Aberdeen Angus and Hereford cattle, reported that cows spent 7–8 h grazing. Hughes and Reid (1951) found that Hereford steers in England grazed 7–9 h. At Entebbe, Uganda, East African short-horned Zebu (*B. indicus*) grazed *Paspalum notatum* (some areas of *Cynodon dactylon* and *Eragrostis mildbraedii*) from 7 to 8 h each day and ruminated between 4.5 and 5.0 h (Harker *et al.*, 1954). The remainder of the time was spent standing and lying without ruminating, walking when not grazing, drinking and licking minerals. At Muguga in Kenya, Zebu steers averaged 7.35 h on fairly good pasture (Lampkin and Quarterman, 1962). When animals were kraaled at night grazing time varied from 5.93 h on leafy pastures and 9.06 h on more mature pastures with stemmy forage. Smith (1959) grazed *Hyparrhenia* veld with different breeds of East African *B. indicus* in northern Rhodesia. The total grazing time varied seasonally between a minimum of about 8 h and over 13 h. The shorter grazing period corresponded to the wet season when forage was abundant and the longer period to the dry season when forage was stemmy and more lignified or scarce. Increased grazing time was accompanied by decreased ruminating time. The longer grazing period suggested extensive selectivity or search for herbage. Under these conditions a longer period of time is needed to attain the level of reticulo-rumen fill that inhibits feeding and prompts regurgitation and chewing of the cud. The effect of reduced rumination is longer retention of rumen contents.

In northern Nigeria, Haggar (1968) observed the grazing behaviour of four groups of White Fulani cattle – young bulls (18 months of age), mature bulls (30 months), dry cows (66 months) and milking cows (66 months). Each herd consisted of about 40 animals with the activities of 2 animals/day observed for a 15-day period in the wet and dry seasons. The herds were released for grazing at 07.00 and allowed to roam over 250 ha of rangeland. In the dry season they were fed hay and silage in the morning and milking cows received concentrates in both seasons. They were penned for the night at 17.30. Observations were made at 2.5 min for a period of 12 h/day. The mean number of hours grazing and ruminating were (1) young bulls, 6.90 and 2.86, respectively, (2) mature bulls, 7.18 and 1.98, (3) dry cows, 7.32 and 1.96 and (4) milking cows 5.57 and 2.82. The total grazing time was lower in the dry season than the wet season because of supplementary feeding, reduced quantity and quality of herbage and increased heat stress. Milking cows spent less time grazing compared to the others because of increased level of supplements. The reduced grazing time caused by supplementation was also noted by Cowan (1975) and Stobbs (1970).

Patterns of grazing

An overall impression formed by periods of cattle-watching or automatic recordings is the regular pattern of behaviour under a set of
conditions. Investigators, however, have noted different numbers of
grazing periods in a 24 h period. In the temperate zone two major
peaks occurred, one after each milking and each period lasted 2–4 h
(Castle and Halley, 1953; Hancock, 1950; Waite *et al.*, 1951).
Hughes and Reid (1951) in England were able to identify five other
less prominent, but clearly defined, peaks in their periodicity charts.
Cowan (1975) noted that lactating Friesians in Queensland showed a
persistent diurnal pattern, one about 06.00 after the morning milking
and the other between 16.00 and 17.00 following the afternoon milking. A third peak, but less prominent, was observed just before midnight. In Kenya at about 1 700 m Goldson (1963) found three main
grazing peaks with lactating Jerseys, one after each milking and a
third between 01.00 and 04.00. The longest period occurred after
the morning milking and continued for most of the herd until the
afternoon milking. Stobbs (1970) used vibracorders on lactating
Jerseys in southeastern Queensland and found four grazing peaks
when animals were kept continuously on pasture. At first light about
05.30 cows began grazing within 15 min of each other and continued
until morning milking. After returning to pasture, there were up to
three intermittent grazing periods before the evening milking. A
further period of intensive grazing of about 1.5 h was noted just before and after sunset. During the night animals grazed individually
for periods up to 3 h with the greatest concentration prior to and following midnight.

The grazing pattern of beef cattle appears to be somewhat similar
to that of dairy cattle. Animals graze throughout most of the day,
depending on climatic conditions, and continue during the night
when given access to pasture. Lampkin *et al.* (1958) recorded two
major peaks with steers at Muguga in Kenya. The first came at
07.00 with high intensity for about 2 h and the second at about
13.00, continuing until dusk. In northern Nigeria, the peak grazing
of mature Fulani bulls and dry cows in the wet season took place between 09.00 and noon and 15.00 to 17.00 (Haggar, 1968). Young
bulls grazed more uniformly throughout the day. In the dry season,
the more intensive periods occurred earlier in the morning and later
in the afternoon. A very marked reduction was noted during midday
in the young bull and dry cow herds, both of which responded to
heat stress. Behaviour of free-grazing West African Shorthorns in
Ghana showed four grazing peaks – three during the day and one at
night (Rose Innes, 1963). During 24 h 43 per cent of the time was
spent grazing, 26 per cent ruminating and 31 per cent idling. Zebu
steers in Uganda grazed during most of the daylight hours but at

east four distinct periods were discernable (Harker *et al.*, 1954). The morning grazing extended over the longest period of time. Some individuals grazed at night, but there was no regular herd pattern.

There were usually two periods of intense day-grazing by steers on *Hyparrhenia* pastures in northern Rhodesia (Smith, 1959). One started at dawn and extended to about 11.00. This was followed by a rest and watering until about 13.00, then afternoon grazing continued to about 1 h after sunset. Minor breaks were observed for salt supplementation. Grazing also occurred between midnight and 04.00 but night grazing did not follow a routine pattern. The initiation of night grazing was gradual. One or two animals would begin with half an hour passing before the entire herd started grazing. The night-grazing comprised about 20 per cent of the total grazing time.

Animal performance in the tropics and subtropics may be limited by nutritional stress, climatic stress and poor sward structure. Harker *et al.* (1954) and Musangi (1965) noted that daytime grazing of steers exceeded 90 per cent of the total grazing time. Under adverse conditions of low-quality herbage or scarcity of forage grazing time extends into the night. Also, high-producing lactating cows graze for longer periods during the night. Goldson (1963) in Kenya and Mugerwa *et al.* (1973) in Uganda reported 33 and 40 per cent night-grazing. Stobbs and Hutton (1974) found about 50 per cent night-grazing in southeastern Queensland (Table 10.7). Night enclosure of cattle is a common practice in parts of Africa. Joblin (1960) in Uganda found that restriction of night-grazing led to a significant decline of 30 per cent in liveweight gain. The difference was attributed to marginal day-grazing due to a shortage of pasture herbage. In Tanzania steers with access to night-grazing made greater gains than those enclosed (Owen, 1968). Smith (1961) and Wilson (1961) emphasized the losses in grazing time and animal performance with the practice of night-kraaling.

Climate, particularly temperature, has a significant influence on grazing time and pasture utilization. Under environmental stress the reaction of *Bos taurus* is more pronounced than *B. indicus*. In Louisiana (about 30° N latitude) under high temperature and humidity conditions, Aberdeen Angus cattle grazed 54 per cent of the day and Brahmin cattle 71 per cent (Rhoad, 1938). Later at the same research station Seath and Miller (1946) noted that grazing time under hot weather conditions (30 °C) was reduced by 1 h compared to cooler conditions (22 °C). A comparison of beef breeds in South Africa where temperatures reached almost 40 °C Aberdeen Angus, Shorthorn, Hereford (*B. taurus*) and Afrikander (*B. indicus*) spent 75, 78, 79 and 89 per cent, respectively, of the day grazing (Bonsma *et al.*, 1940). On the Fiji Islands, less than 20° S latitude, Payne *et al.* (1951) observed that Friesian dairy cows spent 67 per cent of their grazing in the cool hours of the night. Larkin (1954) working with grade Shorthorn steers in Queensland, and Seath and Miller (1946) observing Jersey and Friesian lactating cows in Louisiana, reported that daytime grazing was reduced on hot days and night-time grazing increased. Dairy cattle suffer from hyper-thermal stress that reduces grazing activity when subjected to high ambient temperatures. In the hot, humid tropics this is accentuated by the heat increment that follows consumption of roughages, espe-cially in the dry season. Thus, provision of shade for exotic breeds is essential.

On pastures of ample quantity and quality forage, cattle walk from 2 to 3 km in each 24 h (Larkin, 1954). The actual difference varies with size of paddock, availability of forage or other feed, quality of herbage and density of the sward. During the dry season when the amount of forage and quality decrease cattle spend more time grazing and a concomitant time in walking (Hutchinson *et al.*, 1962; Lampkin and Quarterman, 1962; Smith, 1959). Excessive walking is undesirable and cattle should not travel more than about 1 km for water as this reduces grazing time and leads to non-uniform utilization of pasture.

Biting (Grazing action)

Cattle graze by combined action of the tongue, mouth and head movements. Ingestion of herbage is more of a plucking than biting action. Herbage is seized or grasped by a sweep of the tongue (pre-hended), drawn into mouth and then literally torn or ripped from the plant by a sideward and upper movement of the head. Move-ment of the lower jaw and teeth, as well as the dental pad, assist in separating the 'bite' of herbage from the plant.

There are three types of bites involved in utilization of forage by cattle grazing a pasture sward:
1. Harvest bite – the animal prehends a mass of herbage, separates it from the plant and draws it into the mouth. This occurs with the head lowered or in a grazing position.
2. Mastication bite – occasionally an animal lifts its head after har-vesting a large mouthful to chew the herbage into smaller pieces before ingestion. This is more likely to occur on pastures of high leaf density where an animal draws a large amount of herbage into the mouth. Cows on mature tropical pastures take in small amounts of herbage and mastication bites often account for less than 5 per cent of grazing bites (cited by Stobbs, 1974b).
3. Rumination bite – after obtaining a reticulo-rumen fill, cattle stand or lie, regurgitate the ingested herbage and chew it several times before reswallowing. This is sometimes referred to as 'chewing the cud'. Rumination bites provide some indication of

herbage quality. Long rumination time is related to low-quality, fibrous forage and short rumination to high-quality forage with greater digestibility (Balch, 1971).

Number of bites

Rate of biting indicates the ease with which herbage is harvested by the grazing animal. Stobbs (1974a, 1974b) recorded automatically the number of bites taken by non-lactating Jersey cows grazing temperate and tropical grass species. The biting frequency declined linearly on all pastures throughout the grazing period. Cows on kikuyu grass with 3 weeks of regrowth started grazing at about 65 bites/min but this dropped to 50 after 2 h. On Rhodes grass having 8–10 weeks regrowth rate of biting averaged 75 bites/min at the beginning of grazing with a decline to 56 when grazing ceased. Biting rate of cows on oats with 5–8 weeks regrowth began at 65 bites/min and declined more rapidly than those on Rhodes grass. Rate of chewing during rumination was fairly constant on both oats and Rhodes grass, averaging 48 bites/min. The total number of bites (grazing and ruminating) recorded within a 24-hour period varied among cows from about 25 000 when grazing oats to 75 000 on mature Setaria pasture (Table 10.7). The number of harvesting bites rarely exceeded 36 000 per day. Hancock (1950) noted that lactating cows in New Zealand made about 50 bites/min and 24 000 bites/day when grazing ryegrass–white clover pastures.

Size of bites

The size of bite by grazing cattle is a major factor in determining the amount of herbage intake. The quantity of organic matter ingested per bite is determined by using oesophageal-fistulated animals fitted with vibracorders to measure jaw movements. A foam-rubber plug placed in the lower oesophagus directs ingested feed through the fistula and into a collecting bag hung around the cow's neck. The samples of consumed herbage can be used to calculate average bite size and to determine dry matter, organic matter, chemical analyses and *in vitro* digestibilities. Using this system, Stobbs (1973a, 1974a) found that bite size varied from 0.05 to 0.80 g of organic matter, depending on herbage availability and accessibility (Table 10.7). Cows grazing N-fertilized *Setaria anceps* ingested 0.39 g/bite of organic matter and those on sparse unfertilized *Setaria* averaged 0.13 g/bite. Mean bite size of cows grazing 5-week regrowth of Siratro was 0.24 g organic matter per bite compared to 0.34 and 0.38 g for N-fertilized Pangola and Setaria, respectively, of the same age. A smaller bite from the trailing legume was probably due to difficulty in grasping and harvesting the larger leaves. Hendericksen and Minson (1980), working with *Lablab purpureus*, reported that bite size declined from 0.41 g on day 1 to 0.09 g on day 12. The decrease was

caused by a reduction in quantity of leaf and a lack of desire of cat
tle to consume the stem fraction.

Bite size varied significantly among animals and may be related to
mouth size (Stobbs, 1973b). An interaction existed between pasture
type and stage of maturity (Fig. 10.10). Cows grazing 2- and 4-week
regrowth of Rhodes grass harvested a larger quantity of herbage per
bite than those grazing Setaria. Difference in plant structure contri
buted to the variation in bite size. Rhodes grass is more leafy and
leaves are smaller and more accessible in the early stages of growth
then Setaria. Similar amounts of herbage were prehended and har
vested from the two grasses after 6 weeks of regrowth, with a mean
intake of about 0.15 g/bite. At this level of grazing the intake of dry
matter is inadequate for sustained animal performance. Stobb
(1973a) calculated that an intake of 0.3 g of organic matter per bite
is needed for a 400 kg animal to achieve adequate dry matter when
grazing at 36 000 bites/day. Under conditions of low quality and

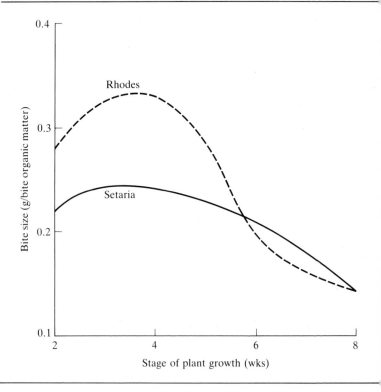

Fig. 10.10   Mean bite size of Jersey cows grazing Rhodes grass and Setaria pastures
at four stages of regrowth (Stobbs, 1973b).

sparse pasture animals graze for a longer period of time in an attempt to compensate for small bite size.

Effect of sward structure on bite size

Grazing cattle prefer the leaf component of the sward rather than the stem fraction. Cows perhend and harvest a bite of maximum size when the sward contains a high proportion of accessible leaf. Factors that influence bite size were noted by Stobbs and Hutton (1974) as follows:

1. Leaf yield – a high leaf : stem ratio is desirable. Size of bite is determined largely by the amount of leaves and their position in the canopy structure.
2. Accessibility of leaves – leaves that are readily accessible, i.e. not mixed with stems within the canopy, are more easily prehended. As plants develop the percentage of stems increase and interfere with grasping a full bite. Fibrosity of the leaf, and petiole in the case of legumes, affect the ease of harvest. Thus, older leaves will be more difficult to remove because of higher lignification and increased cell wall structure.
3. Sward and leaf density – the density per unit area of the leaf–stem mixture and the leaf fraction vary widely with different pastures. Bulk densities increase with higher forage yields but bite size may be low because of a larger proportion of stems and older leaves. Thus, the grazing animal has difficulty in selecting the more nutritious plant parts. Considerable potential exists to improve sward and leaf density and increase herbage intake by selecting leafy cultivars in compositing pasture mixtures, following a regular maintenance fertilizer programme and adjusting stocking rates for optimal carrying capacity.

References

Addison, K. B. (1970) Management systems on spear grass country, *Proc. 11th Int. Grassld Cong.*, pp. 789–93.

Aronovich, S., A. Serpa and **H. Riberio** (1970) Effect of N-fertilizer and legumes upon beef production of pangola grass pasture, *Proc. 11th Intl. Grassld Cong.*, pp. 796–800.

Ayala, H. J. (1969) *El Ganado de Carne y Los Forrajes en Colombia. Curso Corto Sobre Ganado de Carne*, ICA Misc. Publ. No. 11, Palmira: Colombia.

Balch, R. C. (1971) Proposal to use time spent chewing as an index of the extent to which diets of ruminants possess the physical property of fibrousness characteristics of roughages, *Brit. J. Nutr.* **26,** 383–92.

Barnard, C. and **O. H. Frankel** (1964) Grass, grazing animals and man in historic perspective. In C. Barnard (ed.), *Grasses and Grasslands*, Macmillan: New York, Ch. 1.

Bennett, D., F. H. W. Morley, K. W. Clark and **M. L. Dudinski** (1970) The effect of grazing cattle and sheep together, *Aust. J. Exptl Agric. Anim. Husb.* **10,** 694–709.

Bisset, W. J. and **G. W. C. Marlowe** (1974) Productivity and dynamics of two Siratro based pastures in the Burnett Coastal foothills of southeast Queensland, *Trop. Grassld* **8**, 17–24.

Blaser, R. E., R. C. Hammes, Jr., H. T. Bryant, C. M. Kincaid, W. H. Skrdla, T. H. Taylor and **W. L. Griffith** (1956) The value of forage species and mixtures for fattening steers, *Agron. J.* **48**, 508–13.

Blaser, R. E., H. T. Bryant, C. Y. Ward, R. E. Hammes, Jr., R. C. Carter and **N. H. Macleod** (1959) Symposium on forage evaluation: VII. Animal performance and yields with methods of utilizing pasturage, *Agron. J.* **51**, 238–42.

Blaser, R. E., H. T. Bryant, R. C. Hammes, Jr., and **R. L. Boman** (1969) *Managing Forages for Animal Production*, Virginia Polytechnic Inst. Res. Div., Bull. 45, Blacksburg.

Blount, C. G. and **R. J. Jones** (1977) Steer liveweight gains in relation to the proportion of time on Leucaena leucocephala pastures, *Trop. Grassld* **11**, 159–64.

Bonsma, J. C., G. D. J. Scholtz and **F. G. J. Badenhorst** (1940) The influence of climate on cattle, *Fmg in S. Afr.* **15**, 7–12, 16.

Brendon, R. M., D. T. Torell and **B. Marshall** (1967) Measurement of selective grazing of tropical pastures using oesophageal fistulated steers, *J. Range Mgmt* **20**, 317–20.

Brumby, P. J. (1959) The grazing behaviour of dairy cattle in relation to milk production, liveweight and pasture intake, *N. Z. J. Agric. Res.* **2**, 797–807.

Bryan, W. W. and **T. R. Evans** (1971) A comparison of beef production from nitrogen fertilized Pangola grass and from a Pangola grass-legume pasture, *Trop. Grassld* **5**, 89–98.

Burns, J. C., R. D. Mochrie, H. D. Gross, H. L. Lucas and **R. Teichman** (1970) Comparison of set-stocked and put-and-take systems with growing heifers grazing Coastal bermuda grass (Cynodon dactylon (L.) Pers.), *Proc. 11th Intl. Grassl. Cong.*, pp. 904–9.

Caro-Costas, R., J. Vicente-Chandler and **C. Burleigh** (1961) Beef production and carrying capacity of heavily fertilized, irrigated guinea, Napier and Pangola grass pastures on the semi-arid south coast of Puerto-Rico, *J. Agric. Univ. P. R.* **45**, 32–6.

Caro-Costas, R., J. Vicente-Chandler and **J. Figarella** (1965) Productivity of intensively managed pastures of five grasses on steep slopes in the humid mountains of Puerto Rico, *J. Agric. Univ. P. R.* **49**, 99–111.

Caro-Costas, R., J. Vicente-Chandler and **F. Abruña** (1972) Effect of four levels of concentrate feeding or milk production by Holstein cows grazing intensively managed tropical grass pastures, *J. Agric. Univ. P. R.* **56**, 97–104.

Caro-Costas, R., J. Vicente-Chandler (1974) Milk production of young Holstein cows fed on grass from steep, intensively managed tropical grass pastures over three successive lactations, *J. Agric. Univ. P. R.* **53**, 18–52.

Castle, M. E. and **R. J. Halley** (1953) The grazing behaviour of dairy cattle at the National Institute for Research in Dairying, *J. Anim. Behaviour* **1**, 139–43.

Colman, R. L. and **A. G. Kaiser** (1974) The effect of stocking rate on milk production from kikuyu grass pastures fertilized with nitrogen, *Aust. J. Exptl Agric. Anim. Husb.* **14**, 155–60.

Combella, J., R. D. Baker and **J. Hodgson** (1979) Concentrate supplementation, and the herbage intake and milk production of heifers grazing Cenchrus ciliaris, *Grass and Forage Sci.* **34**, 303–10.

Cowan, R. T. (1975) Grazing time and pattern of grazing of Friesian cows on a tropical grass-legume pasture, *Aust. J. Exptl Agric. Anim. Husb.* **15**, 32–7.

Cowan, R. T., I. F. R. Byford and **T. H. Stobbs** (1975) Effects of stocking rate and energy supplementation on milk production from tropical grass-legume pastures, *Aust. J. Expt, Agric, Anim. Husb.* **15**, 740–6.

Cowlishaw, S. J. (1969) The carrying capacity of pastures, *J. Brit. Grassld Soc.* **24**, 207–14.

Dale, A. B. and **J. M. Holden** (1968) Milk production from tropical legume-grass pastures, *Proc. Aust. Soc. Anim. Prod.* **7**, 86–91.

Davies, J. G. (1946) *A Note on Pasture Management,* CISRO Bull, 201, Brisbane, pp. 97–104.

Dirven, J. G. P. (1965) Milk production on grassland in Surinam, *Proc. 11th Intl. Grassld Cong.*, pp. 997–9.

Dirven, J. G. P (1970) Yield increase of tropical grassland by fertilization, *Proc. 9th Cong. Intl. Potash. Inst.*, Antibes, pp. 403–9.

Dradu, E. A. A. and **G. N. Harrington** (1972) Seasonal crude protein content of samples obtained from a tropical range pasture using oesophageal fistulated steers, *Trop. Agric. (Trin.)* **49**, 15–21.

Eng, P. K., L. 't Mannetje and **C. P. Chen** (1978) Effects of phosphorus and stocking rate on pasture and animal production from a guinea grass-legume pasture in Johore, Malaysia, Animal liveweight change, *Trop. Grassld* **12**, 198–207.

Evans, T. R. and **J. B. Hacker** (1973) Comparison of animal production for six tropical grasses, *Aust. CSIRO, Div. Trop. Agron. Ann. Rept. 1972–73*, pp. 11–12.

Firth, J. A., W. W. Bryan and **T. R. Evans** (1974) Updated budgetary comparisons between Pangola grass/legume pasture and nitrogen fertilized Pangola grass for beef production in the southern Wallum, *Trop. Grassld* **8**, 25–32.

Flores, J. F., T. H. Stobbs and **D. J. Minson** (1979) The influence of the legume Leucaena leucocephala and formalin casein on the production and composition of milk from grazing cows, *J. Agric. Sci. (Camb.)* **92**, 351–7.

Gardener, C. J. (1980) Diet selection and liveweight performance of steers on Stylosanthes hamata – native grass pastures, *Aust. J. Agric. Res.* **31**, 379–92.

Goldson, J. R. (1963) Observations on the grazing behaviour of grade dairy cattle in a tropical climate, *E. Afr. Agric. For. J.* **29**, 72–7.

Grof, B. and **W. A. T. Harding** (1970) Dry matter yield and animal production of guinea grass (Panicum maximum) on the humid tropical coast of north Queensland, *Trop. Grassld* **4**, 85–95.

Haggar, R. J. (1968) Grazing behaviour of Fulani cattle at Shika, Nigeria, *Trop. Agric. (Trin.)* **45**, 179–85.

Haines, C. E., H. L. Chapman, R. J. Allen, and **R. W. Kidder** (1965) *Roselawn St. Augustine Grass as a Perennial Pasture Forage for Organic Soils of South Florida,* Univ. Florida Agric. Expt. Sta., Bull. 689, Gainesville.

Hamilton, R. I., L. J. Lambourne, R. Roe and **D. J. Minson** (1970) Quality of tropical pastures for milk production, *Proc. 11th Int. Grassld Cong.*, pp. 860–4.

Hancock, J. (1950) Studies in monozygotic cattle twins. IV. Uniformity trials: grazing behaviour, *N. Z. J. Sci. Tech.* **32** (Sec. A., No. 4), 22–59.

Hancock, J. (1954) Studies of grazing behaviour in relation to grassland management. I. Variations in grazing habits of dairy cattle, *J. Agric. Sci. (Camb.)* **44**, 420–9.

Hancock, J. and **C. P. McMeekan** (1955) Studies of grazing behaviour in relation to grassland management. III. Rotational compared with continuous grazing, *J. Agric. Sci. (Camb.)* **45**, 96–103.

Hardison, W. A. (1966) *Chemical Composition, Nutrient Content and Potential Milk Production Capacity of Fresh Tropical Herbage,* Dairy Trg. Res. Inst., Res. Bull. No. 1, Laguna: Philippines.

Harker, K. W., J. D. Taylor and **D. H. L. Rollinson** (1954) Studies on the habits of Zebu cattle. I. Preliminary observations on grazing habits, *J. Agric. Sci. (Camb.)* **44**, 193–8.

Harker, K. W. and **A. D. McKay** (1962) A stocking rate trial on rough grazing in Buganda, *E. Afr. Agric. For. J.* **27**, 220–2.

Harker, K. W., D. T. Torell and **G. M. Van Dyne** (1964) Botanical examination of forage from esophagael fistulas in cattle, *J. Anim. Sci.* **23**, 465–9.

Harrington, G.N. and **D. Pratchett** (1973) Cattle diet on Onkole rangeland at different seasons, *Trop. Agric. (Trin.)* **50**, 211–9.

Harrington, G. N. and **D. Pratchett** (1974) Stocking rate trials in Onkole, Uganda. I. Weight gain of Onkole steers at intermediate and heavy stocking rates and different managements, *J. Agric. Sci. (Camb.)* **82**, 497–506.

Hart, R. H. (1972) Forage yield, stocking rate and beef gains on pasture, *Herb. Abst.* **42**, 345–53.

Heady, H. F. and **G. M. Van Dyne** (1965) Prediction of weight composition from point samples, *J. Range mgmt.* **18**, 144–8.

Heath, M. E., D. S. Metcalfe and **R. E. Barnes** (1973) *Forages. The Science of Grassland Agriculture*, Iowa State Univ. Press (3rd edn), Ch. 12.

Hendericksen, Q. and **D. J. Minson** (1980) The feed intake and grazing behaviour of cattle grazing a crop of Lablab purpureus cv. Rongai, *J. Agri. Sci. (Camb.)* **95**, 547–54.

Hughes, G. P. and **D. Reid** (1951) Studies on the behaviour of cattle and sheep in relation to the utilization of grass, *J. Agric. Sci. (Camb.)* **41**, 350–66.

Humphreys, L. R. (1972) *The Interaction of Pasture Stocking Rate and Fertilizer Needs*, ASPAC Food Fert. Tech. Bull. No. 7: Taiwan.

Humphreys, L. R. (1977) *Tropical Pastures and Fodder Crops*, Longman: London.

Hutchinson, H. G., R. O. Woof, R. M. Marion, I. Salehe and **J. M. Robb** (1962) A study of the habits of Zebu cattle in Tanganyika, *J. Agric. Sci. (Camb.)* **59**, 301–17.

Jackson, J. J. (1972) Some observations on the comparative effects of short duration grazing systems and continuous grazing systems on the reproductive performance of ranch cows, *Rhod. Agric. J.* **69**, 95–102.

Jeffery, H. J., R. L. Coleman, A. B. Dale and **J. M. Holden** (1970) Milk production from unimproved and an improved grazing system with and without a grain supplement, *Proc. Aust. Soc. Anim. Prod.* **8**, 476–81.

Joblin, A. D. H. (1960) The influence of night grazing on the growth rates of Zebu cattle in East Africa, *J. Brit. Grassld Soc.* **15**, 212–15.

Joblin, A. D. H. (1963) Strip grazing versus paddock grazing under tropical conditions, *J. Brit. Grassld Soc.* **18**, 69–73.

Joblin, A. D. H. (1966) The response of Teso Zebu milking stock to moderate levels of supplementary feeding, *E. Afr. Agric. For. J.* **31**, 368–74.

Johnstone-Wallace, D. B. and **K. Kennedy** (1944) Grazing management practices and their relationship to the behaviour and grazing habits of cattle, *J. Agric. Sci. (Camb.)* **34**, 190–7.

Jones, R. J. (1974) The relation of animal and pasture production to stocking rate on legume based and nitrogen ferilized subtropical pastures, *Proc. Aust. Soc. Anim. Prod.* **10**, 340–3.

Jones, R. J. and **R. L. Sanderland** (1974) The relation between animal gain and stocking rate. Derivation of the relation from results of grazing trials, *J. Agric. Sci. (Camb.)* **83**, 335–42.

Kennen, T. C. D. (1969) A review of research into the cattle–grass relationship in Rhodesia, *Proc. Veld Mgmt. Confn. Byo.*, May 1969, 21–5.

Kidder, R. W. (1952) Ton of beef per acre of grass in 12 months, *Breeders Gazette* **117**, 8.

Lampkin, G. H. and **J. Quarterman** (1962) Observations on the grazing habits of Grade and Zebu cattle. II. Their behaviour under favourable conditions in the tropics, *J. Agric. Sci. (Camb.)* **8**, 119–23.

Lampkin, G. H., J. Quarterman and **M. Kidner** (1958) Observations on the grazing habits of Grade and Zebu steers in a high altitude temperate climate, *J. Agric. Sci. (Camb.)* **50**, 211–18.

Larkin, R. M. (1954) Observations on the grazing behaviour of beef cattle in tropical Queensland, *Queensld J. Agric. Sci.* **11**, 115–41.

Leeuw, P. N. de (1971) *The Prospects of Livestock Production in the North Guinea Zone Savannas*, Seminar on Forage Crops Research in West Africa, Ibadan (mimeo).

Lucas, H. L. (1962) Determination of forage yield and quality from animal responses. In *Range Research Methods*. USDA Misc. Publ, No. 940: Washington, pp. 43–54.

Magadan, P. B., E. Q. Javier and **J. C. Madamba** (1974) Beef production on native (Imperata cylindrica (L.) Beauv.) and pará grass (Brachiaria mutica (Forsk.) Stapf) pastures in the Philippines, *Proc. 12th Intl. Grassld Cong.*, pp. 293–9.

Mannetje, L. 't, D. F. Nicholls and **D. B. Coates** (1975) Cattle breeding performance from pastures on granitic soils, *Aust. CSIRO Div. Trop. Agron. Ann. Rept 1974–75*, pp. 20–1.

Mannetje, L. 't, R. J. Jones and **T. H. Stobbs** (1976) Pasture evaluation by grazing experiments. In N. H. Shaw and W. W. Bryan (eds) *Tropical Pasture Research Principles and Methods*, Commonw. Bur. Past. Fld Crops, Bull. 51, Hurley: England, Ch. 9.

Marshall, B., M. I. E. Long and **D. D. Thornton** (1969) Nutritive value of grasses in Onkole and the Queen Elizabeth Park, Uganda. III. *In vitro* dry matter digestibility, *Trop. Agric. (Trin.)* **46**, 43–6.

Matches, A. G. (1970) Pasture research methods. *Proc. Nat. Confn. Forage Qual. Eval. Util.*, Univ. Nebraska, 1969, Sec. I. pp. 1–32.

McDowell, R. E. (1972) *Improvement of Livestock Production in Warm Climates*, W. H. Freeman: San Francisco, pp. 490–2.

McDowell, R. E., H. Cestero, J. E. Rivera-Anaya, F. Román-Garcia, J. A. Arroyo-Aguilú, C. M. Berrocal, M. Soldevila, J. C. Lopez-Alberty and **S. W. Metz** (1975) *Tropical Grass Pastures with and without Supplement for Lactating Cows in Puerto Rico*, Univ. P. R. Agric. Expt. Sta., Bull. 238.

McDowell, R. E., H. Cestero, J. D. Rivera-Anaya, M. Soldevila, F. Román-Garcia and **J. A. Arroyo-Aguilú** (1977) Value of supplementary feeding for lactating cows grazing fertilized grass pastures in Puerto Rico, *J. Agric. Univ. P. R.* **61**, 204–16.

McKay, A. D. (1968) Rangeland productivity in Botswana, *E. Afr. Agric. For. J.* **34**, 178–93.

McKay, A. D. (1971) Seasonal and management effects on the composition and availability of herbage, steer diet and liveweight gains in a Themeda trianda grassland in Kenya. II. Results of herbaga studies, diet selected and liveweight gains, *J. Agric. Sci. (Camb.)* **76**, 9–76.

McLean, R. W., W. H. Winter, J. J. Mott and **D. A. Little** (1981) The influence of superphosphate on the legume content of the diet selected by cattle grazing Stylosanthes-native grass pastures, *J. Agric. Sci. (Camb.)* **96**, 247–50.

McMeekan, C. P. and **M. J. Walshe** (1963) The inter-relationships of grazing method and stocking rate in the efficiency of pasture utilization by dairy cattle, *J. Agric. Sci. (Camb.)* **61**, 147–63.

Mears, P. T. and **L. R. Humphreys** (1974) Nitrogen response and stocking rate of Pennisetum clandestinum pastures. II. Cattle growth, *J. Agric. Sci. (Camb.)* **83**, 469–78.

Mellor, W., M. J. Hibberd and **B. Grof** (1973a) Performance of Kennedy ruzi grass on the wet tropical coast of Queensland, *Queensld J. Agric. Anim. Sci.* **30**, 53–6.

Mellor, W., M. J. Hibberd and **B. Grof** (1973b) Beef cattle liveweight gains from mixed pastures of some guinea grasses and legumes on the wet tropical coast of Queensland, *Queensld J. Agric. Anim. Sci.* **30**, 259–66.

Miller, C. P. and **J. T. van der List** (1977) Yield, nitrogen uptake and liveweight gains from irrigated grass–legume pasture on a Queensland tropical highland, *Aust. J. Exptl. Agric. Anim. Husb.* **17**, 949–60.

Minson, D. J. (1967) The voluntary intake and digestibility in sheep of chopped and pelleted Digitaria decumbens (Pangola grass) following a late application of fertilizer nitrogen, Brit. J. Nut. **21**, 587–97.

Mott, G. O. (1960) Grazing pressure and the measurement of pasture production, *Proc. 8th Intl. Grassld Cong.*, pp. 606–11.

Mott, G. O. and **H. L. Lucas** (1952) The design, conduct and interpretation of grazing trials on cultivated and improved pastures, *Proc. 6th Intl. Grassld Cong.* pp. 1380–5.

Morley, F. H. W. (1966) The biology of grazing management, *Proc. Aust. Soc Anim. Prod.* **6**, 127–36.

Morley, F. H. W. and **C. R. W. Spedding** (1968) Agricultural systems and grazing experiments, *Herb. Abst.* **38**, 279–87.

Mugerwa, J. S., D. A. Christensen and **S. Ochetim** (1973) Grazing behaviour of exotic dairy cattle in Uganda, *E. Afr. Agric. For. J.* **39**, 1–11.

Murtagh, G. J. (1975) The need for alternative techniques of productivity assessmen in grazing experiments, *Trop. Grassld* **9**, 151–8.

Musangi, R. S. (1965) Feed intake studies in ruminants. II. The grazing behaviour o Friesian and Nganda steers on tropical pastures, *Afr. Soils* **10**, 321–9.

NAS (1978) *Nutrient Requirements of Domestic Animals, No. 3, Dairy Cattle,* Repor Comm. Anim. Nut., Natl Acad. Sci., Natl Res. Council (5th edn): Washington.

Norman, M. J. T. (1960) Grazing and feeding trials with beef cattle at Katherine N. T., *Aust. CSIRO Div. Land. Res. Surv.,* Tech. Paper, No. 12.

Norman, M. J. T. (1970) Relationship between liveweight gain of grazing beef steer and availability of Townsville lucerne, *Proc. 11th Intl. Grassld Cong.,* pp. 829–32.

Norman, M. J. T. (1974) Beef production from tropical pastures. Part 1, *Aust. Mea Res. Comm. Rev.,* No. 16, Canberra, pp. 1–23.

Nuthall, P. L. and **P. C. Whiteman** (1972) A review and economic evaluation of bee production from legume-based and nitrogen fertilized tropical pastures, *J. Aust Inst. Agric. Sci.* **38**, 100–8.

Oakes, A. J. (1960) Pangola grass (Digitaria decumbens Stent.) in the Carribbean *Proc. 8th Intl. Grassld Cong.,* pp. 386–9.

Olaluku, E. A. (1972) The effect of level feed, nutrient intake and stage of lactatior on the milk of White Fulani cows in Ibadan, Ph.D. Thesis, Univ. Ibadan, Nigeria

Olayiwale, M. B. (1974) Feeding and management of dairy cows at Shika, Zaria. I J. K. Loosli, V. A. Oyenuga and G. M. Babatunde (eds.) *Animal Production i the Tropics,* Heineman (Nigeria): Ibadan, pp. 137–51.

Osbourn, D. F. (1975) Beef production from improved pastures in the tropics, *Worlc Rev. Anim. Prod.* **11**, 23–13.

Owen, M. A. (1968) Studies with beef steers on the Kongwa Plain, Central Tanzania *Trop. Agric. (Trin.)* **45**, 159–71.

Patridge, I. J. (1979) Improvement of Nadi blue grass (Dicanthium carioscum) pas tures on hill land in Fiji with superphosphate and Siratro: effects of stocking rate on beef production and botanical composition, *Trop. Grassld* **13**, 157–64.

Patridge, I. J. and **E. Ranacou** (1974) The effects of supplemented Leucaena leucocephala browse on steers grazing Dichanthium caricosum in Fuji, *Trop. Grassl* **8**, 107–12.

Payne, W. J. A. (1963) The potential for increasing efficiency of feed utilizatior through newer knowledge of animal nutrition, *Proc. 1st World Cong. Anim. Nut. Rome* **3**, 204–5.

Payne, W. J. A. (1969) Problems and advances under humid conditions, *Proc. Worl Confr. Anim. Prod.* **29**, 52–60.

Payne, W. J. A., W. I. Laing and **E. N. Raivoka** (1951) Grazing behaviour of dair cattle in the tropics, *Nature (Lond.)* **167**, 610–1.

Peterson, R. G. and **H. L. Lucas** (1968) Computing methods for evaluating pasture by means of animal response, *Agron. J.* **60**, 686–7.

Quinn, L. R., G. O. Mott, W. V. A. Bisshoff and **M. B. Jones** (1965) Beef produc tion of six tropical grasses in central Brazil, *Proc. 9th Intl. Grassld Cong.,* pp 1015–20.

Ramirez, A., F. Rodriguez, J. Lotero, H. Chaverra and **N. S. Raun** (1968) El pas torero continuo en el pasto Puntero, *Agric. Trop. (Colombia)* **24**, 657–63.

Rhoad, A. O. (1938) Some observations on the response of purebred Bos taurus and

Bos indicus cattle and their crossbred types to certain conditions of the environment, *Proc. Amer. Soc. Anim. Prod.* **31**, 284–95.

Richards, J. (1965) Productivity of guinea grass (Panicum maximum), *Proc. 9th Intl. Grassld Cong.*, pp. 1033–5.

Riewe, M. E. (1961) Use of the relationship of stocking rate to gain of cattle in an experimental design for grazing trials, *Agron. J.* **53**, 309–13.

Rodel, M. G. M. (1971) Pasture management at Henderson, *Rhod. Agric. J.* **68**, 114–15. Rose Innes, R. (1963) The behaviour of free-grazing cattle in the West African humid tropics: studies on a herd of West African Shorthorns on the Accra Plains, Ghana. I. Rainy season. *Emp. J. Exptl Agric.* **31**, 1–13.

Scateni, W. J. (1966) Effect of variations in stocking rate and conservation on the productivity of subtropical pastures, *Proc. 10th Intl. Grassld Cong.*, pp. 947–51.

Seath, D. M. and **G. D. Miller** (1946) Effects of warm weather on grazing performances of milking cows, *J. Dairy Sci.* **29**, 199–206.

Shultis, A. (1962) *Estimating Range Land Values and Rents*, Univ. California Agric. Extn. Ser., Publ. AXT-70.

Smith, C. A. (1959) Studies on northern Rhodesia Hyparrhenia veld. I. The grazing behaviour of the indigenous cattle grazed at light and heavy stocking rates, *J. Agric. Sci.* (*Camb.*) **52**, 369–75.

Smith, C. A. (1961) Studies on the northern Rhodesia Hyparrhenia veld. III. The effect of growth and grazing behaviour of indigenous cattle of restricting their daily grazing by night kraaling, *J. Agric. Sci.* (*Camb.*) **56**, 243–8.

Smith, C. A. (1966) Studies on the Hyparrhenia veld of Zambia. IV. The effect of cattle grazing Veld and Dambo at different stocking rates, *J. Agric. Sci.* (*Camb.*) **66**, 49–56.

Smith, C. A. (1970) The feeding value of tropical grass pastures evaluated by cattle weights, *Proc. 11th Intl. Grassld Cong.*, pp. 839–42.

Sparks, D. R. and **J. C. Malechek** (1968) Estimating percentage dry weight in diets using a microscope technique, *J. Range Mgmt* **21**, 264–5.

Stobbs, T. H. (1969a) The use of liveweight gain trials for pasture evaluation in the tropics. III. The measurement of large pasture differences, *J. Brit. Grassld Soc.* **24**, 177–83.

Stobbs, T. H. (1969b) The effect of grazing management upon pasture productivity in Uganda. I. Stocking rate, *Trop. Agric.* (*Trin.*) **46**, 187–94.

Stobbs, T. H. (1969c) The effect of grazing management upon pasture productivity in Uganda. III. Rotational and continuous grazing, *Trop. Agric.* (*Trin.*) **46**, 195–200.

Stobbs, T. H. (1970) Automatic measurement of grazing time by dairy cows on tropical grass and legume pastures, *Trop. Grassld* **4**, 237–44.

Stobbs, T. H. (1971a) Quality of pasture and forage crops for dairy production in the tropical regions of Australia. I. Review of the literature, *Trop. Grassld* **5**, 159–70.

Stobbs, T. H. (1971b) Production and composition of milk from cows grazing Siratro (Phaseolus atropurpureus) and Greenleaf Desmodium (Desmodium intortum), *Aust. J. Exptl Agric. Anim. Husb.* **11**, 268–73.

Stobbs, T. H. (1973a) The effect of plant structure on the intake of tropical pastures. I. Variation in bite size of grazing cattle, *Aust. J. Agric. Res.* **24**, 809–18.

Stobbs, T. H. (1973b) The effect of plant structure on the intake of tropical pastures. II. Differences in sward structure, nutritive value and bite size of animals grazing Setaria anceps and Chloris gayana at various stages of growth, *Aust. J. Agric. Res.* **24**, 821–9.

Stobbs, T. H. (1974a) Components of grazing behaviour of dairy cows on some tropical and temperate pastures, *Proc. Aust. Soc. Anim. Prod.* **10**, 299–302.

Stobbs, T. H. (1974b) Rate of biting by Jersey cows as influenced by the yield and maturity of pasture swards, *Trop. Grassld* **8**, 81–6.

Stobbs, T. H. (1975a) Factors limiting the nutritional value of grazed tropical pastures for beef and milk production, *Trop. Grassld* **9**, 141–50.

Stobbs, T. H. (1975b) Beef production from improved pastures in the tropics, *World Rev. Anim. Prod.* **11**, 58–65.

Stobbs, T. H. (1978) Milk production, milk composition, rate of milking and grazing behaviour of dairy cows grazing two tropical grass pastures under a leader-follower system, *Aust. J. Exptl Agric. Anim. Husb.* **18**, 5–11.

Stobbs, T. H. and **L. J. Cowper** (1972) Automatic measurement of the jaw movement of dairy cows during grazing and rumination, *Trop. Grassld* **6**, 107–12.

Stobbs, T. H. and **E. M. Hutton** (1974) Variations in canopy structure of tropical pastures and their effects on the grazing behaviour of cattle, *Proc. 12th Intl. Grassld Cong.*, pp. 510–7.

Stobbs, T. H. and **A. D. H. Joblin** (1966) The use of liveweight-gain trials for pasture evaluation in the tropics, *J. Brit. Grassld Soc.* **21**, 181–5.

Stobbs, T. H. and **P. A. C. Thompson** (1975) Milk production from tropical pastures, *World Anim. Review FAO*, No. 13, pp. 27–31.

Theurer, C. B. (1970) Determination of botanical and chemical composition of the grazing animal's diet, *Proc. Nat. Confr. Forage Qual. Eval. Util.* Univ. Nebraska, Sec. I, 1–17.

Thornton, D. D. and **G. N. Harrington** (1971) The effect of different stocking rates on the weight gain of Ankole steers on natural grassland in Western Uganda, *Uganda Agric. Sci.* **76**, 97–106.

Thurbon, P. N., G. A. Chambers, R. Sibbick and **J. Stokoe** (1973) Progress report on milk production from cows grazing irrigated, fertilized Digitaria decumbens as influenced by stocking rate and a molasses/biuret supplement, *Proc. 3rd World Cong. Anim. Prod.*, pp. 12–14.

Van Dyne, G. M. and **H. F. Heady** (1965) Botanical composition of sheep and cattle diets on a mature annual range, *Hilgardia* **36**, 465–92.

Veitia, J. (1971) Milk production of Holstein × Brahman cows given free access to elephant grass forage and either a high protein concentrate or molasses/urea, *Rev. Cub. Cienc. Agric.* **5**, 171–4.

Vicente-Chandler, J., L. Rivera-Brenes, E. Boneta, R. Caro-Costas, P. Rodriguez and **W. H. Garcia** (1953) *The Management and Utilization of the Forage Crops of Puerto Rico*, Univ. P. R. Agric. Expt. Sta., Bull. 116.

Vicente-Chandler, J., R. Caro-Costas, R. W. Pearson, F. Abruña, J. Figarella and **S. Silva** (1964) *The Intensive Management of Tropical Forages in Puerto Rico*, Univ. P. R. Agric. Expt. Sta., Bull. 187.

Vicente-Chandler, J., F. Abruña, R. Caro-Costas, J. Figarella, S. Sevando, and **R. W. Pearson** ((1974) *Intensive Grassland Management in the Humid Tropics of Puerto Rico*, Univ. P. R. Agric. Expt. Sta., Bull. 233.

Waite, R., W. B. McDonald and **W. Homes** (1951) Studies in grazing management. III. The behaviour of dairy cows grazed under close-folding and rotational systems of management, *J. Agric. Sci.* (*Camb.*) **41**, 163–73.

Walker, B. (1968) Grazing experiments at Ukiriguru, Tanzania. II. Comparison of rotational and continuous grazing systems on natural pastures of hardpan soils using an 'extra-period latin-square change-over design', *E. Afr. Agric. J.* **34**, 235–44.

Walker, B. and **G. D. Scott** (1968) Grazing experiments at Urikiguru, Tanzania. I. Comparison of rotation and continuous grazing systems on natural pastures of hardpan soils, *E. Afr. Agric. For. J.* **34**, 224–34.

Walton, P. D. (1981) A comparison of continuous and rotational grazing, *J. Range Mgmt.* **34**, 19–21.

Wheeler, J. L. (1962) Experimentation in grazing management, *Herb. Abst.* **32**, 1–7.

Wheeler, J. L., J. C. Burns, R. D. Mochrie, and **H. D. Cross** (1973) The choice of fixed or variable stocking rates in grazing experiments, *Exptl Agric.* **9**, 289–302.

Willoughby, W. M. (1970) Feeding value and utilization of pasture, *Proc. Aust. Soc. Anim. Prod.* **8**, 415–21.

Wilson, P. N. (1961) The grazing behaviour and free-water intake of East African short-horned Zebu heifers at Serere, Uganda, *J. Agric. Sci.* (*Camb.*) **56**, 351–64.

Winks, L. (1975) Integration of native and sown pastures for increased animal production, *Trop. Grassld* **9**, 159–64.

Chapter 11

Range management

Range and rangeland mean different things to different people. They may be vast arid and semi-arid land areas with sparse vegetation, immense regions over which nomadic herds wander in search of water and grass as dictated by the season, grassy plains devoted to wild-life preserves which are used for safaris and at attractions or large ranches with cowboys who herd and brand cattle and hold annual round-ups.

Actually, the two terms are comparable and refer to land areas covered with natural and semi-natural vegetation that are suitable habitats for extensive grazing of both domestic and wild animals (Heady, 1967; Pratt *et al.*, 1966). They include various types of grasslands and wooded, bush and shrub savannas used for grazing cattle, sheep, goats, donkeys and camels. The grazing area may be restricted by fences or open range free to all cattle-keepers and herders. Rangelands are frequently characterized as receiving less than 750 mm average rainfall, but also include areas with higher amounts where soils and topographies are unsuitable for cultivation. Additionally, some produce timber and provide watersheds.

All continents of the earth have extensive arid and semi-arid lands classified as rangeland (Gonzalez, 1969): Australia, 69 per cent; Africa, 51 per cent; Eurasia, 29 per cent; and America, 15 per cent. In North America, Canada has nearly 5 per cent of such lands up to the 60° parallel, the USA 34 per cent, and Mexico, 52 per cent. A similar situation exists in South America: Argentina, Brazil, Peru and Venezuela have among them around 300 million ha of arid and semi-arid lands. In Africa more than 45 per cent of the area between the equator and 20° N latitude is arid or semi-arid, and between 20° N and 40° N 94.7 per cent of the land falls in this category. In East Africa, Kenya has about 87 per cent of the land area available for pastures and rangelands, Uganda 60 per cent and Tanzania 50 per cent (Starnes, 1968). About 90 per cent of Southern Rhodesia (now Zimbabwe) is comprised of natural velds (rangelands) (Kennan, 1955), and approximately 80 per cent of South Africa with less than 650 mm of annual rainfall is separated into various types of veld (Meredith and Rose, 1955).

Rangelands are limited with regard to agricultural production, but they are capable of producing a great variety and quantity of herbage plants that serve as the basis for the extensive types of livestock industry characteristic of these regions. In many areas shifting agriculture and arable cropping are carried out within the total rangeland territory, but the naturalized fallows generally serve as grazinglands while crop residues provide supplemental feedstuffs for range cattle.

Range management in its simplest form is the care of natural and semi-natural grazinglands. More technically, it is the manipulation and utilization of rangeland soils, vegetation and animals for the production of goods and services needed by man (Heady, 1967, 1975). The major products of the range are meat, milk, hides, skin and wool. Planning and administering the use of rangeland is needed to obtain maximum output of herbage and animal product consistent with conservation of the range resources. This implies the application of ecological knowledge to practices involved with range management.

Principles of range management

Rangeland research and management over the past 50 years have led to the development of certain principles, and these can be separated into two groups (1) those associated with animals and (2) those involved with vegetation (Heady, 1967). Details of their application will vary among environments, but they provide basic guidelines for rangelands around the globe.

Principles related to animal practices

1. Grazing intensity (i.e. stocking rate) to balance the productive capacity of the land; otherwise herbage is wasted or the rangeland destroyed.
2. Control over kinds of animals using the range; they have varying feeding habits and influences on their habitat and different values for man.
3. Animal distribution over the range for proper herbage utilization; otherwise heavy concentration in localized areas causes overgrazing, deterioration of the vegetation and soil erosion.
4. Movement of animals according to seasonal production of herbage plants and their capability to withstand grazing so as to allow regeneration of plant growth and reseeding.

Principles related to range vegetation

1. Elimination or suppression of undesirable plant species by proper grazing management and brush control, particularly the judicious use of fire.
2. Seeding improved forage species to supplement those existing in the natural vegetation.
3. Fertilizing to correct soil deficiencies.
4. Soil-conserving practices to control water and erosion.

Grazing management refers to the care and handling of livestock on the range. It is primarily concerned with the selection of livestock and their distribution so as to utilize the forage in a given area effectively (Sampson, 1952; Stoddart *et al.*, 1975). Attention must be given to their protection from injury and death losses. Grazing management constitutes a part of range management and is a determining factor in regard to favourable economic returns.

Vegetation development and soil conservation

Rangelands constitute an ecosystem and have evolved under the influences of the climatic, edaphic and biotic factors discussed in Chapter 1. The vegetative cover of many rangelands, however, developed as a fire subclimax because of repeated burning, but also strongly influenced by soil conditions and the effects of overgrazing (Bartlett, 1956; Phillips, 1965; West, 1955; Le Houerou, 1980).

Rangeland deterioration

The grazing capacity of many rangelands has declined perceptibly on all continents and less desirable plants have rapidly replaced the more palatable and nutritious species (Harlan, 1965; Sampson, 1952). Notable abuse of rangelands has occurred in low rainfall areas under inadequate systems of grazing management and where periodic droughts hastened the effects of excessive grazing (Renner, 1948). Outstanding examples of grassland devastation are visible in North and South Africa, India, Pakistan, Australia and western United States. The range decline of the USA prairielands and the degradation of formerly productive, open grasslands into denuded 'thornbush deserts' in the Karamoja region of East Africa typify areas that have undergone overgrazing and misuse (Naveh, 1966). A similar deterioration of the Orinoco flood plain grassland savannas of Colombia and Venezuela took place after the introduction and increase of livestock (Roseveare, 1948). During the dry season much of the plains area is severely overstocked and wide margins of the water sites and streams are so heavily grazed and trampled tha the soil is left bare of all vegetation.

Decline of the well-grassed countryside in South Africa has beer well documented and serves as an example of rangeland mismanage- ment (Meredith and Rose, 1955). Early records show that the in- digenous stock-keepers followed a definite system of grazing as they migrated with their cattle and sheep, and grazing systems were care- fully manipulated within tribal boundaries. As colonizing farmers and stock numbers increased they moved into new areas, creating clashes with the Bantu herdsmen and overcrowding the grazing- lands. Within 100 years after the first settlement, signs of grassland degeneration became evident and the decline rapidly continued. The onset of a severe drought, political and administrative difficul- ties, along with development of the Merino sheep industry, upset farmers and placed greater demands on available grazing. It was ob- vious that available grass was insufficient for the increasing numbers of sheep and cattle. The lack of adequate grazing led to one of the most important events in the history of South Africa: the so-called 'Great Trek', a mass migration into new territory.

In the northern savanna region of West Africa, where cattle are heavily concentrated, nomadic and semi-nomadic pastoral tribes keep and hoard cattle for prestige and bride price, as a display of wealth and as a means of livelihood and survival. In recent years cattle numbers steadily climbed as land available for grazing de- creased and stocking capacity declined (Putman, 1969; Shepherd and Ba'ashers, 1966). The increasing number of agriculturists broadened the need for more arable land and frequently led to con- flicts with the cattle-keepers and herders. In addition, the encroach- ment of the desert presented a problem for those concerned with both livestock and crop production. The reasons were twofold (1) traditional cultivation practices leaving large areas bare so that they were continually eroded by wind and rain and (2) overstocking causing destruction of the range.

A comparable situation is encountered in regions of East Africa where range development is hampered by low and erratic rainfall, low-quality grasses growing on poor soils, lack of water, incidence of disease and sociological limitations (Naveh, 1966). Low-grade livestock are hoarded in surplus quantities and kept on impover- ished rangelands formerly held in balance by recognized tribal graz- ing and burning patterns. A vicious circle of deterioration–drought– starvation has been induced through the encroachment of better grazinglands by cultivators, the decimation of browsing game herds, and the extension of uncontrolled grazing brought on by increased watering sites and improved veterinary facilities. Without strict con- trol of land use and livestock management, such practices some- times contribute to more rapid devastation of the rangeland.

The principal causes of rangeland degradation are as follows:

1. Climatic changes – largely severe and prolonged drought which causes gradual replacement of the more palatable grasses by undesirable species, especially when combined with mismanagement of the range.
2. Overgrazing – due to overstocking and probably the most acute problem of range management.
3. Uncontrolled fire – whether ignited by man or lightning, a detriment to arid and semi-arid rangelands.
4. Cultivation – destruction of range vegetation by agriculturists who then abandon the land because climatic conditions do not favour arable farming. This has been a serious problem in Mexico and was responsible for the 'dust bowl' in the western USA.
5. Communal grazing – a factor in regions where no restrictions are placed on cattle numbers and owners or keepers have no responsibility for range management.
6. Socio-economic – applied to societies where cattle hoarding is a symbol of wealth and prestige.
7. Rodents and predators – an underestimated but prevalent damage by rabbits, hares, kangaroos, rats and other small rodents which consume forage.
8. Soil erosion – caused by wind and rain as a consequence of other processes which reduce vegetative cover.

In general, deterioration of the range in terms of vegetation changes and soil erosion can be traced to the systems of management. Frequently, the systems of grazing are based on the calendar instead of a knowledge of growth patterns of the range plants and their critical stress periods. By recognizing a decline in the herbage potential and grazing capacity, steps can be taken at an early stage of range deterioration to correct or modify the adverse factors. Most ranges will then rapidly regain their productivity. If advanced deterioration has occurred however, radical management changes are needed and recovery will be slow.

Classification of range condition

Guidelines have been established to indicate the trend in the condition of the range (Sampson, 1952, Stoddart *et al.*, 1975). Range condition refers to departures in the botanical composition of the range, i.e. an assessment of the herbage production and species components of the current year, season or month as compared to the average production established by previous records. It depicts the amount and stability of the herbage and reflects quantity (a measure of stocking rate) and quality (a measure of output) of livestock products. In describing range condition, four classes are employed:

1. Excellent – the vegetative cover is normal with productive, vigorous plants, and a good mulch cover between grass tufts. This denotes 75–100 per cent carrying capacity of the range potential.
2. Good – the better herbage species predominate but less desirable plants are noted; bare spots may be visible because fewer seedlings survive and some soil erosion takes place; range producing at 50–75 per cent of capacity.
3. Fair – undesirable species are dominant in many places and brush encroachment may be serious in other places; the palatable grasses and herbs are weakened, not highly productive and develop few seeds; there are many bare spots with gully erosion; 25–50 per cent of normal stocking capacity.
4. Poor – sparse and unstable vegetative cover, little mulch and heavy soil and wind erosion; severe encroachment of weedy species, especially woody plants; the desirable grasses, herbs and forbs found in protected places; no more than 25 per cent of potential stocking capacity; cattle likely to be in poor condition.

Trend refers to the direction of change in the range condition, i.e. improving or declining. It can be noted by the density and composition of the vegetation, as well as the vigour and growth characteristics of the plants. Adverse trends can be noted by (1) replacement of desirable species by inferior types, (2) decreased herbage production of individual plants and the sward in general, (3) reduced litter on the soil surface with signs of soil erosion and (4) decline in stocking capacity and decrease in animal gains.

Appraisal of range condition

In classifying rangeland for potential productivity under good management, it is necessary to know whether the vegetation is improving or deteriorating. This requires a knowledge of the desirable natural and naturalized plants, their competitiveness and desirable densities, acceptability to animals, tolerance of drought and trampling, etc. Weedy species can serve as indicators of the degree of deterioration or depletion. Information about soil characteristics, particularly inherent fertility, is important in regard to maintainning vegetative cover for sustained herbage growth and as a safeguard against erosion. Long-term climatic patterns are needed as well as the weather of the current year in making estimates and predictions of the range-stocking capacity.

Quantitative-climax approach

This method of rating range conditions is applicable to perennial grasslands and is based on a comparison of the present vegetation with that of the previous composition at a given interval of time (Dykaterhius, 1949). Range plants, whether desirable or undesirable, can be classified as 'increasers', 'decreasers' and 'invaders'. The first two can be valuable herbage types but the latter are mostly undesirable and are generally abundant on overgrazed, unstable rangeland. In developing the base from which future comparisons are made, individual species are placed in one of the three groups and relative percentages of each group (inclusive of all species) are recorded. This is usually tabulated so as to rate the range condition as excellent, good, fair or poor. Such data should be taken periodically throughout the season to provide records under varying conditions. Thus, by knowing the current range condition, and comparing it with past situations under similar circumstances, adjustments in stocking rates and distributions can be made. This scheme is more satisfactory where bunch-type grasses comprise the vegetation rather than creeping and trailing types.

Palatability-rating approach

In this system used for annual-type rangelands, ratings are made of plants highly acceptable to the livestock (Sampson, 1952). Data are collected early in the season before moving animals on to the range. Excellent-to-good conditions indicate a large proportion of highly preferred plants, there being a relatively dense cover, a thin mulch on the soil surface and no active erosion. Ranges in fair-to-poor condition are dominated by less palatable species and a greater number of undesirable plants. Those in poor condition consist of sparse soil cover, plants of poor growth, many weedy species and heavy soil erosion.

Range-potential method

This approach attempts to express the current herbage production in relation to the ultimate potential (Humphrey, 1949). It requires prior knowledge of the range vegetation and output, with emphasis given to ratings of botanical composition and density of cover, plant vigour (potential production), quantity of mulch and degree of soil erosion. In predicting potential herbage production and stocking capacity, it is assumed that (1) range condition is not a temporary state but repeatable under comparable environmental circumstances, (2) a current rating differing from the expected normal would not necessitate reclassifying the range condition on an annual basis and (3) excellent-to-good range will produce more than fair-to-poor condition, even though the current rating might suggest a modification.

Use of score-card

The range evaluator has before him a list of important factors such as (1) general growth and vigour of the desirable herbage species, (2) density, composition and overall grazing value of the vegetation, (3) indicator plants, including annual grasses, weeds and poisonous plants, (4) soil erosion indicators, such as quantity of mulch, extent of erosion and formation of gullies and (5) animal indicators, which include weight gains of the livestock and appraisal of the rabbit and rodent population. Numerical values assigned to the various items are scored and summarized to provide a rating of the range condition (Parker and Woodhead, 1944).

The three-step method

This scheme appraises the condition of the vegetation and gives an index of trends over a period of years (Parker, 1950). The method employs the desirable features of several quantitative measures for evaluating the range potential.

Step 1 is concerned with the establishment of permanently located line transects on the range. At regular intervals along the transect records are taken of the presence and identity of plants and the proportion of ground mulch.

Step 2 consists of a field analysis of the data, classification of the condition at the time of evaluation and estimation of the current range trend. This is usually done with a trend score-card.

Step 3 provides for taking photographs along the transect as a permanent record of the elements observed and measured. These procedures are repeated and the results compared in subsequent examinations during the year and in later years.

It should be noted that these techniques for measuring the condition and trend of the range have been developed and tested primarily in the western USA. They have application, perhaps with modifications, in the development of range management in the tropics, but their value lies in the strict control of animals.

Practices for range management and conservation

A knowledge of the range condition is basic to development of improvements in management and to application of the aforementioned principles. In planning for range improvement and maintenance the range manager must (1) have control over the animals, i.e. their numbers and distribution in space and time, (2) make adjustments in grazing pressure, i.e. stocking rate, through the use ɔf adequate grazing systems, (3) utilize appropriate means of brush ɔontrol, of which fire is the most common, (4) make adequate provisions for water, (5) give consideration to renovation practices, such as seeding, reseeding, fertilizing and (6) have access to supplemental feeds by means of herbage conservation *in situ*, artificial storage of hay or silage, and concentrates.

Intensity of grazing and stocking rate

Excessive numbers of livestock, insufficient area and lack of grazing in the dry season is the common cry of herdsmen on many rangelands (Cozzi, 1965; Edwards, 1942; Frenchou, 1966; Hendy, 1975; Kennen, 1971; Mabey and Rose Innes, 1964; Maule, 1954; Sullivan *et al.*, 1980; Trochain and Koechlin, 1958; Walker and Scott, 1968). One of the most difficult features of range management, especially in the arid and semi-arid regions, is the variability of herbage production from year to year. This makes it difficult to estimate stocking capacity. Grazing pressures in years of high rainfall completely differ from those during prolonged drought periods. Since rainfall is so unreliable in the dry tropics, stocking at rough estimates of average herbage production is likely to lead to deterioration of the range over a period of time. A stocking rate of about 80 per cent of the average carrying capacity maintains the range condition, except in excessively dry years. A rule of thumb sometimes suggested is not to overgraze more than 1 year out of four.

One approach might be the adjustment of stocking rates based on the annual herbage production and has merit in areas where a ready market is available for animal buying and selling. This seldom exists in the tropics, so that seasonal flexibility of balance between cattle numbers and forage resources is difficult or impossible.

A more practical solution is recognizing vegetational changes which lead to range decline and making adjustments of stocking rates before deterioration becomes severe. The importance of range condition and trend studies have been stressed as guides in the determination of carrying capacity and assessment of management. However, there is a lack of information for most tropical rangelands concerning productivity of major range types and their stocking capacities under different treatments. Information on potential herbage production is needed rather than preconceived concepts of vegetation climax and succession. This could be obtained by use of methods and techniques as outlined above, along with small-scale grazing experiments on selected grassland types.

In many places the cattle numbers already exceed the potential stocking capacity, especially in localities of communal grazing. Thus overgrazing is the common practice and destocking becomes difficult to achieve, because it demands a radical change in the pastoralists' way of life. Under such conditions completely new approaches must be devised to modify the land tenure system and the social structure before range management practices can be imposed.

Systems of grazing management

Grazing systems can be classified into the following categories:
1. Continuous – livestock are placed on the range and allowed to remain indefinitely, as is the case of year-round grazing even with seasonal herbage growth. Animals have free access to any part of the range.
2. Deferred – the range is divided into camps or paddocks (sometimes called ranges) so that a long period of rest is systematically alloted to each, the deferment falling at different times of the year over a predetermined number of years. The resting period coincides with a fixed time (by calendar or season) before and after burning and may include an interval to allow for seed set and maturity.
3. Rotational (sometimes called divisional rotation) – separation of the range into 2, 3, or 4 equal, or nearly so, areas and rotating the entire herd from one division to another at systematic intervals.
4. Deferred rotation – a division of the range with a given portion deferred at some critical period of the year, generally during seed set and maturity, with sufficient time allowed for seedling establishment.

Deferred grazing

Variations of the deferred grazing system have been employed in various parts of Africa in order to provide fodder reserves during the dry season and to accumulate flammable material for late and effective burns. The combinations of herd numbers, paddocks, grazing, resting and growing periods could run into the hundreds. Some are too complicated, of course, for practical and flexible management. In East Africa, Rhodesia (Zimbabwe) and South Africa 16 different schemes were found on experimental stations (Heady, 1960, 1967). A number have been described, such as one herd in four paddocks and three herds in four paddocks (West, 1955); one herd in four paddocks with three grazing periods, one herd in four paddocks with four grazing periods, three herds in four paddocks and yearly grazing (Heady, 1960); six camp and one herd system, three camp and one herd (Rains, 1963). Howell (1978) described a multi-camp grazing system in South Africa which was useful in counteracting varying climatic conditions and providing grass cover under adverse growing conditions. Two systems with burning and one system with no burning will be discussed.

Three-camp and one-herd deferred grazing with burning
(Table 11.1)

This system has shown promise in the northern Guinea savanna of West Africa (Rains, 1963) and in the drier coastal areas of Kenya (Naveh, 1966). Although herbage available for grazing during the dry season is not highly nutritive, it is far superior to that commonly encountered where no system of management is imposed and the range is burned annually by uncontrolled fires. A disadvantage of this plan is the movement of cattle on a calendar basis which leaves little flexibility for adjustments, especially in dry years. The cycle of grazing, resting and burning is completed in 3 years. A four camp, one herd system with grazing during the growing season in year 1, a rest for burning in year 2, burning and grazing during the dry season in year 3, and grazing during the dry season in year 4 would provide for greater flexibility than the three camp, one herd system and allow heavier stocking rates when they are needed. The advantages of the latter system are the liberal provision for rests in the growing season and adjustments in stocking during prolonged dry periods. Neither of these systems, however, makes provision for young stock in terms of nutritive requirements or rotational grazing for helminth (parasite) control.

Table 11.1 *Three-camp (paddock) and one-herd system of deferred grazing (Rains, 1963)*

Year*	Camp I	Camp II	Camp III
First	B. April G. Oct.–Mar.	G. Apr.–June	G. July–Sept.
Second	G. July–Sept.	B. April G. Oct.–Mar.	G. Apr.–June
Third	G. Apr.–June	G. July–Sept.	B. April G. Oct.–Mar.

* A new cycle is begun in the fourth year; B. and G. refer to burning and grazing, respectively.

One herd in four paddocks and three grazing periods with no burning (Table 11.2).

This scheme was used in the Rift Valley of Kenya to allow improvement of the grass cover (Heady, 1960). Each paddock is grazed for 4 months and rested for 1 year on a calendar basis, as established by the average growing season. For the first year grazing includes part of the dry season plus the first month of the growing season; second year, 4 months in the middle of the growing season; third year, the last month of the growing season plus part of the dry season; fourth year, no grazing. There is no provision for burning and animals nev-

Table 11.2 *One-herd in four paddocks and three grazing periods during the year* (*Heady, 1960*)

Paddock	Year 1			Year 2			Year 3			Year 4		
	1	2	3	1	2	3	1	2	3	1	2	3
A	G				G				G			
B		G				G				G		
C			G				G				G	
D				G				G				G

Note. G indicates the grazing periods; lines are the growing seasons:
(1) = Jan.–April,   (2) = May–Aug.,   (3) = Sept.–Dec.

er enter a paddock with completely fresh herbage. With fixed dates of animal movement and a constant stocking rate every paddock should have the same grazing capacity at the beginning of the scheme. Drought and accidental fires usually force flexibility into the fixed schedule. After the desired improvement of vegetation has occurred, it would probably be wise to consider a more flexible system.

Factors to consider in choosing the system of grazing

The choice of a grazing system depends on such factors as the condition of the range and the trend of condition, i.e. toward improvement or decline; rainfall and its distribution, length of the dry season; vegetative cover and the potential production of grasses, forbs and browse plants; objectives of the livestock enterprise; the land tenure system; type of livestock and quality of desired product; and above all managerial skill. The system should be simple in design and implementation since the more complicated grazing schemes require closer supervision and greater attention. If a deferred plan is employed, a small number of paddocks and infrequent moving of cattle simplify range and grazing management. There is need for flexibility to allow for adjustments when prolonged droughts occur and when accidental and uncontrolled fires destroy reserve forage. Some means of bush control must be included as a part of the overall plan, and in many areas this means controlled burning. Consideration of the growth cycle of range plants is highly important, so that root reserves can be accumulated for subsequent regeneration of plant growth after the dry season. No single system is applicable to all situations and the selected scheme must be geared to local ecological and management conditions.

One factor which limits the year-round utilization of vast areas of grazinglands in much of tropical Africa is the incidence of tsetse fly, the vector of trypanosomiasis (sleeping sickness). For example, in West Africa, *Glossina moristans* is found in the southern savanna vegetation under tree and shrub cover, while *G. longipalpis* is restricted to heavy woodland. *Glossina palpalis* and *G. trachinoides* are riparian-dependent, being found in vegetation along streams. It is only during the dry season, when fly species recede, that cattle are herded into the infested regions. A number of measure can be applied for meeting the trypanosomiasis problem, such as exploitation of tolerant and resistant livestock, chemotherapy and proper animal husbandry, but more important is a higher plane of animal nutrition. Methods directed against the vector include vegetation clearing, along with application of persistent insecticides (such as DDT and dieldrin) using prescribed spraying techniques (Davies, 1971; MacLennan and Aitchinson, 1963, 1969).

Comparison of grazing system

A number of studies show that in the arid and semi-arid regions requirements of the livestock and the range are met by continuous grazing for about 3 years followed by 1 year of rest to provide herbage and root reserves and to accumulate fuel for burning (Bogdan and Kidner, 1967; McKay, 1968, 1970; Norman, 1963; Walker and Scott, 1968).

This is illustrated by data in Table 11.3 taken from the northern Guinea savanna of Nigeria where annual rainfall is less than 1 000 mm. Under low grazing pressure herbage is kept in a more nutritive stage during the growing season with continuous rather than rotational grazing. With less available moisture the value of continuous grazing becomes more pronounced. In regions with less than 250–375 mm yearly rainfall the formalized deferred and rotational schemes are not appropriate, so that traditional nomadic and transhumance rotations between wet- and dry-season grazing, under controlled stocking rates, are probably more desirable.

Table 11.3 *The effect of grazing systems on performance of bulls on range stocked at 2.7 ha/head. North Guinea Savanna, Nigeria, 1961–65 (Leeuw, 1971)*

Grazing system	Liveweight gain per year		Supplementary concentrates*
	(kg/head)	(kg/ha)	(kg)
Continuous	43	15.8	138
2 – paddocks, deferred	39	14.4	136
3 – paddocks, deferred	33	12.1	136
3 – paddocks, rotational	30	11.2	144

* Whole cottonseed fed during dry season of 5–6 months.

In Mexico deferred rotation grazing increased carrying capacity of three rangeland types under moderate or heavy stocking as compared to continuous grazing (Gonzales, 1969). The deferred system has also been used in southwestern USA (Harlan, 1965) and in Australia (Nunn and Suijdendorp, 1954) for improved seedling establishment and development to vegetation for ground cover.

Rotational grazing is generally more beneficial in the more humid areas where there is a wider range of species and greater differences in animal acceptability of plants. For example, in the West African southern Guinea savanna where annual rainfall exceeds 1 000 mm young bulls were grazed on *Andropogon gayanus* during the rainy season of 6–7 months. Daily gains were 0.33 kg/head for paddocks rotationally grazed and 0.25 kg/head for those continuously grazed at stocking rates of one animal per hectare (Leeuw, 1971). Adequate rest periods between grazings allowed the desired species to recover from repeated defoliation, whereas intensive continuous stocking reduced their regrowth potential and encouraged inferior and less acceptable grasses.

Fodder and feed from trees and shrubs

Herbage for livestock taken from trees and shrubs is known as browse. It may be eaten directly from the natural growth of the plants or from regrowth of sprouts after cutting near ground level (known as coppice). In addition, woody branches can be cut or lopped from taller shrubs and trees, thus falling to the ground, where the twigs, seeds, pods and even the bark are eaten (known as pollarding). In rangelands of seasonal rainfall areas browse plants may be of equal or greater value as stock feed than grasses and other herbaceous plants, especially during the dry season.

It is likely that as many animals feed on plant associations of which trees and shrubs comprise a large part, as on grazinglands of pure grass or mixed grass and other herbs (Semple, 1970). The extensive use of woody–grassy vegetation is found in Africa, Australia, Asia, western USA, Mexico, Central and South America and in many regions of the subtropical and temperate zones of the earth. In India and Pakistan the problems of grazing are especially acute due to extreme overstocking. Here, a great scarcity of herbage exists throughout the long dry season between the monsoons so that browse is the major source of livestock feed. In some areas the cutting of branches from trees is inhibited or strictly regulated, as indiscriminate lopping destroys species that are valuable for lumber and fuel. The fodder from trees and shrubs represents the only available feed during the dry season in the arid northeast of Brazil and in parts of El Salvador, Nicaragua and Indonesia where cattle live principally on the fruits and leaves of trees for a considerable part of each year. In Mexico some cattle obtain their entire nourishment from Mesquite (*Prosopis iuliflora*) and from cacti (*Opuntia* spp.). The latter are also important for livestock in the arid and semi-arid states of Brazil. In this region several thousand hectares of spineless cactus have been planted and are in production as fodder plants. Cattle feeding on cacti lose weight rapidly but are able to survive long periods without water because of the thick, fleshy stems.

As estimated 75 per cent of the trees and shrubs are browsed to some extent by domestic animals and game in Africa (Whyte, 1947). In Tanzania, all but four of 100 species offered to goats were eaten and most were accepted by cattle when grasses were scarce or not available. An incomplete list from Kenya named 57 species of browse plants, most of which were eaten by cattle and some only by goats (Dougall and Bogdan, 1958). It was noted that 202 species were grazed or browsed in East Africa (Glover *et al.*, 1966) and 385 trees, shrubs and other browse plants were listed as being eaten by cattle in South America (Roseveare, 1948). Other lists have been compiled showing that livestock frequently browse on some species and occasionally on others (Everist, 1958; Kadambi, 1963; Lawton, 1968; Van Reusburg, 1948; Wilson and Brendon, 1963).

Browse plants are less subject to seasonal variation than grasses in terms of nutrient content. Furthermore, they leaf out at the end of the dry season, before the rains and before other forage plants appear. This occurs at a time when animal need is maximal for feed of a high nutrient content, as they are grazing on low-quality grasses. Browse plants alone keep healthy animals in fair condition, but may be inadequate as the sole feedstuff. A mixture of several species for browsing is superior to a single species. Much of the material is high in protein, ranging from about 10 to more than 25 per cent on a dry weight basis, and is high in most minerals except P which may drop to 0.12 per cent (Lawton, 1968; Mabey and Rose Innes, 1964, 1966; Rose Innes and Mabey, 1964). The crude protein usually exceeds that of grasses by two- to four-fold, especially in the dry season when the content of the latter may drop to 3 per cent or lower. Crude fibre content of most browse plants may be high, but no more than dry season grass or hay. Dry matter content ranges between 30 to 60 per cent while trees and shrubs are growing compared to 60 to 80 per cent for the dry, parched grass. Digestibility trials have shown that the feeding value of fodder from some browse plants is higher than that of grasses during the dry season, and equal to or greater than *Centrosema pubescens* (Table 11.4). Intake values of browse plants range from about 9.0 to 12.2 per cent overall and from 1.8 to 2.3 per cent of liveweight.

The browsing of cattle depends on (1) availability of grasses and browse plants – if browse is limited a greater percentage of time is

Table 11.4 *Nutritive value of some browse plants, legumes and native grass in Ghana* (*Mabey and Rose Innes, 1964, 1966*)

Species	Digestibility and chemical composition*				
	DM	OM	CP	CF	NFE
Antiaris africana	64(40)	67	77(12)	39(21)	74
Baphia nitida	57(37)	55	72(21)	32(27)	58
Grewia carpinfolia	70(42)	71	78(18)	54(18)	80
Griffonia simplicifolia	69(41)	70	81(16)	59(29)	70
Centrosema pubescens	53(25)	53	62(20)	40(30)	61
Native grass	54(37)	58	40(5)	66(34)	56

* Figures are percentages, DM per cent and chemical composition in parentheses; data for browse plants are averages of monthly samples, centro prior to flowering, and grass cut in June and July; local bullocks used for browse plants and grass, sheep for centro.

spent in search of grazing material and if plentiful it is eaten readily, (2) length of grazing time – if short there is less browsing, (3) time spent walking, i.e. in search of herbage and (4) adjustment to browse plants, i.e. whether accustomed to browsing (Payne, 1963). When browse plants are plentiful, up to 25 per cent of the grazing time may be spent browsing and if they are limited from 10 to 16 per cent may be spent browsing. When cattle are accustomed to browsing, and if grasses are scarce, they may browse as much as 45 per cent of the grazing time. They usually do not browse indiscriminately but concentrate on particular species if given a choice.

The pods of some trees are highly nutritive and readily consumed by livestock and game. In Rhodesia 10–20 large *Acacia albida* trees per hectare produced from 1 100 to 2 200 kg of pods without serious yield reduction of surrounding grass (West, 1950). Smaller trees, such as *A. subulata*, when spaced at 25–50 per hectare yielded from 550 to 1 100 kg of pods per year. Cattle grazing dry and mature grass and supplemented with edible pods were able to maintain their weight. Feeding 4.5 kg of veld hay and 4.5 kg of pods made a satisfactory daily maintenance ration for a 450 kg animal.

Similar types of pod-bearing trees are found in other tropical regions and make considerable contribution to livestock feed. More attention should be directed toward their care and utilization. It would also be good range management to leave useful browse plants when land is cleared to stumps and to plant strips of browse species.

Burning as a management practice

Large areas of rangelands are repeatedly burned every year either by design, mistake or natural causes. Because of its seeming ɔmnipresence during the dry season, fire has a greater and more direct influence on bush encroachment and herbage productivity of many grazinglands than any other method used for bush control. Fire in vegetation antedates man by millions of years and played a role in the development of plant associations and plant–animal relations before his appearance (Bartlett, 1956). The full impact of burning on changing vegetation, and especially on producing and maintaining fire subclimax grasslands (rangelands), was not realized until man discovered how to make fire and use it as a part of his activities (Komark, 1971; West, 1965). It is likely that man has always discussed ways in which fire influenced his life and affected his surroundings. Even today controversy exists as to the place and usefulness of burning in regard to range management. At one time scientific opinion almost universally opposed burning of grazinglands as being wasteful and destructive. In recent times, however, research directed towards understanding the effects of fire on vegetation has shown that controlled burning of natural and semi-natural grazinglands serves a useful purpose for the grazier. There is still concern, however, that fire which originates in the grassland or savanna spreads into and destroys the forest, which in turn produces derived savanna and eventually semi-natural grassland (West, 1971).

Extent of fire

The relative amount or percentage of area burned in the subhumid and arid rangelands has not been accurately determined (Fig. 11.1). In East Africa it has been said that:

the extent of fire is not easy to assess. Most European ranchers either do not burn and take precautions against fire sweeping across their properties or they follow a planned programme of burning to keep the bush open. They have found by experience that forage cannot be grazed and burned and also that bush invades without burning. Only the nomadic herdsmen can still be extravagant with fire (Heady, 1960).

Similar statements are still applicable to a vast majority of the earth's rangelands. One exception might be that many ranchers in other areas of the tropics do not follow a schedule of burning and are less cautious about fire protection of their properties. Aerial observations in Kenya during the 1958 dry season indicated that no more than 25 per cent of the land surface of a transect from Nairobi to Mombasa (some 500 km) had been burned (Heady, 1960). Flights made by one of the authors over the *llanos* of Colombia during the 1950s and 1960s confirm this figure. Such estimates, however, have little meaning in a localized area and for small graziers. Some areas are burned annually where seasonal grazing occurs and sufficient flammable material accumulates. This is especially true in regions of shifting agriculture where fires escape from land-clearing operations. It also occurs on communal grazinglands where stocking

Fig. 11.1 Burning of native and naturalized grazinglands, as shown here in the *llanos* of Colombia, is a means of stimulating growth during the dry season, eliminating excess and unused forage and is an aid in weed control. It must be used judiciously to avert decline of plant nutrient reserve and deterioration of the pastureland.

is uncontrolled. In very dry regions, and in overgrazed areas, burning is usually light and sporadic, or does not occur, because of insufficient fuel.

Fires are usually visible throughout the dry season and until after the first rains. Once an area is burned, it is generally not fired a second time in the same season due to lack of flammable material. In Uganda, most burning was done in the afternoon and fires were always greater in number during the week than on Sunday (Masefield, 1948).

Reasons for burning

Research on the use of fire as a range management practice was greatly expanded in Africa and western USA during the 1930s and in Australia during the 1950s. Investigation has been less extensive in other tropical regions. A large number of the studies centered around the beneficial or detrimental effects of fire on plant succession and the plant responses to time of burning, with relatively few being concerned with burning–grazing relationships.

Many reasons are given for burning. Among these can be listed several which fall outside of range management practices.

1. Carelessness of honey hunters who burn so as to see bee trees

and hives hung in trees (to escape raids by animals and ants) and who smoke the trees and hives before collecting honey.

2. Land preparation under shifting agriculture when fires get out of control.

3. Burning to drive out game or to attract game (when regrowth occurs) for hunting.

4. Social and religious festivities, i.e. prior to the rainy season to assure bountiful harvests.

5. Revenge to destroy someone's house, crops and grazing.

The more pertinent reasons for burning, of course, can be related to range and land management practices:

6. Control the encroachment of undesirable plants, mainly bush types. This is the foremost reason put forth for burning, and experimental evidence shows that burning retards the establishment and growth of trees and bushes but does not completely eliminate them. It maintains a low percentage of bushes and shrubs in terms of numbers as compared to no burning (Edwards, 1942; Kennan, 1971; Norman, 1963; Rose Innes and Mabey, 1971; Thomas and Pratt, 1967; Trapnell, 1959; Van Reusburg, 1971).

7. Remove the old stemmy and fibrous growth which would not be eaten by livestock. The effect of accumulated material and subsequent deterioration may be temporary or permanent, depending on environmental conditions. Too much litter tends to suppress the desirable grasses, but if large quantities accumulate and are burned, many grass tussocks may be killed. In some regions, a reason given for burning of the previous season's growth is to prevent the build-up of material which would be burned anyway, thus reducing the danger of accidental fires.

8. Obtain more desirable species composition. Grass species respond differently to burning and their reaction is strongly influenced by time and intensity of burn, the type of grass cover, and local environmental conditions (Kennan, 1971; Skovlin, 1971; West, 1965). Fire favours the development of *Themeda trianda* over several other grasses and its density is improved by burning. When completely protected the species disappears. In parts of Queensland, Australia, *Heteropogon contortus*, an undesirable species, predominates with annual burning and replaces *T. australis* (Shaw, 1957). If an area is burned, but protected from grazing, the latter becomes dominant, indicating that *H. contortus* develops in response to burning and grazing. Many desirable species such as *Panicum maximum, Hyparrhenia rufa* and *Andropogon gayanus* show a wide tolerance of burning but others such as *Melinis minutiflora* are eliminated.

9. Stimulate growth out of season and improve herbage quality. Fire promotes regrowth with a fairly constant 'period of sprout-

ing' which varies from 6 to 10 days for grasses and 18 to 30 days for trees (Hopkins, 1963, 1965). It has been postulated that high temperatures initiate a sprouting promoter or destroy a sprouting inhibitor, thus allowing the regeneration of new tillers even during the dry season. The rapid growth following burning is associated with an increase in the population of nitrifying organisms (Vine, 1968), but may also be related to a modification of the availability of carbohydrates for their utilization.

10. Facilitate the movement of livestock and aid the distribution of animals on the range. Livestock tend to concentrate in favourite areas, but with judicious burning they can be attracted to regrowth in outlying areas (Campbell, 1960).

11. Constraint of tsetse fly, other biting flies and ticks which transmit diseases. The tsetse fly renders over 10.4 million km^2 of land in Africa unavailable for year-round cattle grazing (Pratt, 1969). This represents 37 per cent of the total land area of the continent. Fire to reduce bush is used as one means of suppressing the fly population (MacLennan and Aitchinson, 1969).

12. Establish fire breaks in developing a system of protection from uncontrolled fire.

13. Prepare a seedbed for natural reseeding and artificial seeding of desired forage species.

14. The ash remaining after burning is thought to have a fertilizer effect, but this factor is questionable since the same minerals would eventually be released by decomposition of the organic matter (Egunjobi, 1970).

Disadvantages of burning

The disadvantages and deleterious effects of burning have been given but few have been substantiated by scientific investigation.

1. Burning injures range plants by removing top growth and depletes food reserves available for regrowth. Repeated burning during the dormant period stimulates regrowth for out-of-season grazing and causes a drastic reduction of root and crown reserves (Van Reusburg, 1971; West, 1965). This practice leads to invasion of weedy species, and a good example is the prevalence of *Imperata cylindrica* found in many parts of the tropics where annual burning occurs.

2. Burning causes deterioration of the vegetation. Burning too late in the dry season may injure browse plants as they usually leaf out ahead of grasses. The damaging effect on vegetative cover and encroachment of undesirable species, however, is related to time and frequency of burning.

3. It bares the soil to erosion and water runoff, deprives the soil of litter and mulch and reduces the effectiveness of rainfall. Again,

these deleterious effects can be considerably reduced by attention to burning management.
4. There is loss of organic nitrogen, carbonaceous material and organic matter. Nitrogen in the above-ground portion of the plants is lost on burning, but the organic matter content of a soil changes little except under cultivation, unless there is a deep burn which damages rhizomes, e.g. an extended burn such as might occur when swampy areas dry out (Van Reusburg, 1971).
5. Uncontrolled fires which spread beyond the intended area of burning may destroy forests and other properties (but this is an indictment against the practice rather than the use of burning *per se*).

Time, frequency and intensity of burning

Burning every third or fourth year in a properly managed grazing system is sufficient to check bush encroachment and provide optimal forage production (Guillotean, 1958; Pratt and Knight, 1971; Ramsey and Rose Innes, 1963; Savage, 1980; West, 1965). The grasses should be dormant, or nearly so, for a rapid effective burn without damage to living parts. Late burning just prior to the onset of rains combats bush encroachment because new leaves have emerged, but may injure low-growing browse plants. Early burning at the end of the growing season weakens perennial grasses because food reserves have not been completely transported to the crowns and roots. Should they be induced to tiller again out of season the root reserves are further depleted. Annual fires set early in the growing season, usually just after the first rains when trees are sprouting but grasses are still dormant, effectively reduce brushland to open grassland suitable for grazing in 4–5 years. The same result can be achieved over a longer period of time by burning and grazing in alternate years. Effective control of woody growth is influenced by the frequency of burning and the severity of each fire, which is related to the quantity of flammable material.

A rest period is needed before burning to allow an accumulation of combustible material. At least 1 000 kg/ha of dry matter is required for an effective fire to control woody scrub of 1 m height and 1 500–2 500 kg to control shrubs and taller trees (Rains, 1963; Thomas and Pratt, 1967). In deferred grazing systems this factor is taken into consideration.

An intense but rapid fire is most desirable. Temperatures at soil level rise sharply, depending on the season, height and density of fuel, and wind speed, but usually return to ambient levels within a few minutes (Hopkins, 1965; Masson, 1948; Norton and McGarity, 1965; Pitot and Masson, 1951; Tothill and Shaw, 1968). Ground level temperatures frequently exceed 500 °C but are less severe above 1.0 m height. Early-season burning generates temperatures of 60–70 °C above a height of 3 m but late-season fires cause them to exceed 500 °C up to 3 m and 100 °C at heights of 6 m. Tall grasses produce higher temperatures than short grasses, especially at ground level. At 2 cm soil depth the temperature may fluctuate as much as 14 °C above normal, but often as little as 3–4 °C. The percentage of herbage burned indicates the severity of a given fire. Light burns from early-season fires may consume 25 per cent of the material, but late-season burns range from 65 to 95 per cent.

A post-burn rest is needed to enable grasses to recover and build up food reserves before being grazed, but the optimal period is difficult to establish. If herbage productivity is high and rainfall distribution ensures vigorous regrowth, grazing within 4–6 weeks does not harm the grass cover. Animal performance under such conditions is superior to that of unburned range. With reduced regrowth, owing to irregular rainfall and limited plant nutrient reserves, a longer rest period is required or less intensive stocking must be practised to prevent further depletion and deterioration.

The success of a pasture legume in the tropics is partly related to its fire tolerance. Annual legumes, e.g. Townsville stylo, avoid the effects of fire because they survive as seeds in the dry season, but perennating types are subject to damage. A study in northern Australia (Gardener, 1980) showed that species which are adapted to more arid conditions had greater fire tolerance than those adapted to more humid conditions. Cultivars and lines of *Stylosanthes guianensis* succombed to fire drainage, excepting cv. Oxley. Plants of resistant lines regenerated growth from buds formed more than 14 mm below the soil surface. Those of *S. scabra* and *S. viscosa* developed from root tissue as deep as 38 mm. Oxley is a fine-stemmed, perennating and semi-prostrate cultivar with underground crown development (Bowen, 1980). Genetic diversity was noted within species studied by Gardener (1980), which provides an opportunity for selection of fire-tolerant lines within the desirable grazing types.

Bush control

Once a proper balance between rangeland productivity and stocking rate has been established the principal and universal method of vegetation control should be that of manipulating grazing pressure. Since this is seldom attained and vast areas are already covered by undesirable woody species, some other means is necessary for suppression or control of bush. A notable increase in stocking capacity can be achieved by clearing land of unwanted woody growth so as to allow the growth of native and naturalized grasses. Bush-control operations should first be carried out on land of high production potential and these followed by measures to prevent reinfestation. Whatever the system used, the costs and expected economic returns should be carefully calculated before the decision is made to launch a bush-control programme.

1. Use of fire

This is the most common and presently the most practical and economical method of bush control on most rangelands. Controlled grazing in combination with burning ensures a more effective control of bush, desirable spread of grass cover and improved output of animal product. In a planned system of grazing each succeeding burn is more uniform and effective than the preceding one, i.e. an accumulative effect occurs with repeated burning.

2. Hand-slashing

Slashing of all above-ground woody stems and stumping is an age-old practice for suppressing growth of woody species.. In the more humid regions rapid regrowth occurs from stems and roots, so that other control measures are often needed or else the benefits may not last even for 1 year. Slashing with sufficient time for drying of woody material prior to burning generates more intense and uniform fires when heavy bush covers the area than burning without slashing.

3. Mechanical control methods

These include: (a) Bulldozer – with blade and attached tines to clear bush and small trees and to push the debris into wind-rows for burning. Much of the soil is also moved into the wind-rows, which tends to destroy grasses and grass seeds. (b) Holt Breaker – a heavy roller with four blades, drawn by a powerful tractor or crawler for flattening and smashing bushes and small trees. There may be little killing and seedling trees and shrubs may actually increase, unless follow-up burning and reseeding are used. Virtue of the Holt Breaker lies in mechanical simplicity, capacity to reduce bush canopy to ground level and, on some types of land, it leaves a seedbed for direct sowing. (c) Chaining – an anchor chain of 50–60 tonnes, often with a heavy steel ball to hold the back loop close to the ground, is dragged between two tractors. Large trees are broken down and some are pulled out, but shrubby growth usually bends and little damage is done. (d) Cutters – tractor-drawn, take-off power machines with heavy knives which turn horizontally to the ground level, such as the Gyramon or rotary mower, or with a chain slasher instead of knives. They are effective in cutting bushes of less than 5 cm, but do not disturb the stubble and root system. Repeated mowing of thinly wooded areas and removal of tall-growth, tussocky grasses tends to favour development of low-growing grass cover under some conditions. All of these operations are costly and they are generally used on land of high productive potential along with other practices for range improvement.

4. Chemical control

Studies have been made in many regions of tropical rangelands and shown to be of practical value in conjunction with other grassland improvements, provided that cost–benefit ratios are favourable (Little and Ivens, 1965). Mineral oils and arsenic compounds were used in early studies, and followed by a vast number of trials with 2,4-D and 2,4,5-T formulations and mixtures. These are applied as sprays directly on to the foliage, slashed stems or stumps by using knapsack and mechanically powered equipment and airplanes. Granulated products, such as Picloram and urea compounds, have also been spread on the ground in close proximity to bushes, shrubs and trees (Harrington, 1972). Results from herbicides and arbocides have been variable, with kills up to 80 per cent or more for first applications. Repeated treatments, and combinations of other control measures such as burning, are needed for more complete control. These products are selective in their action so that varying dosages and times of application are important. Woody species respond differently and many show resistance and tolerance to some compounds. Browse plants are likely to be seriously damaged or killed. Before considering the use of chemicals, the range manager should consult with an expert who is knowledgeable with products and their application, and with the local flora and botanical composition of the rangeland.

5. Grazing of goats

The observation that goats prefer to browse led to trials for their use in bush control. With proper control, goats effectively reduce the percentage of certain woody types e.g. *Acacia* spp., but do not sufficiently browse others, e.g. *Tarchonanthus* and *Dichrostachys* to ensure control (Pratt and Knight, 1971). Through repression of the woody species, grass cover is improved, but allowing the goats to feed indiscriminately in large numbers leads to deterioration of the vegetation. In terms of practical range management, goat-keeping should be first for livestock value and second to check bush invasion into grazingland.

Provision of water

On arid and semi-arid rangelands it is rare for adequate water supplies to be available for cattle to have ready access to daily watering. Thus, water sites are provided or else the livestock are moved to water at fixed times. The daily water capacity of range stock reaches about 8.2 litres per 100 kg of liveweight (1.0 gal per 100 lb) in the dry season and around one-half this amount in the wet season (Wilson, 1961). Actual intake of water varies with the moisture content of herbage and climatic conditions. Slightly more will be consumed daily than when watered at less frequent intervals. In Kenya 3½ year-old Zebu drank about 19 litres (approx. 5 gal) when watered daily but 88 per cent of this quantity when watered every other day (French, 1956). An average figure of 15.2 litres/head per day should be used in calculating the need for water (Heady, 1960; Rains, 1963).

Animal output is greater with daily watering than every second or third day and the distance walked between grazing and the watering site affects productivity. For daily watering the distance walked should not be more than 8–10 km. If the two-way journey exceeds 15 km watering should be done every second day. Thus, the water supply determines to some extent grazing patterns. With widely spaced watering points the area around the supply is seriously trampled and overgrazed, while the more distant herbage is not utilized and generally of low quality.

Range reseeding and fertilizing

Reseeding is not a substitute for poor range management. Before a decision is made to reseed, the system of grazing should be critically evaluated to determine whether modifications will bring about an improvement in the natural vegetation. If clearing and cultivation are involved, range reseeding probably cannot be justified in economical returns of cost inputs, unless as a part of an intensive and reclamation scheme (Bogdan and Pratt, 1966; Rains, 1963). Other factors to be considered include the amount and distribution of rainfall (a minimum of about 150 mm during the growing season is needed for seedling establishment), soil fertility level, topography, availability of seeds, method of seeding and appropriate management practices to assure success of seedling establishment. If reseeding seems to be an economic enterprise, the practices discussed in Chapter 8 would be applicable.

It is unlikely that economic benefit will be derived from applying fertilizer to arid and semi-arid rangeland. Responses of natural and semi-natural grasslands to added fertilizer nutrients have been demonstrated (Ch. 9), but more favourable results occur when combined with other improvement practices.

Fencing

In terms of livestock-keeping, fences are used either to enclose animals or to enclose crops. In areas where ranching is practised and where livestock-keeping is a part of land ownership, or land rental, fences are used to mark boundaries and prevent trespass. Within a recognized property they are erected around grazinglands to reduce open herding, allow separation and control of animals and enhance the employment of grazing management schemes. Fencing is then an essential part of range management. In some areas of the tropics free range is allowed and communal grazing is practised. Livestock-herding is a way of life and most fences belong to the agriculturists, not the pastorialists. Under such conditions they are used by the animal husbandryman to construct kraals for protection of cattle at night and as holding enclosures for diseased animals and young stock.

Various types of fences are used, depending on the kind of live-stock-keeping. Wire is the common form employed with improved range management, but may be of the so-called living type, such as hedges of *Euphorbia* spp., and various thorny or woody plants. These are also used by the agriculturists to enclose croplands. Herders frequently utilize branches of trees and palms, bamboos and stalks of crop residue to erect small enclosures for temporary holdings.

References

Barlett, H. H. (1956) Fire, primitive agriculture and grazing in the tropics. In W. L. Thomas, *Man's role in Changing the Face of the Earth*, Univ. Chicago *Press*, pp. 697–721.

Bogdan, A. V. and **D. J. Pratt** (1966) *Reseeding Denuded Pastoral Land in Kenya*, Govt Printer: Nairobi.

Bogdan, A. V. and **E. M. Kidner** (1967) Grazing natural grassland in western Kenya, *E. Afr. Agric. For. J.* **33**, 31–4.

Bowen, E. J. (1980) Oxley stylo: a small plant for big beef production, *Queensld Agric. J.* **106**, 24–30.

Campbell, R. S. (1960) *Use of Fire in Grassland Management*, FAO Party. Past. & Fodder Dev. Trop. Amer.: Maracay, Venezuela.

Cozzi, P. (1965) L'allevamento del Restiame en Somalia, *Riv. Agric. Subtrop.* **59**, 123–39.

Davies, H. (1971) Further eradication of tsetse in the Chad and Gongolia River systems in northeastern Nigeria, *J. Appl. Ecol.* **8**, 563–78.

Dougall, H. W. and **A. V. Bogdan** (1958) Browse plants of Kenya with special reference to those occurring in South Baringo, *E. Afr. Agric. For. J.* **23**, 236–45.

Dykaterhius, E. J. (1949) Condition and management of rangeland based on quantitative ecology, *J. Range Mgmt*, **2**, 104–15.

Edwards, D. C. (1942) Grass burning, *Emp. J. Exptl Agric.* **10**, 219–31.

Egunjobi, J. K. (1970) Savanna burning, soil fertility and herbage production in the derived savanna zone of Nigeria, *Wildlife Consr. W. Afr. Proc. 7th Bienn. Conf.*, pp. 52–8.

Everist, S. L. (1958) Our best fodder trees, *Queensld Agric. J.* **84**, 581–2.

French, M. A (1956) The importance of water in the management of cattle, *E. Afr. Agric. For. J.* **21**, 171–81.

Frenchou, H. (1966) L'élevage et le commerce du bétail dans le nord du Cameroun, *Cah. ORSTOM, Ser. Sci. Humaines* **3**, 1–125.

Gardener, C. J. (1980) Tolerance of perennating Stylosanthes plants to fire, *Aust. J. Exptl Agric. Anim. Husb.* **20**, 587–93.

Glover, P. E., J. Stewart and **M. D. Gwynne** (1966) Masai and Kipsigis notes on East Africa plants. Part I. Grazing, browse, animal associated and poisonous plants, *E. Afr. Agric. For. J.* **32**, 184–91.

Gonzales, M. N. (1969) Optimizing livestock production through range management, *Livestock Dev. Dry Intermediate Savanna Zones Conf. Proc.*, Ahmadu Bello Univ.: Zaria, Nigeria.

Guillotean, J. (1958) The problem of bush fires and burns in land development and soil conservation in Africa south of the Sahara, *Afr. Soils* **4**, 64–102 (Fr. & Eng.)

Harlan, J. R. (1965) *Theory and Dynamics of Grassland Agriculture*, van Nostrand: Princeton, New Jersey, Ch. 5.

Harrington, G. N. (1972) The effects of spraying Acacia Lockii de Wild with picloram and 2,4-D/picloram mixtures and additional treatment with picloram granules, *E. Afr. Agric. For. J.* **37**, 197–200.

Heady, H. F. (1960) *Range Management in East Africa*, Govt Printer: Nairobi.

Heady, H. F. (1967) Range management in East Africa, *Symp. E. Afr. Range Problems*, Ville Serbelloni: Italy, pp. 78–9.

Heady, H. F. (1975) *Rangeland Management*, McGraw-Hill: New York.

Hendy, K. (1975) Review of natural pastures and their management problems on the north coast of Tanzania, *E. Afr. Agric. For. J.* **41**, 52–7.

Hopkins, B. (1963) The role of fire in promoting sprouting of some savanna species, *J. West Afr. Sci. Assoc.* **7**, 162–4.

Hopkins, B. (1965) Observations on savanna burning in the Olokemeji Forest Reserve, *Niger. J. Appl. Ecol.* **2**, 367–81.

Howell, L. N. (1978) Development of multi-camp grazing systems in the Southern Orange Free State, Republic of South Africa, *J. Range Mgmt* **31**, 459–61.

Humphrey, R. R. (1949) Field comments on the range condition method of forage survey, *J. Range Mgmt* **2**, 1–10.

Kadambi, K. (1963) Useful fodder trees and grasses for cultivation in Ghana, *Ghana Farmer* **7**, 75–80.

Kennan, T. C. D. (1955) Veld management in Southern Rhodesia, *Rhod. Agric. J.* **52**, 4–21.

Kennan, T. C. D. (1971) The effects of fire on two vegetation types at Matapos, Rhodesia, *Proc. Tall Timbers Fire Ecol. Conf.* **11**, 53–98.

Komark, E. V. (1971) Lightning and life ecology in Africa, *Proc. Tall Timbers Fire Ecol. Conf.* **11**, 473–511.

Lawton, R. M. (1968) The value of browse in the dry tropics, *E. Afr. Agric. For. J.* **33**, 227–30.

Leeuw, P. N. de. (1971) The prospects of livestock production in the northern Guinea zone savannas, *Seminar Forage Crops in W. Afr. Univ. Ibadan, Nigeria* (mimeo).

Le Houerou (1980) The rangelands of the Sahel, *J. Range Mgmt* **33**, 41–5.

Little, E. C. S. and **G. W. Ivens** (1965) The control of brush by herbicides in tropical and subtropical grassland, *Herb. Abst.* **35**, 6–12.

Mabey, G. L. and **R. Rose Innes** (1964) Studies on browse plants in Ghana. II. Digestibility. (a) Digestibility of Griffonia simplicifolia, (b) Digestibility of Baphia nitida, *Emp. J. Exptl Agric.* **32**, 125–30, 274–8.

Mabey, G. L. and **R. Rose Innes** (1966) Studies on browse plants in Ghana (c) Digestibility of Antiaris africana, (d) Digestibility of Grewia carpinfolia, *Expt. Agric.* **2**, 27–32, 113–17.

MacLennan, K. J. R. and **R. J. Aitchinson** (1963) Simultaneous control of three species of Glossina by the selective application of insecticide, *Bull. Ent. Res.* **54**, 199–212.

MacLennan, K. J. R. and **R. J. Aitchinson** (1969) Trypanosomiasis, *Livestock Dev. Dry Intermediate Savanna Zones, Confr. Proc.*, Ahmadu Bello Univ.: Zaria Nigeria.

Masefield, G. B. (1948) Grass burning: Some Uganda experiences, *E. Afr. Agric For. J.* **13**, 135–8.

Masson, H. (1948) *La Température du Sol au Cours d'un Feu de Brousse au Sénégal.* Comm. No. 12, Conf. des sols, Léopoldville: Congo Belge, pp. 1933–40.

Maule, J. P. (1954) Meat and milk for Africa, *New Commonwealth* **28**, 22–5.

McKay, A. D. (1968) Rangeland productivity in Botswana, *E. Afr. Agric. For. J.* **34** 179–93.

McKay, A. D. (1970) Range management research at EAAFRO, *E. Afr. Agric. For J.* **35**, 346–9.

Meredith, D. and **C. J. Rose** (1955) Grasslands in South African agriculture. In D Meredith (ed.), *The Grasses and Pastures of South Africa*, Central News Agency Cape Town, Part 2. Ch. 1.

Naveh, Z. (1966) Range research and development in the dry tropics with special reference to East Africa, *Herb. Abst.* **36**, 77–85.

Norman, M. J. T. (1963) The short term effect of time and frequency of burning native pastures at Katherine, N. T., *Aust. J. Exptl Agric. Anim. Husb.* **3**, 26–9.

Norton, B. E. and **J. W. McGarity** (1965) The effect of burning of native pasture on soil temperature in northern New South Wales, *J. Brit. Grassld Soc.* **20**, 101–5.

Nunn, W. M. and **H. Suijdendorp** (1954) Station management – the value of deferred grazing, *J. Dept. Agric. W. A.* **3**, 585–7.

Parker, K. W. (1950) *Report on 3-step Method for Measuring Condition and Trend Forest Ranges*, USDA For. Ser.: Washington.

Parker, K. W. and **P. V. Woodhead** (1944) What's your range condition? *Amer. Cattle Producer* **26**(6), 3.

Payne, W. J. A. (1963) A brief study of cattle browsing behaviour in a semi-arid area of Tanganyika, *E. Afr. Agric. For. J.* **29**, 131–3.

Phillips, J. (1965) Fire – as master and servant: its influence on the bioclimatic regions of Trans-Saharan Africa, *Proc. Tall Timbers Fire Ecol. Conf.* **4**, 9–109.

Pitot, A. and **H. Masson** (1951) Données sur la température en cours des feux de brousse aux environs de Dakar, *Bull. IFAA* **13**, 711–32.

Pratt, D. J. (1969) Management of arid rangeland in Kenya, *J. Brit. Grassld Soc.* **24** 151–7.

Pratt, D. J., P. J. Greenway and **M. D. Gwynne** (1966) A classification of East Africa rangeland, with an appendix of terminology, *J. Appl. Ecol.* **3**, 369–82.

Pratt, D. J. and **J. Knight** (1971) Bush control studies in the drier areas of Kenya. V Effect of controlled burning and grazing management in Tarchonanthus/Acacia thicket, *J. Appl. Ecol.* **8**, 217–37.

Putman, W. C. (1969) Livestock in West Africa, *Livestock Dev. Dry Intermed Savanna Zones, Conf. Proc.*, Ahmadu Bello Univ., Zaria, Nigeria. pp. 7–21.

Rains, A. B. (1963) *Grassland Research in Northern Nigeria*, Samaru Misc. Pape No. 1, IAR, Zaria, Nigeria.

Ramsey, J. M. and **R. Rose Innes** (1963) Some quantitative observations on the effect of fire on the Guinea savanna vegetation of northern Ghana over a period o eleven years, *Afr. Soils* **8**, 41–86.

Renner, F. G. (1948) Range condition: A new application to the management o natural grazinglands, *Proc. Inter-Am. Conf. of Nat. Resources*, USDA Publ. 3382 pp. 527–35.

Rose Innes, R. and **G. L. Mabey** (1964) Studies on browse plants in Ghana. I. Chemical composition, *Emp. J. Exptl Agric.* **32**, 115–24.

Rose Innes, R. and G. L. Mabey (1971) Fire in West African vegetation, *Proc. Tall Timbers Fire Ecol. Conf.* **11**, 147–73.

Roseveare, G. M. (1948) *The Grasslands of Latin America*, Imp. Bur. Past. & Fld Crops, Bull. 36, Wilson-Lewis: Cardiff.

Sampson, A. W. (1952) *Range Management Principles and Practices*, Wiley: New York.

Savage, M. J. (1980) The effect of fire on the grassland microclimate, *Herb. Abst.* **50**, 499–503.

Semple, A. T. (1970) *Grassland Improvement*, Leonard Hill Books: London.

Shaw, N. H. (1957) Bunch spear grass dominance in burnt pastures in southeastern Queensland, *Aust. J. Agric. Res.* **8**, 325–34.

Sheperd, W. O. and M. M. Ba'ashers (1966) Plant introduction and potential, including range and pastures management, *Report Mtg Savanna Dev., Khartoum, Sudan, UNDP/FAO*, pp. 136–49.

Skovlin, J. M. (1971) The influence of fire on important range grasses of East Africa, *Proc. Tall Timbers Fire Ecol. Conf.* **11**, 201–17.

Starnes, O. (1968) East African rangelands, *Symp. E. Afr. Range Problems*, Ville Serbelloni: Italy, pp. 93–101.

Stoddart, L. A., A. D. Smith and T. W. Box (1975) *Range Management*, McGraw-Hill: New York.

Sullivan, G. M., K. W. Stokes, D. E. Farris, T. C. Nelson, and T. C. Cartwright (1980) Transforming a traditional forage/livestock system to improve human nutrition in tropical Africa, *J. Range Mgmt* **33**, 174–9.

Thomas, D. B. and D. J. Pratt (1967) Bush control in the drier areas of Kenya. IV. Effect of controlled burning on secondary thicket in upland Acacia woodland, *J. Appl. Ecol.* **4**, 325–35.

Tothill, J. C. and H. H. Shaw (1968) Temperatures under fires in bunch speargrass pastures of southeast Queensland, *J. Aust. Inst. Agric. Sci.* **34**, 94–7.

Trapnell, C. G. (1959) Ecological results of woodland burning experiments in northern Rhodesia, *J. Ecol.* **47**, 129–68.

Trochain, J. L. and J. Koechlin (1958) Les pasturages naturels de sud de l'Afrique Equatoriale Francaise, *Bull. Inst. Centraf. No. 15–16*, pp. 59–63.

Van Rensburg, H. J. (1948) Notes on some browse plants, *E. Afr. Agric. J.* **13**, 164–6.

Van Rensburg, H. J. (1971) Fire: Its effect on grasslands, including swamps – southern, central and eastern Africa, *Proc. Tall. Timbers Fire Ecol. Confr.* **11**, 175–99.

Vine, H. (1968) Developments in the study of soils and shifting agriculture in Tropical Africa. In R. P. Moss (ed.), *The Soil Resources of Tropical Africa*, Cambridge Univ. Press, Ch. 5.

Walker, B. and G. D. Scott (1968) Grazing experiments at Ukiriguru, Tanzania. III. A comparison of three stocking rates on the productivity and botanical composition of natural pastures of hard pan soils, *E. Afr. Agric. For. J.* **34**, 245–55.

West, O. (1950) Indigenous tree crops for southern Rhodesia, *Rhod. Agric. J.* **47**, 214–7.

West, O. (1955) Veld management in the dry, summer-rainfall bushveld. In D. Meredith (ed.), *The Grasses and Pastures of South Africa*, Central News Agency: Cape Town, Part 2. Ch. 5.

West, O. (1965) *Fire in Vegetation and its Use in Pasture Management with Special References to Tropical and Subtropical Africa*, Commonw. Agric. Bur. Mimeo. Publ. No. 1/1965.

West, O. (1971) Fire, man and wildlife as interacting factors limiting the development of vegetation in Rhodesia, *Proc. Tall Timbers Fire Ecol. Conf.* **11**, 121–45.

Whyte, R. O. (1947) *The Use and Misuse of Shrubs and Trees as Fodder*, Jt Publ. Commonw. Agr. Bur. No. 10, Hurley, England.

Wilson, P. N. (1961) The grazing behaviour and free water intake of East African shorthorn Zebu cattle at Serere, Uganda, *J. Agric. Sci. (Camb.)* **56**, 351–64.

Wilson, J. G. and R. M. Brendon (1963) Nutritional value of some common cattle browse and fodder plants of Karamoja, Uganda, *E. Afr. Agric. For. J.* **28**, 204–8.

Chapter 12

Supplemental feed, preservation and storage

The native and natural grasslands provide the basic diet of the vast majority of ruminant livestock in the tropics. A majority of these grasslands are located in regions characterized by erratic rainfall patterns and varying periods of extreme drought and often on soils of poor fertility unsuited to cropping. Seasonal distribution of rainfall and soil fertility conditions impose a direct influence on the amount and quality of forage available during the year and indirectly affect animal performance. Though dry matter yields of 30–130 tonnes/ha have been obtained from cultivated tropical forages when nutrients and moisture are not limiting, the normal dry matter yields of natural grasslands are considerably lower and range from 2 to 6 tonnes/ha in the humid tropics, 1 to 4 tonnes/ha in the subhumid tropics and less than 0.5 tonnes/ha in the arid tropics (Crowder, 1977; Humphreys, 1977). The nutritive value of herbage in the grass-based grazinglands is inherently lower than the temperate grasslands and declines rapidly with advancing maturity. Thus, apart from very few exceptions in the wet tropics, animals in tropical environment live for a considerable period during the year on a sub-maintenance diet.

Need for supplementation in the tropics

The productivity and chemical composition of tropical grasslands vary annually depending on the length and intensity of wet and dry seasons. Herbage quality is generally satisfactory during 3–6 months of the growing season. Voluntary intake, digestibility and nutrient content decline rapidly as grasses become mature, dry and fibrous.

Chemical analyses of the herbage from a large number of tropical grasslands have indicated the protein content to be around 3–4 per cent in the dry months and occasionally as low as 1–2 per cent during the latter part of a long dry season. Bredon and Horrell (1962) studied the chemical composition and nutritive value of several common grasses in Uganda. They reported that the starch equivalent appeared to be generally sufficient for maintenance requirement of cattle throughout the year. Digestible crude protein values, however, were low and generally inadequate for animal maintenance over 5–6 months of the year. This lack of protein in the animal's diet is one of the primary factors limiting animal performance in the tropics. Low protein content in the herbage results in low apparent protein digestibility and approaches a zero value when the crude protein content declines to around 3 per cent. Thus, animals subsisting solely on dry season herbage are often on a diet almost void of digestible crude protein and in many cases may even show a negative nitrogen balance. Low protein content also causes a reduction in (1) the overall digestibility of herbage dry matter, (2) total digestible energy content and (3) voluntary intake of dry matter. The end result of reduced voluntary intake and lower digestibility is a sharp decline in the total intake of digestible energy. Thus animals on low protein herbage suffer not only from protein deficiency but also from lack of energy (Zemmelink, 1974).

Poor performance of livestock subsisting on low protein herbage is further accentuated by widespread mineral deficiencies, particularly phosphate, and a wide Ca : P ratio. In widely scattered areas of the tropics dietary deficiencies of copper, cobalt, chlorine, iodine and sodium have been reported. In areas with a long and profound dry season, animals also suffer from vitamin A deficiency because of low carotene content in the bleached mature herbage.

Livestock grazing the native and natural tropical grasslands find adequate quantities of a fair-to-good quality herbage during the initial months of the wet season and gain weight. They start losing weight, however, as soon as the grasses reach maturity and their condition deteriorates progressively through the dry months. Often by the end of the dry season they are emaciated and deaths in large numbers may occur if rains are delayed. Johnson (1975) discussed the status of pastoral nomadism in the Sahelian zone of Africa and described the catastrophic animal losses that occur during a large-scale regional drought. There are few compensatory devices or remedies in this situation except drastic reduction in herd size (Table 12.1). Under normal conditions, in most of the tropics the animals exhibit a characteristic 'stop-go' cycle of growth (Fig. 12.1), resulting in retarded body development, late maturity, low reproductive rates and ultimately poor and uneven output in terms of animal products. Dirven (1970) estimated average liveweight gains of about 20 and 90 kg/ha from cattle grazing natural grasslands in monsoonal and humid tropical climates, respectively. Lansbury (1960) in Ghana and Leeuw (1971) in northern Nigeria recorded losses of over 15 per cent of cattle bodyweight during the dry seasons. In East Africa, French (1951) reported an average of 10–20 kg/ha liveweight losses during the dry season and in years of prolonged

310   *Supplemental feed, preservation and storage*

Table 12.1 *Scenario of nomadic response to large-scale regional drought (Johnson, 1975)*

Stage 1. *Deepening drought* Restricted movement around permanent water sources; overgrazing near wells; death of weakest animals; raiding; some animal sales if near a market

Stage 2. *Intense drought* Crucial move/stay decisions; massive herd die-offs, especially of water-demanding species; depletion of animal and material capital reserves; livestock sales wherever possible, but poor price since stock condition is poor

Stage 3. *Total, prolonged drought* Drought of several years' duration; most of animals consumed, dead or sold; capital reserves expended; involuntary sedentarization

Stage 4. *Drought recovery* Use of remaining capital reserves, surviving stock and labour sales to rebuild the herds

drought up to 50 kg/animal. Similar nutritional problems are encountered by livestock keepers in Asia, South America and Australia, where fast growing native grasses provide nutritious grazing for 3–5 months of the year and animals suffer from temporary starvation during the dry months.

In some parts of the tropics, inundated conditions of grasslands during the rainy season, e.g. the plains of Venezuela and Colombia, lowlands of Bolivia, Brazil and monsoon Asia, do not permit grazing for considerable periods.

An adequate supply of high-quality feed throughout the year is a primary basis for increasing animal performance. The existing

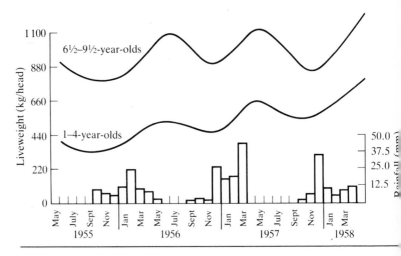

Fig. 12.1 Seasonal changes in liveweight of two age groups of cattle (1955–58) in the Northern Territory of Australia (Norman and Stewart, 1964).

vegetation in most tropical grazinglands will not support a high level of animal performance. Large amounts of energy, however, are stored in the form of dry and mature herbage on these grasslands. This energy can be utilized to improve livestock performance significantly during periods of stress by providing supplements to raise the overall nutritive value of the feed ingested by animals. In formulating supplementary feeding programmes it will be necessary to correct protein and mineral deficiencies.

Types of supplementation

Animal producers often attempt to mitigate the effects of prolonged drought by selling part of the herd, moving animals to public or rented lands, feeding edible shrubs or tree branches or hand-feeding young stock. Such practices are usually uneconomical and do not raise the standards of the animal industry much above the existing subsistence level.

The development of a successful supplementary feeding programme will depend on a knowledge of (a) nutritional requirements of livestock for maintenance and production, (b) available level of these nutrients in the herbage and (c) how the deficient nutrients can be economically supplied to the animal. Other factors include type of grazingland and species components, rainfall pattern and distribution, soil nutrient status, availability of fertilizers, type of livestock operation and genetic potential of the livestock.

Improved animal performance during periods of scarcity can be achieved by use of (1) conserved forage, (2) biological residues, (3) drought-tolerant species, (4) industrial by-products, (5) chemical and concentrate supplementation and (6) improved management practices. The operational level of livestock farms and economic factors will influence decision-making. They vary from region to region and affect the costs of inputs as well as returns from outputs.

Conservation processes

Herbage availability during the wet season often exceeds animal requirements. The accumulated forage loses most of its nutritive value with maturity. It could be conserved for supplementary dry season feeding. Forage conservation basically aims to produce, at low cost, a stable product suitable for animal feeding with minimum loss of nutritive value. Deterioration due to internal chemical changes and external microbial action of cut herbage are prevented either by dehydration or acidification. In the temperate countries hay-making and ensilage are accepted practices of forage conservation. In recent years an increasing quantity of forage, particularly lucerne, is conserved by high temperature dehydration and subsequently processed into wafers, cobs or pellets. In the tropics conservation of natural pasture surpluses is rarely practised because of inherent inefficiencies in the feeding system, low nutritive value of products (Table 12.2) and unfavourable weather conditions (Miller, 1969; Humphreys, 1977). These could be offset to some extent by proper choice of plant species, good management and adequate fertilization of sown pastures. Such practices increase herbage yields, improve chemical composition, enhance digestibility and improve the quality of conserved products.

Bush foggage

Conservation *in situ* consists of leaving excess herbage and browse plants as standing vegetation in the grazing area. Animals grazing such material show marked selectivity and make small gains in

Table 12.2 *Digestible organic matter (DOM), digestible crude protein (DCP) and ad libitum intake levels of fresh and conserved Nigerian grasses, conserved forages and crop residues (Miller, 1969)*

	DOM (% DM)	DCP (% DM)	Intake level (% maintenance)
Bush foggage (C)*	52.0	−1.0	57
Andropogon gayanus			
herbage (C)	54.0	1.4	85
silage unwilted (C)	53.8	2.7	98
silage wilted (C)	47.3	1.2	77
herbage (S)	52.4	1.3	122
silage (S)	40.7	2.5	59
hay (S)	44.9	0.7	87
Hyparrhenia rufa			
herbage (S)	39.6	0.8	72
silage (S)	34.3	−0.9	56
Panicum maximum			
silage unwilted (S)	46.2	4.0	95
silage wilted (S)	50.6	5.7	122
Sorghum + mucuna silage (C)	50.2	2.6	97
Maize silage (C)	56.7	2.4	80
Maize + cowpea silage (C)	60.1	3.8	92
Soybean hay (S)	58.0	7.9	198
Soybean hay (C)	59.0	7.7	151
Groundnut haulms, leaf loss (C)	44.2	3.3	63
Groundnut haulms, leaves retained (C)	56.1	6.3	107
Groundnut haulms, early harvested (C)	53.8	8.4	153

* Cattle
† Sheep

liveweight in the early dry season (Miller, 1969). As the dry period progresses the herbage loses most of its nutritive value owing to maturity, bleaching and loss of leaf (Table 12.3). Animal performance declines sharply when this occurs. Nitrogen fertilizer applied at the end of the wet season has been used in an attempt to sustain crude protein of accumulated forages. The practice is of doubtful value in the humid tropics and of marginal success in the cooler subtropics.

Table 12.3 *Comparative chemical analysis of Pennisetum pedicellatum hay and foggage (Rains, 1963)**

	CP	CF	EE	Ash	P_2O_5
Hay	4.78	37.58	1.24	6.73	0.23
Foggage	1.91	33.76	0.63	10.22	0.11

* Figures are percentages based on a dry matter basis. CP = crude protein, CF = crude fibre, EE = ether extract.

Animals grazing bush foggage maintain bodyweight if provided with salt−mineral block licks and supplemented with small quantities of protein feeds. For example, Zemmelink (1974) in Nigeria found that a supplement of 75 g/day cottonseed and groundnut cake in equal mixture was adequate to maintain weight of 215 kg steers. Without supplement they lost about 300 g daily. The concentrate provided 11 per cent of the energy required for maintenance with the remainder coming from grazed bush foggage. At higher levels of concentrate feeding, cattle may actually gain weight. In Australia, beef cattle fed dry roughage *ad libitum* and a concentrate supplement of 1.4 kg/head per day made daily gains of 0.75−1 kg in a trial of 108 days (Whyte *et al.*, 1959).

In South Africa, spraying dry standing herbage with about 400 litres/ha of a solution made from 15.5 kg of urea, 14 gal of molasses and 34 gal of water improved palatability and protein content. Mature steers grazing the treated material gained 0.19 kg/head per day, while those grazing untreated material lost weight without supplementary feed (Altona, 1966). Under the prevailing conditions of cattle management in many parts of the tropics, placing urea−salt−mineral blocks in open ranges is practicable. These block-licks enhance utilization of bush foggage and considerably alleviate protein and mineral deficiencies of the roughage.

Standing hay

In parts of northern Australia Townsville stylo 'hays off' after seeding and remains as standing hay with about 12 per cent protein until the first rains of the following wet season (Shaw and Norman, 1970;

Gillard and Fisher, 1978). The potential of this free-seeding summer annual was recognized in the 1920s and has become widespread in tropical and subtropical Australia. This legume led to a 5- to 10 fold increase in carrying capacity over native grasslands of northern regions with 760 mm of rainfall or more. It grows in dense stands and varies from prostrate to semi-prostrate in growth habit. Plants develop many branching stems. The legume requires a growing season of about 12 weeks. Flowering occurs over a prolonged period and seed yields range from 450 to 785 kg/ha (Davies and Hutton, 1970). Dry matter yields of 5.0 tonnes/ha or more accumulate during the growing season. The standing hay is readily grazed by cattle. Seeds are also eaten but sufficient fall to the soil for regeneration of dense stands in the following wet season. At Katherine, Northern Territory, average liveweight gains of steers on stand-over Townsville stylo during the dry season were 68 kg/head at a stocking rate of 2.5 animals/ha. On unimproved grazingland steers lost about 20 per cent of their initial liveweight when stocked at 40 ha/head. Steers maintained continuously on Townsville stylo from weaning at 1.2 ha/head reached slaughter weight at less than 3 years of age, as compared with 5–7 years on native grasses (Norman, 1968).

In some areas the Townsville stylo standing hay gradually deteriorates because of dew and light rainfall. Under these conditions cattle lose weight but the amount and duration is less than on native grasses.

Attempts have been made to utilize other legumes in a similar manner but with less success than Townsville stylo. The recent introduction of *Stylosanthes hamata* cv. Verano into Australia appears to have potential as a legume for standing hay as well as a component of pasture mixtures, (McKeagne, *et al.*, 1978; Gillard *et al.*, 1980). Townsville stylo has been introduced into other countries for use in overseeding native and naturalized grasslands for improvement of dry-season grazing. It appears to have potential in the northern Guinea and Sudan savannas of Nigeria (Leeuw and Brinckman, 1974).

Herbage of sown grass pastures is frequently accumulated as standing hay for dry season reserve. With decline in quality, however, it seldom provides maintenance requirements without supplementation. An experiment was carried out in Malawi to compare (1) Rhodes grass foggage, (2) Rhodes grass foggage with stylo and Siratro sown in alternate pure strips so as to occupy about one-third of the total area and (3) Rhodes grass foggage supplemented daily with 0.34 kg/head of cottonseed cake (0.10 kg crude protein) to provide about one-third of maintenance requirements (Addy and Thomas, 1976). Cattle grazed the pastures in the dry, cool period from June to November at 2.85 units/ha. The Rhodes grass was topdressed with 45 kg/ha of N at the beginning of the wet period and grazed until about 6 weeks before the onset of the dry season. Steers on Rhodes grass alone lost weight, but liveweight of those on the other two treatments increased (Table 12.4). Crude protein of the grass leaves dropped to 2.4 per cent during the experimental period, while that of the legumes remained above 11.0 per cent.

Table 12.4 *Liveweight changes (kg/head) in the cool, dry season of cattle grazing Rhodes grass foggage with or without protein supplement* (Addy and Thomas, 1976 – Malawi)*

Treatment	Malawi Zebu	Friesan × Malawi Zebu	Mean	Response to protein
Rhodes grass alone	−3.2	+0.7	−3.2	–
Rhodes grass and legumes[†]	+19.0	+30.0	+24.5	+25.8
Rhodes grass plus cottonseed cake[‡]	+18.2	+28.6	+23.4	+24.7

* Six weeks' accumulation of herbage before the dry season; grazing from June to November.

[†] Alternate strips of stylo and Siratro in Rhodes grass allowed to accumulate herbage along with the grass.

[‡] 0.34 kg/head daily cottonseed cake (0.10 kg crude protein to provide about one-third of maintenance requirements).

Hay

A high quality hay can be defined as a forage dried so as to retain most of the leaves, without deterioration of dry matter and nutrients, without mould development, having its natural green colour and palatability, and capable of being stored over a long period of time (Hodgson *et al.*, 1948). A detailed knowledge of seasonal growth and development and the accompanying chemical changes in the forage is necessary to obtain maximum yield of digestible nutrients in the hay. The early flowering stage appears to be the most appropriate time to harvest grasses for hay.

Unreliable weather conditions and generally poor herbage quality are the two most serious constraints to large-scale hay production in the tropics. Uneven land surface, lack of technical know-how and scarcity of hay-making equipment are also important inhibiting factors. Average losses of dry matter during hay-making and storage are estimated to be around 25–30 per cent and may be greater with adverse weather.

From the time forage is cut to the time it is adequately dried for storage, dry matter losses and deterioration in nutritive value and quality occur due to respiration and fermentation, mechanical and handling damage and weather effects. Hay-making procedures, therefore, should be utilized to keep these losses at a minimum. For production of superior quality hay, conditions must be favourable for rapid removal of moisture from cut herbage. This can be achieved by (1) making hay during periods most likely to have continuous sunshine, (2) drying the herbage partly in the field and partly under artificial conditions, (3) complete artificial drying, (4) preharvest desiccation of forage with chemicals such as formic acid and (5) crushing, bruising or cracking forage stems by mechanical devices to accelerate the drying process. To avoid serious deterioration in quality, loss of nutrients from fermentation, mould development and danger of spontaneous combustion, moisture content should be reduced to 20 or 25 per cent or less for loose storage in stacks. Baled or chopped hay should be drier than long hay for storage.

The usual method of hay-making is to cut the forage, allow it to dry in a wind-row in the field and stack it when dry. To minimize losses due to rain and to facilitate drying, the forage can be stacked on tripods, hurdles and racks in the field. Kilns and smokehouses have been occasionally used to complete the drying process in parts of Asia and Africa. In the wet tropics, hay-making must be confined to the latter part of the rainy season or to the early dry season when the forage is mature and has poor nutritive value. In the semi-arid tropics, conditions are more conducive to hay-making.

Chemical analyses and nutritive-value studies of hays made from tropical forages permit the following generalizations to be made.

1. Although hay made from natural grasslands is rarely of high quality, it has a better feeding value than the standing hay or foggage left in the field (Table 12.3 p. 313).
2. Hay made from the first growth after rains or from the aftermath is considerably superior to hay made from the whole season's growth or hay made during the dry season (Table 12.5).
3. Crude protein and digestible crude protein contents of tropical grass hay range between 4–10 and 1–6 per cent, respectively. Tropical grass hay is generally inadequate as a sole feed to supply the animal's nutrient requirements for maintenance. Some species such as *Pennisetum clandestinum, Panicum maximum, Cenchrus ciliaris, Cynodon* spp. and *Digitaria decumbens*, when grown under good fertility conditions and carefully managed, offer opportunities for making hay of acceptable quality. This hay should be adequate for animal maintenance requirements and even allow liveweight gains.
4. Crude protein and digestible crude protein contents of legume hay are high and range between 11–17 and 7–13 per cent, respectively. This indicates their potential as a cheap source of protein for supplemental feed.

Silage

Silage can be described as a moist succulent feed produced as a result of controlled fermentation of fresh forage when stored in a silo

Table 12.5 *Nutritive value of some grass and legume hays*

	Crude protein (%)	Digestible crude protein (%)	Crude fibre (%)	Starch equivalent (unit)	Total digestible nutrients (%)
*Native grass**					
First growth	8.8	5.5	34.5	38	
Aftermath growth	8.5	4.8	34.2	31	
Whole season's growth	6.1	1.6	33.5	23	
*Cynodon plectostachyus**					
First growth	10.5	6.1	31.4	36	
Aftermath growth	8.0	4.6	33.4	29	
Whole season's growth	7.3	3.8	37.1	24	
Dried *in situ* and					
mid-season	5.2	1.2	39.7	20	
Chloris gayana[†]					
Young and leafy	14.7	9.6	28.9		58
In full flower	7.0	3.1	33.1		56
Pennisetum clandestinum[†]					
Young and leafy	14.1	9.1	28.4		58
Setaria anceps [‡]					
Young and leafy	15.1	10.0	28.7		58
Stizolobium deeringannum[†]					
Without much loss of					
leaves and seeds	17.9	10.9	23.8	51	
With much loss of					
leaves and seeds	13.0	7.1	33.7	33	
Medicago sativa[†]					
Flowering	18.9	13.6	28.9	57	
Vigna sinensis[‡]	10.9	7.4	27.4		

* French (1943).
[†] Dougall (1960).
[‡] Foster and Mundy (1961).

under anaerobic conditions. Since climatic conditions in many parts of the tropics make hay-curing difficult, silage-making offers a more reliable means of forage conservation. Grasses are the most common material for silage-making. Tall-growing fodder grasses such as *Panicum maximum, Pennisetum purpureum, Setaria anceps* and *Tripsacum laxum*, when well managed and fertilized, are suitable for silage-making. In addition, crops such as maize, sorghum and millet with or without legumes are also used. The prostrate, stoloniferous grasses and trailing legumes are less desirable.

The quality of silage depends on the stage of maturity, dry matter and nutritive content, particularly the carbohydrate fraction. To produce good quality silage, most grasses should be harvested in the vegetative stage of growth and not later than the early bloom stage. Cereal crops can be cut and ensiled at pre-bloom or boot stage to early dough stage, but preferably at the early milk stage (Table 12.6).

Silos

The containers used for silage-making range from the sophisticated and expensive gravity self-feeding vertical silos to the simple and inexpensive do-it-yourself trench or pit silos commonly used by farmers all over the world. Silage can even be made by stacking, although losses are likely to be high. Large-diameter concrete pipes can serve as good containers for silage. The choice of silo and its size are largely dependent on the magnitude of the livestock enterprise, facilities available and various economic factors.

Good-quality silage has been made in pit or trench silos in many parts of the tropics. Such silos should be located on well-drained, high grounds of the farm, preferably on hillsides. The walls should be plastered with mud or cement to minimize loss. Consolidation of the cut and usually chopped herbage can be effected by tractors or heavy rollers in the larger open-ended silos or manually in the smaller silos.

Production of silage

Plant respiration continues after the silo is filled and until the oxygen present in the air and trapped in the forage is used up. Respiration is associated with the breakdown of carbohydrates and production of carbon dioxide, water and heat. Presence of excessive amounts of air results in a greater rise in temperature. Heavy losses can occur when consolidation of forage materials is poor or if the silo is improperly sealed. This occurs because of continuing aerobic respiration and increased losses due to surface spoilage.

Proper compaction restricts carbohydrate losses due to respiration. The fermentation process is initiated by enzymatic activity and presence of yeasts, moulds and aerobic bacteria. This results in the breakdown of structural carbohydrates and sugars and in the production of various acids such as acetic, propionic and lactic. Degradation of protein into simpler substances such as amino acids and ammonia also occurs. As oxygen is exhausted and acidity increases, moulds and yeasts cease to grow or disappear and only the anaerobic bacteria remain active. These produce further amounts of acid from soluble carbohydrates. The type of acid produced depends on the organisms present. Clostridia are mainly responsible for the production of butyric acid. Undesirable changes in the protein fraction

Table 12.6 *Effect of harvest age on the silage quality of sorghum (Catchpoole, 1962)*

Age (wks)	Moisture (%)	Crude protein* (%)	Yield of silage (t/ha)	Silage protein (kg/ha)	Total loss[+] (%)	pH	Non-volatile acid / Volatile acid
4	80.8	17.2	6.9	247	63	5.31	0.3
6	82.9	12.6	31.8	762	57	5.70	0.4
8 (Boot)	82.7	9.1	52.2	897	42	4.09	1.3
10 (Early milk)[‡]	80.5	6.7	76.9	1 110	14	3.49	3.2
12 (Early dough)	78.3	4.8	80.7	953	4	3.43	3.6
14	73.5	5.4	80.4	1 087	–	3.74	2.7
16	71.9	5.1	69.7	1 110	4	3.81	3.5

* Based on dry matter.
[+] Percentage of original weight.
[‡] Recommended stage of ensiling.

are associated with the formation of butyric acid (Murdock, 1966; Catchpoole, 1970). The aim in silage-making is to prevent this group of bacteria from becoming dominant and to encourage those that produce lactic acid. Conversion of sugar to lactic acid involves only a small loss of energy. Increasing acidity and high osmotic pressure control the activity of anaerobic bacteria and their growth is almost completely inhibited when the pH value falls below 4.2. Further anaerobic decomposition during storage is prevented and the silage remains in a stable condition for a considerable period of time if kept properly sealed and covered to prevent entry of air and water.

A well-preserved, unwilted silage has a pH of 4.2 or below, butyric acid concentration of less than 0.2 per cent, ammoniacal nitrogen content of less than 11 per cent of total N and lactic acid concentration of 3–13 per cent of dry matter and exceeding that of volatile acids (Catchpoole, 1970; Catchpoole and Henzell, 1971).

Losses in silage-making and their control

Losses in dry matter and nutritive value of silage normally range from 10 to 20 per cent but may rise higher. They occur from (1) field spoilage, (2) type of fermentation (increasing with rising pH values and greater activity of clostridia), (3) seepage or effluent (higher when the dry matter content of herbage is low or under excessive compaction in the silo) and (4) spoilage or wastage during storage and after the silo is opened for feeding due to entry of air into the exposed surfaces resulting in fungal and bacterial growth (Murdock, 1966; Takano, 1972). The efficiency of silage-making can be significantly improved by judicious use of harvesting, storage and handling equipment, proper compaction and careful sealing of silos, mechanical treatments such as fine chopping and laceration of herbage before ensiling, wilting and use of additives.

Wilting of herbage to a moisture level of 70 per cent or below prior to ensiling results in (1) significant reduction or elimination of effluent flow, (2) reduction in dry matter losses due to aerobic respiration, (3) production of more palatable silage with higher pH and higher concentration of sugar and (4) increased stability due to higher osmotic pressure which suppresses the growth of clostridia organisms. Spraying of forage crops with formic acid at a rate varying from 0.25 to 1 per cent of the crop fresh weight can successfully reduce the moisture content of the herbage to levels optimum for ensilage (Tetlow *et al.*, 1975).

Low sugar content and high moisture in the herbage cause delayed or unsatisfactory fermentation and anaerobic decomposition during storage. Additives either inhibit anaerobic decomposition by significantly reducing the growth of bacteria or stimulate the natural fermentation processes during silage-making (Watson and Nash,

1960). A number of nutritive additives have been used, such as molasses, cereal grains and citrus pulp, as well as non-nutritive additives such as sodium metabisulphite, calcium formate, formic and other mineral acids and antibiotics. Inoculation of herbage with lactic acid bacteria has given promising results. Aerobic stability of silage has been increased by the addition of pimaricin, an antimycotic antibiotic which has no effect on bacteria or ruminant animals (Javier, 1976). The choice of additive will depend on availability, cost facilities required for mixing and handling, and the farmer's knowledge of silage-making.

Silage from tropical forage species

Silage made from tropical pasture species does not compare favourably with that of temperate species (Table 12.2, Miller, 1969; Catchpoole and Henzell, 1971; Holm, 1974). Tropical grasses are generally coarse and stemmy, higher in crude fibre, lower in soluble carbohydrates. A combination of these factors results in greater retention of cell contents, delayed fermentation and proliferation of clostridial organisms. The latter cause production of butyric acid, greater loss of energy, protein degradation and increased effluent losses. Low levels of soluble carbohydrates and high buffering capacity, combined with woody stems and soft leaf tissues, generally result in unsatisfactory ensilage of tropical legumes. Cereals alone and cereal–legume mixtures appear to be ideally suitable for silage in the tropics.

The following conclusions can be made about silage and silage-making from tropical herbage species in the tropics (Catchpoole and Henzell, 1971; Tuah and Okyere, 1974; Wylie, 1975):

1. The densities of silage from tropical species can be low. Unwilted tropical pasture species often fail to settle or settle only slowly and slightly during ensilage (McWilliams and Duckworth, 1949; Davies, 1963). Low-density silage can suffer extensive spoilage due to aerobic decomposition during storage (Vera Cruz, 1967; Davies, 1963). Special precautions, therefore, should be taken to exclude air during storage. Losses in dry matter and N during natural fermentation of tropical pasture species in efficiently sealed silos are comparable to those reported from temperate species.

2. Production of stable silage from tropical pasture species without the use of additives has been reported by many workers in the tropics. Such silage generally has high pH (4.8 or more), low to medium concentration of volatile acids (below 6% of dry matter), and moderate to high amounts of ammoniacal N. Decreased water activity and increased osmotic pressure of the cell sap appear to play a significant role in maintaining the stability of

high pH tropical silages by limiting the growth of clostridia. Thus, *Setaria anceps* grown under a range of conditions in Australia produced a remarkably stable acetic acid silage with pH of around 4.8, lactic acid concentration of less than 1 per cent, volatile acid concentration of 2–6 per cent of dry matter and ammoniacal N content of around 18 per cent of the total N (Catchpoole, 1968). *Chloris gayana*, on the other hand, offered little resistance to decomposition with final contents of volatile acids and bases being high. *Lotononis bainesii* and *Desmodium* produced well-prepared silage but *Macroptilium atropurpureum* decomposed badly during ensilage, probably due to the low content of sugar (Catchpoole, 1970).

Production of lactic acid silage in the temperate countries is associated with sugar contents of plant materials that range between 13 to 16 per cent of the dry matter. Tropical grasses, on the other hand, have low concentration of soluble carbohydrates, often below 6 per cent. Forage crops such as maize and sorghum, however, have higher sugar contents and therefore produce good lactic acid silage in tropical environments (Miller *et al.*, 1964).

3. Molasses has been used in the tropics to inhibit anaerobic decomposition during fermentation and enhance production of good lactic acid silage. Since the sugar content of tropical grasses is low, larger applications are necessary (up to 80 kg molasses per tonne wet weight of plant material). Use of 40 kg/tonne or less produced silage of unsatisfactory quality with significant concentration of butyric acid.

Sodium metabisulphite has rarely been used as an additive to reduce butyric and proteolytic fermentations and effluent losses in the tropics, and results appear to be inconclusive. There is a need for further investigation of other additives for silage-making in the tropics.

4. Wilting prior to ensiling has improved silage preservation in the tropics (Ferreira *et al.*, 1974). Wilting is associated with low butyric acid concentrations and reduction in dry matter and N losses. Silage made from unwilted *Chloris gayana* with 80 per cent moisture lost 28 per cent of its dry matter and 44 per cent of its N. The losses were reduced to 20 and 24 per cent, respectively, when the material was wilted to 70 per cent moisture level prior to ensiling (Davies, 1963). Wilting, however, could be difficult in the tropics because of high humidity and abrupt changes in the weather. Mould growth in wilted silage also can be a serious problem.

5. Most tropical grasses are inherently low in their feeding value and changes during ensiling reduce their feeding values even further.

Dehydration

High-temperature dehydration with minimum loss of dry matter and nutrients has been commercially practised to conserve forage crops in some temperate countries of Europe and in the USA and the Soviet Union. The common practice is to use fuel oil and produce temperatures of 635–745 °C with exhaust air temperature around 135–190°C.

Freshly cut forage is taken from the field with minimum delay, chopped and passed through the dehydrator chamber either in trays or by a continuous flow system. The time of exposure of forage to heated air depends on the temperature in the dryer and usually in about 5–15 min the forage is sufficiently dehydrated for storage. It can be stored as such or made dense by further processing for storage, transport and feeding. This can be done by compressing it into bales, wafers (extruding chopped forage from a ram press) or cobs (extruding from a rotary die), or it can be milled and pelleted.

Artificial dehydration is expensive and will be economical only when fuel costs are low. The material to be dehydrated must be of high quality, so that the final product compares favourably with concentrate supplements in feeding value for poultry and pig rations and as a supplement for young ruminant animals. Lucerne is a good example. Losses in dry matter rarely exceed 5 per cent when forage crops are artificially dehydrated and most of the carotene present in the green material is also preserved. In the present circumstances artificial drying of tropical forage crops is unlikely to be encouraged because of high cost of fuel and low nutritive value of the final product.

Feeding value and livestock production potential of conserved forages

Herbage quality and nutritive value of native tropical grasses are usually low compared to high-yielding grasses. Furthermore, techniques used for forage conservation in most of the tropics are relatively less advanced and inefficient than in temperate zones. Therefore, the feeding value of conserved products as measured in terms of voluntary intake of dry matter, digestibility of dry matter digestible crude protein content is relatively low and marginal for animal maintenance requirements.

Dehydration *per se* does not significantly affect the nutritive value of a forage. Though the digestibility and metabolizable energy are slightly depressed by drying, the efficiency with which metabolizable energy of the dried crop is used is enhanced because of reduced heat losses when animals ingest dried rather than fresh material (Wilkins, 1974). Due to inherent deficiencies in the hay-making process, however, considerable losses occur in dry matter and nutritive value. Significantly larger leaf losses can often reduce the protein content of the field crop by 40 per cent (Raymond, 1969). Under slow-drying conditions in the field, 6–8 per cent reductions in digestibility of hay can be expected and still further reductions can occur in storage due to losses in sugars and overheating. Poor nutritive value and reduced digestibility result in greater retention time of the ingested materials in the rumen and consequently reduced voluntary intake of hay by the animals. Low protein content in the hay reduced voluntary intake and digestibility due to insufficient nitrogen for proliferation and sustained activity of rumen microorganisms. In Nigeria, consumption of *Andropogon gayanus* hay by cattle decreased from 76 to 46 g/kg metabolic weight with a decrease in crude protein content from 5.4 to 2.4 per cent (Zemmelink *et al.*, 1972). Reduction in the crude protein content from 7 to 2 per cent in *Hyparrhenia* foggage in Rhodesia resulted in a decline in dry matter intake at the rate of about 10 g/kg metabolic weight for every 1 per cent decrease in crude protein content (Smith, 1962).

The grinding or milling of the dried herbage causes breakdown of material into fine particles. Processing of this material into pellets significantly alters nutritive value by increasing the voluntary intake and efficiency of utilization digested nutrients, even though percentage digestibility is decreased (Javier, 1976). When fed *ad libitum*, the pelleted herbage has a higher bulk density than unprocessed hay and is consumed in larger quantities because of higher rate of passage through the digestive tract. Greater intake results in an increased rate of bodyweight gain. A smaller proportion of herbage is required for maintenance, resulting in a significant improvement in the feed conversion efficiency as noted in Tables 12.7 and 12.8 (Hogan *et al.*, 1962; Holmes *et al.*, 1966). The magnitude of response to pelleting varies with the quality of the herbage pelleted, and is usually larger when herbage of poor nutritive value is pelleted (Heaney *et al.*, 1963; Holmes *et al.*, 1966).

Ensiling has a slight depressing effect on digestibility and the efficiency with which energy in the forage is utilized, but causes greater reduction in the efficiency of utilization of crude protein (Wilkins and Wilson, 1971). Some reports from the tropics have indicated larger reductions in digestibility as a result of ensiling (Catchpoole and Henzell, 1971). The major factor that limits the production potential of animals fed silage is the level of voluntary intake which can be as low as 70 per cent of dried forage (Wilkins and Wilson, 1971). The extent of reduction in intake varies and depends on (1) the moisture content of the ensiled material, (2) amounts of free organic acids in the silage that may affect pala-

Table 12.7 *Voluntary feed intake (dry matter) and gains of steers given various bermudagrass (Cynodon dactylon) feeding treatments for 60 days (Hogan et al., 1962)*

	Grazed	Soilage	Hay	Pellets
No. of steers	10	10	10	10
Ave. initial weight (kg)	243	242	245	245
Ave. final weight (kg)	280	270	274	301
Ave. daily gain (kg/head)	0.62	0.47	0.48	0.95
Ave. daily feed intake (kg DM)		4.95	5.18	7.05
Ave. daily feed intake per 100 kg bodyweight (kg DM)		1.95	2.00	2.58
Feed kg DM per kg gain		12.41	12.05	7.44

Note. Bermudagrass 3–4 weeks old and originating from an adequately fertilized pasture was used for the various feeding treatments or conventionally grazed. The quality of the grass was considered excellent, except during the first 3 weeks of the study when the herbage grazed or conserved contained residual oat straw.

Table 12.8 *Voluntary intake and weight gains (kg/day) of heifers fed Setaria anceps and Chloris gayana hay, chaff or pellets (Holmes et al., 1966)*

	Setaria		Chloris	
	Voluntary intake (air dry)	Weight gain	Voluntary intake (air dry)	Weight gain
Hay	3.63	0.05	3.31	0.16
Chaff	3.90	0.12	3.86	0.12
Pellets	5.40	0.49	5.13	0.22

Note. Experimental period of 28 days; *Setaria anceps* had 9.6 per cent CP and *Chloris gayana* had 10.6 per cent CP; dry matter digestibility of both grasses was 53 per cent; heifers were 15 months old and each weighed 160–190 kg.

tability and disturb the pH in the rumen and (3) the ammonia content in the silage which indicates the extent of protein degradation. A review of dairy and beef production in the tropics when grass silage and hay were used to supplement dry season grazing indicated relatively poor animal output (Javier, 1976). For further improvement, the animal diet must be supplemented with protein and energy concentrates as well as minerals.

Rains (1963) stated that 'it must always be recognized that while bulky foodstuffs may be desirable for ruminant animals, the only criterion for assessing the value of hay or silage is the cost per unit of digestible constituents and to compare this cost with that from others'. He concluded that the cost/benefit ratio would be much higher with some concentrates and that conservation in terms of hay and silage of low quality herbage would be highly uneconomical. Careful choice of species, harvesting at the appropriate stage of growth (Table 12.6 p. 319) and adoption of efficient procedures for conservation will result in a better quality product. More favourable cost/benefit ratios could then be achieved by using conserved products in livestock-feeding. For example, Couper (1971) demonstrated that the cost of producing good quality maize silages could be reduced by over 60 per cent with doubling the plant population and using the herbicide Atrazine for weed control.

Biological residues

The value of crop residues as an important source of fodder to livestock has been recognized for a long time. In recent years attempts have also been made to use animal waste as a component of animal diet.

Crop residues

After a crop is harvested the plant materials left in the field constitute the crop residue. Spilled or unthreshed grains, voluntary weeds, grasses and other forage materials growing in fields and along boundaries, irrigation channels and roadsides often form a significant component of the crop residues when animals are allowed to graze after harvest.

The nutritive value of crop residues and quantity available from a unit area are highly variable and depend on the species and variety of crop, farming methods and environmental conditions prevailing in the area. Often a considerable portion of the residue may be used by the farmer as fuel for cooking, thatching material, erection of fences or may be sold for other uses such as brick-making or manufacture of paper. Table 12.9 shows the range of variation in the nutritive value of some straws and stovers commonly used for ruminant feeding. Residues from other crops such as cotton, cassava, sugar-cane, sweet potato tops, etc. are fed to animals in many countries. Papaya rinds, citrus and mango peels, pumpkin and melon rinds when fed to animals during unfavourable seasons provided adequate supplementary nutrients for maintenance of bodyweight in East Africa (French, 1951). Feeding cattle with succulent residues such as banana trunks considerably reduces their drinking water requirements and has some value during periods of moisture stress.

In some densely populated areas of the tropics, particularly in Asia where animal husbandry is more closely integrated with agriculture, many of the non-productive livestock are maintained for much of the year on various crop residues carefully collected and conserved by the farmers. The productive livestock such as working oxen and milk animals are provided with varying amounts of green fodder and supplements. In the State of Madras, India, the total fodder production was estimated to be 34 million tonnes annually and crop residues accounted for 21 million tonnes (Whyte, 1964).

Table 12.9 *Average nutritive values of some straws and stovers (Dougall, 1960)*

	Digestible			Total digestible nutrients (%)	Gross digestible energy (%)	Crude fibre (%)	Ca (%)	P (%)
	Crude protein (%)	Crude protein (%)	Nutritive ration (%)					
Dolichos*, leaves and vines	19.9	15.1	3.2	63	69	20.5	1.44	0.15
Dolichos, bean pods containing some seed	14.1	9.0	5.4	58	62	29.6	0.40	0.14
Maize stover	3.7	1.0	37.9	49	49	39.6	0.11	0.04
Oat straw	5.3	1.8	25.7	48	49	41.9		
Sorghum stover	2.8	0.5	74.1	38	38	43.1		
Soybean stover	6.0	2.4	22.4	56	57	33.7		
Sunflower stover	2.0	0.1	120.0	12	12	56.2	0.40	0.02

Lablab purpureus (syn. *Dolichos lablab*).

Over 2 million tonnes of rice straw are produced annually in Taiwan. More than 0.5 million tonnes are of the soft-straw type, which has considerable potential as a major roughage component in the ruminant diet (O'Donovan and Chen, 1972). In many African countries where animal husbandry and crop-farming are not integrated, crop residues are not fully utilized for animal production. They are either lost due to burning or incompletely utilized because of their non-availability until the associated crops are harvested in a mixed cropping system. The residues from the early maturing components of mixed cropping must be collected and conserved or utilized immediately. Otherwise, they deteriorate rapidly due to leaf loss, insect and termite activity, leaching and bleaching effects of the weather, and consequently lose most of their nutritive value.

Sugar-cane leaves are burned or otherwise separated from the stalks at harvest. In many areas they are fed to livestock along with the leafy portion at the top of the stalk. To maintain high feeding value they should be collected on the day of harvest. They are often fed green but could be cured and stored (Preston and Leng (1978)). O'Donovan (1970) fed chopped leaves containing 5–6 per cent crude protein to dairy and beef cattle. This feed provided maintenance requirements of dairy cattle producing 2 kg head of milk daily. Maintenance requirements of beef cattle were met and an average daily liveweight gain of 0.25 kg/head was obtained. To achieve higher rates of gain it was necessary to feed additional protein and energy foodstuff. In Australia, Droughtmaster steers gained 0.61 kg/day when fed chopped sugar-cane supplemented with meat meal or herbage of *Leucaena leucocephala* (Siebert *et al.*, 1976).

Among the various crop residues available for animal feeding, legume residues generally provide a satisfactory supplement to animals that subsist on poor quality herbage (Miller, 1969). This material provides a maintenance ration and minimizes weight loss. Thus, animals make less compensatory gains when pasture growth resumes in the following wet season.

The highly lignified cereal straws and stovers are poor in nutritive value and their usefulness as supplement is mainly limited to providing additional bulk to the ruminant diet. When animals are fed poor-quality crop residues, it is necessary to provide additional protein or non-protein N, energy supplementation and minerals, particularly sulphur and phosphorus. Urea and molasses can be profitably used in many tropical countries as the major components of supplementary ration (O'Donovan and Chen, 1972). Undersowing crops with forage legumes offers another possibility for efficient utilization of crop residues with low protein and high fibre. The undersown legumes provide a useful supplement during the dry season. With proper management they can be successfully grown as companion crops with cereals such as sorghum, maize and rice without significantly affecting the yield of the cereal crops. The crop yields can be maintained at, or near, normal levels by manipulating crop and legume density, crop maturity type, sowing dates of the crop and legume and fertilizer applications.

Production systems using molasses (Preston, 1975), cassava (Müller *et al.*, 1975, Nestel and Graham, 1978 and Ahmed 1980), alkali-treated roughages and other fibrous materials have been developed in an effort to improve the digestibility and density of diets.

Treatment with hydroxides such as sodium hydroxide offers the additional possibility of improving the feeding quality of poor-quality crop residues as well as hay (Bula *et al.*, 1977; Perez, 1976 Rexen and Vestergaard–Thomsen, 1976; Capper *et al.*, 1977; Jackson, 1978; McManus *et al.*, 1979). Such treatment reduces the effect of lignin on digestibility, increases intake by as much as 33 per cent and digestibility from 5 to 16 per cent. Animals fed on such treated crop residues have shown comparable daily weight gains as those fed on maize silage. Development of simple and economic methods of treating roughages are likely to pay rich dividends in terms of efficient animal production from poor-quality crop residues. The value of lime-treated rice straw was shown by a study with grade Zebu bulls in the Philippines (Table 12.10). The treatments consisted of untreated straw and lime-treated (2% CaO) straw comprising 60 per cent of the total ration, with concentrates making up 40 per cent. Animal performance in terms of daily liveweight gain and efficient feed utilization was about 40 per cent in favour of the lime-treated straw.

Rice straw in Bangladesh was treated for 20–40 days with wine and urea in earthen pits, bamboo baskets and stacks covered with materials such as banana and coconut leaves, gunny bags, bamboo

Table 12.10 *Feed-lot growth performance of grade Zebu bulls on lime-treated and untreated rice straw-based rations in the Philippines (Perez, 1976)**

	Rice straw treated	Rice straw untreated
Initial weight (kg)	153.0	152.8
Final weight (kg)	234.5	210.5
Ave. daily gain (kg)	0.73	0.52
Daily feed intake (kg)		
Concentrate	2.31	2.12
Rice straw	3.17	3.18
Feed efficiency (kg/kg gain)		
Concentrate	3.22	4.22
Rice straw	4.46	6.34

* Four bulls used for each treatment; concentrate fed at 40 per cent of ration; trial period of 112 days.

mats, plastic and soil. Urea was mixed at the rate of 3–5 per cent by weight of straw and wine at 1.0 litre/kg of straw. *In vitro* analyses showed an increased digestibility of dry matter from 35 to 52 per cent. Nitrogen content increased from 0.6 to 1.0 per cent. Dry matter intake of urea-treated straw increased from about 2.8 to 5.0 per cent, as percentage of bodyweight, and wine-treated straw from about 2.8 to 3.2 per cent. Trials also indicated that an improvement occurred in the nitrogen balance of sheep (Dolberg *et al.*, 1980).

The margin of safety between bare survival and death due to starvation of animals subsisting on natural pastures is narrow, and a cause of great concern to livestock keepers in much of the tropics. Supplementary grazing provided by crop residues during periods of stress has contributed to this margin of safety in many areas. Without it the traditional industry in these areas would have either collapsed or suffered great reductions in the numbers of animals kept.

Raay and Leeuw (1970) studied the importance of crop residues as fodder in the Guinea and Sudan savanna zones of the Katsina province of Nigeria. These areas receive from 750 to 1 000 mm of rainfall during April–September. Guinea corn or sorghum, millets, cowpeas, groundnuts and cotton are the common crops grown. Approximately 0.8 million animal units are maintained on the native upland shrub savanna grasslands of the region. If kept entirely on the grazinglands during the dry months, the animals are completely emaciated towards the end of the dry season and many die before the next wet season. Crop residues are grazed intensively during the dry season. Millet residues are generally lost due to early harvest of the crop and non-availability of these residues to the animals because of intercropping. Guinea corn, groundnuts and cowpeas are harvested by October and the grazing of these residues starts in November. More areas are released for grazing when the cotton harvest is completed by January. The estimated production of the edible crop residues is 0.75–1 million tonnes annually (guinea corn 1 738 kg/ha, 2.2 per cent CP; cotton 281 ka/ha, 8.0 per cent CP; groundnuts 515 kg/ha, 9.2 per cent CP; cowpeas 258 kg/ha, 10.0 per cent CP). Taking into consideration the number of animal units and their daily dry matter consumption (7.5 kg/animal unit) gives a fodder potential of 121–159 grazing days for the entire livestock population of the area. The crop residues and browse plants provide animals with a varied and more nutritious diet during the dry season and ensure their survival until the next rainy season. Animals largely avoid grasses during these months and spend most of their grazing time on crop residues and browse (Fig. 12.2). They return to grass only after these sources of food are completely exhausted or after the rains begin. The Fulani word *nyali* meaning 'rush and finish' aptly indicates the value of crop residues to the graziers.

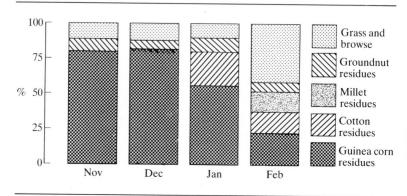

Fig. 12.2 Time spent on grazing different fodders by livestock in Katsina province of northern Nigeria, November 15 1958–March 1 1969 (Raay and Leeuw, 1970).

Animal residues

Recycling of animal waste, particularly poultry litter, as a source of nitrogen in the ruminant diet has been found potentially worth while in several studies (Noland *et al.*, 1955; Bhattacharya and Fontenot, 1966; Jacobs, 1976). Broiler litter is a waste product of poultry houses and contains bedding materials, excreta, wasted feed and feathers. The litter is high in nitrogen and ash content. *In vitro* studies of poultry manure have shown that rumen micro-organisms can utilize the uric acid and it is comparable to soybean as a source of N (Leibholz, 1969). Inclusion of 30–45 per cent unprocessed poultry litter in the concentrate supplement can increase animal intake of roughages and overall general performance. Jacobs (1976) studied the effects of feeding poultry litter to young calves. With rations containing 30–50 per cent poultry litter, 6–12-week-old calves did not develop satisfactorily. Calves 12–18 weeks old grew at a slower rate than those on control concentrate diet but gave a favourable cost/benefit ratio. The costs per kilogram of liveweight gain were 33 per cent less when poultry litter was included in the concentrate ration as compared a high concentrate ration.

The usefulness of animal waste as a supplementary feed will be limited until the danger of potential hazards to human and animal health from the likely presence of various drug residues and pathogenic organisms can be overcome. Pasteurization by heating the litter or addition of antibiotics have been recommended. Salmonella organisms in unsterilized poultry litter produced abnormal effects in adult animals, particularly under conditions of poor management (Leibholz, 1969).

Industrial by-products

By-products of the various food-packing and processing industries offer yet another source of supplementary feed for animals. Cereal brans, oil cakes, cottonseeds, molasses, citrus pulp and brewery by-products are a few of the numerous by-products used for livestock-feeding. These products are capable of providing energy, protein, minerals and vitamins in varying amounts and many of them are often reprocessed, enriched and sold as concentrate feeds. When a by-product is bulky or of poor keeping quality its usefulness will be limited unless animal houses are situated near the industrial centres. Thus, pineapple bran from the fruit-canning industry is a good source of energy. This product, however, is of high moisture content and uneconomical for long distance transport (O'Donovan *et al.*, 1972; O'Donovan, 1975).

Waste paper has been considered as a potential source of energy for ruminant production in some countries. *In vitro* organic matter digestibility of various grades of paper ranging from 22 to 64 per cent has indicated that high-quality paper and brown paper are valuable sources of energy and most other types except newspapers are digestible enough to form at least the roughage basis for maintenance rations (Coombe and Briggs, 1974). Waste paper is more digestible than the roughage used for its manufacture because of almost complete delignification of the roughage during the paper-manufacturing process.

Grazing under tree crops

Coconuts and livestock

Livestock graze at low stocking rates under coconuts, but the practice is seldom exploited as a coconut–cattle enterprise. A study in the Philippines showed that 10–50 per cent of coconut plantations were grazed (Barker and Nyberg, 1968), figures which correspond to those of other countries. A number of other studies revealed that improved grass and legume species can be successfully introduced under coconuts (Ferdinandez, 1973; Hill, 1969; Reynolds, 1978; Salting, 1980; Thomas, 1978), but there is a paucity of quantitative information regarding long-term associations. Controversy exists regarding light intensity effects of the palm canopy on pasture plants. The undersown forage plants compete for available nutrients unless fertilizers are regularly applied. They also reduce available moisture in areas of low rainfall. Heavy grazing causes compaction of some soil types, but this can be minimized by rotational grazing and occasional disking. Feed supplies can be substantially increased by use of appropriate herbage species, improved management practices and grazing intensity under tall and unimproved coconuts. Whether livestock can be integrated with high-yielding hybrid coconuts is less certain.

Other tree crops

A cover crop is important to soil conservation and tree productivity in tree crop plantations such as rubber and palm oil. Natural grasses and herbaceous cover vary widely in composition, yield and nutritive value, depending on tree density and plantation management. Trailing-type legumes, such as calapo, centro and puero, are frequently used as cover crops during the establishment and growth of young trees, but other legumes, such as stylo, Siratro and greenleaf desmodium, have been grown successfully. Grazing under these tree crops has been less successful than under coconuts because of (a) damage to young trees during their establishment, (b) difficulty in managing the trailing legumes which are usually grown in pure stands as a cover crop, (c) diminishing feed supply as trees reach maturity and (d) damage to the plantation tree product.

Chemical and concentrate supplementation

Growth and productivity of animals maintained entirely on the available herbage from grazinglands and crop residues are erratic, and the characteristic 'stop-go' cycle of growth can be favourably modified to a considerable extent by judicious supplementary feeding of concentrates and minerals.

Concentrate supplementation

Concentrates are rich sources of energy and/or nitrogen for the animal and contain little or no fibre and varying amounts of minerals and vitamins. They can be classified as carbohydrate concentrates, vegetable or animal protein concentrates and mixed concentrates. Some of the important economic factors that have placed constraints on the widespread use of concentrates in developing countries of the tropics are (1) low genetic potential of tropical livestock, (2) high prices of grain or its greater value as human food, (3) scarcity of oil cakes and oil seeds and their large-scale export overseas at high prices, (4) unattractive market prices and (5) lack of consumer appreciation for better-quality livestock products due to generally low standards of living. Additionally, there is a lack of accurate information on the feed requirements of various classes of tropical livestock for maintenance and production, as well as on the nutritional value of a wide variety of locally available feeds. Supplementary feeding of concentrates in many of the tropical countries is therefore primarily restricted to productive animals such as working oxen, milking animals (particularly those kept around urban centres) and the exotic, sick and very young animals.

Lansbury (1960) in West Africa estimated that a 225 kg bullock with an assumed dry matter appetite of 2 per cent bodyweight would require feed containing about 3 per cent digestible crude protein for maintenance and a liveweight gain of 0.45 kg/day. Conventional forage sources in West Africa do not provide these requirements during a considerable part of the year. Cattle grazing native pastures in the Northern Territory of Australia at about 6 ha/head in the dry season responded well to a daily protein supplement of 0.91 and 1.82 kg of guar seeds (*Cyamopsis tetragonploba*, 25% CP) with daily liveweight gains of 0.19 and 0.46 kg, respectively. The unsupplemented animals lost 0.5 kg daily (Norman and Stewart, 1964). Kidner (1966) in East Africa recommended 0.45–0.90 kg/head daily of protein supplement in the form of an oil cake during the 3 months of dry season for beef animals with unrestricted access to pasture. This would allow liveweight gain to progress during the dry season and consequently reduce the period of finishing by about 20 per cent. Rapid growth with fattening cattle could only be achieved by restricted access to the pasture (a maximum of 7 h daily) and very high levels of supplementary feeding (5.5–11 kg of complete ration daily for 275–450 kg animals) to provide the high nutritive requirements of the animals. In this instance feedstuffs cannot be considered supplements but instead are substitutes for the basal diet. Under most situations in the tropics this scheme is likely to be uneconomic. Supplementary feeding to animals on higher-quality pastures causes reduction in grazing intake. The immediate effects of the supplement on bodyweight gains are likely to be small unless grazing intake is limited by low availability of herbage (Corbett, 1976).

Effects of supplements on the feeding value of the diet

Supplements fed to the grazing animals increase the feeding value of the entire diet by direct addition of nutrients and energy over and above that supplied by the pasture and other roughages. They often increase the supply of energy from the roughage due to increased intake and/or digestibility of the herbage on offer. Holmes and Jones (1964) calculated the equation $I = 2.8 - 0.034\ D$, where I = increase in total feed intake per pound of concentrate fed and D = organic digestibility (OMD) of the roughage. Thus, when D is 53 per cent I will be 1 and the increase in intake is due to concentrate with no increase in the consumption of the roughage. Above 53 per cent OMD roughage intake is depressed but total intake increases. Below 53 per cent OMD roughage intake increases.

Low protein content in the grazing diet, i.e. 6–7 per cent, depresses voluntary intake but this can be offset by protein or non-protein N supplement. This is illustrated by data presented in Table 12.11 (Morris, 1966). Lucerne and meat meal were utilized as supplements and not as replacement feed for the basal diet. Energy contained in the supplement, extra energy from additional intake of

Table 12.11 *Effect of protein supplements on the ad libitum consumption of native grass hay and bodyweight changes of heifers* (Morris, 1966)*

	Intake per head per day			Bodyweight change (kg)
Supplement (kg)	Hay (kg)	Total (kg)	N (g)	
Nil	5.95	5.95	43.7	− 9.1
0.59 lucerne	6.41	7.00	65.7	+ 9.6
1.36 lucerne	6.55	7.91	91.0	+37.7
0.18 meat meal	6.91	7.09	67.2	+17.7

* Feeding period of 26 weeks; native grass hay with 0.74 per cent N; initial bodyweight of heifers 250 kg.

the basal diet and additional N resulted in significant liveweight gains. The unsupplemented animals lost weight. Protein and non-protein N supplements can also increase the dry matter digestibility of low protein roughages fed to the animals by sustaining active microbial digestion in the rumen (Musangi, 1966; Fishwick *et al.*, 1978). In some instances N supplements do not increase the digestibility and occasionally may slightly decrease the digestibility of poor-quality conserved forages (Morris, 1958; Haggar, 1970). Generally, when the herbage has 4 per cent or less crude protein content N supplementation increases both the level of digestibility and intake. At higher crude protein levels, little or no digestibility increases are observed with non-protein nitrogen supplementation (Raymond, 1969).

Non-protein nitrogen supplementation

In 1891, Zuntz indicated the possibility of providing the protein requirement of ruminants by simple nitrogenous substances such as ammonia and amides. Since then non-protein N sources such as urea or complexed forms of urea-like biuret, di-ureido isobutane or urea-phosphate have received considerable attention as cheap sources of N for the ruminant animal. The data from a large number of experimental studies in temperate and tropical countries have clearly indicated their usefulness as supplement to the low-protein, high-fibre roughage feeds (Altona, 1966; Morris, 1966; Pieterese, 1967; Loosli and McDonald, 1968; Raymond, 1969; Morris and Gulbransen, 1970; Romero *et al.*, 1976).

The principal use of urea in the diets of ruminants is to provide adequate N to the rumen micro-organisms for their growth and proliferation. They in turn provide the necessary amino acids needed by the animal. Addition of urea will increase the voluntary intake and possibly the digestibility of poor-quality, low-N roughage feeds such as dry-season standing grass, hay, silage or straws and stovers. Urea deaminates rapidly within the rumen causing an increase in rumen pH and a rise in the level of ammonia in the rumen and blood. These can lead to ammonia toxicity which could be lethal.

Adequate amounts of carbohydrates in the diet in the form of low-protein cereal grains, starch or molasses reduce the toxicity hazard associated with urea feeding. Rapid hydrolysis of the highly digestible carbohydrate concentrates provides for the immediate energy needs of rumen micro-organisms before the energy from the slowly hydrolysable roughages is available. The rapid rise in the rumen and blood ammonia level is therefore controlled and kept below toxic levels. Urea can be efficiently utilized by the ruminant animal when sufficient energy and about one-half of the total dietary nitrogen requirements are present in the diet (O'Donovan *et al.*, 1972). A period of adaptation by gradually increasing the concentration of urea in the diet is necessary for its efficient utilization. Virtanen (1966) showed that by using such a procedure dairy cows were capable of giving high milk yields when urea and ammonium salts were the sole source of N in a synthetic diet. The presence of some protein, however, increases the response from added urea. Lower rumen ammonia levels and more efficient N utilization occur when urea is fed to animals in small, frequent amounts rather than in larger quantities (Burns, 1965; Raymond, 1969; Romero *et al.*, 1976).

Complex forms of urea, such as biuret, urea-phosphate or di-ureido isobutane, when fed to animals give lower concentrations of ammonia in the rumen and blood than urea *per se*. Fishwick *et al.* (1978) compared voluntary intake by beef animals and digestibility of diets based on cereal straw (2.2% CP) and grains (10.8% CP) when supplemented with different forms of non-protein N. The organic matter digestibility of the diet was 53 per cent when urea was fed along with cereal grains one to four times daily in the ration. The digestibility increased to 56.7 per cent when di-ureido isobutane replaced urea. A solution containing urea + P + Ca + Na included with the straw increased digestibility to 59 per cent.

Although considerable advances have been made in the development of safe methods of urea supplementation, the production of slow-release urea seems to have great potential for future use, particularly under extensive grazing conditions, where visits to water and supplement by the stock are likely to be infrequent. Under such conditions the desired release rate would permit the hydrolysis of urea in the rumen to in-phase with fermentation rates required for efficient microbial activity (Romero *et al.*, 1976).

The chemical structure of biuret gives it certain physical and chemical characteristics that favour its use as a nitrogen supplement for low-quality roughages. It has the following advantages compared with urea (1) less soluble, (2) slower rate of ammonia release, (3) less toxic and (4) more palatable. Feed-grade biuret, however, is more expensive per unit of N than urea. This form of N has been used in Florida as a supplement to low-quality Pangola grass hay containing 4.6 per cent crude protein (Ammerman, 1974). Feeding 25 or 50 g of biuret per steer daily in combination with corn meal increased voluntary intake of the hay by 16 per cent over the non-N supplement (Table 12.12). Another study with sheep showed cellulose digestibility of 44 per cent of Pangola hay plus minerals. This value was increased to 55 per cent by adding 10 g/head per day of N in the form of biuret. Furthermore, consumption of hay was increased from 698 g/day to 851 g/day for the nil and 10 g biuret treatments, respectively.

Compensatory growth

Supplementation with energy or N during long periods of nutritional inadequacy is of considerable importance for animal survival. It results in improved body status of the animals compared to unsupplemented animals, either by minimizing liveweight losses or by increasing bodyweight during periods of stress. The long-term benefits of supplementary feeding in some instances may be small or absent because of subsequent compensatory gains by the unsupplemented animals (Morris, 1966; Owen, 1968; Corbett, 1976). Animals surviving on a low plane of nutrition in the dry season gain much more bodyweight on native pastures in the following wet season than animals in good condition at the start of the rains (Norman and Stewart, 1964). Under such conditions it can be suggested that supplementation during the dry season should be restricted to older ani-

Table 12.12 *Effect of biuret and corn meal supplementation on the consumption of low-quality Pangola grass hay by steers (Ammerman, 1974)*

Item	Treatment*			
	A	B	C	D
Supplement/steer, daily				
Biuret[†](g)	0.0	0.0	25[‡]	50[‡]
Corn meal (g)	0.0	254.0	254	254
Consumption/steer, daily				
Hay (kg)	4.15	4.09	4.82	4.80
Hay + corn meal (kg)	4.15	4.55	5.27	5.26

* Adequate salt, Ca, P, S and trace minerals provided.
† Kedlor 230 feed-grade biuret.
‡ Provided 9.2 and 18.4 g of N per steer daily for the 25 and 50 g biuret, respectively.

mals destined for slaughter and animals pregnant or likely to conceive during the period. Provision of supplementary feed to young stock may not yield permanent liveweight benefits because of compensatory growth in the following rainy season. Zemmelink (1974) reported no compensatory growth on pastures in northern Nigeria. Thus, advantages of supplementary feeding in the dry season were still present at the end of the following wet season (Haggar *et al.*, 1971).

Mineral and vitamin supplementation

Mineral deficiencies are common in animals grazing tropical pastures, particularly native grasslands. Nutrient deficiency in the diet is generally manifested in a reduction of feed intake and animal productivity. This is often associated with the animal's tendency to chew bones, metal, wood or soil and is followed by distinctive clinical symptoms indicative of specific deficiency. Iodine (I) deficiency results in goitre or swelling of the thyroid gland. Copper (Cu) deficiency results in anaemia, and cobalt (Co) deficiency may result in anaemia as well as diarrhoea in young animals. Unhealthy appearance, rough coat and intense craving for salt indicate Na deficiency. Phosphorus or Ca deprivation have similar symptoms, consisting of improper calcification of cartilage and often resulting in disabling fractures, poor growth rates and emaciation. Animals require relatively larger quantities of P, K, Ca, Mg, Na and S and small amounts of Fe, Mn, Zn, Cu, Co, I, Mo and Se in their daily diet.

Mineral deficiencies need to be identified and corrected, either indirectly by fertilization of the pastures or directly by providing the necessary minerals as supplements to the animals. The adequacy of a diet in essential nutrients can be studied by chemical analysis of the diet or of body tissues and comparing the levels of various elements with published standards (e.g. Table 13.2, p. 351). The diet can also be evaluated by feeding supplements of a nutrient suspected to be deficient and determining the effects on animal performance. For example, Long *et al.* (1972) studied the mineral status of grass–legume pastures in eastern Uganda by determining N, Na, K, Ca, Mg, P, S, Mn, Cu, and Zn in the pasture and Ca, Mg and P in bovine plasma. They concluded that the pastures contained adequate amounts of N, K, Mg, S and Cu for the dairy cows of the region. Sodium, Zn and P were deficient or marginally deficient and supplementation of these elements was recommended. Increased use of K fertilizers in the region was deemed likely to result in Mg deficiency. Calcium and Mn were considered adequate only in pastures containing a high proportion of legumes.

Mineral requirements of various classes of livestock under different tropical conditions are not precisely known, and mineral content of herbage during different times of the year in many areas has not yet been determined. Thus, it would be advisable to provide standard mineral licks or supplements as a first step towards the improvement of animal output in these areas. For example, in the eastern plains of Colombia, a base supplement of salt, bone meal and trace elements was provided to cattle grazing native grasslands. Supplementation increased liveweight gain of cows (302 to 330 kg), calf weaning weight (132 to 172 kg) and tended to increase calf weaning percentage (36.5 to 57.8%) (Raun, 1968; Soekanto *et al.*, 1980)

Mineral supplements correct specific nutritional deficiencies in the diet of animals. It also increases voluntary intake and, in some cases the digestibility of the basal diet, thus resulting in considerable improvement in animal performance. For example, intake of Co deficient pastures by sheep was increased by 60 per cent as a result of oral administration of 1 mg Co per day (Marston *et al.*, 1938). Playne (1969) studied the effects of sodium sulphate on the intake and digestibility of a 4 : 1 mixture of spear-grass and Townsville stylo hay by sheep. Sheep receiving a daily supplement of 14 g of sodium sulphate consumed 48 per cent more feed than unsupplemented sheep during the 6-week experimental period. Dry matter digestibility of the diet increased by 5.3 percentage units. The combined effects of increased dry matter intake and digestibility caused a 62 per cent increase in digestible dry matter intake. Animals receiving supplement gained weight, while those not receiving supplement lost considerable weight during the experimental period. The beneficial effects of the supplement were largely attributed to the sulphur in the sodium sulphate. This element caused more efficient utilization of N in the roughage. Animals on supplement had reduced nitrogen in the urine and maintained a zero N balance, while the unsupplemented animals showed a negative nitrogen retention of – 0.45 g/day.

Phosphorus deficiency is widespread in the tropics and is commonly associated with low protein content in the herbage. The deficiency becomes more acute when animals are fed mature standing grass. Advancing plant maturity depresses P content in the herbage because of translocation to the seeds and roots. Application of phosphatic fertilizers in P-deficient soils not only raises the P level in the herbage but may also improve herbage yield or its overall nutritive value. Mineral supplements containing P in the form of bone meal increased the liveweight gains of grazing cattle by 12 kg in the wet season during a 2 year study in the Northern Territory of Australia (Norman, 1960). Winks *et al.* (1977) in Queensland also reported significant increases in liveweight gains of steers grazing Townsville stylo-based pastures due to P supplementation in the wet season. Significant response was not obtained with P supplementation in the dry season.

Provision of common salt to animals grazing pastures deficient in Na content improves animal performance. Murphy and Plasto (1973) detected Na deficiency in cattle grazing native pastures in parts of Queensland on the basis of salivary Na and K levels. Common salt was offered *ad libitum* to cows and their suckling calves grazing these pastures. Supplemented cows gained 0.35 kg/head per day, whereas the unsupplemented cows gained 0.14 kg/head per day. The respective gains for suckling calves were 0.77 and 0.62 kg/ head per day.

Mineral supplements are normally provided in the feed or in drinking water or in form of salt licks. Some concentrate supplements are also rich in minerals. For example, molasses not only provides energy but also contains significant amounts of K, Ca, Mg, S, Cl, Co and Fe. Simple treatments are now available for some trace element deficiencies, e.g. Co and Se 'bullets' and injectable Cu (Corbett, 1976).

The above examples emphasize the need for proper mineral nutrition for optimum livestock productivity, and indicate that livestock performance can be radically altered by relatively small changes in the nutrient supply. It is therefore necessary to consider various important nutritional animals subsisting on native or naturalized grasslands as well as poorly managed pastures. Addition of an element to a diet results in the addition of an accompanying anion or cation. Diets must than be balanced with the accompanying ion or all other nutrients supplied in adequate amounts (Minson *et al.*, 1976).

Grazing ruminants seldom suffer from vitamin deficiencies except vitamin A. The fat-soluble vitamin A is synthesized in the liver from the carotene present in the green plant materials ingested by the animal. Deficiency of this vitamin results in inflammation of the eye, night blindness, hardening of the skin and lowering of the animal's resistance to diseases. Considerable quantities of vitamin A are stored in the liver and carotene in the body fat of the animal. When the animal's diet is low in carotene for a considerable period of time deficiency develops. In the arid and semi-arid areas of the tropics when animals subsist on bleached mature grass that has lost most of its carotene due to oxidation, a deficiency of vitamin A can occur during the latter part of the dry season. Prolonged feeding of bleached and weathered hay or crop residues also results in vitamin A deficiency. Well-preserved green hay and silages or substances such as cod-liver oil and red palm oil are good sources of carotene and can be effectively used in the animal diet to correct vitamin A deficiency.

References

Addy, B. L. and **D. Thomas** (1976) *Rhodes Grass Pastures. Some Aspects of Management and Utilization*, Ves. Bull. 3/76, Min. Agric. Natl Resources: Malawi.

Ahmed, F. A. (1980) Feeding cassava as an energy supplement to dried grass, *E. Afr. Agric. For. J.* **46**, 368–72.

Altona, R. E. (1966) Urea and biuret as protein supplements for range cattle and sheep in Africa, *Outl. Agric.* **5**, 22–7.

Ammerman, C. B. (1974) Improved utilization of low quality roughages by nitrogen supplementation, *Down to Earth* **29**(4), 14–17.

Barker, R. and **A. J. Nyberg** (1968) Coocnut-cattle enterprises in the Philippines, *Phil. Agric.* **52**, 49–60.

Bhattacharya, A. N. and **J. P. Fonetenot** (1966) Protein and energy value of peanut hull and wood shaving poultry litters, *J. Anim. Sci.* **25**, 367–71.

Bredon, R. M. and **C. R. Horrell** (1962) The chemical composition and nutritive value of some common grasses in Uganda, *E. Afr. Agric. For. J.*, **39**, 13–17.

Bula, T. J., V. L. Lechtenberg and **D. A. Holt** (1977) Potential of temperate zone cultivated forages. In *Potential of the World's Forages for Ruminant Animal Production*, Winrock Intl Livestock Res. Trg. Center, Arkansas, pp. 7–28.

Burns, M. A. (1965) Urea block licks, *Queensld Agric. J.* **91**, 12–15.

Capper, B. S., D. J. Morgan and **W. H. Parr** (1977) Alkali-treated roughages for feeding ruminants: a review, *Trop. Sci.* **19**, 73–88.

Catchpoole, V. R. (1962) The ensilage of sorghum at a range of crop maturities, *Aust. J. Exptl Agric. Anim. Husb.* **2**, 101–5.

Catchpoole, V. R. (1968) Effects of season, maturity and rate of nitrogen fertilizer on ensilage of Setaria sphacelata, *Aust. J. Exptl Agric. Anim. Husb.* **8**, 569–73.

Catchpoole, V. R. (1970) The silage fermentation of some tropical pasture plants, *Proc. 11th Intl Grassld Congr.*, pp. 891–4.

Catchpoole, V. R. and **E. F. Henzell** (1971) Silage and silage making from tropical herbage species, *Herb. Abst.* **41**, 213–21.

Coombe, J. B. and **A. L. Briggs** (1974) Use of paper as a feedstuff for ruminants, *Aust. J. Exptl Agric. Anim. Husb.* **14**, 292–301.

Corbett, J. L. (1976) Usefulness to the animal producer of research findings. Nutritional aspects of the growth of grazing animals, *Proc. Aust. Soc. Anim. Prod.* **11**, 251–88.

Couper, D. C. (1971) The cost of silage production at Shika agricultural research station, *Niger. Agric. J.* **8**, 77–84.

Crowder, L. V. (1977) Potential of tropical zone cultivated forages. In *Potential of the World's Forages for Ruminant Animal Production*, Winrock. Intl. Livestock Res. Trg Center, Arkansas, pp. 49–78.

Davies, J. G. and **E. M. Hutton** (1970) Tropical and sub-tropical pasture species. In R. M. Moore (ed.), *Australian Grasslands*, Aust. Natl Univ. Press: Canberra, Ch. 19.

Davies, T. (1963) Fodder conservation in Northern Rhodesia, *J. Agric. Sci. (Camb.)* **61**, 309–28.

Dirven, J. G. P. (1970) Yield increase of tropical grassland by fertilization, *Proc. 9th Intl. Potash Inst. Congr. (Antibes)*, pp. 403–9.

Dolberg, F., M. Sadullah, M. Hague, R. Ahmad and **R. Hague** (1980) Methods of keeping urea-treated straw as a source of ammonia using indigenous materials, *Proc. 1st Asia Australian Anim. Sci. Cong.*, Serdang, Malaysia (mimeo).

Dougall, H. W. (1960) Average nutritive values of Kenya feeding stuffs for ruminants, *E. Afr. Agric. For. J.* **26**, 119–28.

Ferdinandez, D. E. F. (1973) Utilization of coconut lands for pasture development, *Ceylon Cocon. Plrs Rev.* **7**, 14–19.

Ferreira, J. J., J. F. C. DaSilva and **J. A. Gomide** (1974) Efeito do estado de desolvimento, do emurchecimento e da adiçao de raspa de mandioca sobre o valor nutritivo do silagem do campin-elefante (Pennisetum purpureum Schum.), *Experientae (Brazil)* **17**, 85–108.

Fishwick, G., J. J. Parkins, R. G. Hemingway and **N. S. Ritchie** (1978) A comparison of the voluntary intake and digestibility by beef cows of diets based on oat straw and supplemented with different forms of non-protein nitrogen, *Anim. Prod.* **26**, 135–41.

Foster, W. H. and **E. J. Mundy** (1961) Forage species in northern Nigeria, *Trop. Agric. (Trin.)* **38**, 311–18.

French, M. H. (1943) The composition and nutritive value of Tanganyika feeding stuffs, *E. Afr. Agric. For. J.* **8**, 126–32.

French, M. H. (1951) Factors affecting animal nutrition in Tanganyika, *E. Afr. Agric. For. J.* **16**, 198–203.

Gillard, P. and **M. J. Fisher** (1978) The ecology of Townsville stylo-based pastures in northern Australia. In J. R. Wilson (ed.), *Plant Relations in Pastures*, CSIRO: Melbourne, Ch. 22.

Gillard, P., L. A. Edye and **R. L. Hall** (1980) Comparison of Stylosanthes humilis with Stylosanthes hamata and Stylosanthes subsericea in the Queensland dry tropics. Effects on pasture composition and cattle liveweight gains, *Aust. J. Agric. Res.* **31**, 205–20.

Haggar, R. J. (1970) The intake and digestibility of low quality Andropogon gayanus hay supplemented with various nitrogenous feeds as recorded by sheep, *Niger. Agric. J.* **7**, 70–6.

Haggar, R. J., P. N. de Leeuw and **E. Agishi** (1971) The production and management of Stylosanthes gracilis at Shika, Nigeria. II. In savanna grassland, *J. Agric. Sci. (Camb.)* **77**, 437–44.

Heaney, D. P., W. J. Pigdon, D. J. Minson and **G. I. Pritchard** (1963) Effect of pelleting on energy intake from forages cut at three stages of maturity, *J. Anim. Sci.* **22**, 752–7.

Hill, G. D. (1969) Grazing under coconuts in the Morobe district, *Papua New Guinea Agric. J.* **21**, 10–2.

Hodgson, R. E., R. E. Davis, W. H. Hosterman and **T. E. Hienton** (1948) Principles of making hay. In *Yearbook of Agriculture*, USDA: Washington, D.C., pp. 161–7.

Hogan, W. H., O. L. Brooks, E. R. Beaty and **R. A. McCreery** (1962) Effect of pelleting Coastal bermudagrass on livestock gains, *Agron. J.* **54**, 194–5.

Holm, J. (1974) Nutritive value and acid contents of silages made from tropical forages at Chiang Mai, Thailand, *Thai. J. Agric. Sci.* **7**, 11–21.

Holmes, W. and **J. G. W. Jones** (1964) The efficiency of utilization of fresh grass, *J. Brit. Grassld Soc.* **19**, 119–29.

Holmes, J. H. G., M. C. Franklin and **L. J. Lambourne** (1966) The effects of season, supplementation and pelleting on intake and utilization of some sub-tropical pastures, *Proc. Aust. Soc. Anim. Prod.* **6**, 354–63.

Humphreys, L. R. (1977) Potential of humid and subhumid rangelands. In *Potential of the World's Forages for Ruminant Animal Production*, Winrock Intl Livestock Res. Trg Center, Arkansas, pp. 29–48.

Jackson, M. G. (1978) Treating straw for animal feeding, *FAO Anim. Health and Prod., Paper No. 10*, Rome.

Jacobs, G. J. L. (1976) Feeding poultry litter to ruminants, *Proc. Aust. Soc. Anim. Prod.* **11**, 417–20.

Javier, E. Q. (1976) *Techniques of Intensive Forage Production*, ASPAC Food Fert. Techn. Cent., Extn Bull. 81, Taiwan.

Johnson, D. L. (1975) The status of pastoral nomadism in the Sahelian zone. In *The Sahel: Ecological Approaches to Land Use*, MAB Tech. Notes, UNESCO, pp. 75–88.

Kidner, E. M. (1966) Beef production. Part. II. Production from unrestricted access to pasture and from restricted access but with supplementary feeding, *E. Afr. Agric. For. J.* **32**, 91–5.

Lansbury, T. J. (1960) A review of some limiting factors in the nutrition of cattle on the Accra plains, Ghana, *Trop. Agric. (Trin.)* **39**, 185–92.

Leeuw, P. N. de (1971) The prospects of livestock production in the Northern Guinea Savanna zones, *Samaru Agric. Newsl.* **13**, 124–33.

Leeuw, P. N. de and **W. L. Brinckman** (1974) Pasture and rangeland improvement in the Northern Guinea and Sudan zone of Nigeria. In J. K. Loosli, V. A. Oyenuga and G. M. Babatunde (eds), *Animal Production in the Tropics*, Heinemann (Nigeria): Ibadan, pp. 124–36.

Leibholz, J. (1969) Poultry manure and meat meal as a source of dietary nitrogen for sheep, *Aust. J. Exptl Agric. Anim. Husb.* **9**, 589–93.

Long, M. I. E., B. Marshall, W. K. Ndyanabo and **D. D. Thornton** (1972) Mineral status of dairy farms in eastern Uganda. II. Nitrogen and mineral content of grasses and some mineral contents of bovine plasma, *Trop. Agric. (Trin.)* **27**, 227–34.

Loosli, J. K. and **I. W. McDonald** (1968) *Non-protein Nitrogen in the Nutrition of Ruminants*, FAO Agric. Studies, Rome.

Marston, H. R., R. G. Thomas, D. Murnane, E. W. L. Lines, I. W. McDonald, H. O. Moore and **L. B. Bull** (1938) *Studies on Coast Disease of Sheep in South Australia*, Bull. 13, CSIRO: Australia.

McKeague, R. J., C. P. Miller and **P. Anning** (1978) Verano, a new stylo in the dry tropics, *Queensld Agr. J.* **104**, 31–5.

McManus, W. R., L. L. Grout, U. N. E. Robinson, P. Southwell-Keely and **P. N. Woodhart** (1979) Ensilage from alkali-treated roughages, *Aust. J. Exptl Agric. Anim. Husb.* **19**, 354–61.

McWilliams, A. P. and **J. Duckworth** (1949) The preparation of elephant grass silage and its feeding value for tropical dairy cattle, *Trop. Agric. (Trin.)* **26**, 16–23.

Miller, T. B. (1969) Forage conservation in the tropics, *J. Brit. Grassld Soc.* **24**, 158–62.

Miller, T. B., A. B. Rains and **R. J. Thorpe** (1964) The nutritive value and agronomic aspects of some fodders in Northern Nigeria. 3. Hay and dried crop residues, *J. Brit. Grassld Soc.* **19**, 77–90.

Minson, D. J., J. H. Stobbs, M. P. Hegarty and **M. J. Playne** (1976) Measuring the nutritive value of pasture plants. In N. H. Shaw and W. W. Bryan (eds), *Tropical Pasture Research Principles and Methods*, Commonw. Bur. Past. and Fld Crops, Bull. 13, Hurley: England, Ch. 13.

Morris, J. G. (1958) Drought feeding studies with cattle and sheep. I. The use of native grass hay (bush hay) as the basal component of a drought fodder for cattle, *Queensld J. Agric. Sci.* **15**, 161–6.

Morris, J. G. (1966) Supplementation of ruminants with protein and non-protein nitrogen under northern Australia conditions, *J. Aust. Inst. Agric.* **32**, 185–95.

Morris, J. G. and **B. Gulbransen** (1970) Effect of nitrogen and energy supplementation on the growth of cattle grazing oats and Rhodes grass, *Aust. J. Exptl Anim. Husb.* **10**, 379–83.

Müller, D. J., K. C. Chou and **K. C. Nah** (1975) Cassava as a total substitute for cereals in livestock and poultry rations, *Proc. Confr. Anim. Feeds Subtrop. Trop. Origin*, Trop. Prod. Inst.: London, pp. 58–95.

Murdock, J. C. (1966) Grass silage, *Outl. Agric.* **5**, 17–21.

Murphy, G. M. and **A. W. Plasto** (1973) Liveweight response following sodium chloride supplementation of beef cows and their calves grazing native pastures, *Aust. J. Exptl Agric. Anim. Husb.* **13**, 369–74.

Musangi, R. S. (1966) The influence of supplementary feeding and grazing restriction on the voluntary intake of roughages and liveweight gain by beef cattle, *E. Afr. Agric. For. J.* **31**, 271–5.

Nestel, B. and **M. Graham** (eds) 1978. *Cassava as an Animal Feed*, IRDC Workshop Proc., University of Guelph.

Noland, P. R., B. F. Ford and **M. L. Ray** (1955) The use of ground chicken litter as source of nitrogen for gestating-lactating ewes and fattening steers, *J. Anim. Sci.* **14**, 860–5.

Norman, M. J. T. (1960) *Grazing and Feeding Trials with Beef Cattle in Katherine N.T.*, CSIRO Aust. Div. Land Res. Surv., Tech. Paper No. 12.

Norman, M. J. T. (1968) The performance of beef cattle on different sequences of Townsville lucerne and native pasture at Katherine, N. T., *Aust. J. Exptl Agric. Anim. Husb.* **8**, 21–5.

Norman, M. J. T. and **G. A. Stewart** (1964) Investigations on the feeding of beef cattle in the Katherine region, N. T., *J. Aust. Inst. Agric. Sci.* **30**, 39–46.

O'Donovan, P. B. (1970) *Livestock Production on Marginal Land in Taiwan*, FAO: Rome.

O'Donovan, P. B. (1975) Potential for by-product feeding in tropical areas, *World Anim. Rev. (FAO)*, No. **13**, pp. 32–7.

O'Donovan, P. B. and **M. C. Chen** (1972) Performance of dairy heifers fed different levels of cane molasses with rice straw as roughage, *Trop. Agric. (Trin.)* **49**, 125–34.

O'Donovan, P. B., M. C. Chen and **P. K. Lee** (1972) Conservation methods and feeding value for ruminants of pineapple bran mixtures, *Trop. Agric. (Trin.)* **49**, 135–42.

O'Donovan, P. B., S. P. Liand and **M. C. Chen** (1972) Composition of urea with soyabean meal in concentrates for milking cows, *Trop. Agric. (Trin.)* **49**, 311–20.

Owen, M. A. (1968) Studies with beef steers on the Kongwa Plain, Central Tanzania, *Trop. Agric. (Trin.)* **45**, 159–71.

Perez, C. B. (1976) *Fattening Cattle on Farm By-products.* ASPAC Food and Fert. Techn. Cent., Extn. Bull. No. 83, Taiwan.

Pieterese, P. J. S. (1967) Urea in winter rations for cattle. In M. H. Briggs (ed.), *Urea as a Protein Supplement*, Pergamon Press: Oxford.

Playne, M. J. (1969) Effect of sodium sulphate and gluten supplement on the intake and digestibility of mixture of spear grass and Townsville lucerne hay by sheep, *Aust. J. Exptl Agric. Anim. Husb.* **9**, 393–9.

Preston, T. R. (1975) Sugarcane as the basis for intensive animal production in the tropics, *Proc. Confr. Anim. Feeds Subtrop. Origin*, Trop. Prod. Inst.: London, pp. 64–83.

Preston, T. R. and **R. A. Leng** (1978) Sugarcane as cattle feed, *World Anim. Rev.* **27**, 7–12 and 28, 44–8.

Raay, J. G. T. van and **P. N. de Leeuw** (1970) The importance of crop residues as fodder. A resource analysis in Katsina Province, Nigeria, *Tijdschrift voor Econ. En Soc. Geografie.* **61**, 137–47.

Rains, A. B. (1963) *Grassland Research in Northern Nigeria.* Samaru Misc. Publ. 1, Inst. Agric. Res., Ahmadu Bello Univ., Zaria.

Raun, N. S. (1968) Producción de ganado de carne en los Llanos Orientales, *Agric. Trop. (Colombia)* **24**, 641–8.

Raymond, W. F. (1969) The nutritive value of forage crops, *Adv. Agron.* **21**, 1–108.

Rexen, F. and **K. Vestergaard-Thomsen** (1976) The effect on digestibility of a new technique for alkali-treatment of straw, *Anim. Feed Sci. Tech.* **1**, 78–83.

Reynolds, S. G. (1978) Evaluation of pasture grasses under coconuts in Western Samoa, *Trop. Grassld* **12**, 146–51.

Reynolds, S. G. (1980) Grazing under coconuts, *World Anim. Rev.* **35**, 40–5.

Romero, V. A., B. D. Siebert and **R. M. Murray** (1976) A study on the effect of frequency of urea ingestion on the utilization of low quality roughage by steers, *Aust. J. Exptl Agric. Anim. Husb.* **16**, 308–13.

Salting, D. D. (1980) Integrating coconut, beef and sugarcane production, *Greenfields* **10**, 12–5.

Siebert, B. D., R. A. Hunter and **P. N. Jones** (1976) The utilization by beef cattle of sugarcane supplemented with animal protein, plant protein or non-protein nitrogen and sulphur, *Aust. J. Agric. Anim. Husb.* **16**, 790–4.

Shaw, N. H. and **M. J. T. Norman** (1970) Tropical and sub-tropical woodlands and grasslands. In R. M. Moore (ed.), *Australian Grasslands*, Aust. Natl Univ. Press: Canberra, Ch. 8.

Smith, C. A. (1962) The utilization of Hyparrhenia for the nutrition of cattle in the dry season. III. Studies on the digestibility of the produce of mature veld and veld hay and the effect of feeding supplementary protein and urea, *J. Agric. Sci. (Camb.)* **53**, 173–8.

Soekanto Lebdosoekojo, C. B. Ammerman, N. S. Raun, J. Gomez, and **R. C. Littell** (1980) Mineral nutrition of beef cattle grazing native pastures on the eastern plains of Colombia, *J. Anim. Sci.* **51**, 249–60.

Takano, N. (1972) *Grassland Farming-Silage.* ASPAC Food and Fert. Techn. Cent. Extn. Bull. 23, Taiwan.

Tetlow, R. M., R. J. Wilkins and **W. E. Evans** (1975) *Pre-harvest Dessication with Formic Acid*, Grassld Res. Inst.: Hurley, England.

Thomas, D. (1978) Pastures and livestock under tree crops in the humid tropics, *Trop. Agric. (Trin.)* **55**, 39–44.

Tuah, A. K. and **O. Okyere** (1974) Preliminary studies on the ensilage of some species of tropical grasses in the Ashanti forest belt of Ghana, *Ghana J. Agric. Sci.* **7**, 81–7.

Vera Cruz, N. C. (1967) Making grass silage, *Phil. Agric.* **35**, 266–71.

Virtanen, A. I. (1966) Milk production of cows on protein-free feed, *Science* **153**, 1603–14.

Watson, S. J. and **M. J. Nash** (1960) *The Conservation of Grass and Forage Crops*, Oliver and Boyd: Edinburgh and London.

Whyte, R. O. (1964) *The Grassland and Fodder Resources of India*, ICAR: New Delhi, pp. 150–1.

Whyte, R. O., T. R. G. Moir and **J. P. Cooper** (1959) *Grasses in Agriculture*, FAO Agric. Studies 21, Rome.

Wilkins, R. J. (1974) Scientific and technical progress in forage crop dehydration, *Proc. 12th Intl Grassld Congr.*, pp. 289–306.

Wilkins, R. J. and **R. F. Wilson** (1971) Silage fermentation and feed value, *J. Brit. Grassld Soc.* **26**, 108.

Winks, L., F. C. Lamberth and **P. K. O'Rourke** (1977) The effect of a phosphorus supplement on the performance of steers grazing Townsville-stylo based pasture in north Queensland, *Aust. J. Exptl Agric. Anim. Husb.* **17**, 357–66.

Wylie, P. B. (1975) Silage in Queensland, *Queensld Agric. J.* **101**, 708–18.

Zemmelink, G. (1974) Utilization of poor quality roughages in Northern Guinea savanna zone. In J. K. Loosli, V. A. Oyenuga and G. M. Babatunde (eds) *Animal Production in the Tropics*, Heinemann (Nigeria): Ibadan, pp. 167–76.

Zemmelink, G., R. J. Haggar and **J. H. Davies** (1972) A note on the voluntary intake of Andropogon gayanus hay by cattle as affected by level of feeding, *Anim. Prod.* **15**, 85–88.

Zuntz, N. (1891) Pflugers Arch. Ges., *Physiol.* **49**, 483.

Chapter 13

Herbage quality and nutritive value

Ruminants are peculiarly adapted to feed on forages and a good forage source is indispensable for their survival. In much of the tropics annual and seasonal variations in forage availability impose a 'stop-go' growth pattern characterized by alternate gains and losses in animal bodyweight. Luxuriant growth of a forage does not always indicate its capability adequately to supply the nutritional demands of animals. Studies by nutritionists in earlier times considered the different dietary components in the forage in an estimation of its overall nutritive value.

Presence of a component in the plant does not necessarily mean that it is readily and completely available to the animal ingesting the herbage. Its level of consumption and digestion, as well as an estimation of the actual need by the various classes of animals, are the major factors that have to be considered while determining the usefulness of a forage to the animal. Therefore, any attempt to describe the quality of a forage should take into account (1) the feed components in forage species and factors affecting their variability and (2) the ability of animals to utilize the feed components and factors affecting the variability in such utilization.

Evaluations of herbage quality are necessary for the attainment of high animal productivity goals, either in terms of output per animal or output per hectare or both. It must, however, be realized that productivity of grazing animals not only depends on herbage quality but also on animal potential, carrying capacity, grazing pressure and provision of other supplements.

Nutritive value

The nutritive value of a forage refers to its chemical composition, digestibility and the nature of digested products. However, the amount of forage consumed by the animal is very important, as it affects total nutrient intake and therefore the animal response. Various factors such as acceptability, presence of undesirable substances, rate of passage and availability of forage influence consumption or intake by animals. Thus, assessment of herbage quality involves an integrated evaluation of its nutritive value and its level of consumption by the animal.

Chemical composition

Since various elements are connected with body functions of both plants and animals, early nutritionists attempted to characterize the nutritive value of a forage by studying its chemical composition. The Weende proximate analytical scheme as modified in its present form by Henneberg in 1860 was introduced. The scheme resolves a given feedstuff into five fractions: crude protein (CP), fat or ether extract (EE), crude fibre (CF), ash and nitrogen-free extract (NFE) by subjecting it to (a) determination of Kjeldhal nitrogen and multiplying the N value by 6.25 (CP), (b) extractions with anhydrous ether (EE), (c) extractions with ether, sulphuric acid and sodium hydroxide (CF) and (d) subtraction of CP, EE, CF and ash contents from the sample weight (NFE).

Initially, it was believed that the entire CF fraction represented the indigestible part of a forage and NFE constituted the digestible portion of the carbohydrates. Forages having low CF content were considered high in nutritive value. Crude fibre is an empirical substance and its composition varies within and between forage species. Increase of lignin with advancing maturity does not necessarily increase CF content even though digestibility declines (Phillips *et al.*, 1954). Realization of the fact that part of CF is digestible, and at times even more than the NFE, along with inherent difficulties of chemically separating the CF and NFE fractions, reduced the usefulness of herbage nutritive value evaluations based on proximate analyses.

Alternate methods have been proposed to partition the carbohydrate fraction of forages. Crampton and Maynard (1938) resolved carbohydrate into lignin, cellulose and other carbohydrates to predict the feeding value. Van Soest (1966) proposed that forages are made up of two basic dietary fractions: cell contents (CC) and cell wall contents (CWC). The division was based on the extent of availability of these fractions to the animal (Table 13.1).

Therefore, for a clear understanding of digestibility and nutritive value, knowledge of the following dietary components will be useful.

Cell contents (CC) and cell wall contents (CWC).

Lipids, sugars, starch, non-protein nitrogen, soluble protein, pectin, organic acids and water-soluble matter comprise the cell contents. They are soluble in neutral detergents and therefore almost com-

Table 13.1 *Classification of forage fractions according to nutritive characteristics (from Van Soest, 1966 and 1967)*

Class	Fraction	Nutritional availability	
		Ruminant	Non-ruminant
Category A (Cellular contents)	Sugars, soluble carbohydrates, starch	Complete	Complete
	Pectin	Complete	High
	Non-protein N	High	High
	Protein	High	High
	Lipids	High	High
	Other soluble	High	High
Category B (Cell wall contents)	Hemicellulose	Partial	Low
	Cellulose	Partial	Low
	Heat-damaged protein	Indigestible	Indigestible
	Lignin	Indigestible	Indigestible
	Keratin	Indigestible	Indigestible
	Silica	Indigestible	Indigestible

pletely degraded by enzymes secreted in the animal digestive tract. They can be extracted by using disodium dihydrogen ethylene di-aminetetracetate (EDTA) in borate buffer containing sodium lauryl sulphate (Van Soest, 1966). Earlier attempts to extract forage solubles did not effect complete separation of CC from CWC. The cell wall contents are resistant to enzymatic action and degradation is possible through microbial intervention in the rumen. Their breakdown depends on the proportion of different constituents such as lignin, hemicellulose, cellulose and silica. Van Soest (1966, 1967) separated the CWC into (1) a fibre fraction made up of an insoluble portion in neutral detergent (NDF) and another insoluble portion in acid detergent (ADF) and (2) lignin. The overall digestibility of fibre depends on the amount of lignin and other physical incrustations included in it. Fonnesbeck and Harris (1970) suggested a modification of the Van Soest procedure by using additional pepsin digestion to remove the protein completely from the cell wall contents. The relationships of the various components of forage dry matter as determined by various procedures have been summarized by Harris (1970) as seen in Fig. 13.1.

Crude protein (CP).

Crude protein is estimated by multiplying Kjeldhal nitrogen by 6.25. This is not an estimate of true protein since estimation of total N also includes other nitrogenous compounds. Since rumen micro-

Dry matter						
Organic matter					A s h	Organic matter system
Crude fibre	Nitrogen-free extract		Ether extract	Crude protein	A s h	Weende or proximate system
Cell walls (neutral detergent fibre)			Cell contents (neutral detergent solubles)			Van Soest system
Cell walls (Fonnesbeck and Harris)			Cell contents			
Non-nutritive matter	Partially nutritive matter		Nutritive matter			
Lignin and acid insoluble ash	Cellulose	Hemi-cellulose	Soluble carbohydrate Protein Ether extract Soluble ash			

Fig. 13.1 Various systems of partitioning the dry matter of forage (Harris, 1970 – taken from Javier, 1975).

organisms are able to assimilate elemental nitrogen, estimation of CP or total nitrogen in forages is useful. Such estimations, however, will be of only limited value without determining the extent of availability to animals.

Fibre.

The crude fibre (CF) fraction obtained in the Weende proximate scheme does not represent the true fibre portion of the forages. Forages with similar CF content do not necessarily have similar digestibilities. For example *Panicum coloratum* and *Panicum maximum*, both having CF content of 43 per cent, showed digestibility values of 48.3 and 56.6 per cent, respectively (Ademosun 1974). Digestibility differences have been reported even within the same species.

The fibre fraction in forages varies in degree of development, chemical composition and structural complexity. The nature of fibre will depend on the type of deposition on primary cell walls. Substances such as cellulose, hemicellulose, lignin, suberin, cutin, waxes and salts are normally incorporated during the formation of secondary cell walls. Cellulose and hemicellulose are the major components of the cell wall and account for a large proportion of energy obtained from forages. They are both closely associated in plants and digested in the same way in ruminants and are often described in literature under a common term 'holocellulose' (Ely and Moore, 1955). The extent of their utilization by animals is influenced by other associated materials such as lignin and silica.

Lignin.

Lignin is always associated with holocellulose and its amount is insignificant in very young forages. Structurally it is a phenylpropane and is untouched by rumen micro-organisms. Lignin is not only indigestible but also reduces the digestibility of other components possibly because of their encrustation. Ball-milling of lignified tissues increases digestibility. Lignin contents of tropical forages are high as compared to their temperate counterparts.

Minerals.

Ash left after heating a forage sample at 600 °C provides an approximation of the total minerals. Carbon(C), O, H and N are lost due to combustion. The mineral requirements of plants and animals are not fixed; variations occur according to species, maturity and level of production. For example, K requirements of plants are much higher than animal requirements and animals fed on forages rarely suffer from K deficiency (Reid and Jung, 1974). Plants accumulate some elements such as Mo, Se and Fe to levels that may be toxic to animals. Animal requirements of some elements may exceed those available by feeding on forages or animals may require some elements such as Na, Cl, I, Co and Se that though not useful to plants, are found in them. Magnesium and Ca requirements of milking cows can be higher than the amounts forages could normally supply, particularly in areas where legumes are scanty. The legumes generally contain higher amounts of N, Ca, P, K and Mg and lower amounts of Na and chlorine (Cl) when compared with grass species (Underwood, 1966).

   Determinations of mineral composition in forages will be only of limited value in the overall estimation of herbage quality and nutritive value. Such determinations are particularly useful to establish critical levels of nutrients required by forage plants and to serve as a guide in pasture fertilization programmes, as well as aid the formulation of mineral supplementation programmes necessary for optimum animal productivity. Some estimates of animal mineral requirements based on work done in temperate regions are presented in Table 13.2.

Table 13.2 *Estimates of animals' mineral requirements (recommendations are from the National Research Council and adapted from Reid and Jung, 1974)**

| Mineral | Animal requirement | |
	Dairy cow giving 20–30 kg milk (%)	Beef cattle, growing and finishing steers and heifers (%)
Phosphorus	0.35	0.18–0.43
Potassium	0.70	0.60–0.80
Calcium	0.47	0.18–0.60
Magnesium	0.10	0.04–0.10
Sodium	0.18	0.10
Sulphur	0.20	0.10
	(ppm)	(ppm)
Iron	100	10
Manganese	20	1–10
Zinc	40	10–30
Copper	10	4
Cobalt	0.10	0.05–0.10
Iodine	0.60	
Selenium	0.10	0.05–0.10
Molybdenum	1.0	

* See also Cohen (1980).

Factors affecting chemical composition

Agronomic and management practices geared towards increasing the total yields of digestible nutrients from a pasture and therefore conducive to high animal productivity are based on our understanding of the various factors which influence chemical composition of the forages and their manipulation. Soil and climatic conditions, stage of growth and genotype have predominant influence on chemical composition of forages.

Soil and climatic conditions

Tropical areas supporting large ruminant populations can experience distinct wet and dry seasons of 2–10 months' duration. Wide seasonal fluctuations in chemical composition are common in tropical forages (Denium, 1966; Mohamed Saleem, 1972; Crowder and Chheda, 1977). Nutrient absorption is hampered under high moisture stress, and high heat intensity induces rapid physiological maturation accompanied by the formation of highly lignified tissues (French, 1957). Yields of green and dry matter, per cent crude protein, silica, free ash and NFE are directly related to the amount of precipitation (Oyenuga, 1960).

The physical, chemical and biological properties of soil, rates at which nutrients are supplied and renewed in the rooting zone and fertilizer practices affect forage chemical composition. Tropical soils are generally low in N. Deficiencies of other nutrients are also common, depending on the extent of land utilization, and addition of them is necessary for optimum productivity. In highly weathered tropical soils forages have a greater tendency to absorb large quantities of silica (D'Hoore and Coulter, 1972) which significantly depresses digestibility of the herbage.

Fertilizer application alters yield and chemical constituents in the forage depending on whether its growth is limited by the nutrients derived from the fertilizer or not. Any nutrient below the critical level in the soil would limit forage yield. Increasing forage yields will be accompanied by increases in nutrients in the plant tissues if they are not limiting in the soil. Addition of N influences yield and chemical composition of herbage when all other essential elements are available at an optimum level. Increases in crude protein and lignin and inconsistent effects on P, Ca, K and Mg in herbage have been reported when tropical pastures were fertilized with N (Vicente-Chandler, 1974; Mohamed Saleem, 1972). Phosphorus and S deficiencies are common in Utisols (Vicente-Chandler, 1974). Application of P in deficient soils will increase forage yield and P content in the tissues (Mohamed Saleem, 1972; Little *et al.* 1977). The ability of grasses to utilize excess amounts of K can depress Mg uptake and show symptoms of its deficiency (Vicente-Chandler, 1974). Sulphur deficiencies in forages have been reported in tropical America, resulting in low animal performance (Reid and Jung, 1974). Grazed pastures respond differently to fertilizer applications as compared to ungrazed pastures, due to variations in the amount of nutrients removed by grazing animals, trampling effects and recycling of nutrients through animal excreta (Mohamed Saleem, 1972).

Stage of growth

Advancing maturity is accompanied by increase in dry matter which is reflected in increases in cell wall contents, and a decrease in cell contents. In most tropical grasses, CF and NFE continue to increase with age, and in the case of CF most of the increase occurs during the first month of growth. The percentage increase of NFE is more pronounced during later stages of growth. A linear increase in CF content amounting to 1.58 per cent per week was observed in several *Digitaria* species in Trinidad when cut at weekly intervals up to 8 weeks (Miller and Cowlishaw, 1976).

Cell contents of herbage decrease with maturity. This does not mean that elemental absorption has ceased. Rather, the rate of dry matter accumulation through photosynthetic activity is greater than the rate of mineral absorption. This causes a dilution of mineral contents in proportion to the increase in bulk and results in lower crude protein, P and K percentages. Other nutrients tend to follow a similar pattern or remain inconsistent with increasing maturity of forage tissues (Table 13.3). It might thus appear that frequent clipping or grazing at very short intervals would provide animals with higher levels of digestible nutrients. Frequent defoliation, however, also produces a cumulative destructive effect on forage yield because of rapid exhaustion of carbohydrate reserves in the stubble and roots. Such pastures would require prolonged recovery periods and at times may be completely ruined by excessive defoliation.

Differences in chemical composition of different forage species with advancing growth should be considered while determining the optimum stage of utilization to derive maximum benefit without adversely affecting subsequent herbage productivity. Much greater care should be exercised particularly when the pasture contains a mixture of grass and legume species.

Chemical composition also varies vertically in a sward due to endogenous variations in maturity along this axis. The top of the sward, therefore, would remain young and contain low amounts of cell wall components. Per cent dry matter, ADF and lignin increased downwards along the axis while CP decreased in Coastal bermudagrass (Wilkinson *et al.*, 1970). Similar trends have been observed in elephant grass and guinea grass grown in West Africa (Mohamed Saleem and Aken 'Ova, personal communication).

Coward-Lord *et al.* (1974) suggested that remarkable changes in chemical composition occur only after extensive tissue differentiation has taken place between 30 and 60 days of growth. The overall quality of a forage depends on the relative proportion of high-quality fractions. Leaves and stems are nutritionally of equal value in beginning growth. The different dietary components in them decline, however, at different rates with age. The cell wall contents increase at a faster rate in stems as compared to the leaf (Reid *et al.*, 1973).

Genotype

Variations in chemical composition also arise as a result of genetic diversity of forage plants. Legumes are inherently superior to grasses in their feeding value, particularly more so in tropical environments. When different species of forage plants and different genotypes within a species are grown in a common environment and under uniform management, estimations of their chemical composition and feeding value often reveal significant difference (Reid *et al.*, 1973; Klock *et al.*, 1975). Using such a procedure a number of *Cynodon*, *Pennisetum americanum* × *P. purpureum* F_1 hybrids, and *Panicum maximum* clones were evaluated at the University of Ibadan, Nigeria. Considerable differences in per cent crude protein,

Table 13.3 *Influence of plant age on the chemical composition of some tropical grasses*

Grass species	(wks)	CP (%)	CF (%)	P (%)	K (%)	Ca (%)	Mg (%)	Cu (ppm)	Mn (ppm)	Fe (ppm)	Zn (ppm)
Napier grass*	4	23.8	24.6	0.33	2.38	0.61	0.42	17.4	138	708	39.8
	12	10.2	36.8	0.15	1.20	0.38	0.28	12.4	111	524	27.5
	20	8.6	40.4	0.11	0.34	0.43	0.36	14.1	128	355	32.6
	28	6.3	38.0	0.10	0.47	0.40	0.31	11.2	96	148	25.6
	36	–	42.2	0.08	0.24	0.30	0.30	13.5	144	192	36.4
Molasses grass*	4	17.2	25.5	0.31	1.18	0.46	0.50	18.4	137	981	47.2
	12	10.4	38.8	0.20	0.67	0.40	0.39	14.4	102	384	29.0
	20	8.5	41.7	0.16	0.32	0.38	0.36	19.5	126	342	34.1
	28	6.6	41.2	0.13	0.30	0.40	0.33	16.0	116	188	27.4
	36	–	39.5	0.12	0.20	0.48	0.33	15.2	122	217	29.9
Pangola grass*	4	13.4	33.6	0.16	1.32	0.56	0.39	24.9	192	426	34.9
	12	7.4	37.7	0.11	0.74	0.50	0.38	14.8	188	179	22.4
	20	7.6	36.4	0.12	0.37	0.66	0.39	15.6	317	274	30.9
	28	5.6	36.6	0.11	0.38	0.61	0.39	12.4	246	198	18.9
	36	–	34.8	0.11	0.38	0.76	0.38	16.0	298	217	25.0
Cynodon IB.8†	3	15.1	19.8	0.51	2.36	0.50	0.38				
	6	10.8	28.4	0.32	1.38	0.42	0.29				
	10	8.0	38.0	–	–	–	–				
	15	5.8	39.8	0.25	1.41	0.39	0.28				
	17	5.0	38.9	0.25	0.66	0.36	0.26				
Themeda trianda‡											
May	Growth	6.7	35.2	0.12		0.24					
June	Growth	4.6	35.4	0.09		0.38					
July	Growth	4.8	35.5	0.11		0.43					
Aug.	Growth	3.2	35.4	0.08		0.43					
Sept.	Growth	2.7	35.2	0.07		0.44					

* Gomide *et al.* (1969).

crude fibre and digestibility were observed (Chheda, 1974; Aken 'Ova, 1975; Crowder and Chheda, 1977; Mohamed Saleem and Aken'Ova, personal communication). Chemical composition of 107 grass species out of about 450 that occur in Kenya was studied by Dougall and Bogdan (1958) at early flowering stage. Crude protein varied from 5 to 20 per cent and in the majority of species it ranged from 8 to 16 per cent. Crude fibre varied between 14 and 43 per cent. Calcium and P contents also varied considerably and the values for these elements were between 0.09 and 1.05 per cent and 0.05 and 0.37 per cent, respectively. Within *Digitaria* species variations between 0.12 to 0.32 per cent for P, 0.44 to 0.25 per cent for K, 0.12 to 1.10 per cent for Na and 0.45 to 0.83 per cent for Ca were observed when cut at monthly intervals (Strickland, 1974). Differences in maturity dates, leaf : stem ratio, growth habit, responsiveness to fertilizer and management practices and acceptability are some of the important genotypic factors that indirectly influence the chemical composition of herbage. Their manipulation in a desired direction can lead to improvement in the chemical composition of the herbage and ultimately its feeding value.

Sampling and processing.

Herbage nutritive value estimations based on chemical composition are often unreliable, owing to difficulties in obtaining representative samples from the field. Animals tend to graze pastures selectively and laboratory estimates of herbage nutritive value may be only of limited value. Hand-picked samples in the field reportedly narrow the error margin as compared to mowed samples, since they simulate grazing conditions more closely (Miller and Cowlishaw, 1976).

Chemical analyses involve some pretreatment of the samples. Drying is the most important aspect. Improper drying can bring about irreversible changes in the sample and obscure the actual herbage quality. Soluble carbohydrates are underestimated in oven-dried samples as compared to freeze-dried samples. Significant increases in lignin and fibre have been reported in samples dried at high temperatures, possibly due to non-enzymatic browning reactions and by reactions causing hemicellulose to be estimated as part of lignin (Van Soest, 1965). In legumes and to some extent in grasses high temperature causes leaf shattering, and if proper care is not taken errors in estimation are likely. Freeze drying at −15 °C can minimize the destructive effects caused by heat drying of samples (Minson *et al.*, 1976). The capital and recurrent costs, however, may be prohibitive in much of the tropics. More reliable results for *Cynodon* were obtained by drying at 57 °C for 48 h (Wilkinson *et al.* 1969). There is, therefore, a need to standardize laboratory techniques so as to allow for valid comparisons of herbage analytical results originating from different experiments, locations and seasons and to draw meaningful conclusions regarding the actual nutritive value of forages under study.

Toxic substances

Occasionally, forages that are rated high in their major dietary components also contain substances which cause deleterious effects in the livestock that feed on them. Some of the important substances are cyanogenetic glucosides, organic acids such as oxalic acid, amino acids, alkaloids, oestrogenic isoflavones and saponin. Legumes generally contain a wider range of deleterious substances than grasses. Additionally, in localized areas forages tend to accumulate excessive amounts of minerals such as Mo that causes scouring, Se that causes the 'alkali disease' and fluorides that cause mottling and wear of the teeth.

Some grasses during periods of rapid growth are high in nitrates. These can be dangerous to livestock, since they are often reduced to nitrites which are toxic. Warm and dry weather conditions favour accumulation of nitrates in the leaf sheath of *Pennisetum purpureum* beyond lethal levels of nitrites (Harker and Kamau, 1961).

Prussic acid is one of the most prevalent and serious toxic principles in plants. The cyanogenetic glucosides commonly found in *Sorghum* species and other grasses such as *Cynodon plectostachyus* when hydrolysed with appropriate enzymes liberate hydrocyanic acid. The acid inhibits the oxidative enzymes in the body tissues and thus causes internal asphyxiation. Death occurs very rapidly. Agronomic practices such as heavy application of N, and climatic conditions, such as drought, significantly increase the concentration of glucosides.

Tropical grasses also contain varying amounts of oxalic acid. A study in Puerto Rico by Garcia-Rivera and Morris (1955) revealed oxalic acid content in *Paspalum plicatulum* to be as low as 0.02 per cent, while in *Pennisetum purpureum* cultivars it was as high as 2.6 per cent. *Melinis minutiflora*, *Cenchrus ciliaris* and *Panicum maximum* had intermediate oxalic acid contents. In Australia, *Setaria anceps* contained oxalic acid at levels high enough to cause acute oxalate toxicity in grazing animals (Jones and Ford, 1972). Formation of calcium oxalate ties up blood calcium and thus causes poor coagulation of the blood.

Ergot (*Claviceps* spp.) infection of forages produces alkaloids that cause ergotism and can result in abortion and sloughing of hooves.

Indospicine, a hepatotoxic amino acid, is found in *Indigofera spicata*, a pantropic pasture legume, and in *I. domini*, a native legume of Central Australia. It causes liver degeneration and abortion in cattle and sheep, and staggers and even death in horses, due to interference with metabolism (Hutton, 1970; Minson *et al.*, 1976).

Bloating is commonly observed in temperate countries when animals feed on large quantities of legumes. However, in the tropics *Lablab purpureus* is the only legume known to cause bloat at a young and rapidly growing stage. It is likely that tropical legumes do not normally contain the bloat-inducing 18 S protein (Hutton, 1970). High contents of tannin found in *Desmodium* spp. can cause reduction in the palatability and digestibility of the herbage.

Leucaena leucocephala, a pantropic legume shrub with considerable potential for forage, contains up to 0.5 per cent of the nitrogen as a toxic amino acid, mimosine. Diets containing large quantities of this legume can result in wool shedding and abortion in sheep, and loss of hair on the rump and tail and loss of weight in cattle, due to incomplete metabolization of the substance in the rumen. Heifers kept in pens and fed solely a diet of *Leucaena* grew normally but produced small calves with enlarged thyroid glands (Hamilton *et al.*, 1971). Prolonged grazing of cattle on pasture containing *Leucaena* cv. Peru in Queensland resulted in enlarged thyroid glands, low liveweight gains, excessive salivation and hair loss (Jones *et al.*, 1976). This indicated a goitrogenic substance in or developed from *Leucaena*. Cattle previously unaccustomed to *Leucaena* made rapid gains of 1.0 kg/day as compared to poor gains of 0.35 kg/head per day of animals on *Leucaena* for 8–12 months, suggesting a cumulative effect on animal performance. Sheep could be adapted to the mimosine toxin by gradually increasing the *Leucaena* content in their diet (Hegarty *et al.*, 1964; Reis *et al.*, 1975).

In temperate countries, varieties of lupins low in alkaloids and white clover low in cyanogenetic glucosides have been developed through plant breeding. A similar approach is likely to pay rich dividends in tropical forages.

Digestibility

Proximate analyses give a quantitative estimation of the various constituents in forage species but do not indicate their precise nutritive value. Though the cell contents are highly digestible and lignin indigestible by higher animals, ruminants have an advantage over non-ruminants in the utilization of fibrous forage crops because of their ability partly to digest other structural carbohydrates (holocellulose) through the intervention of rumen micro-organisms.

Digestibility is an important measure of the nutritive value of forages and can be defined as the difference in value between the feed eaten and materials voided by the animals, expressed as percentage of feed eaten. Thus, the overall digestibility of a forage will be the summation of the content × digestibility of different chemical components of the forage (Javier, 1975).

Digestibility measurements

A direct method (*in vivo*) that involves feeding experiments with cattle, sheep or goats and indirect laboratory estimation methods (chemical methods; *in vitro*; nylon bag technique) have been developed to measure digestibility of forage crops. These are described in brief below. For detailed procedures the reader should consult Raymond (1969) and Minson *et al.* (1976), Beaty and Engel (1980).

In vivo method.

The technique involves determination of forage DM or OM or chemical constituents eaten by animals confined to digestion stalls and the respective amounts voided in the faeces. This can be represented by the equation:

$$\text{Digestibility \%} = \frac{C_{\text{feed}} - C_{\text{faeces}}}{C_{\text{feed}}} \times 100$$

where C_{feed} and C_{faeces} refer to the amount of forage DM or constituent in forage eaten and faecal excretion respectively.

The actual procedure is laborious, time-consuming and expensive. Animals are fed measured amounts of feed and the rejected feed and faeces voided are carefully measured. Animals require a pre-trial period of at least 7 days followed by a collection period of 14 days (Ademosun, 1970). Chemical composition, and therefore digestibility, can vary from day to day, since forages have to be cut and fed daily during the experimental period. Feeding with hay or grass cut and dried at 100 °C or stored at −15 °C has been suggested. Special harness devices may be needed to collect faeces to avoid contamination with urine and animals have to get accustomed to these devices before the experiment is initiated. Animals are unable to express their selective eating habits fully when fed indoors. Digestive efficiency of animals harbouring intestinal parasites may be seriously hampered.

Thus to replace, as far as possible, the laborious and expensive *in vivo* method, inexpensive and reliable laboratory techniques to predict *in vivo* digestibility of a forage have been developed.

Chemical methods.

As a result of a large number of studies on chemical composition and *in vivo* digestion trials in temperate zones, regression equations have been computed to predict forage digestibility from the chemical composition data obtained from Weende proximate analyses. Their usefulness, however, is restricted, and when applied to a wide range of forages grown in different locations the relationships become considerably less precise. It may even by misleading if used directly to predict the nutritive value of tropical forages (Duckworth, 1946; Brendon *et al.* 1963a). It is therefore necessary to obtain reliable regression equations for tropical pastures (Oyenuga, 1968). Brendon *et al.* (1963a, 1963b) computed regression equations based on digestion trials carried out with cattle in East Africa. They employed 13 forages and reported highly significant correlation coefficients between digestibility and chemical components such as crude fibre, crude protein and total carbohydrate.

Until recently, inadequate methods of separating the nutritional components of forages and their improper interpretations limited the usefulness of digestibility predictions based on proximate analyses. With improved methods available to separate the various nutritional components and with a greater understanding about their availability to animals, it is now possible to predict digestibility more accurately. Thus, Ademosun (1974) studied the relationship between dry matter digestibility and certain chemical constituents of improved tropical forages and found high correlation between dry matter digestibility and acid detergent fibre, lignin and crude fibre (Table 13.4).

Van Soest (1967) suggested that digestibility predictions can be made from the nutritional components of a forage more accurately if the following factors are taken into account (1) indigestible lignin in the forage, (2) endogenous materials, similar to cell contents of plants that are passed with the faeces of the animal and (3) indigestible silica in the forage. After having accounted for these factors, Van Soest and Jones (1968) produced a summative equation for estimating total apparent digestibility (TAD) of dry matter:

$$TAD\% = 0.98S + W(1.473 - 0.789 \log X) - 3.0(SiO_2) - E$$

where S = cell contents, 98 per cent digestible; W = neutral detergent fibre, digested to an extent depending upon the lignification of the acid detergent fibre fraction (X); SiO_2 = silica; E = endogenous excretion in faeces equal to 12.9 per cent of dry matter.

Evidence regarding the applicability of this summative equation in tropical forages is limited. There may be need for refinement be-

Table 13.4 *Relationship between chemical components and digestibility of some forages (Ademosun, 1974)*

Y	X	Regression equation	Correlation
DMD	ADF	$Y = 83.2 - 0.59x$	−0.87
DMD	L	$Y = 57.5 - 0.18x$	−0.76
DMD	CF	$Y = 82.2 - 0.76x$	−0.71
DCP	CP	$Y = 1.14x - 6.0$	0.93
CPD	L	$Y = 61.6 - 0.14x$	−0.85
CWCD	CP	$Y = 62.3 + 0.19x$	0.91

Note. DMD = Dry matter digestibility; DCP = Digestible crude protein; CPD = Crude protein digestibility; CWCD = Cell wall constituent digestibility; ADF = Acid detergent fibre; L = Lignin; CF = Crude fibre; CP = Crude protein.

cause of the peculiarities of tropical forages in terms of fibre, lignin and cell wall constituents (Ademosun, 1970), as well as silica.

In vitro method.

This method attempts to approximate digestion in an artificial environment where rumen conditions are simulated in a test-tube by (1) incubating the dried forage sample with rumen fluid obtained from a fistulated animal, (2) buffering the fluid medium with an artificial solution of saliva and (3) maintaining the temperature at the approximate rumen temperature of 39 °C during the 48 h incubation period. The method may underestimate digestibility when compared with *in vivo* estimations. Close correlation between *in vitro* and *in vivo* estimations are now possible by subjecting the residue of the first microbial digestion to a second enzymatic digestion for 48 h, involving acid pepsin (Tilley and Terry, 1963). Analysis of 146 herbage samples by the authors resulted in a regression equation, *in vivo* digestibility = 0.99 × *in vitro* digestibility −1.01 with SE = 2.31. Rogers and Whitmore (1966) have modified the procedure to achieve greater efficiency in terms of laboratory output.

The *in vitro* method is basically simple and quite efficient. It does not require extensive laboratory facilities and many samples can be analysed simultaneously. Success will depend on correcting the various sources of error that originate due to variations in (1) microbial population, (2) sample preparation and sortage, (3) pH of the medium during incubation and (4) procedures. Large errors can be introduced through the inoculum due to variations in the diet of fistulated animal, feeding systems, time of rumen fluid removal and methods of handling and processing the rumen fluid before use.

Goto and Minson (1977) used a pepsin–cellulose assay instead of rumen–liquid pepsin in predicting the digestibility of tropical grasses. The validity of both procedures has been examined by Terry *et al.* (1978), Rees and Minson (1979) and Peña and Paladines (1979). Thomas *et al.* (1980) noted that the pepsin–cellulose method is less accurate but cheaper than the rumen–liquid pepsin method.

Minson *et al.* (1976) considered the *in vitro* method to be the most accurate of all laboratory techniques for the purpose of prediction of *in vivo* digestibility and recommended its use in tropical forage research programmes. They further suggested that chemical methods should be used only when large differences in samples are expected or when establishment of an *in vitro* laboratory is not practicable.

Donefer *et al.* (1963) suggested the use of cellulase as an alternative for determining organic matter digestibility. The digestion technique was described by Jones and Hayward (1973, 1975) using acidic pepsin to remove the cell wall and crude cellulase from *Trichoderm viride* to dissolve dry matter. Goto and Minson (1977) successfully used the pepsin–cellulase assay instead of rumen–liquid pepsin to predict the digestibility of tropical grasses. *In vitro* rumen liquor–pepsin, pepsin–cellulase and fibre analysis were compared by Thomas *et al.* (1980) for predicting the digestibility of *Setaria anceps, Chloris gayana* and *Pennisetum purpureum*. They found that dry matter digestibility by rumen pepsin was more closely related to *in vivo* determination than that dissolved by cellulase from two sources: *Basidiomycetes* and *Trichoderm viride*. The enzyme from the latter dissolved 33 per cent more dry matter than that from the former. These procedures have also been examined by Jarrije *et al.* (1970), Terry *et al.* (1978), Rees and Minson (1979) and Peña and Paladines (1979).

Nylon bag technique.

The method was first described by Burton *et al.* (1967) and has undergone several modifications since then. Six to ten grams of ground forage samples are placed in small bags made from nylon satin cloth or parachute nylon. Up to 48 of these bags are then tied to each of two circular steel weights and placed in the ventral sac of the rumen of a fistulated cow. The bags are removed after 48 h, washed thoroughly under running water and dried in a forced air oven. The contents are weighed to determine dry matter loss for computation of digestibility. The rate of digestion of dry matter or chemical constituents can be determined by removing the bags at varying intervals.

In spite of some limitations, such as loss of small solid dry matter particles and entry of rumen particles through the bag pores, the method can be useful where establishment of an *in vitro* laboratory is not feasible. It is also beneficial in plant-breeding programmes where small quantities of forage samples emanating from a large number of introductions and breeding lines needed to be evaluated on a comparative basis for their herbage quality.

Factors affecting digestibility

It is generally believed that tropical forages have a lower digestibility, with wider variations among species, than their temperate counterparts at all stages of growth (Milford and Minson, 1966; Denium and Dirven, 1975). A mean difference of 12.8 percentage units in digestibility was calculated when tropical and temperate forages were compared (Minson and McLeod, 1970). This difference was associated with high temperature and transpiration rates. Temperate and tropical species had similar dry matter digestibility values when grown under similar climatic conditions. This led to the conclusion that the observed digestibility differences between tropical and temperate forage species may be environmental rather than genetic.

The dry matter digestibility of a forage, during the grazing stage, can vary considerably, and is related to changes in the chemical composition particularly in fibre, lignin and silica contents and to some extent in crude protein content of the plant, arising as a result of differences in species or genotype, stage of growth, environmental conditions and cultural and management practices. The primary cause of large variations in the digestibility of various forages is the amount of lignin present. A unit increase in lignin can result in three to four unit decrease in digestibility of a forage (Bula *et al.*, 1977).

Stage of growth and genotype

Forages are highly digestible at young and immature stages of growth. High *in vitro* digestibility values ranging from 75 to 85 per cent after 1 week's regrowth have been reported in several improved varieties of tropical pastures (Reid *et al.*, 1973; Mohamed Saleem, 1972). Digestibility declines with advancing maturity. Digestibility, and the rate of its decline with age vary considerably between genera, species and varieties (Table 13.5). The differences in digestibility, as well as the differences in the rates of decline of digestibility, are much larger among tropical grasses than among temperate grasses. In temperate forages the rate of decline in digestibility varies between 0.3 and 0.5 percentage units each day after the boot stage in grasses and bud stage in legumes (Bula *et al.*, 1977). In tropical grasses Milford and Minson (1966) observed the rate of decline to be 0.1–0.2 per cent daily. However, workers in East Africa (Reid *et al.*, 1973; Ogwang and Mugerwa, 1976), West Africa (Crowder and Chheda, 1977), Trinidad (Miller and Cowlishaw, 1976) and Queensland (Wilson and Mannetje, 1978) reported up to 0.5–0.6 per cent daily decline in digestibility of several grasses. Mohamed Saleem (1972) observed approximately 1 per cent daily decline in digestibility in *Cynodon nlemfuensis*, a commonly grown pasture species in southern Nigeria. Grasses, particularly improved varieties and selections, and legumes have comparable digestibilities to begin with but the rate of decline is slower in most of the legumes (Reid *et al.*, 1973). This may be due to the relatively lower contents of ADF and hemicellulose in legumes (Garcia and Ferrer, 1974).

Reid *et al.* (1973) studied *in vitro* dry matter digestibility of 42 tropical grass species and varieties and 11 legumes in Uganda under good fertility conditions. Three definable patterns of digestibility were identified in the grasses:

Group I – characterized by high initial digestibility (70–85%) followed by a high rate of decline in digestibility with advancing maturity e.g. *Brachiaria*, *Setaria* and *Panicum* spp., *Chloris gayana*, *Cen-*

Table 13.5 Influence of age on in vitro dry matter digestibility (%) of some tropical grasses and legumes (Reid et al., 1973)

Species	Weeks of growth								
	1	2	4	6	8	10	12	14	16
Grasses									
Andropogon gayanus	67.9	64.9	63.9	60.7	53.4	60.9	50.8	43.5	46.0
Brachiaria ruziziensis	82.5	81.2	73.9	72.1	66.2	64.0	54.8	54.5	49.8
Cenchrus ciliaris	78.6	75.3	70.5	63.7	59.0	56.6	45.0	40.6	38.6
Chloris gayana cv. Masaba	78.9	79.3	69.2	60.0	49.9	50.4	44.7	41.0	35.6
Cynodon dactylon		68.3	65.5	60.5	53.8	60.7		52.5	
Digitaria decumbens		77.0	75.8	70.4	59.5	59.8	56.3	55.2	56.3
Hyparrhenia rufa	60.1	59.6	63.2	56.2	57.5	59.2	48.7	42.0	39.5
Panicum maximum cv. likoni	84.7	79.9	67.8	53.1	51.8	52.3	50.0	43.0	47.7
Paspalum urvillei	61.9	62.0	64.7	66.8	64.2	61.3	63.5	59.9	59.0
Pennisetum purpureum	74.8	68.6	70.2		63.2	63.1	63.8	58.1	48.9
Setaria anceps		70.7	71.4	63.8	53.7	49.3	36.9	39.3	30.5
Themeda trianda	74.3	58.2	57.5	60.6	55.6	54.4	43.5	40.0	
Legumes									
Centrosema pubescens				60.2	61.4	56.2	54.8	55.3	53.0
Desmodium intortum			60.9	58.9	52.1	54.3	51.7	50.8	48.6
Glycine wightii			69.4	63.6	62.4	62.0	65.3	60.7	62.6
Trifolium semipilosum			76.5	79.9	78.1	77.3	77.1	75.7	75.0
Macroptilium atropurpureum cv. Siratro				73.9	62.8	67.2	61.8	58.3	60.1
Stylosanthes guianensis					62.0	63.8	65.3	61.4	61.7
Medicago sativa	77.2	73.1	65.1	62.5	63.8	63.0	58.3	55.4	57.5

chrus ciliaris, Digitaria decumbens, Melinis minutiflora, Pennisetum purpureum and *Sorghum sudanense.*

Group II – characterized by intermediate digestibility (60–70%) and variable decline in digestibility with advancing maturity e.g. *Andropogon gayanus, Cynodon dactylon, Dactylis glomerata* and *Digitaria uniglumis.*

Group III – characterized by low initial digestibility (50–60%) and a generally low rate of decline in digestibility e.g. *Paspalum, Cymbopogon, Hyparrhenia* and *Themeda* spp.

Group I contained the genera and species which are commonly planted or have undergone breeding or selection work, while Group III mostly contained the indigenous grasses found as components of natural swards.

A knowledge of the changes in chemical composition and digestibility as a result of maturity of a given forage is essential for understanding the efficient animal utilization of forage crops. The cultural and management practices of pastures should be geared to obtaining high-quality forage for efficient animal production by striking an appropriate balance between forage yield and digestibility. Poor pasture management can result in low dry matter yields and low-quality forage.

Considerable genetic differences in forage digestibility at the species and variety levels have been observed in several tropical forages such as those belonging to the genera *Setaria* (Fig. 13.2), *Hyparrhenia, Paspalum, Cynodon, Pennisetum, Panicum, Digitaria* and *Stylosanthes* (Hacker and Minson, 1972; Brolman and Kretschmer, 1972; Burton and Monson, 1972; Reid *et al.*, 1973; Chheda, 1974; Burton *et al.*, 1973; Klock *et al.*, 1975; Crowder and Chheda, 1977). The results indicate a strong possibility of improving the digestibility of tropical forage species through plant-breeding.

Plant fractions

The leaf and stem fractions of a forage plant are equally highly digestible because of high level of cell contents and high digestibility of the cell wall fractions at young stage of growth. With advancing maturity, however, considerable differences in the digestibility between the two fractions arise. As the stem matures the cell contents decrease rapidly. The cell wall contents increase and the cell wall fraction is also now less digestible because of increased lignification. The cell contents of leaves remain at higher levels and rate of lignification is much slower with increase in age than stems. This results in a rapid decline in digestibility of the stem fraction compared to the leaf fraction with advancing maturity (Raymond, 1969).

Samples of leaf and stem plus sheath did not differ in digestibility until the sixteenth week of sampling, and thereafter marked decrease in *in vitro* digestibility was observed in stem plus sheath fraction (Reid *et al.*, 1973). Haggar and Ahmed (1970) reported

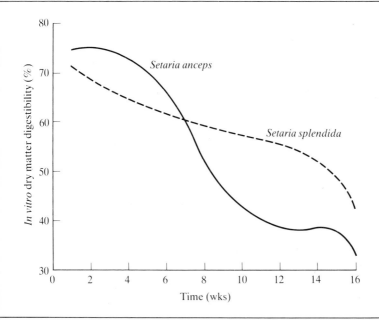

Fig. 13.2 *In vitro* digestibility of Setaria (Reid *et al.*, 1973).

the daily *in vitro* digestibility decline in leaves and stems of *Andropogon gayanus* to be 0.5 and 0.4 per cent, respectively, in the vegetative phase, but after seed-head emergence the decline in digestibility of stems was 0.75 per cent daily.

The digestibility of species and varieties of *Digitaria* that develop stolons declined at a faster rate than those which did not (Miller and Cowlishaw, 1976). Herbage of *Digitarias* found to be low in digestibility also had more secondary thickening in stem nodal regions than those high in digestibility (Klock *et al.*, 1975).

Vertical differences in digestibility are also encountered in tropical pastures. Within the mass of herbage, digestibility decreases from the top downwards. Rapid maturation of stems and senescence of leaves in the basal layers can reduce the overall forage digestibility. Leaf : stem ratio at the stage of forage utilization and date of flowering are therefore important factors associated with digestibility. Species with short internodes may have an advantage since they would contain higher proportions of leaf. Milford and Minson (1968) and Minson (1971), however, observed that in *Chloris gayana*, *Panicum maximum* and *P. coloratum* digestibility was not related to leafiness or floral development. This suggested that morphological characteristics of a forage may not always be indicative of its digestibility.

Climate

Reduced digestibility and poor nutritive value of tropical forages can be attributed to high heat intensity that causes rapid growth, enhanced maturity with decreasing leaf : stem ratio and increased crude fibre content, particularly higher lignification, as well as increased levels of acid detergent fibre and cell wall contents.

Climatic variations, particularly temperature changes during the growing season, can cause differences in the digestibility of a forage. In regions that experience wet equatorial or wet monsoonal climates, and where daily temperatures are comparable during the wet and dry seasons, forage digestibility during the two seasons does not differ markedly. Miller and Cowlishaw (1976) recorded similar patterns of digestibility in *Digitaria* spp. during the two growing seasons in Trinidad. On the other hand, Olubajo and Oyenuga (1970), who studied digestibility of various grass–legume mixtures in Nigeria found consistent 8–10 per cent differences in digestibility between the wet and dry seasons. *In vitro* organic matter digestibility of 24 *Cynodon* genotypes cut at 6-week intervals was found to be on an average 8.7 per cent lower during the late dry season as compared to early wet season in Ibadan, Nigeria (Chheda, unpublished). Deinum and Dirven (1975) studied *in vitro* organic matter digestibility of temperate and tropical grasses grown at different temperatures and reported 0.79–1.02 per cent decrease in digestibility for each 1 °C increase in temperature in *Brachiaria ruziziensis* at vegetative and stem elongated stages of growth, respectively. The average decline in digestibility with 1 °C increase in temperature for a large group of forages was calculated to be 1.14 per cent by Minson and McLeod (1970).

Protein content

In many parts of the tropics under natural fertility conditions, and up to about 2 months of growth after the onset of rains, the crude protein content in forage grasses is well above 7 per cent. With uninterrupted growth, due to high heat intensity, the grasses rapidly become mature and crude protein content drops drastically, reaching values of 4–6 per cent after 3–5 months. During dry months, crude protein content often drops below 4 per cent and in later parts of the dry season values of 1–2 per cent have been recorded (Crowder and Chheda, 1977).

Crude protein content in forages can have a significant effect on digestibility. When it exceeds 7 per cent in the herbage, digestibility does not appear to be affected (Milford and Minson, 1966). If herbage with a crude protein content below 7 per cent is fed to animals, however, microbial activity in the rumen is depressed by lack of nitrogen. This causes an incomplete utilization of structural carbohydrates in the ingested forage and a slow rate of passage of the digesta. Therefore, forage digestibility and voluntary intake are significantly reduced. Glover and Dougall (1960) observed marked decline in total carbohydrate digestibility when crude protein content in the forage fell below 6 per cent. It is therefore important to maintain crude protein level of the animal's diet above 6–7 per cent for maintenance and production. Grass–legume mixtures, high-protein feed supplementation or feeding of urea are some of the ways by which digestibility of a low-protein forage can be improved considerably.

Mineral content

Intake and digestibility are adversely affected when mineral content of plants drops below the level required for animal growth. Thornton and Minson (1973b) reported that dry matter digestibility of Pangola grass–legume pastures increased from 41.6 per cent to 44.9 per cent when P concentration increased from 0.11 to 0.15 per cent. Increasing the P content of feed through P fertilization of pastures resulted in higher liveweight gain of sheep in Australia (Ozanne *et al.*, 1976). Liveweight gain was linearly related to both P and Ca levels in guinea grass in Malaysia (Eng *et al.*, 1978), increasing from 280 to 379 g/head per day with a mean concentration of 0.13 and 0.19 per cent P, and 0.18 and 0.26 per cent Ca, respectively. The liveweight gain response was supported by Rees and Minson (1979), who showed that dry matter digestibility of pen-fed sheep increased by 2.1 per cent and intake by 11.3 per cent when Ca concentration of forage increased from 0.22 to 0.38 per cent. Silica is absorbed by forages along with water and reduces forage digestibility, much in the same way as lignin (Van Soest and Jones, 1968). Tropical grasses appear to contain more Si than legumes (D'Hoore and Coulter, 1972) and can reach concentrations of over 5.0 per cent on a dry weight basis (Oyenuga, 1957; Miller and Blair-Rains, 1963).

Digested products

Animals in the tropics and subtropics derive their energy requirements primarily from the herbage they consume. Therefore, for an accurate measurement of the nutritive value of a given forage, knowledge of the quantity of digestible energy and net energy available to the animal from that forage and the efficiency of energy utilization for maintenance and production is essential.

Microbial activity within the rumen results in the production of a number of steam-volatile fatty acids that are the major sources of the energy supply of ruminants. Quantitative determinations of these acids and their relative proportions, as well as the efficiency of their utilization by the animal, are also used to estimate the nutritive value of forages.

Digestible energy and net energy

Of the total energy available from a forage (between 4.19 and 4.90 cal/g on a dry matter basis in tropical pastures, Butterworth, 1964) a portion is lost as faecal energy and the remainder is considered as the usable digestible energy. About 20 per cent of this digestible energy is lost as combustible gases, mostly methane during digestion and in the production of urine, leaving metabolizable energy available for the animal's metabolic and productive activities. Additional energy is lost as heat when the animal digests and metabolizes the forage. The energy that remains is the net energy of the forage available to meet maintenance and production needs of the animal.

The net energy content of a forage depends on the concentration of digestible energy in that forage. When forages of low digestibility are consumed by the animal, increased amounts of energy are given off in the form of heat during the digestion process. Thus, a larger proportion of the metabolizable energy is wasted during eating and digestion of low-quality forage as compared to forage of high digestibility. The efficiency of utilization of net energy depends on the class of animals such as cattle, sheep or goats and on animal requirements in terms of maintenance, meat or milk production. For example, animals use the energy available from forages much more efficiently for maintenance or milk production than for meat production. Thus, results of research in temperate countries have indicated that at 70 per cent digestibility forages are 65–75 per cent as efficient as grain for meeting animal maintenance and production requirements. At 50 per cent digestibility, though the forages are 50 per cent as efficient as grain in terms of energy provisions for maintenance and milk production, their efficiency is only 20 per cent of grain for meat production (Bula *et al.* 1977).

The various energy estimations in forage plants require specialized and expensive equipment and therefore cannot be routinely carried out. As digestible energy is highly correlated with dry matter digestibility, energy requirements of ruminants are often expressed as total digestible nutrients (TDN) or starch equivalents (SE), and for tropical pastures these can be predicted from dry matter digestibility (DMD) or organic matter digestibility (OMD) data (Minson *et al.*, 1976):

$$\text{TDN} = 0.99 \, \text{DMD} + 0.96 \pm 2.6 \text{ or } 0.95 \, \text{OMD} + 1.49 \pm 2.5,$$
$$\text{SE} = 1.24 \, \text{DMD} - 33.8 \pm 3.3 \text{ or } 0.19 \, \text{OMD} - 33.0 \pm 3.2.$$

It is important to remember that with maturity the decline in net energy available to animals from forages will be more pronounced than the decline in digestibility, even though forage digestibility and net energy are related.

Volatile fatty acids

The wide variety of feeds and forages consumed by the animals are converted into a small number of steam-volatile fatty acids (VFA) as a result of microbial fermentation in the rumen, which in turn provide most of the energy requirements of these animals.

A review of pertinent literature on the subject by Raymond (1969) indicated that:

1. Acetic, propionic and butyric acids are the most important.
2. Their relative proportions in the total amount differ and, in general, as the forages become less digestible the proportion of acetic acid increases, while the combined proportion of propionic and butyric acids decreases.
3. While all three acids are used with equal efficiency for maintenance purposes, acetic acid is much less efficiently used for fattening.
4. For milk production acetic acid appears to be utilized more efficiently than propionic acid.

These findings have led to the belief that the various fatty acids produced in the rumen play different roles in animal production. Therefore, any decision regarding the most effective use of forages of differing qualities for animal production should be based on an understanding of the shift in the relative proportions of these acids in the rumen as caused by different forages.

Forage intake

An animal will consume forages in varying amounts. Therefore, assessment of forage quality depends not only on the nutritive value of the forage but also on the quantity of that forage voluntarily eaten or, in other words, on the total quantity of digestible nutrients consumed by the animal. Crampton (1959) stated that 'the feeding value of a forage depends primarily on the magnitude of its contribution towards the daily energy need of an animal. Differences in this respect between forages are almost completely a consequence of the relative amounts in which they are voluntarily eaten'.

In monogastric animals intake is controlled mainly by the levels of blood metabolites. In ruminants intake depends largely on the capacity of the digestive tract, particularly the rumen. The animal stops eating when a certain degree of 'fill' is reached and starts to eat again when the 'fill' is reduced as a result of digestion and movement of the residue through the digestive tract. Even though with forages of very high digestibility (over 70–75%) blood metabolite level can control intake by ruminants, under most conditions, particularly in the tropics where forages generally have lower digestibility, the gastro-intestinal fill controls intake. Thus, involuntary physiological reflexes, rather than subjective preference, control voluntary forage intake by the ruminants (Raymond, 1969; Cordoba *et al.*, 1978).

Animal intake of tropical forages, particularly the grasses, is in most cases considerably lower than temperate species. This low level of intake is generally considered to be the major cause of low animal productivity in tropical environments. Ingalls *et al.* (1965) suggested that 70 per cent of the variation in animal productivity can be accounted for in terms of voluntary intake differences, as compared to 30 per cent accounted for by digestibility differences. A higher rate of intake is directly related to the shorter time that ingesta is retained in the rumen (Poppi *et al*, 1980). Ademosun (1974) studied the intake of several tropical grasses by sheep and goats in Nigeria and obtained dry matter intake values (g/ kg $LW^{0.75}$) ranging from 25 to 48. These values are about one-half to two-thirds of those reported in Australia for similar species of tropical grasses at comparable stages of growth (Minson, 1971, 1972; Thornton and Minson, 1973a), and only about one-third of the figures recorded for temperate forages. These low intake figures place serious limitations on forage utilization in the tropics. Intake of tropical legumes, however, is considerably higher than that of tropical grasses.

Measurement of intake

Voluntary intake is defined as the amount animals will ingest when an excess of 15 per cent is offered (Blaxter *et al.*, 1961). For reliable estimates of intake, an aequate number of pen-fed animals (8–10 sheep or goats) must be used because animals vary considerably in their eating capacity. Since voluntary intake is affected by the selective eating habit of the animal, it is necessary to specify the amount of herbage available for each animal. Forage in the fresh, dried or frozen form should be presented to the animal at least twice daily. Uneaten residues are removed to prevent fouling. Accurate weight records of the amount presented and that removed are essential. Voluntary intake can be expressed as kilograms of forage dry matter eaten per 100 kg of liveweight or g/kg LW. In order to correct the differences in intake that arise due to species or size of animals, however, most data are reported in terms of metabolic weight of the animal expressed as liveweight raised to 0.75 power ($LW \ kg^{0.75}$), and for comparative purposes these are then converted to g/kg $LW^{0.75}$.

Factors affecting intake

Among the various factors influencing voluntary intake of forage by animals the most important ones appear to be stage of growth, digestibility and genotype. Low protein content and mineral deficiencies can also affect intake. Under grazing conditions, availability of forage and environmental fluctuations can significantly alter intake. Furthermore, animals differ in their selection of the proportion of a species in the diet (Jones, *et al.*, 1979). Bite size, which is influenced by sward density, stem content and proportion of leaf in the basal and top layers of the herbage mass, can place limitations on forage intake under grazing (Stobbs, 1973). High levels of nitrogen fertilization have been reported to have an adverse effect on intake (Reid *et al.* 1966). Supplements can also alter the voluntary intake (Rees and Minson, 1976).

Stage of growth

Advancing maturity adversely affects the nutritive value of a forage and may affect its voluntary intake by animals (Tables 13.6 and 13.7).

Investigations carried out in Australia (Milford and Minson, 1966, 1968; Minson, 1971, 1972) and in East Africa (Gihad, 1976) on the daily animal intake (g/kg $LW^{0.75}$) of several tropical forages at various stages of growth, indicated that intake of grasses declines with advancing maturity. The rate of decline, however, varies among species. When intake of six tropical grasses was examined under uniform conditions Minson (1972) observed two distinct groups of species, one represented by *Chloris gayana, Panicum maximum, Paspalum dilatatum* and *Setaria splendida* where intake declined rapidly with age (0.23–0.32 g/kg $LW^{0.75}$ per day), and another represented by *Digitaria decumbens* and *Pennisetum clandestinum* with a very slow rate of decline (0.04–0.09 g/kg $LW^{0.75}$ per day). Grasses belonging to the latter group will be more desirable in areas with a long dry season because of their ability to be utilized more efficiently with advancing maturity. The average intake values of these grasses at 4, 10 and 14 weeks of regrowth were 55, 49 and 41 g/kg $LW^{0.75}$. Ademosun (1974) in West Africa and Johnson *et al.* (1968) in the Philippines, however, observed that intake was independent of stage of maturity.

Minson and Laredo (1972) studied the available information regarding intake and digestibility of several tropical grasses and concluded that differences in voluntary intake among *Panicum* varieties were closely related ($r = 0.86$) with leafiness. Differences among species could not be attributed to differences in leaf percentage. Flowering behaviour was also a relatively poor guide of the intake of tropical forages.

Digestibility and genotype

Within a particular forage variety or cultivar, intake is related to its

Table 13.6 Per cent dry matter digestibility (DMD) and g/kg LW$^{0.75}$ voluntary dry matter intake (IDM) of six grasses harvested after 4, 10 and 14 weeks of regrowth (Minson, 1972)

Age at harvest (wks)	Setaria splendida		Paspalum dilatatum		Chloris gayana		Panicum maximum		Pennisetum clandestinum		Digitaria decumbens		Mean	
	DMD	IDM	DDM	IDM	DDM	IDM	DDM	IDM	DDM	IDM	DDM	IDM	DDM	IDM
4	66.8	52.4	56.7	58.6	60.6	55.0	63.5	62.0	62.0	48.0	67.2	56.1	62.8	55.4
10	55.8	42.6	49.5	48.7	55.3	51.9	49.9	51.8	49.1	44.4	57.1	53.4	52.8	48.8
14	51.4	30.0	42.5	43.0	47.5	39.0	48.0	39.8	47.4	45.2	49.1	49.1	47.7	41.0
Mean	58.0	41.6	49.6	50.1	54.5	48.6	53.8	51.2	52.8	45.7	57.8	52.9	54.4	48.4

Table 13.7 Per cent dry matter digestibility (DMD) and g/kg LW$^{0.75}$ voluntary dry matter intake (IDM) of six Panicum varieties harvested after 4, 9 and 13 weeks of regrowth (Minson, 1971)

Age at harvest (wks)	Kabulabula*		Burnett*		C.P.I. 13372*		Green panic†		Coloniao†		Hamil†		Mean	
	DDM	IDM	DDM	IDM	DDM	IDM	DDM	IDM	DDM	IDM	DDM	IDM	DDM	IDM
4	62.5	60.9	62.4	63.0	59.3	57.6	62.0	69.2	62.2	69.2	59.3	68.8	61.3	64.8
9	49.4	40.5	54.9	53.0	50.3	43.5	47.4	47.6	53.4	51.6	50.3	57.1	51.0	48.9
13	45.7	35.7	49.0	46.3	43.2	36.4	47.7	42.3	50.2	52.1	48.2	62.2	47.3	45.8
Mean	52.3	45.7	55.4	54.1	50.9	45.8	52.4	53.0	55.3	57.6	52.6	62.7	53.2	53.2

* P. coloratum varieties.
† P. maximum varieties.

digestibility. As herbage matures its digestibility declines, and when forage of lower digestibility is consumed by the animal it occupies a greater volume, remains within the rumen for a longer period of time and produces a larger quantity of indigestible residue to be passed down the hind tract as compared to forage of high digestibility. Thus, dry matter intake of low digestible forage is poor. It increases as digestibility increases, until the animal's requirements are satisfied. Above a digestibility level of 65–70 per cent, intake is not affected by digestibility. In fact, it may decrease with increasing digestibility since the needs of the animal are likely to be satisfied with lower intake (Conrad *et al.*, 1964; Raymond, 1969). In the absence of feed supplementation of concentrates, it is only on very rare occasions that animals have access to forages of very high digestibility in tropical environments.

The relationship between digestibility and intake is highly variable. Inherent differences among families, species and varieties significantly affect voluntary intake. It has long been recognized that animals consume considerably larger amounts of legumes than grasses when both have similar digestibility. Milford and Minson (1966) observed the intake of *Glycine wightii* was more than twice that of *Setaria anceps*, *Digitaria decumbens* and *Chloris gayana*, all having about 55 per cent greater intake of six legumes than that of eight equally digestible grasses. The difference in intake was due to a 17 per cent shorter retention time in the rumen required by the legumes, as well as a 14 per cent higher organic matter in the rumen digesta. A close relationship between the daily intake of digestible organic matter and retention time was observed. Van Soest (1965) observed a higher proportion of cell contents and a lower proportion of cell wall constituents in legumes than in grasses of the same level of digestibility.

Inter- and intraspecific differences in voluntary intake occur (Tables 13.6 and 13.7). These data are a part of those reported by Minson (1971, 1972). *Setaria splendida* had the highest digestibility but lowest voluntary intake. The decline in intake with decreasing digestibility in *Pennisetum clandestinum* and *Digitaria decumbens* was approximately one-third the rate of *S. splendida*, *Chloris gayana* and *Panicum maximum* for each unit decrease in dry matter digestibility. At a uniform dry matter digestibility level of 50 per cent the predicted intake values ranged from 33.0 for *S. splendida* to 50.4 for *D. decumbens*, and at 60 per cent digestibility the values ranged from 45.9 for *S. splendida* to 58.1 for *P. maximum*.

Intraspecific intake variation patterns in *Chloris gayana* (Milford and Minson, 1968) and in *Panicum maximum* and *P. coloratum* (Minson, 1972), represent the two extreme possibilities. In *C. gayana* the diverse varieties tested showed rather small and only marginal differences in digestibility and similar dry matter intake

values. This suggested that a variety of *C. gayana* with superio
feeding value will be difficult to obtain. On the other hand, result
in *Panicum* (Table 13.7) indicated considerable potential for selec
tion of types with superior feeding value. In this case, the averag
differences in digestibility between varieties were only about 3 pe
cent, but differences in voluntary intake were considerable
Panicum maximum cv. Hamil had a voluntary intake of 50 and 2
per cent greater than *P. coloratum* cv. Kabulabula when both ha
50 and 60 per cent digestibility, respectively. At 50 per cent digesti
bility, even among *P. maximum* varieties, cv. Hamil had 22 per cen
greater voluntary intake than cv. Green panic. Among *P. coloratum*
varieties, C.P.I. 13372 showed 23 per cent superiority over Kabula
bula. These and similar reports have encouraged researchers to in
tensify their search for superior-quality genotypes within alread
identified and widely grown species of forage plants.

Joint estimation of intake and digestibility

Digestibility and intake are the two basic components that gave
measure of herbage quality. Crampton *et al* (1960) proposed tha
they should be combined into one joint measure called nutritiv
value index (NVI) so as to evaluate more efficiently the feedin
value of a forage.

$$NVI = relative\ intake\ (RI) \times \frac{per\ cent\ forage\ energy\ conten\ digestibility}{100}$$

where

$$RI = \frac{actual\ intake/kg\ LW^{0.75}}{80} \times 100$$

(80 is a theoretical standard forage of dry matter intake 80 g
kg $LW^{0.75}$).

A similar measure is 'daily digestible energy intake' (DEI) ex
pressed as kcal/kg $LW^{0.75}$ and is the product of digestible energ
content × measured voluntary intake. Minson and Milford (1966
reported a close correlation ($r > 0.998$) between NVI and DEI.

Under most circumstances when pure stands of a pasture ar
evaluated or within-species comparisons are made for forage qualit
as NVI determined by DEI (digestible energy content) or dry matte
intake are equally reliable guides. DEI is considered to be more re
liable in evaluating forage mixtures and making comparisons amon
forage species (Javier, 1975). For purposes of uniformity, Minso
and Milford (1966) suggest that DEI be used with feeding value
of pastures.

Nutritive value and herbage quality of grazed pastures

Estimations of chemical composition, digestibility and intake of a forage based on cut samples can differ markedly from those based on the forage actually grazed by the animals. Data obtained from in-house studies are quite reliable for stored materials such as hay and silage, or for forages that are cut and fed to animals kept in confinement. Grazing, however, imposes several modifying influences on the basic components of forage nutritive value and quality because of selective eating habits of the animal, trampling and return of animal excreta.

When allowed to graze freely, animals tend to consume relatively larger proportions of the more nutritious fractions of the forage. Pastures, particularly grass–legume mixtures, are rarely uniform in their botanical composition. Browse plants are an important component under range conditions and provide a considerable proportion of the dietary requirements of the animals, particularly protein, during periods of herbage scarcity of the grazingland. Therefore, chemical analyses and digestibility estimates from cut samples are likely to differ considerably from those of the actually grazed forage. Forage intake under grazing conditions depends on the nutritive value of that forage and on its availability, as well as environmental conditions. Trampling and soiling of the forage due to urine and dung influence forage quality, and animals avoid eating the affected portions of the forage.

Methods have therefore been devised to give a better indication of the nutritive value and herbage quality of grazed forages. Some of these are discussed below. For detailed explanation of the procedures and for information on additional methods, the reader should consult Bulletins 45, 47 and 51, Commonwealth Bureau of Pastures and Field Crops, Hurley, Berkshire, England.

Fistulated animals

Animals with an oesophageal fistula can be used to obtain samples of grazed forage from pure swards, mixed pastures or grazinglands under range conditions. Such samples can then be utilized for analyses of chemical constituents and for digestibility estimations. Errors can arise because of sample contamination with the animal's saliva which is rich in P, Na and K, and also sample preparation for the analyses. Digestibility estimates are depressed by 3–4 per cent if saliva contamination is ignored (Minson *et al.*, 1976). Nevertheless, this method of sampling has received attention in the tropics. The following example illustrates the usefulness of the fistulated technique.

Dradu and Harrington (1972) studied the seasonal crude protein content of a tropical rangeland using oesophageal fistulated animals. Species of perennial grasses *Themeda, Brachiaria, Cymbopogon, Digitaria, Hyparrhenia, Chloris* and *Loudetia* were the main sward components, along with the browse plant *Acacia hockii.* The data showed wide variations in crude protein content of the samples and followed a bimodal pattern through the year, being in phase with liveweight changes of the steers and with rainfall pattern (Fig. 13.3). The researchers also noted the potential usefulness to estimate quickly the amount of protein supplement required to maintain animal growth.

Faecal index methods

These methods are used to estimate digestibility of grazed forages and do not require samples of forages eaten by the animals. Instead, digestibility is estimated from regressions derived with stall-fed animals relating indoor digestibility to the concentration of an in-

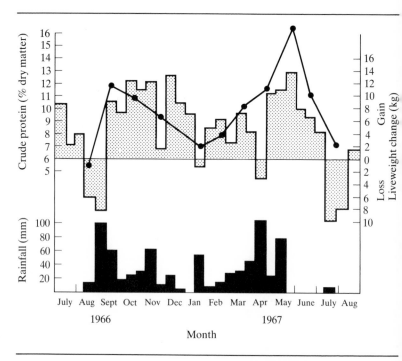

Fig. 13.3 Crude protein content of monthly fistula samples (graph), fortnightly mean liveweight changes of five steers on the same pasture (shaded histogram) and fortnightly rainfall (solid histogram) (Dradu and Harrington, 1972).

dicator in the faeces. Faecal contents of chromogen, N, fibre and lignin as well as Cu, Mg and silica have been reported to relate to forage digestibility (Raymond, 1969). Faecal nitrogen is one of the most commonly used indicators because of the ease with which it can be determined using standard laboratory procedures. Minson *et al.* (1976) emphasized the existence of a prediction error of about ± 3 digestible units in using regressions to relate digestibility to faecal composition. Also, they noted an 'application error', due to applying a regression derived from stall-fed animals to animals that graze selectively.

Measurement of forage intake by grazing animals

Clipping technique

A number of measured areas within the pasture available to the animal are harvested immediately prior to grazing and the weight of dry matter (or constituents) offered for grazing is determined. Immediately after grazing a similar number of measured areas are again harvested and appropriate determinations are similarly made. The difference between the two sets of data gives an estimate of the quantity of forage dry matter (or constituents) consumed by the grazing animals. Intake estimations by this method generally have quite large errors and individual animal intake can only be determined if the animals are grazed separately. Olubajo (1974) reported 14–25 per cent overestimation of daily dry matter intake from this technique, when compared to the nitrogen–chromic oxide faecal indicator method.

During the dry season the magnitude of overestimation was much greater (up to 100%, Olubajo, 1970). The method has proved useful with strip-grazing and when there are large differences between pregrazing and post-grazing yield estimates.

Indirect methods

Herbage intake of grazing animals can also be determined from estimates of faecal production and dry matter digestibility using the equation:

Daily herbage intake
$$= \text{Daily faecal output} \times \frac{100}{100 - \text{per cent dry matter digestibility}}$$

Daily faecal output can be measured accurately by the total collection of faeces voided using harnesses and faecal collection bags. The method is laborious, difficult under grazing conditions and interferes with the animal's grazing behaviour, particularly cattle. Most researchers prefer an indirect method for estimation of faecal production, one that uses non-toxic indigestible tracers that pass unchanged through the animal's digestive tract and can be completely recovered in the faeces.

The most commonly used faecal tracer is chromic oxide. It can be administered in powder form, capsules or preferably as paper strips impregnated with Cr_2O_3. Bulked samples of 40–50 defecations collected from the pasture are analysed for greater accuracy. Collection periods of about 5 days with four animals give a reasonably accurate estimation of faecal output. Dosing of animals with polystyrene particles helps in identifying the faeces of individual animals used in the study. Daily faecal output is estimated by the formula:

$$\text{Daily faecal output} = \frac{\text{grams marker } (C_2O_3) \text{ fed/day}}{\text{grams marker/g faeces}}$$

Though not perfect, the method is widely used and in almost all cases has proved to be far more reliable than the conventional clipping technique.

Animal productivity

The measurement of animal products to which a forage can be converted is the most significant measure of a forage, and all other methods are eventually referred to it as the ultimate standard of measurement (Harlan, 1956). For the purpose of assessment of herbage quality, the pasture is evaluated in terms of production per animal and under conditions where the amount of forage available is not limiting.

Measurement of animal production from forages requires extensive facilities in terms of land, livestock, fencing, water supply, field staff and considerable expertise in experimental design and statistical analyses of data. Basically the procedure involves establishing small paddocks of forages, placing test animals such as lactating cows, beef animals or sheep on the paddocks and measuring the relevant animal products. The following example of a simple grazing study shows the value of measuring animal product as an index of forage quality.

Aken'Ova (1975) produced a large number of *Pennisetum americanum* × *P. purpureum* F_1 hybrids and studied the forage potential of many of these hybrids in comparison with several promising elephant grass selections. Though the dry matter yields of the F_1 hybrids were less than that of elephant grass, they appeared to be better utilized by the grazing animals. Chemical composition and *in vitro* organic dry matter digestibility data were comparable, even though a few hybrids had superior digestibility when compared to the most promising elephant grass selections.

A grazing trial was conducted to measure beef production potential of elephant grass and F_1 *Pennisetum* hybrid pastures under a rotational grazing system. The 12 paddocks, alternately containing each of the two forages, were grazed during the later part of the wet season by equal numbers of White Fulani Zebu heifers of comparable weight and age. The results clearly indicated the superior herbage quality of the F_1 hybrid pastures (Fig. 13.4). In spite of 20 per cent lower total dry matter production, the F_1 hybrids produced 43 per cent more liveweight gains, largely due to better utilization of herbage on offer and increased dry matter intake. Animals on elephant grass pastures required 33 per cent more dry matter than those on F_1 *Pennisetum* pastures for each kilogram of liveweight gain.

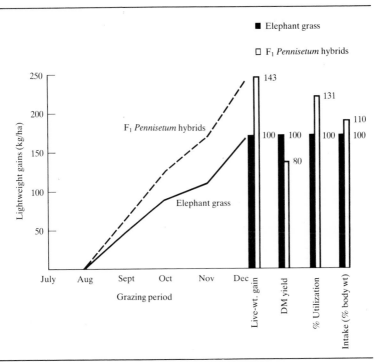

Fig. 13.4 Liveweight gains from rotationally grazed F_1 *Pennisetum* hybrid and elephant grass pastures. Vertical bars compare the F_1 hybrids and elephant grass performances on a percentage basis. University of Ibadan, Nigeria.

References

Ademosun, A. A. (1970) Nutritive evaluation of Nigerian forages. II. The effect of stage of maturity on the nutritive value of Stylosanthes gracilis, *Niger. Agric. J.* **7**, 164–73.

Ademosun, A. A. (1974) Utilization of poor quality roughages in the derived savanna zone. In J. K. Loosli, V. A. Oyenuga and G. M. Babatunde (eds), *Animal Production in the Tropics*, Heinemann (Nigeria): Ibadan, pp. 152–66.

Aken'Ova, M. E. (1975) Improvement of Pennisetum purpureum Schum. for forage in the low-altitude humid tropics, Ph. D. Thesis, Univ. Ibadan, Nigeria.

Beaty, E. R. and **J. L. Engel** (1980) Forage quality measurements and forage research. A review, critique and interpretation, *J. Range Mgmt* **33**, 49–54.

Blaxter, K. L., F. W. Wainmann and **R. S. Wilson** (1961) The regulation of food intake by sheep, *Animal Prod.* **3**, 51–62.

Brendon, R. M., K. W. Harker and **B. Marshall** (1963a) The nutritive value of grasses grown in Uganda when fed to Zebu cattle. I. The relation between the percentage crude protein and other nutrients, *J. Agric. Sci. (Camb.)* **61**, 101–4.

Brendon, R. M., K. W. Harker and **B. Marshall** (1963b) The nutritive value of grasses grown in Uganda when fed to Zebu cattle. II. The relation between crude fiber and nitrogen-free extract and other nutrients, *J. Agric. Sci. (Camb.)* **61**, 105–8.

Brolman, J. B. and **A. E. Kretschmer** (1972) Agronomic and morphological evaluation of some Stylosanthes introductions, *Abst., Amer. Soc. Agron. J*, Madison, Wisconsin, p. 42.

Bula, T. J., V. L. Lechtenberg and **D. A. Holt** (1977) Potential of temperate zone cultivated forages. In *Potential of the World's Forages for Ruminant Animal Production*, Winrock Intl Livestock Res. Trg Cent., Arkansas, pp. 7–28.

Burton, G. W. and **W. G. Monson** (1972) Inheritance of dry matter digestibility in bermudagrass (Cynodon dactylon (L.) Pers.), *Crop Sci.* **12**, 375–8.

Burton, G. W., R. H. Hart and **R. S. Lowrey** (1967) Improving forage quality in bermudagrass by breeding, *Crop Sci.* **7**, 329–32.

Burton, G. W., J. C. Millot and **W. G. Monson** (1973) Breeding procedures for *Panicum maximum* Jacq. suggested by plant variability and mode of reproduction, *Crop Sci.* **13**, 717–20.

Butterworth, M. H. (1964) The digestible energy content of some tropical forages, *J. Agric. Sci. (Camb.)* **63**, 319–22.

Chheda, H. R. (1974) Forage crops research at Ibadan. I. Cynodon spp. In J. K. Loosli, V. A. Oyenuga and G. M. Babatunde (eds) *Animal Production in Tropics*, Heinemann (Nigeria): Ibadan, pp. 79–95.

Cohen, R. D. H. (1980) Phosphorus in rangeland and ruminant nutrition, *Livestock Prod. Sci.* **7**, 25–37.

Conrad, H. R., A. D. Pratt and **J. W. Hibbs** (1964) Regulation of feed intake in dairy cows. I. Change in importance of physical and physiological factors with increasing digestibility, *J. Dairy Sci.* **47**, 54–62.

Cordoba, F. J., J. D. Wallace and **R. D. Pieper** (1978) Forage intake by grazing livestock: A review, *J. Range Mgmt* **31**, 430–8.

Coward-Lord, J., J. A. Arroyo-Aguilú and **O. Garcia-Molinari** (1974) Proximate nutrient composition of ten tropical forage grasses, *J. Agric. Univ. Puerto Rico* **58**, 305–11.

Crampton, E. W. (1959) Interrelations between digestible nutrient and energy content, voluntary dry matter intake, and the overall feeding value of forages. In H. B. Sprague (ed.), *Grasslands*, AAAS. Publ. 13: Washington, pp. 205–12.

Crampton, E. W., E. Donefer and **L. E. Lloyd** (1960) Nutritive value index for forages, *J. Anim. Sci.* **19**, 538–52.

Crampton, E. W. and **L. A. Maynard** (1938) The relation of cellulose and lignin content to the nutritive value of animal feeds, *J. Nutrition* **15**, 508–12.

Crowder, L. V. and **H. R. Chheda** (1977) Forage and fodder crops. In C. L. A. Leakey and J. B. Wills (eds), *Food Crops of the Lowland Tropics*, Oxford Univ. Press: London, pp. 127–59.

Denium, B. (1966) Influence of some climatological factors on the chemical composition and feeding value of herbage, *Proc. 10th Intl. Grassld Cong.*, pp. 415–18.

Denium, B. and **J. P. G. Dirven** (1975) Climate, nitrogen and grass. 6. Comparison of yield and chemical composition of some temperate and tropical grass species grown at different temperatures, *Neth. J. Agric. Sci.* **23**, 69–82.

D'Hoore, J. and **J. K. Coulter** (1972) Silicon and plant nutrition. In *Soils of the Humid Tropics*, Nat. Acad. Sci.: Washington D.C., pp. 163–73.

Donefer, E., E. W. Crampton, and **L. E. Lloyd** (1963) The prediction of digestible energy intake potential (NVI) of forages using a simple in vitro technique, *Proc. 10th Intl. Grassld Cong.*, pp. 442–5.

Dougall, H. W. (1960) Average nutritive value of Kenya feeding stuffs for ruminants, *E. Afr. Agric. J.* **26**, 119–28.

Dougall, H. W. and **A. V. Bogdan** (1958) The chemical composition of the grasses of Kenya, *E. Afr. Agric. J.* **24**, 17–23.

Dradu, E. A. A. and **G. N. Harrington** (1972) Seasonal crude protein content of samples obtained from a tropical range pasture using oesophageal fistulated steers, *Trop. Agric. (Trin.)* **49**, 15–21.

Duckworth, J. (1946) A statistical comparison of the influence of crude fibre on the digestibility of roughage by Bos indicus (Zebu) and Bos tarus cattle, *Trop. Agric. (Trin.)* **23**, 4–9.

Eng, P. K., L. 't Mannetje and **C. P. Chen** (1978) Effects of phosphorus and stocking rate on pasture and animal production on a guinea grass-legume pasture in Johore, Malaysia. 2. Animal liveweight change, *Trop. Grassld* **12**, 198–207.

Ely, R. E. and **L. A. Moore** (1955) Hollocellulose and the summatic analysis of forages, *J. Anim. Sci.* **14**, 718–24.

French, M. H. (1957) Nutritional value of tropical grasses and fodders, *Herb. Abst.* **27**, 1–9.

Fonnesbeck, P. V. and **L. E. Harris** (1970) Determination of plant cell-walls in feeds, *Proc. Amer. Soc. Anim. Sci., West. Sec.* **21**, 153–5.

Garcia, R. and **F. Ferrer** (1974) Relative digestibility (in vitro) of some tropical grasses and legumes, *Proc. 12th Intl. Grassld Cong.*, pp. 187–93.

Garcia-Rivera, J. and **M. P. Morris** (1955) Oxalate content of tropical forage grasses, *Science* **122**, 1089–90.

Gihad, E. A. (1976) Studies on the nutritional value of pasture species in Zambia, *E. Afr. Agric. For. J.* **41**, 335–9.

Glover, J. and **H. W. Dougall** (1960) The apparent digestibility of the non-nitrogenous components of ruminant feeds, *J. Agric. Sci. (Camb.)* **55**, 391–4.

Gomide, J. A., C. H. Noller, G. O. Mott, J. H. Conrad and **D. L. Hill** (1969) Effect of plant age and nitrogen fertilization on the chemical composition and in vitro cellulose digestibility of tropical grasses. Mineral composition of six tropical grasses as influenced by plant age and nitrogen fertilization, *Agron. J.* **61**, 116–28.

Goto, I. and **D. J. Minson** (1977) Prediction of the dry matter digestibility of tropical grasses using a pepsin-cellulose assay, *Anim. Feed. Sci. Tech.* **2**, 247–53.

Hacker, J. B. and **D. J. Minson** (1972) Varietal differences in in vitro dry matter digestibility in Setaria and the effects of site and season, *Aust. J. Agric. Res.* **23**, 959–67.

Haggar, R. J. and **M. B. Ahmed** (1970) Seasonal production of Andropogon gayanus. II. Seasonal changes in digestibility and feed intake, *J. Agric. Sci. (Camb.)* **75**, 369–73.

Hamilton, R. I., L. E. Donaldson and **L. J. Lambourne** (1971) Leucaena leucocephala as a feed for dairy cows: direct effect on reproduction and residual effect on the calf and lactation, *Aust. J. Agric. Res.* **22**, 681–92.

Harker, K. W. and **A. K. Kamau** (1961) A preliminary survey of the nitrate content of various grasses, *E. Afr. Agric. J.* **27**, 57–60.

Harlan, J. R. (1956) *Theory and Dynamics of Grassland Agriculture,* Van Nostrand: Princeton, New Jersey, Ch. 10.

Harris, L. E. (1970) *Nutrition Research Techniques for Domestic and Wild Animals*, vol. I, Int. Agric. Services, Logan: Utah.

Hegarty, M. P., P. G. Schinkel and **R. D. Court,** (1964) Reaction of sheep to the consumption of Leucaena glauca Benth. and its toxic principle mimosine, *Aust. J. Agric. Res.* **15**, 153–67.

Hutton, E. M. (1970) Tropical pastures, *Adv. Agron,* **22**, 1–73.

Ingalls, L. R., J. W. Thomas, E. J. Benne and **M. Tesar** (1965) Comparative response of wether lambs to several cuttings of alfalfa, birdsfoot trefoil, bromegrass and reed canarygrass, *J. Anim. Sci.* **24**, 1159–64.

Javier, E. Q. (1975) *Breeding for Quality Forage,* ASPAC Food Fert. Techn. Cent., Extn. Bull. 50, Taiwan, pp. 1–29.

Jarrije, P., P. Thivaid and **C. Demarquily** (1970) Development of cellulalytic enzyme digestion for predicting the nutritive value of forages, *Proc. 11th Intl. Grassld Cong.,* p. 762.

Johnson, W. L., W. A. Hardison, A. L. Ordaveza and **L. S. Castillo** (1968) Factors affecting voluntary intake by cattle and water buffaloes, *J. Agric. Sci. (Camb.)* **71**, 67–71.

Jones, R. J. and **C. W. Ford** (1972) The soluble oxalate content of some tropical pasture grasses grown in south-east Queensland, *Trop. Grassld* **6**, 201–3.

Jones, D. I. H. and **M. V. Hayward** (1973) A cellulase digestion technique for predicting the dry matter digestibility of grasses, *J. Sci. Food Agric.* **24**, 1419–26.

Jones, D. I. H. and **M. V. Hayward** (1975) The effect of pepsin pretreatment of herbage on the prediction of dry matter digestibility from solubility in fungal solutions, *J. Sci. Food Agric.* **26**, 711–18.

Jones, R. J., C. G. Blount and **J. H. G. Holmes** (1976) Enlarged thyroid glands in cattle grazing Leucaena pastures, *Trop. Grassld* **10**, 113–16.

Jones, R. J., M. M. Ludlow, J. H. Troughton and **C. G. Blount** (1979) Estimation of the proportion of C_3 and C_4 plant species in the diet of animals from the ratio of 12C and 13C isotopes in the faeces, *J. Agric. Sci. (Camb.)* **92**, 97–100.

Klock, M. A., S. C. Schank and **J. E. Moore** (1975) Laboratory evaluation of quality in subtropical grasses. III. Genetic variation among Digitaria species in in vitro digestion and its relationship to plant morphology, *Agron. J.* **67**, 672–5.

Little, D. A., R. W. McLean and **W. H. Winter** (1977) Prediction of the phosphorus content of herbage consumed by grazing cattle, *J. Agric. Sci. (Camb.)* **88**, 533–8.

Milford, R. and **D. J. Minson** (1966) The feeding value of tropical pastures. In W. Davies and C. L. Skidmore (eds), *Tropical Pastures,* Faber and Faber: London, Ch. 7.

Milford, R. and **D. J. Minson** (1968) The digestibility and intake of six varieties of Rhodes grass (Chloris gayana), *Aust. J. Exptl Agric. Anim. Husb.* **8**, 413–18.

Miller, I. L. and **S. J. Cowlishaw** (1976) Effects of stage of growth and season on the nutritive value of four digit grasses in Trinidad, *Trop. Agric. (Trin.)* **53**, 305–20.

Miller, T. B. and **A. Blair-Rains** (1963) The nutritive value and agronomic aspects of some fodders in Northern Nigeria. 1. Fresh herbage, *J. Brit. Grassld Soc.* **18**, 158–67.

Minson, D. J. (1971) The digestibility and voluntary intake of six varieties of Panicum, *Aust. J. Exptl Agric. Anim. Husb.* **11**, 18–25.

Minson, D. J. (1972) The digestibility and voluntary intake by sheep of six tropical grasses, *Aust. J. Exptl Agric. Anim. Husb.* **12**, 21–7.

Minson, D. J. and **A. Laredo** (1972) Influence of leafiness on voluntary intake of tropical grasses by sheep, *J. Aust. Inst. Agric. Sci.* **38**, 303–5.

Minson, D. J. and **M. N. McLeod** (1970) The digestibility of temperate and tropical grasses, *Proc. 11th Intl. Grassld Cong.,* pp. 719–32.

Minson, D. J. and **R. Milford** (1966) The energy values and nutritive value indices of Digitaria decumbens, Sorghum almum and Phaseolus atropurpureus, *Aust. J. Agric. Res.* **17**, 411–23.

Minson, D. J., T. H. Stobbs, M. P. Hegarty and **M. J. Playne** (1976) Measuring the nutritive value of pasture plants. In N. H. Shaw and W. W. Bryan (eds), *Tropical*

Pasture Research Principles and Methods, Commonw. Bur. Past. Fld Crops, Bull. 51, Hurley, England, Ch. 13.

Mohamed Saleem, M. A. (1972) Productivity and chemical composition of Cynodon IB.8 as influenced by level of fertilization, soil pH and height of cutting, Ph. D. Thesis, Univ. Ibadan, Nigeria.

Ogwang, B. H. and **J. S. Mugerwa** (1976) Yield response to nitrogen application and in vitro dry matter digestibility of elephant grass and bulrush millet hybrids, *E. Afr. Agric. For. J.* **41**, 231–42.

Olubajo, F. O. (1970) Use of chromic oxide in indoor digestion experiments with tropical grass–legume mixtures using White Fulani steers, *Niger. Agric. J.* **7**, 105–10.

Olubajo, F. O. (1974) Pasture research at the University of Ibadan. In J. K. Loosli, V. A. Oyenuga and G. M. Babatunde (eds), *Animal Production in the Tropics*, Heinemann (Nigeria): Ibadan, pp. 67–8.

Olubajo, F. O. and **V. A. Oyenuga** (1970) Digestibility of tropical mixtures using the indicator technique, *J. Agric. Sci. (Camb.)* **75**, 175–81.

Oyenuga, V. A. (1960) Effect of stage of growth and frequency of cutting on the yield and chemical composition of some Nigerian fodder grasses – Panicum maximum Jacq, *J. Agric. Sci. (Camb.)* **55**, 339–50.

Oyenuga, V. A. (1968) *Animal Production in Africa Meeting Nutrient Requirements of Range Cattle for Optimum Yield*, Agric. Res. Priorities Econ. Dev. Afr., The Abidjan Confr., vol. III, pp. 32–40.

Ozanne, P. G., M. D. Purser, E. M. W. Howes, and **I. Southay** (1976) Influence of phosphorus content on feed intake and weight gain in sheep, *Aust. J. Exptl Agric. Anim. Husb.* **16**, 353–60.

Peña, M. and **O. Paladines** (1979) Digestibilidad de la materia seca de forrajes tropicales usando el metodo de solubilidad en pepsina-cellulasa, *Turrialba* **29**, 189–94.

Phillips, T. G., J. T. Sullivan and **M. E. Laughlin** (1954) Chemical composition of some forage grasses. I. Changes with maturity, *Agron. J.* **46**, 361–9.

Poppi, D. P., D. J. Minson, and **J. H. Termouth** (1980) Studies of cattle and sheep eating leaf and stem fractions of grass, *Aust. J. Agric. Res.* **32**, 99–137.

Raymond, W. F. (1969) The nutritive value of forage crops, *Adv. Agron.* **21**, 1–108.

Rees, M. C. and **D. J. Minson** (1976) The influence of supplements on the voluntary intake and digestibility of feed – an alternative in fertilizers? In G. J. Blair (ed.), *Review in Rural Sci. 3. The Efficiency of Phosphorus Utilization*, Univ. New England: Armidale, pp. 165–9.

Rees, M. C. and **D. J. Minson** (1979) The validity of the in vitro technique using rumen fluid or cellulase for predicting change in the dry matter digestibility of grasses caused by fertilizer calcium, sulphur, phosphorus and nitrogen. *Grass and Forage Sci.* **34**, 19–25.

Reid, R. L. and **G. A. Jung** (1974) Effect of elements other than nitrogen on nutritive value of forage. In D. A. Mays (ed.), *Forage Fertilization*, ASA, CSA & SSA: Wisconsin, pp. 66–100.

Reid, R. L., G. A. Jung and **S. J. Murray** (1966) Nitrogen fertilization in relation to the palatability and nutritive value of orchard grass, *J. Anim. Sci.* **25**, 636–45.

Reid, R. L., A. J. Post, F. J. Olsen and **J. S. Mugerwa** (1973) Studies on the nutritional quality of grasses and legumes in Uganda. I. Application of in vitro digestibility techniques to species and stage of growth effects, *Trop. Agric. (Trin.)* **50**, 1–15.

Reis, P. J., D. A. Tunks and **R. E. Chapman** (1975) Effect of mimosine, a potential chemical defleecing agent, on wool growth and the skin of sheep, *Aust. J. Biol. Sci.* **28**, 69–84.

Rogers, H. H. and **E. T. Whitmore** (1966) A modified method of the in vitro determination of herbage digestibility in plant breeding studies, *J. Brit. Grassld Soc.* **21**, 150–2.

Stobbs, T. H. (1973) The effects of plant structure on the intake of tropical pastures

II. Differences in sward structure, nutritive value, and bite size of animals grazing Setaria anceps and Chloris gayana at various stages of growth, *Aust. J. Agric. Res.* **24**, 821–9.

Strickland, R. W. (1974) Performance of Southern African Digitaria spp. in southern Queensland, *Aust. J. Exptl Agric. Anim. Husb.* **14**, 186–96.

Terry, R. A., D. C. Murdett, and D. F. Osbourn (1978) Comparison of two in vitro procedures using rumen-fluid pepsin and pepsin cellulase for prediction of forage digestibility, *J. Brit. Grassld Soc.* **33**, 13–8.

Thomas, C. T., P. K. Njorage, and J. S. Felon (1980) Prediction of digestibility in three tropical grasses, *Trop. Agric. (Trin.)*, **51**, 75–81.

Thornton, R. F. and D. J. Minson (1973a) The relationship between apparent retention time in the rumen, voluntary intake and apparent digestibility of legume and grass diets in sheep, *Aust. J. Agric. Res.* **24**, 889–98.

Thornton, R. F. and D. J. Minson (1973b) Effects of soils, fertilizers and stocking rates on pasture and beef production on the Wallum of southeast Queensland, *Aust. J. Exptl Agric. Anim. Husb.* **13**, 537–43.

Tilley, T. M. A. and R. A. Terry (1963) A two-stage technique for the in vitro digestion of forage crops, *J. Brit. Grassld Soc.* **18**, 104–11.

Underwood, E. J. (1966) *The Mineral Nutrition of Livestock*, FAO/CAB: Aberdeen.

Van Soest, P. J. (1965) Use of detergents in analysis of fibrous feeds, III. Studies of effects of heating on yield of fiber and lignin in forages, *JAOAC* **48**, 785–90.

Van Soest, P. J. (1966) Non-nutritive residues. A system analysis for the replacement of crude fiber, *JAOAC* **49**, 546–62.

Van Soest, P. J. (1967) Development of a comprehensive system of feed analysis and its application to forages, *J. Animal Sci.* **26**, 119–28.

Van Soest, P. J. and L. H. P. Jones (1968) Effect of silica in forages upon digestibility, *J. Dairy Sci.* **51**, 1644–8.

Vicente-Chandler, J. (1974) Fertilization of humid tropical grasslands. In M. A. Mays (ed.), *Forage Fertilization*, ASA, CSA and SSA: Washington, D.C., pp. 277–300.

Vicente-Chandler, J., S. Silva and J. Figarella (1959) The effect of nitrogen and frequency of cutting on the yield and chemical composition of three tropical grasses, *Agron. J.* **51**, 202–6.

Wilkinson, S. R., R. N. Dawson and W. E. Adams (1969) Effect of sample drying procedure on chemical composition and in vitro digestibility of Coastal bermudagrass, *Agron. J.* **61**, 457–63.

Wilkinson, S. R., W. E. Adams and W. A. Jackson (1970) Chemical composition and in vitro digestibility of vertical layers of Coastal bermudagrass (Cynodon dactylon L.), *Agron. J.* **62**, 39–43.

Wilson, J. R. and L. 't Mannetje (1978) Senescence, digestibility and carbohydrate content of buffel grass and green panic leaves in swards, *Aust. J. Agric. Res.* **29**, 503–16.

Chapter 14

Measurement of pasture production

Pasture measurements involve two biological systems – the plant and the animal. Aspects of the systems that interest the animal producer and the pasture investigator include the quantity and continuous supply of forage, the amount of herbage consumed by the grazing animal, nutritive value and digestibility of the herbage and animal performance. The quantity of herbage produced by a pasture or grazingland can be measured by weight, visual estimate and animal output. The most reliable measurement of quality is that of carrying out a well-designed grazing trial, but the relative nutritive value can be determined by *in vitro* digestibility studies and chemical analyses if the data are properly interpreted.

The pasture *per se* can be described in terms of forage yield, botanical composition (proportion of species), proportion of plant parts (leaves, stems, reproductive organs), chemical composition of the total sward or different plant parts, and digestibility of the total herbage mass or component parts. A description of the pasture should also include the flora, i.e. the floristic units present, and the vegetation or the numerical presence and distribution of the floristic units. Pasture studies rarely encompass a full description and seldom depict the changing nature of the pasture complex. Instead, one or more pasture attributes will be evaluated and described. There are a number of methods available for measuring each attribute and no single one can be designated as being the most appropriate under all conditions. Often the deciding factor as to the method selected will be available budget, facilities and labour.

Sampling herbage

Sampling is essential in determining botanical composition and herbage yield of a pasture or grazingland. The sample is thought of as being the small quantity of herbage used for hand separation of species, dry matter determination and chemical or *in vitro* analyses. It has a broader connotation, however, and refers to the units selected in making measurements. These include the plot, quadrats and cage areas, length of transect, selection of components comprising sample, amount of herbage collected, number and frequency of collections, degree of grinding for chemical and *in vitro* analysis, etc.

Sampling units

The units used for sampling herbage in pasture studies consist of areas, lines or points.

Areas

The area to be sampled may be oblong, square or circular. In theory an oblong unit has a distinct advantage. Since plant associations occur in circular fashion, a long and narrow strip will traverse a greater number of plants and species in a mixed population (Davies, 1931). In addition, a more diverse soil gradient is covered within an oblong area than a square or circle.

An area may be delineated by a plot that is either temporary or permanent. The plot may be the length of swath cut by a mower and is moved at each harvest. It may be permanently located and identified by fixed corners set at ground level. Cages, either movable or fixed, are also plots.

The sampling area is frequently determined by a quadrat of square, oblong or circular shape. These units vary in size, e.g. 12 in^2, 1.0 yd^2 or 1.0 m^2 and are constructed of metal or wood. Some are flexible for folding. In some instances the quadrat is subdivided in a grid fashion to facilitate observations. Rings or hoops of varying dimensions are used for circular sampling. Plots are sometimes referred to as quadrats, but more commonly the latter is used for small movable units.

Lines

The line may also be called a transect. It can be a wire stretched for a given distance over the sward or a visual sighting from one position in the pasture to another position. The line functions as a sampling unit *per se* or marks the course along which sampling units are located.

Points

Herbage may be sampled by point contact and each provides a measurement. A number of systems have been devised to make point recordings, e.g. knots on a string, plumb bob suspended from a line, probes positioned along a line and supported by a frame, a spot selected at the toe of a shoe.

The type of sampling unit employed will depend on the kind of information desired, the type of pasture or grazingland being studied, materials and facilities available for collecting and handling information or samples, and to some extent personal preference of the investigator. Size and shape of the sampling unit for weight determination will depend on the type of cutting implement utilized and the purpose for which the sample is required. Hand shears or hedge-trimmers are appropriate for areas less than 1.0 m². Motorized mowers are best suited for long and narrow plots.

Sampling errors

Two types of errors are common to sampling, namely random sampling error and bias. Random sampling error is inherent because samples are estimates of the population. Bias arises from the human aspect of personal preference, either conscious or unconscious.

Random sampling error

The positioning of plots, quadrats and cages should be done by strict randomization so that all parts of the population have an equal chance of being included in the sample. This necessitates the use of random numbers taken from a table or some other objective method such as pulling numbered cards from a box. Even then sampling error is present, because the herbage samples provide only an average of the measurements recorded. They do not represent the true mean because the entire population is not measured.

There are several ways to reduce random sampling error. This begins with planning the study when attention is given to the accuracy desired in the experiment. At this time an appropriate design is selected, number of replications determined and treatments arranged at random. Other means of reducing random sampling error include the following (Brown, 1954):
1. Increased number of units sampled. More accurate results are obtained from a greater number of small samples than from fewer large ones.
2. Increased size of sample so as to include a greater proportion of the population.
3. Stratification of the sample by dividing the pasture into sections and sampling within these divisions. The strata may represent areas of recognized moisture difference, contours, gradients, etc.
4. Use of supplementary information. For example, it may be known that one portion of the pasture received a different fertilizer treatment that influences yield. This area can be sampled separately and a correction factor calculated to adjust yield estimates.

Source of bias

Bias usually occurs because of inappropriate methodology. For example, a common practice in taking measurements with quadrats is to toss the sampling unit (a metal or wooden frame) at random on-to the sward. Often this is done over the shoulder so as not to bias the direction of toss. This is not a random method of demarcating the sampling area. A system to assure random selection of their location is to begin at a pre-chosen site, perhaps a corner of the pasture, and to move a predetermined distance to position the first sampling unit. The next would then be placed at a given distance and in a given direction, both selected at random. More appropriately, fence posts can be numbered and randomly selected as the point of reference from which to measure randomly chosen distances for positioning the sampling units.

The investigator is sometimes inclined to select a so-called representative area for locating quadrats, transects or cages. Again, this represents a bias. Another source of bias is that of cutting herbage samples in a pasture species trial at the same level above the soil. Upright growing and trailing types are likely to respond differently to such treatment. Estimates of species composition is sometimes made by the investigator and enumerators instead of sampling by weight. Individuals tend to favour certain species in their evaluation and thus introduce unconscious bias. Boundaries of quadrats may not be well defined under some conditions and can affect the quantity of herbage cut or estimated. Some persons are more meticulous than others in collecting and handling harvested material. These are only a few examples of how bias can modify pasture measurements. The only way to reduce or remove bias is through improvement of sampling method.

Yield measurements

The most important aspect of pasture herbage is its value as animal feed. Pasture productivity can be quantitatively measured by weight of herbaceous material. Knowledge of forage yield, however, is important only as related to an animal product. It can be used to calculate stocking rate and to estimate the amount of herbage consumed by grazing animals. The methods utilized for calculating herbage yields can be grouped into categories as follows (Brown, 1954; Reppert *et al.*, 1962):

1. Agronomic methods in which the yield of herbage is determined by (a) cutting and weighing samples from known areas and at appropriate intervals of time, (b) weight estimates based on visual observations and (c) indirect determinations using relationships between selected plant factors and production.
2. Animal production methods that measure (a) animal production and (b) the quantity of herbage consumed and digested by the grazing animal.

Weight from clipped plots

Harvesting herbage from an area of known dimension has been the standard approach for calculating the amount of forage available for grazing or consumed by grazing animals. The method of harvest must be integrated into the system of grazing to arrive at a valid measurement. Regardless of method, the procedure is that of cutting the herbage by hand or mechanically, collecting clippings, weighing fresh material from large plots, drying a sample of known fresh weight (sometimes the entire amount of material from a quadrat is dried without recording fresh weight) and accurately weighing the dried sample. The dry weight of samples from a given treatment are then used to calculate the production of dry matter per unit area (hectare, acre or other unit of measure). Even though the procedure is relatively simple and straightforward, certain precautions should be taken to ensure accuracy (Reppert *et al.*, 1962). The following points should be considered:

1. Positioning of sampling units in random fashion.
2. Well-defined plot or quadrat boundaries and accurate harvesting within the delineated area.
3. Clear understanding of whether the sampling represents total yield or separation of species into component parts.
4. Selection and adherence to a predetermined height of cutting above ground level.
5. Careful identification of sample and accuracy in weight records.
6. Appropriate intervals between harvests.

The cage method for determining herbage yield and consumption

Agronomic measurement of herbage yield on permanently grazed pastures is usually accomplished by harvesting areas protected by cages (Brown, 1954; Carter, 1962; Klingman *et al.*, 1943; Linehan *et al.*, 1952; Rodriguez *et al.*, 1979). Cages are randomly distributed throughout the pasture and the herbage within the cage clipped from time to time during the year. A comparable nearby area is harvested outside the cage at the same time. This allows the determination of herbage consumed by the so-called 'difference' or 'cage-difference' method. It is assumed that consumption equals the herbage yield from the caged area minus the yield of the outside area. The cage is moved to another area after clipping to avoid the modified environmental effects within the cage. Wind velocity and transpiration are lower and humidity higher in the caged area than outside. This may lead to an increased yield of 10 per cent or more (Cowlishaw, 1951). Furthermore, soil compaction and trampling of grazing animals may reduce herbage yields as compared to the ungrazed area. A system of calculating herbage yields that accounted for plant growth during the grazing period was proposed by Linehan *et al.* (1952). They noted that yield as determined by the cage method tends to overestimate forage consumption of the grazing animal. A system was proposed that made allowance for differences in growth of pasture plants in protected and unprotected areas. The following formula was employed to make adjustment for growth:

$$\text{Amount of herbage consumed} = (c - f) \frac{(\log d - \log f)}{(\log c - \log f)}$$

where c is the quantity of herbage present at the beginning of grazing; d – the quantity within the caged area at the end of grazing and f – the uneaten quantity outside the caged area at the end of grazing. This formula was designed for long rotational grazing periods in the temperate zone, but would have application to seasonal grazing or sampling periods in the tropics.

The total herbage produced per hectare could be calculated by knowing the amount on offer at the beginning of the grazing period, that consumed by grazing animals and that uneaten at the end of the grazing period.

The size of cages has varied with different trials, and includes dimensions of 0.91×0.91 m, 1.2×1.2 and 1.21×2.73 m with a height of 0.45 or 0.60 m. They are usually constructed of heavy wire or small metal rods. Considerable difficulty is encountered in maintaining the position of the cage, even though pegged down with metal or wooden stakes. Grazing animals frequently push against them in an effort to obtain the lush herbage inside. A more stable cage 1.21×2.73 m is used at the Grassland Research Institute at Hurley (Institute Staff, 1961). The framework consists of metal tubing covered with mesh chain-link netting. It is constructed as an ellipse about 0.65 cm at the centre.

Information about the number of cages to use per hectare is limited. A study in North Carolina showed that 12–13 per hectare of 1.2×1.2 m dimension were insufficient to measure total herbage yields or botanical composition accurately and more than 24 would be needed to obtain reliable results (Joint Comm. Rept., 1952). In studying temperate zone pastures, Davis and Bell (1958) used 25 cages/ha of 1.2×1.2 m in continuously grazed pastures and Naylor (cited by Carter, 1962) recommended 127 and 32 cages/ha of the same size to obtain coefficients of variation of 10 and 20 per cent, respectively. These large numbers involve considerably more expense and labour than most investigators wish to invest and usually they use from 12 to 15 cages/ha.

Mower strip method

Two schemes of clipping have been utilized in estimating herbage yield by this technique. Wagner *et al.* (1950) in the USA compared herbage yields from cages and mowed strips, with calculated yields based on nutrient needs of dairy cows for maintenance, milk production and milk gains. Herbage yields were obtained by the difference method using six cages (1.2 × 1.2 m in size) per pasture of about 1.0 ha and six replications. The mower strip method consisted of cutting four swathes 0.91 × 4.5 m with a sickle-bar mower in the same pastures. Strips were mowed just before allowing cattle into the rotationally grazed paddocks. Fresh material was collected and sampled for dry weight determinations. Each pasture was grazed for 10–14 days and rested for 3–4 weeks. Herbage yields obtained from the cage and mower strip methods exceeded that of the yield calculated from nutrient needs by 24 and 6.0 per cent, respectively. The herbage residue after grazing was not considered in the mower strip technique. Advantages cited for this method included (a) samples taken more rapidly, (b) less labour and capital required, (c) more representative sample due to the size and shape of sampling and (d) danger of losing a sample from animals moving a cage eliminated. It must be recognized, of course, that the mower strip method cannot be used on continuously grazed pastures. It may also have limitations on unproductive and weedy pastures.

Another way of sampling in the mower strip method is to make cuttings before and after the rotational grazing period. In theory this approach would more accurately estimate the quantity of herbage consumed by grazing animals.

Leguminous shrubs and trees, browse plants

Production of edible material for browse is based on current leaf and twig growth either available to grazing animals or cut and fed to them. Arriving at an estimate of yield is difficult. First of all, it is not meaningful to sample the entire plant due to inedible woody portions. Then, in harvesting leaves and twigs a judgement must be made as to the proportion that animals would select. A guide for sampling *Leucaena leucocephala* is to collect all leaves plus stems less than 6 mm in diameter (Shaw *et al.*, 1976). A decision must also be made regarding the choice of sampling unit, whether by land area or single plants. The latter is usually preferred as the smallest sampling unit.

Visual estimates of weight

Measuring herbage yield by cutting, weighing and drying is time-consuming and costly. Furthermore, each measurement represents a single sample in a variable population. Even though great accuracy can be obtained, the number of samples are limited. Several methods have been described to estimate the quantity of herbage quickly and with a high degree of accuracy.

A weight-estimate method for determining herbage yield was described by Pechanec and Pickford (1937) under range conditions of the western USA. The technique involved making estimates of forage weight within a delineated area, checking the estimate by harvesting and weighing the sample and using covariance analysis of the data to test the degree of association between estimated and harvested yields. They concluded that weight estimates were related to yield with reasonable accuracy, rapidly determined, readily checked by actual weights, suitable for use with replication and easily learned with minimum instruction. The need for training and double sampling, i.e. checking visual estimates by actual weights, were stressed by Wilm *et al.* (1944) in the USA.

Morley *et al.* (1964), working with large grazing experiments in Australia, pointed out that visual estimates can be transformed to actual yields by use of a calibration procedure. Linear regression of actual yield values (obtained from harvested samples) on the estimated values were fitted for calibration of each observer. The visual estimation technique was evaluated by Campbell and Arnold (1973) in Western Australia. They found a significant (linear) relation between visual estimates and actual yields over a range of pastures examined. The training technique of Pechanec and Pickford (1937) was simply that of estimating yield within a quadrat of given size then checking the fresh weight with a small spring-type scale sensitive to 10 g. Campbell and Arnold (1973) located quadrats in different paddocks to cover a range of yields grouped into five classes of low-to-high herbage yields. Their results indicated that novice observers with limited training in estimating and checking weights could predict yields with a high degree of accuracy and minimal bias. They concluded that:

(a) Trained observers are needed to give consistently accurate predictions.
(b) Observers need a prescribed period of training to recognize their biases, especially with tall, thin stands (usually overestimated as opposed to short, dense stands that were underestimated).
(c) Observers need to be aware of the range of yields before making official estimates.
(d) A number of estimates over varying conditions must be made and calibrated against actual weights.
(e) Greatest precision is obtained when observers estimate the whole of each paddock.
(f) The estimates can be entered directly into computer data sheets for immediate processing of predicted yields.

The present authors can substantiate these conclusions having been associated with observers who could estimate yields within ±100 kg/ha of the actual mean dry weight.

A comparative yield method for estimating dry matter production was described by Haydock and Shaw (1975) using tropical pastures in southeast Queensland. A set of reference quadrats were pre-selected to provide a scale for making relative weight estimates rather than absolute weight estimates. Five or nine reference quadrats were located to encompass a range of sward densities. For example, to construct a five-point scale two quadrats were placed on low (1) and high (5) yielding areas. Experienced observers selected these standards. Then novice observers indicated areas of the intermediate standards. Their final selection was made after consultation with experienced supervisors. The quadrats were left in position at each standard and protected by cages in grazed pastures. They served as references for observers to return and refresh their visual imagery. Observers moved through the pasture with a quadrat and equipment to check visual estimates in the training period. Calibrations of each observer were made using regression coefficients and correlations of estimated and actual yield values. These investigators found that more than one calibration line was needed where species of different growth habits comprised pasture swards. For example, Townsville stylo has narrow leaves and tends to grow within the associated grasses, whereas Siratro has broad leaves on trailing stems that tend to grow over the grasses. They also emphasized the vertical development of the sward and differences in sward density of lower layers. Haydock and Shaw (1975) combined their comparative yield estimate and the dry-weight-rank method of Mannetje and Haydock (1963) to rate herbage yield and botanical composition, respectively. Approximately 2 800 quadrats in a 130 ha pasture experiment were scored in 7 man-days.

The comparative yield estimate of Haydock and Shaw (1975) was similar to that of Hutchinson *et al.* (1972) who studied closely grazed pastures of New South Wales, Australia. The latter investigators cut sod cores 10.8 cm in diameter and 6.0 cm in depth to represent differences in quantity of fresh herbage from low to high amounts. The cores fit into a tray with a hole of the same diameter in the centre. By placing the tray with the standards over random sites of the sward observers were trained to make visual estimates of the sward appearing in the centre. By use of regression analysis the estimates were converted to actual dry weights with a high degree of accuracy.

Hutchings and Schmautz (1969) also field-tested the relative-weight-estimate method for determining herbage production of rangeland in the western USA. They used a single standard as a base quadrat to estimate production of other sampling units. These investigators found more reliable estimates for total production than for species components. They also noted less accuracy with a change in light conditions, and fatigue or attitude of the observer. Dependable estimates were closely related to the period of training and experience. Particular emphasis was given to double sampling, i.e. repeated checking of estimated and actual weights.

Indirect methods of estimating herbage yields

Indirect estimation of pasture yield is concerned with the relationship between plant factors and production. Easily measured plant traits or related characteristics such as sward height, density of ground cover or a combination of these are measured. These contribute in some way to herbage yield and can be converted to yield through standards and regression analyses.

Height and cover combination

Plant height and percentage ground cover each show a close relationship to yield. As individual characteristics, they probably are more useful under range conditions than with higher-yielding improved pastures (Brown, 1954). The combination of height and cover bring together depth and area measurements. Their product is a rough index of volume and can be used to predict herbage yields more accurately than either alone. Instead of individual measurements, Alexander *et al.* (1962) in the USA described a method of incorporating both components into a single direct measurement. They dropped a piece of light-weight plywood approximately 0.4 m^2 on to the pasture sward and measured the distance of each corner from ground level. Height in this case represented per cent cover of herbage as well as sward height. The final resting elevation of the board expressed bulk density of the sward. A formula for rapid conversion to yield estimate was from the combined measurements of the four corners and weight of herbage directly below the board. Forage yield of a known pasture was estimated using the following formula:

$$85\ H - 190 = \text{dry matter/ha}$$

where H is the sum of the four corners above ground level. Assume a total of 50 cm, then $80 \times 50 - 190 = 4\ 060$ kg/ha of dry matter. The method was tested on several pasture types including common bermudagrass. Herbage yield estimates were significantly and positively corrected with actual weight measurements. This scheme permitted a rapid estimation of forage production without substantially disturbing the experimental area. A corrected formula would need to be derived for each type of pasture.

A simple disc instrument was used by Castle (1976) to estimate herbage yield in Scotland. The device consists of an aluminium rod (shaft) marked in divisions of 0.5 cm. Two discs, also made of aluminium, are linked at a 20 cm distance by three rods. They have holes in the centre and together weigh 200 g. The lower disc measures 30 cm and the upper 10 cm in diameter. In yield estimates the shaft is thrust through the sward and held upright with the base resting on the soil. The discs are placed over the shaft then dropped on to the sward and settle to a constant position on the herbage. Height of the upper disc from ground level is read directly from the markings on the shaft. A number of actual sampling unit weights are determined to allow calculation of simple correlation coefficients and coefficients of the regression of yield on height. Phillips and Clarke (1971) and Powell (1974) used similar simple mechanical instruments for estimating herbage yields in New Zealand. Such instruments are easily constructed and require little skill in their use. They allow rapid yield estimates, e.g. about 50 readings and recordings in about 15 min within a paddock of 2–3 ha.

Use of electronic equipment

Estimates of herbage yield *in situ* with capacitance meters have been made in several countries. Fletcher and Robinson (1956) in the USA modified a portable field instrument used for determining soil moisture and tested its value for estimating forage weight. The instrument was comprised of a series of probes placed vertically in the sward, and a dial that measured a change in electrical displacement of energy when opposite surfaces of a non-conductor were maintained at a difference of potential (capacitance). It functioned on the principle that air has a low dielectric constant while herbage has a high dielectric constant. Any change that occurs in electrical capacitance when probes are placed within the sward should be shown by the dial reading. In theory different quantities of herbage should give different readings. The instrument actually measures water content of the herbage and not dry matter yield. Inconsistent readings have been obtained because of relative differences in water content of species, their stage of growth as well as arrangement of plant material around the probes (Shaw *et al.*, 1976). Furthermore, the instrument is not sensitive to the quantity of dead material due to the low dielectric constant of dry forage. A number of modified models have been designed (Campbell *et al.*, 1962; Johns *et al.*, 1965; Jones and Haydock, 1970; Neal and Neal, 1973; Angelone *et al.*, 1980) and several are available commercially. None have proven entirely satisfactory over a range of pasture types, especially for legume–grass mixtures of different species and moisture contents.

The beta-attenuation technique has been employed on a limited scale to estimate forage yields (Teare *et al.*, 1966; Mitchell, 1972). It relies on the thickness-gauging principle where absorption or attenuation of beams of radioactive emissions (beta particles) is a predictable function of density of the absorbing material (herbage). The difference in absorption due to herbage density can be measured and transformed to biomass information when measurements are summed over height (vertical development of the sward). Teare *et al.* (1966) used standing plants of wheat, soybean and tall fescue grass as the absorptive material and as the dependent variable in a regression against radioactivity. A linear regression was obtained between herbage density and absorption of beta particles. Mitchell (1972), working with standing forage, found that the method accounted for 90 per cent of the yield variation observed under field conditions. He noted that the instrument is relatively inexpensive, gives accurate results even when the material is compressed, allows repetitive sampling of the same location but would have limited value under heavily grazed conditions.

Photography

Photography is a valuable technique to provide supplemental information in pasture experimentation. The adage that 'a picture is worth a thousand words' has real meaning to describe pasture and range condition. A series of well-planned photographs taken at timely intervals effectively portrays changes in vegetation composition and density. The longevity of plants and species can be depicted and characteristics of individual plants recorded. Effects of grazing and other treatments can be followed by sequential photographs. Differences in stocking rate, effects of fertilizers and comparisons of species lend themselves well for photographic study. Photographs aid in describing equipment and research methods. On rangelands and larger-scale experiments, aerial photographs provide useful information in locating boundaries and experimental areas, delineating vegetation types, planning improvements and other special uses. Coloured photographs are particularly useful in maintaining records and presenting results.

Evaluating pastures by animal performance

Pasture output can be expressed in terms of (1) animal productivity, i.e. animal days per hectare, liveweight gain per hectare, milk or wool production per hectare and (2) the quantity of feed units required for animal maintenance, weight increase, milk production and other products. Animal production is determined by carrying out grazing trials. The number of feed units produced by a pasture are calculated by two methods: (a) indirectly by using animal product and calculating the feed units required to obtain that product and (b) directly from analyses of the herbage nutrient composition and herbage yield.

Two common feed units used are total digestible nutrients (TDN) and starch equivalent (SE). TDN has been criticized as a measure of the useful energy value of forages and concentrates, and net energy (NE) is recommended as the expression for denoting energy requirements and energy value of cattle feeds. Net energy is separated into portions required for maintenance (NE_m), gain (NE_g) and milk production (NE lactating cows).

Total digestible nutrients

These include all of the digestible organic nutrients, namely protein, fibre, nitrogen-free extract and fats. The percentage of each nutrient is obtained by chemical analysis, and a digestibility coefficient is obtained by carrying out digestion trials. The digestibility coefficients of many temperate-zone feeds are known from numerous digestion trials and are given in tables of various publications, e.g. NAS (1976, 1978). There is less information available for tropical feedstuffs, including grasses and legumes.

The TDN content of a feed (herbage) is obtained by multiplying the percentage of each nutrient in feed by its digestibility coefficient and summing the products. This has been done for some tropical forage crops. For instance, elephant grass in the late vegetative stage contains 63 per cent TDN and in the late bloom stage 52 per cent TDN for beef cattle (NAS, 1976). Assume that an elephant grass pasture produces 20 tonnes/ha of dry matter and is maintained in a vegetative stage of growth. The calculated TDN/ha would then be 12.6 tonnes/ha. Knowing the yield of TDN one could calculate a potential stocking rate of beef cattle when making an indicated liveweight gain, or the number of milking cows that could be maintained when producing milk at a given level. Conversely, having at hand the animal weights and liveweight gain per hectare or milk production per hectare one could calculate the amount of TDN required for maintenance, gain and milk production. This method provides an indirect measure of TDN produced by the pasture. Figures given in tables of nutrient requirements of beef and dairy cattle would be needed for this calculation (NAS, 1976, 1978).

Starch equivalent

This is the amount of starch needed to produce the same quantity of body fat (energy) as would 100 units (kg or lb) of herbage. Any nutrient can be expressed in terms of starch, e.g. the value of protein in terms of starch is 0.94 when starch is 1.0. The SE of herbage can be calculated from the percentage composition of a given digestible nutrient as determined by chemical analysis.

Net energy

This unit is defined as the difference between metabolizable energy (amount of gross energy consumed minus faeces, gaseous products of digestion and urine energy) and heat increment (heat of feed utilization resulting from consumption of feed when an animal is in a thermoneutral environment), and includes the amount of energy used for maintenance, gain and milk or wool production. Net energy values are listed in nutrient requirement tables (NAS, 1976, 1978) but are limited for tropical forages. Elephant grass in the vegetative stage of growth has a NE_m of 1.36 M cal/kg and a NE_g of 0.76 M cal/kg for beef cattle.

Net energy values are useful in predicting weight gains or determining whether cattle have gained according to expectations. For example, suppose that steers are grazing elephant grass kept in a vegetative stage of growth. They are placed on pasture when weighing 200 kg and will be sold at 400 kg/head. The mean weight for the grazing period is 300 kg so that NE_m will average 5.55 M cal/day or 4.08 kg (5.55 ÷ 1.36 NE_m for elephant grass) of dry matter for maintenance. Total dry matter consumption will average 7.5 kg/day, based on 2.5 per cent intake of bodyweight. This leaves 3.42 kg (7.50 − 4.08) of dry matter available for weight gain. This portion of the diet will contain 2.6 M cal NE_g (2.60 × 0.76 NE_g). Steers averaging 300 kg/head during the grazing period and having available 2.6 M cal NE_g should gain about 0.65 kg/day. Thus, a period of about 310 days would be needed to reach the indicated marketable age. Figures used in these calculations were taken from tables given in NAS (1976).

Botanical composition

Botanical analysis of the vegetative cover in a pasture or grazingland identifies the species present and their proportion. This determination is difficult because of the highly variable nature of pasture herbage and lack of accurate methods that estimate the animals' diet. Many techniques have been described and utilized for vegetational analyses. They have been classified into five primary categories, namely (1) weight, (2) area covered, (3) number of individuals, (4) frequency of occurrence and (5) pattern (Brown, 1954; Tothill and Peterson, 1962).

Weight of species components

The weight of each species or cultivar is the most precise and objective method for determining botanical composition of the pasture sward. Each component is expressed as a percentage of the total dry weight of a given sample or treatment.

Samples used for botanical analyses may be drawn from herbage cut for yield determination or from sampling units specifically used for measuring species composition. Size of sample varies, but one of about 0.5 kg is considered sufficient for most studies. The weight of samples fluctuate with the size of quadrat used for sampling. Also, labour available for making hand separations and number of samples influences the size of single samples. Determination of botanical composition by weight is a manual operation. Samples are usually taken to the laboratory shortly after harvesting but separations can be made in the field. It is easier and more rapid to separate fresh rather than dry material. A sample is spread cut before the person who identifies each species and manually removes stems and leaves of that type from the mixture. Each species is kept separate for dry weight determination and calculation of its percentage contribution to the total dry weight. The method is laborious and time-consuming. The speed of hand separation can be increased by maintaining stem orientation so as to keep bases together and stems parallel. This is easily accomplished when samples are cut by hand, but not with machine harvesting.

Height of cutting herbage for botanical analysis should correspond to the height of grazing. Even then the sample does not represent the botanical content of ingested herbage since animals graze in a selective manner. Furthermore, grazing is not done at a uniform height even under heavy stocking rates. Any attempt to simulate the pattern of grazing by cutting or plucking samples has not met with great success.

Since hand-sorting is a slow process other methods have been suggested for measuring species composition, namely (1) estimation of relative weight of each species in the standing herbage and (2) estimation of species weight in the cut sample. These are subjective and less precise than actual weights but are useful for preliminary trials, surveys and rough estimates of changes in botanical composition. They are also more meaningful in simple rather than complex mixtures.

A method of estimating botanical composition on a dry-weight basis of species ranking was described by Mannetje and Haydock (1963) and Van Dyne *et al.* (1975). This technique was based on the rank method developed by De Vries (1933) that gave relative species percentage in terms of fresh weight. In the dry-weight-rank method a quadrat was placed on the sward so as to be clearly visible. An observer recorded all of the species present, then ranked them in numerical order based on a visual estimate of dry weight. This was replicated from 50 to 100 times. The data were tabulated by the number of quadrats with ranks of 1, 2, 3 for each species and the proportion of quadrats (based on 100) recorded for each ranking. These proportions were multiplied by factors of 70.19, 21.08 and 8.73, respectively. The multipliers were derived from botanical composition data of samples that were hand-separated into four or five components. Thus, the maximum percentage dry matter of a species could not exceed 70.2. Results of the dry-weight-rank method were closely correlated with hand-separated samples. This technique has the advantage of being rapid so that large numbers of samples can be examined. Furthermore, the sward is left undisturbed. Tothill *et al.* (1978) and Hargreaves and Kerr (1978) noted further details and refinements for comprehensive sampling and described computational procedures for the dry-weight-rank method which has been designated BONTAL. Additional improvements and derivation of multipliers were provided by Jones and Hargreaves (1979). Other sets of multipliers must be derived for divergent pastures in different environments.

Area covered

Cover refers to the proportion of ground covered or occupied by the whole vegetation or individual species. Measurements of plant cover include (1) foliage or leaf spread (plant canopy), (2) crown coverage and (3) basal area or basal coverage. The sward develops in vertical layers and these must be sampled to obtain a valid measure of cover. Types that creep or trail maintain a certain consistency in the amount of cover, even under grazing and with seasonal change. Those that climb show variable cover depending on grazing intensity. These forms may actually have small crown and basal coverage. Bunch grasses develop considerable foliage and crown cover unless heavily grazed. Under conditions of heavy grazing basal area gives a valid measure of their cover. In contrast, it is difficult to define or delimit basal area of spreading grasses and legumes. A measure of foliage or canopy cover is usually made for the dense swards of improved pastures and basal cover for swards of rangelands.

Many methods and techniques have been used to measure and estimate plant cover. Those most frequently used for present or past study of improved and naturalized grazinglands will be discussed here. For more extensive descriptions the reader is referred to surveys by Brown (1954), Hutchings and Pase (1962), Tothill and Peterson (1962) and Tothill (1978). Data are generally expressed in one of the following ways (1) an absolute and cumulative measurement in square units (e.g. point sampling), (2) percentage cover of the ground surface (per cent of area), (3) individual species as a percentage of the total cover (percentage composition by area) and (4) degrees of estimated cover based on a scale.

Point sampling

A point represents the smallest unit of measure in botanical analysis and as a sampling unit has been referred to as the point quadrat. This technique measures botanical composition and percentage of cover. It is objective and rapid, not dependent on random distribution, allows precise identification of species present, can be performed on uneven ground, permits effective studies in persistence of species and seasonal changes as well as effects of fertilizer, other agronomic factors and grazing.

Points used in vegetation sampling consist of pins for contact of herbage. An apparatus described by Levy and Madden (1933) comprised a frame about 50 cm long made of two horizontally spaced bars about 2.5 cm apart and mounted on legs that could be pushed into the soil. The bars were attached with screws for removal and easy transport. Each bar contained 10 holes 5.0 cm apart so that steel pins could be moved vertically. Frame height was about 60 cm but this can be adapted to the vegetation being measured. As the pin moves downward it touches the foliage or bare soil. The types of hits include (1) the first – favours the taller species as they are likely to be touched first, (2) crown or ground – favours prostrate and low growing species and (3) several (all-hits) – gives the greatest accuracy but requires more time.

Tinney *et al.* (1937) bent the frame so that pins rested at a 45° incline. They noted that vertical pins tend to hit broad-leaved plants more frequently than those with narrower leaves. Pins extending through the sward at a 45° inclination cover a wider area than those in a vertical position and allow for a greater number of hits. Furthermore, they are more clearly visible through the vegetation.

Size of the pin point, and movement of the pin and foliage significantly affect precision of measurement (Goodall, 1952). The thicker the pin or point the greater the overestimation of cover as compared to dry weight analysis. More hits are recorded of fine-leaved grasses with larger pins and fewer coarse and bulky species. Thus, it is important to use pins as fine as possible.

The number of samples (hits) to obtain close correlation with weight analysis depends on the heterogeneity of the sward, and whether information is desired on percentage of cover or percentage of species composition. In general, pastures that have a greater number of species or those with sparse vegetation require more points for an adequate sample than those with fewer species and denser swards. Crocker and Tiver (1948) in southeastern Australia found that 200 points were sufficient to detect dominant species and about 500 for the less frequent ones in pastures of 5–50 ha. When a species approaches 50 per cent composition of the sward 100 points will be ample, but if it occupies only 10 per cent of the area up to 1 000 may be required to obtain reliable information (Goodall, 1952). Under conditions of wide variability, test observations should be made using the point quadrat and hand separations for weight analysis of species composition.

There is no standard practice as to the number of hits to record at a given site or the number of sites to sample. In general, sampling efficiency per pin increases when fewer pins or a single pin is used per site, but efficiency of time per pin decreases (Goodall, 1952). Kemp and Kemp (1956) examined the number of sites and different numbers of pins per frame (at the same site) so as to obtain precision of sampling. Their data showed that 300 sites of 5 pins, i.e.

1 500 observations, gave the same precision as 200 sites and 10 pins (2 000 observations). In practice the investigator must balance the time saved in making fewer observations to offset the extra time needed in moving from one site to another. In selecting sites or locations for sampling some investigators distribute the equipment at random. Others lay out line transects across the pasture and follow the line, but stopping at random sites. If change in botanical composition is to be studied over time then permanent sites are suggested to reduce sampling error.

Evans and Love (1957) described the step–point method for use in rangelands of the western USA. This technique comprised a combination of two procedures:

1. Location of a point at the tip of a shoe after a specified number of paces. The tip of the shoe was held at an angle of about 30° and a pin thrust downward into the sward. A record of the first hit was noted by species or bare ground. This gave a measure of species composition and ground cover.
2. A visual estimate of ground cover within a 929 cm^2 quadrat placed on to the sward at the shoe toe after each 10 paces. As noted by other investigators, total plant cover was measured with greater precision than species composition. It was considered, however, that the data gave a reliable estimate of the latter and the method took only one-sixth the time of the conventional point quadrat.

Another variation is the line–point transect method, in which pointed rods are projected downward at intervals of 15 cm along a transect line (Cook and Box, 1961). This technique was more time-consuming and less precise than the point–quadrat method.

Point sampling has many desirable features. It is objective, essentially unbiased if properly used and provides precision with ample replication. It is particularly suitable for use in short vegetation but less so for tall vegetation. The contact site is temporary since points cannot be relocated with accuracy. Line points can be utilized for sampling the same general location, however.

Line intercept

This method consists of the identification and horizontal measurement of plants along the course of a line (wire) stretched over the sward. Length of the line depends on ground cover, being longer where vegetation is sparse. Readings are made in units of length along the line and their total represents the ground surface occupied by the plants. The diameter of each plant directly under the wire is recorded along the linear plane of the wire. In some instances an arbitrary decision must be made as to whether or not a plant lies under the wire. The method is more time-consuming than the point quadrat, detects a greater number of species, but the data are more variable. The technique has been successfully used in rangelands, especially under semi-arid conditions. It has limited usefulness in improved pastures, particularly those containing stoloniferous grasses and trailing legumes.

Loop frequency

This is a combination of the line-interception method and the point quadrat, in which vegetation within a 1.9 cm (approximately ¾ inch) loop is recorded instead of vegetation contacted by a point (Parker and Harris, 1959). In estimating plant cover or species composition, a transect line is laid out and the loop lowered at specified intervals until it contacts foliage (first hit) or allowed to rest on the sward or bare soil (basal coverage). Modifications in manipulation of the loop have been made, one being that of welding the loop on to the end of a rod (Cook and Box, 1961). Since the loop has area the technique overestimates plant cover, detects fewer species and gives more variable data when compared to the point–quadrat and line-intersection methods. It has been used in studies of rangeland, but has little application to improved pastures.

Visual estimates

Area covered can be estimated by visual observation after a period of training or experience. Observers should check their estimates against a more precise method, e.g. point quadrat, at periodic intervals. Even then divergence will occur among samples of each individual as well as among individuals. Several methods have been employed as follows (Brown, 1954):

1. Percentage area estimation – a small area of the sward is delimited by a quadrat and examined from directly above. An estimate is made of the proportion of ground cover by each species. The vegetal cover and bare soil should total 100. To facilitate estimation the quadrat can be subdivided into smaller units and each rated separately. This method has application to dense swards, but is best suited to low-growing species or grazed pastures.

2. At one time rangeland was surveyed by the point observation plot in which circular plots of 100 square feet were first identified. Then the observer(s) used a square foot area as a unit of estimation within the plot. Viewing the vegetation from above, the examiner visualized how much growth of a vegetation class (grass, forb, shrub) or a species covered one square foot of ground. The number of square foot readings by classes or species when totalled gave the percentage cover of each group, since each unit of estimation represented 1 per cent of the area. This method was also known as square-foot density.

3. Basal areas of individual species or of total vegetation can be

estimated within a sampling unit of any size, but usually 1.0 m². These are entered on a chart and in a list until a larger and desig- nated plot area is covered. This method is suitable for semi-arid grassland with bunch or tussock grasses.

Charting procedures

This is a detailed and objective representation of the position and area occupied by species in a given plot. The methods used require time, labour and skill. Charting can be accomplished without special equipment by identifying plants, measuring their diameter and drawing their position to scale on graph paper. The pantograph, a draughtsman's instrument, has been used to record position and area of plants. It requires two people, one to guide an arm of the apparatus around the outline of plants and the other to manipulate the recording needles on to the chart and to note the species by position. A tripod method is sometimes used by fixing a tracing table between the legs of a camera tripod. Outlines of plants as viewed through the hole of the tripod are traced by hand on to transparent paper resting on the table. A camera fixed to a tripod can be used to focus on to a quadrat. An image of the vegetation is projected onto a glass plate that replaces the focusing screen of the camera. This is covered with transparent paper for tracing vegeta- tion. Several other methods of charting described by Brown (1954) have application for rangeland or long-term grazingland, but are not useful in mixed agriculture where pastures alternate with crops in the farming system.

Number of species or plants

The total number of plants or tillers (stems) in a specified area is known as density. It is a quantitative measure of population or abundance. The numbers of individuals may be determined by actual count or estimations. They are expressed as (1) density cal- culation of the number per unit area and (2) percentage composi- tion – ratio of one species to the total number of individuals of all species (Brown, 1954; Strickler and Stearns, 1962).

This method has application to studies of percentage germina- tion, seedling establishment, rate of tillering, regeneration of annual species, persistence of perennial species, weed control, disease sur- veys and estimates of seed yield. The number of individuals as a mea- sure of abundance is related to cover within a given growth habit, but should not be used to estimate vegetative cover in general. Spe- cies that appear most prominent are not necessarily the most abun- dant, e.g. broad-leaved kudzu.

In making counts the investigator must decide whether to count tillers or whole plants. In a dense sward individual plants of bunch-type, rhizomatous and stoloniferous grasses, or trailing legumes cannot be detected.

Counting provides the most accurate measure of number. This is done by using quadrats or transects to delimit an area, then recording the total number of individuals or tillers. In studies of persistence the sites are marked in a permanent manner. Estimation also provides an indication of abundance or plentifulness. This approach should be based on a scale or classes so as to be more descriptive. A scale of 1–5 or 1–10 may refer to numbers or percentages. Similar references can be used for scales such as sparse to abundant or rare to frequent. Estimation is a subjective measure and highly influenced by bias. Furthermore, morphological differences of species add a dimension of confusion. In estimating numbers, particularly on rangelands, accuracy can be increased by including distance (spacings) between plants in the rating.

Frequency of occurrence

Frequency is expressed as the number of samples in which a given species occurs. The terms degree of frequency and percentage frequency are used interchangeably with frequency of occurrence. They all refer to whether a species is present or absent and not to number of plants (density).

Size of the sampling unit influences the probability of a species being noted. The smaller the sample the less likelihood of it being detected. With a small sampling unit, a large number of samples must be examined, otherwise certain species may be recorded as having a low frequency. A heterogeneous pasture, i.e. several to many species and highly mixed, requires more samples than a homogeneous pasture. Sparse vegetation must be sampled in more detail than dense vegetation.

A common method for measuring frequency of occurrence is to toss a quadrat randomly onto the sward a prescribed number of times, making sure that the pasture is well sampled, then to record the presence or absence of a species. Another method is to walk over the pasture, stopping after a predetermined number of paces and noting the presence or absence of a species at the toe of a shoe. The entire pasture must be traversed, since plants of a given species tend to occur in communities.

Features of the frequency method are:
1. It is both rapid and objective. The presence or absence of a species is definite and fairly constant, especially perennials. They are not subject to short-term fluctuation caused by seasonal variation. In contrast, the attributes of weight, number of tillers and area covered are likely to show seasonal differences.
2. An indication of the species composition of the sward is

obtained. When sampled over time a broad estimate of change in botanical content can be noted.

3. There is a linear relationship between frequency and density (number of species) at low-frequency ranges when species are distributed at random. At higher frequencies there is less agreement between the two attributes (cited by Brown, 1954).

4. Frequency reveals that certain species may be localized due to micro-environmental and climatic conditions.

5. Less prevalent species become evident when the pasture is thoroughly sampled.

6. There is no information regarding the quantitative contribution of a species to total herbage production.

Pattern of distribution

The pattern of species distribution, influence of local environment on plant growth and morphological growth habits are not always given recognition in a study of pasture flora (Tothill and Peterson, 1962; Tothill, 1978). Species are not always evenly distributed over a large pasture area. Those found in low-lying wet areas are usually different from those in well-drained sites. In addition, factors such as inherent soil fertility, depth of soil horizon, gradient and topography and exposure (northerly, southerly, etc.) affect species distribution. Before embarking on a study of pasture flora the investigator should note such divergent areas and delineate their boundaries for subsampling. Additional variation also occurs within stratified areas because of localized soil and micro-environmental differences and the general tendency of species to occur in small communities when grown in complex mixtures. These may not be readily appreciated but will contribute to experimental error.

The differences in life cycles and growth patterns of species significantly contribute to variation in botanical composition, ground cover, frequency of occurrence and herbage production. Grasses generally make earlier and more rapid growth and develop flowering stems before legumes. There are also differences among species within both grasses and legumes. Legumes usually continue growth into the dry season and thus comprise a greater proportion of herbage during this period. Such differences in morphologic and phenologic characteristics must be recognized and taken into consideration when planning the method of study and type of sampling unit, as well as time and interval of sampling.

Seedling establishment

The number of seedlings per unit area are usually counted by species but may be estimated. This is done by using a quadrat of known dimension, taking a transect or denoting a specified length of row if seeds are sown by this method. The observer must be able to distinguish species of the sown grasses and legumes as well as weedy types that occur naturally. Counts should be made at a time when individual seedlings are easily identified, i.e. before excessive tillering and stem extension of creeping types take place. A count of seedlings is of general interest in establishment and can be related to later contribution to sward development and eventual persistency. It is of particular interest in determining seed distribution patterns. In addition, a knowledge of seedling emergence and establishment is important in evaluating vegetation cover and herbage production where diverse macro- and micro-environmental conditions exist.

Nutrient reserve

Agronomic and grazing management treatments influence the accumulation of nutrient reserves in roots, rhizomes, stolons and crowns. Their differential effect can be measured by the etiolation technique. Herbage should be cut at soil level and sod cores cut or dug from plots or pastures that received different treatments. Size of the core may vary according to equipment for removal. One of 15 cm diameter and 20–30 cm in depth is convenient for handling. They are placed in pots or plastic bags and kept in absolute darkness. Only water is added as needed to promote development of etiolated tillers and elongated stems. After 3–4 weeks most of the nutrient reserves will have been utilized as the plants grow in the dark. The herbage is then cut for dry weight determinations. These data give a relative measure of nutrient reserve. This technique is suitable for pasture species that develop a dense and uniform sward such as kikuyu, Pangola and *Cynodon* species. Heavily tufted species such as guinea grass, on the other hand, are difficult to sample and the data are likely to be unreliable.

References

Alexander, C. W., J. T. Sullivan and **D. E. McCloud** (1962) A method for estimating forage yield, *Agron. J.* **54**, 468–9.

Angelone, A., J. M. Toledo and **J. C. Burns** (1980) Herbage measurements in situ by electronics. 1. The multiple-probe-type capacitance meter: a brief review, *Grass and Forage Sci.* **35**, 25–34.

Brown, D. (1954) *Methods of Surveying and Measuring Vegetation*, Commonw. Bur. Past. Fld Crops Bull. **42**, Farnham Royal, Bucks: England.

Campbell, A. G., D. S. M. Phillips and **E. D. O'Reilly** (1962) An electronic instrument for pasture yield estimation, *J. Brit. Grassld Soc.* **17**, 89–100.

Campbell, N. A. and **G. W. Arnold** (1973) The visual assessment of pasture yield, *Aust. J. Exptl Agric. Anim. Husb.* **13**, 263–7.

Carter, J.F. (1962) Herbage sampling for yield: tame pastures. In *Pasture and Range Research Techniques*, Comstock: Ithaca, New York, pp. 90–101.

Castle, M. E. (1976) A simple disc instrument for estimating herbage yield, *J. Brit. Grassld Soc.* **31**, 37–40.

Cook, C. W. and **T. W. Box** (1961) A comparison of the loop and point methods of analyzing vegetation, *J. Range Mgmt* **14**, 22–7.

Cowlishaw, S. J. (1951) The effect of sampling cages on the yields of herbage, *J. Brit. Grassld Soc.* **6**, 179–82.

Crocker, R. L. and **N. S. Tiver** (1948) Survey methods in grassland ecology, *J. Brit. Grassld Soc.* **3**, 1–26.

Davies, J. G. (1931) *The Experimental Error of Yield from Small Plots of Natural Pasture*, CSIRO Div. Trop. Past. Bull. **45**.

Davis, R. R. and **D. S. Bell** (1958) A comparison of birdsfoot trefoil-bluegrass and Ladino clover-bluegrass for pasture. II. Yield of herbage and relationships to lamb response, *Agron. J.* **50**, 520–4.

De Vries, D. M. (1933) De rangorde methode. Een schattings methode voor plant-kundig grasland-onderzoek met volgorde bepaling, *Versl. Landb. Onderz.* **39A**, 1–24.

Evans, R. A. and **M. R. Love** (1957) The step-point method of sampling, a practical tool in range research, *J. Range Mgmt* **10**, 208–12.

Fletcher, J. E. and **M. E. Robinson** (1956) A capacitance meter for estimating forage weight, *J. Range Mgmt* **9**, 96–7.

Goodall, D. W. (1952) Some considerations in the use of point quadrats for the analysis of vegetation, *Aust. J. Sci. Res. Ser.* B5(*1*), pp. 1–41.

Hargreaves, J. N. G. and **J. D. Kerr** (1978) *BONTAL – A Comprehensive Sampling and Computing Procedure for Estimating Pasture Yield and Composition. II. Computational Package*, CSIRO. Aust. Div. Trop. Past. Trop. Agron., Tech. Memo. No. 9.

Haydock, K. P. and **N. H. Shaw** (1975) The comparative yield method for estimating dry matter yield of pasture, *Aust. J. Exptl Agric. Anim. Husb.* **15**, 663–70.

Hutchings, S. S. and **C. P. Pase** (1962) Measurements of plant cover-basal, crown, leaf. In *Range Research Methods*, USDA Misc. Publ. No. 940, pp. 22–30.

Hutchings, S. S. and **J. E. Schmautz** (1969) A field test of the relative-weight-estimate method for determining herbage production, *J. Range Mgmt* **22**, 408–11.

Hutchinson, K. J., R. W. McLean and **B. A. Hamilton** (1972) The visual estimation of pasture availability using standard pasture cores, *J. Brit. Grassld. Soc.* **27**, 29–34.

Institute Staff (1961) *Research Techniques in Use at the Grassland Research Institute, Hurley*, Commonw. Bur. Past..Fld Crop Bull. **45**, Hurley: England.

Johns, G. G., G. R. Nicol and **B. R. Watkin** (1965) A modified capacitance probe technique for estimating pasture yield. 1. Construction and procedure for use in the field, *J. Brit. Grassld Soc.* **20**, 212–17.

Joint Comm. Rept (1952) Pasture and range research techniques, *Agron. J.* **44**, 39–50.

Jones, R. M. and **J. N. G. Hargreaves** (1979) Improvements to the dry-weight-rank method for measuring botanical composition, *Grass and Forage Sci.* **34**, 181–9.

Jones, P. J. and **K. P. Haydock** (1970) Yield estimation of tropical and temperate pasture species using an electronic capacitance meter, *J. Agric. Sci. (Camb.)* **75**, 27–36.

Kemp, C. D. and **A. W. Kemp** (1956) The analysis of point quadrat data, *Aust. J. Bot.* **4**, 167–74.

Klingman, D. L., S. R. Miles and **G. O. Mott** (1943) The cage method for determining consumption and yield of pasture herbage, *J. Amer. Soc. Agron.* **35**, 739–46.

Levy, B. E. and **E. A. Madden** (1933) The point method of pasture analysis, *N. Z. J. Agric.* **46**, 267–79.

Linehan, P. A., J. Lowe, and **R. H. Stewart** (1952) The output of pasture and its measurement, *J. Brit. Grassld Soc.* **7,** 73—98.

Mannetje, L. 't and **K. P. Haydock** (1963) The dry-weight-rank method for the botanical analysis of pasture, *J. Brit. Grassld Soc.* **18,** 268–75.

Mitchell, J. E. (1972) An analysis of the beta-attenuation technique for estimating standing crop of prairie range, *J. Range Mgmt* **25,** 300–4.

Morley, F. H. W., D. Bennett and **K. W. Clarke** (1964) The estimation of pasture yield in large grazing experiments, *CSIRO Div. Pl. Ind., Field Sta. Record 3(2),* pp. 43–7.

NAS (1976) *Nutrient Requirement of Domestic Animals. Nutrient Requirements of Beef Cattle,* No. 4, Nat. Acad. Sci.: Washington, D.C.

NAS (1978) *Nutrient Requirements of Domestic Animals. Nutrient Requirements of Dairy Cattle,* No. 3, Nat. Acad. Sci.: Washington, D.C.

Neal, D. L. and **J. L. Neal** (1973) Uses and capabilities of electronic capacitance instruments for estimating standing herbage. Part 1. History and development, *J. Brit. Grassld Soc.* **28,** 81–9.

Parker, K. W. and **R. W. Harris** (1959) The 3-step method for measuring condition and trend of forest ranges: a résumé of its history, development and use. In *Techniques and Methods of Measuring Understory Vegetation,* US Forest Service South and Southeast, Forest Expt Sta. Proc. 1959, pp. 55–69.

Pechanec, J. F. and **G. D. Pickford** (1937) A weight estimate for the determination of range or pasture production, *J. Amer. Soc. Agron.* **29,** 894–904.

Phillips, D. S. M. and **S. E. Clarke** (1971) The calibration of weighted disc against pasture dry matter yield, *Proc. N. Z. Grassld Assoc.* **33,** 68–75.

Powell, T. L. (1974) Evaluation of weighted disc meter for pasture yield estimation on intensively stocked dairy pasture, *N. Z. J. Exptl Agric.* **2,** 237–41.

Reppert, J. N., R. L. Hughes and **D. A. Duncan** (1962) Vegetation measurement and sampling. Herbage yield and its correlation with other plant measurements. In *Range Research Methods,* USDA Misc. Publ. No. 940, pp. 15–21.

Rodriguez, J., E. Rivera and **J. Vicente-Chandler** (1979) Productivity of four intensively managed grasses under grazing management in the humid hills of Puerto Rico, *J. Agric. Univ. P. R.* **64,** 236–40.

Shaw, N. H., L.'t Mannetje, R. M. Jones and **R. J. Jones** (1976) Pasture mesurements. In N. H. Shaw and W. W. Bryan (eds), *Tropical Pasture Research Principles and Methods,* Commonw. Bur. Past. Fld Crops., Bull. 51, Hurley: England, Ch. 10.

Strickler, G. S. and **F. W. Stearns** (1962) The determination of plant density. In *Range Research Methods,* USDA Misc. Publ. No. 940, pp. 30–40.

Teare, I. D., G. O. Mott and **J. R. Eaton** (1966) Beta attenuation – a technique for estimating forage yield in situ, *Radiation Bot.* **6,** 7–11.

Tinney, F. W., P. S. Aamodt and **H. L. Alghren** (1937) Preliminary report of methods used in botanical analysis of pasture swards, *J. Amer. Soc. Agron.* **29,** 835–40.

Toledo, J. M., J. C. Burns, H. L. Lucas and **A. Angelone** (1980) Herbage measurements in situ by electronics. 3. Calibration, characterization and field application of the earth-plate forage capacitance meter: a prototype, *Grass For. Sci.* **35,** 189–96.

Tothill, J. C. (1978) Measuring botanical composition of grasslands. In L. 't Mannetje (ed.), *Measurement of Vegetation and Animal Production of Grasslands,* Commonw. Bur. Past. Fld Crops Bull. **52,** Hurley: England.

Tothill, J. C. and **M. L. Peterson** (1962) Botanical analysis and sampling tame pastures. In *Pasture and Range Research Techniques,* Comstock: Ithaca: New York, pp. 109–34.

Tothill, J. C., J. N. G. Hargreaves and **R. M. Jones** (1978) *BONTAL – A Comprehensive Sampling and Computing Procedure for Estimating Pasture Yield and Composition. I. Field Sampling,* CSIRO Aust. Div. Trop. Crop Past. Trop. Agron., Tech. Memo., No. 8.

Van Dyne, G. M., L. J. Bledsoe, J. D. Gustaffson, J. W. Hughes and **D. M. Swift** (1975) *A Study of Dry Weight Rank Method of Botanical Analysis*, IBP Grassld Biom. Tech. Rept No. 27, Nat. Resources Ecol. Lab., Colorado State Univ.

Wagner, R. E., M. A. Hein, J. B. Sheperd and **R. E. Ely** (1950) A comparison of cage and mower strip methods with grazing results in determining production of dairy pastures, *Agron. J.* **42**, 487–91.

Wilm, H. G., D. F. Costello and **G. E. Klipple** (1944) Estimating forage yield by the double-sampling method, *J. Amer. Soc. Agron.* **36**, 194–203.

Chapter 15

Forage and fodder grasses and legumes in farming systems

Shifting agriculture has been the common means of resting the land where intensive agriculture is not practised. In the forest zones and wooded savanna regions the non-cropping period is appropriately termed 'forest fallow'. After an initial invasion of indigenous and naturalized grasses and herbaceous species there is a regrowth of tree species and vines. In bush savannas, the fallow vegetation consists of grasses and herbs with shrubs and sparse regrowth of trees and bushes. After many cycles of cropping and reduced intervals of fallow, grasses predominate and only a few trees and shrubs appear during the fallow period. In grassland savannas grasses and other herbaceous species constitute the main vegetational cover with sparsely scattered woody plants.

Vegetational cover and its effects

In savanna regions the growth of grasses depends primarily on rainfall, but varies with the grass species. For example, the high-grass Andropogoneae savannas of West Africa yield approximately 8.0 tonnes/ha of dry herbage per year, but a great percentage of the organic matter is lost due to annual burning (Nye and Greenland, 1960). Roots of this grass found in the 0–30 cm soil layer contribute about one-third of the potential production. Leaf and stem litter from scattered trees add another 2.5 tonnes annually, unless burned.

Imperata cylindrica represents a biotic climax purposefully created and maintained by carefully regulated methods of cultivation and herding in Southeast Asia (Burkill, 1935; Spencer, 1966; Seavoy, 1975; Dove, 1980). The above-ground dry matter yield of this grass averages about 3.5 tonnes/ha, but that of roots and stolons is double this quantity.

Sown grasses used during a fallow period will provide higher amounts of dry matter than naturalized grasses. A heavy growth of *Pennisetum purpureum* may exceed 30 tonnes/ha yearly and that of

Panicum maximum 15 tonnes/ha. Roots of both add close to another 30 per cent of the top growth (Vicente-Chandler *et al.* 1964).

Soil humus content

The top 5 cm of fertile soils under tropical forests contain from 2.0 to 5.0 per cent organic matter and the next 15 cm has from 0.5 to 2.0 per cent (Jenny *et al.*, 1948, 1949; Nye and Greenland, 1960). Under conditions of other systems the total soil humus content is lower. Unpublished data from Nigeria illustrate the ranges in organic matter encountered (Table 15.1). Uncut forest had the highest amount and continuous cultivation the lowest. A short-term grass fallow improved the organic matter content of the soil to a measurable extent as compared to continuous cultivation. Use of *Pennisetum purpureum* in a 5-year fallow in Uganda increased the organic matter in the upper 15 cm soil layer from less than 3.0 per cent under a cotton–bean–maize rotation to about 4.0 per cent under grass rotation (Stephens, 1967). Studies in other countries showed that significant increases of soil organic matter occurred with a 3–5-year grass or legume fallow (Adil *et al.*, 1966; Brockington *et al.*, 1965; Bruce, 1965; Nye and Greenland, 1960; Pereira *et al.*, 1954; Popenoe, 1957; Tuley, 1968).

Accumulation of nutrients

An important feature of the fallow period is mobilization of soil nutrients from the deeper layers of the soil profile and their accumulation in organs of plants comprising the vegetational cover.

The accumulation of minerals by grassland savanna is considerably less than that of forest fallow (Nye, 1958). An *Andropogon–Hyparrhenia* grass fallow in Ghana, undisturbed for 20 years, mobil-

Table 15.1 *Organic matter content of a forest soil with different management treatment, Ibadan, Nigeria*[*]

Crop and fallow	0–15 cm (%)	15–30 cm (%)
1. Uncut forest	2.30	0.74
2. Permanent pasture (largely *Cynodon robusta*)	1.38	0.84
3. Continuous cultivation 20 years	0.85	0.59
4. Continuous cultivation 15 years + *Pennisetum purpureum* 3 years	0.94	0.60
5. Continuous cultivation 15 years + *C. nlemfuensis* 4 years	1.46	1.22

[*] Unpublished data from the Department of Agronomy, University of Ibadan.

ized slightly less than 25 per cent as much N, 40 per cent as much K but 80 per cent as much P as a 50-year forest in the Congo (Table 15.2). With annual burning, about two-thirds of the N is lost from the grass and the shrub litter (Nye and Greenland, 1960). Thus, soil nutrients are likely to be inadequate for economic crops after this type of fallow.

Short-term fallow of improved grasses effectively mobilizes soil nutrients due to the more rapid growth of these types as compared to native grasses. The quantities approach that of the forest fallow

Table 15.2 *Nutrients stored in the plant residue of secondary forest regrowth and grassland savanna*

Fallow, location, rainfall	Age of fallow (yrs)	Plant part	Dry matter (kg/ha)	Nutrient content		
				N	P (kg/ha)	K
1. Forest – Yangambi	18	Leaves	6 380	140	7	79
Congo (185 cm)		Wood	130 000	330	62	310
(Bartholomew *et al.*,		Litter	5 500	74	2	8
1953)		Roots	30 800	138	34	187
		Total	172 880	682	105	584
2. Forest – Yangambi	5	Leaves	5 600	123	7	78
Congo (185 cm)		Wood	72 800	183	13	245
(Batholomew *et al.*,		Litter	7 260	78	3	14
1953)		Roots	25 300	176	8	110
		Total	110 960	560	31	447
3. Savanna – Ejura,	20	(Grass)				
Ghana (150 cm)		Aerial	7 480	22	7	40
(Nye and Greenland,		Litter	1 210	4	1	4
1960)		Roots	3 740	13	3	11
		(Trees)				
		Leaves	550	7	1	5
		Stems	52 800	91	13	138
		Total	65 780	127	25	198
4. *Pennisetum purpureum* – Bambesa, Congo (175 cm) (Laudelot *et al.*, 1954)	2	Leaves and litter	28 600	292	20	400
5. *P. Purpureum* – Puerto Rico (175 cm) (Vicente-Chandler *et al.*, 1964)	2–4	Leaves and litter	27 700	332	70	550
6. *Panicum maximum* – Puerto Rico (175 cm) (Vicente-Chandler *et al.*, 1964)	2–4	Leaves and litter	25 300	316	48	400

and exceed that of the savanna regeneration (Table 15.2) (Laudelot *et al.*, 1954). If the grass herbage is cut and removed, however, the nutrients are lost, with the exception of those found in the roots. It has been shown that various grasses and legumes accumulated amounts of 40–60 kg/ha of P and 90–800 kg of K in 4 years (Laudelot, 1961; Vine, 1968). *Pennisetum purpureum* significantly increased the N and exchangeable K content on a deep latosol in Uganda (Stephens, 1967). A natural regeneration (largely *Imperata cylindrica*) showed a slight increase over bare soil fallow. A 4-year fallow of *Chloris gayana* in Uganda, with the residue cut and left to decompose, gave a net accumulation of 220 kg/ha of N (Simpson, 1961). Over time, the grasses gradually absorbed subsoil N and deposited it in the topsoil by leaf and stem litter and root residue. That deep-rooted grasses move nutrients from soil regions below 30 cm has been demonstrated by other investigations (Barrow, 1969; Nye and Hutton, 1957). For example, in Nigeria, roots of *Cynodon nlemfuensis* extracted Ca from below 25 cm and deposited it in the upper soil layer by decomposition of leaves, stems and roots (Saleem *et al.*, 1972). In this way soil pH was increased by about one unit over a 1½-year period. At the same time it decreased in the lower layers of soil.

A legume fallow of *Stylosanthes guianensis* undisturbed for 4 years in Uganda gave an increase of 278 kg/ha of N over bare soil fallow. A portion of the N came from fixation by *Rhizobium* (Simpson, 1961). In Nigeria, the upper 30 cm of soil under a 2-year-old *Cynodon–Centrosema* pasture contained 1 760 kg/ha more organic matter and 715 kg more N than under grass alone (Moore, 1962). The accumulation of N by various legumes in the tropics can be substantial, ranging from less than 100 to over 400 kg/ha (Birch, 1962; Bryan, 1962; NAS, 1977; Williams, 1967).

Use of grasses and legumes

A dense cover of grass affords almost as much protection against soil erosion as does forest cover with a closed canopy. The grass foliage breaks the impact of rainfall and disperses the raindrops. Dead leaves and stems provide surface litter that disrupts the force of falling water. The fibrous root system helps to improve soil structure and enhance infiltration of water into the soil. Dead and decaying roots leave channels to aid percolation of water.

Soil-conserving crops
Erosion in regions of alternating wet and dry seasons generally exceeds that in regions having evenly distributed rainfall, since the

vegetation cover is more likely to be continuously maintained. The first rains after a dry season are likely to cause severe erosion on arable land that is exposed after land preparation and before the crops form an effective cover.

Surface runoff of water may be excessive from a bare soil, with heavy soil loss under high and intense rainfall. A cover of grass or canopy of vegetation, on the other hand, can eliminate runoff and reduce soil erosion considerably (Dugain and Fauck, 1959). On a moderate slope in the Andean region of Colombia 200 tonnes/ha of soil was washed from a bare surface, up to 20 tonnes with dense crop cover, but only 1.0 tonne from ungrazed naturalized grassland (Suarez de Castro, 1952). A study of soil erosion in savanna soils of Senegal (Charreau, 1969) showed annual soil loss of 8.4 tonnes/ha under sorghum and 0.2 tonne/ha under grazed natural bushland composed of grasses, forbs, shrubs and scattered trees. Precipitation totalled about 1 150 mm/year, and occurred within a 5-month rainy period. Similar results were obtained in Uganda (Sperow and Keefer, 1975): bare soil – 40.6 per cent runoff of rainfall and 81.5 tonnes/ ha annual soil loss, maize – 23.4 and 34.0, natural grassland grazed and burned – 10.8 and 2.5, closely grazed pasture – 5.6 and 4.4 and well-managed pasture – 0.5 per cent and 0.1 tonne, respectively.

The areas of central and western São Paulo state have undulating and rolling topography and are prone to erosion (Marques *et al.*, 1961). Soil losses from annual crops and pastures were found to vary with soil type under similar rainfall as follows: sandy Bauru (6% slope, 45% clay) 21.1 tonnes/ha under annual crops and 1.2 tonnes/ ha under sown pastures; Terra Roxa (10% slope, 15% clay) 9.5 and 2.7 tonnes/ha, respectively.

In Rhodesia, under rainfall of around 1 000 mm/year, plots on 8 per cent slope with no cover lost 875 tonnes/ha of soil over a 3-year period, while those covered with grass lost 8.25 tonnes (Hudson, 1957). Using *Pennisetum purpureum* as a cover crop and cut periodically, the soil loss on 4.5 per cent slope totalled 7.0 tonnes/ha. In Nigeria, on 3.0 per cent slope, soil losses from bare soil and from soil covered with maize, cowpea, and *Cynodon nlemfuensis* occurred at the ratio of 11: 6.7: 1.7: 1, respectively, over the growing season of about 5 months.

If soil-conserving crops are grazed, the vegetational cover determines to a great extent the amount of water runoff and soil erosion. With a dense cover they may be nil. Under intense rainfall, surface runoff of water may occur but with little soil loss. If overgrazing occurs, bare soil spots develop so that both water runoff and soil erosion may be quite heavy.

Creeping legumes, such as *Pueraria phaseoloides* and *Centrosema pubescens*, reduce soil erosion when sown so that cover is obtained prior to the onset of the rainy season. In Puerto Rico (Smith and

Abruña, 1955) reported erosion control of 98 per cent under *P. phaseoloides* as compared to bare soil. Heavy erosion may occur, however, if sowing is delayed so that the soil is exposed during early rains. Erect-growing legumes are less satisfactory and relatively ineffective in preventing soil erosion, especially if planted at the beginning of the rainy season.

Mixed agriculture

An alternative for forest or natural regeneration fallow is that of using soil-improving grasses and legumes in a system of ley farming that facilitates the utilization of modern methods of arable cropping and tillage operations. Fields that are deliberately sown with grasses and legumes for rotation with arable crops are considered part of a 'regulated' ley system where fertilization, fencing and grazing management are practised.

In the tropics we can distinguish between this and the 'unregulated' ley. An unregulated ley refers to the regeneration of naturalized and spontaneous-appearing grasses and herbs (Ruthenberg, 1980). Generally, communal grazing is practised but there is little management of grazinglands. Residues from harvested crops also provide sources of fodder and feedstuff. In some instances, a system of systematic folding (enclosing cattle at night) is combined with semi-permanent cultivation, and manure from the enclosure (kraal) is transported to the fields as fertilizer for crops.

Regulated ley systems are less common in the tropics than in the temperate zones. They have been established on the large, mechanized tobacco farms of Zambia and Rhodesia, the wheat and pyrethrum farms in the highlands of Kenya, the maize smallholdings in the medium altitude of Kenya, and the mixed crops smallholdings in the medium altitude of Uganda (Ruthenberg, 1980).

Alternate grazing and arable cropping is practised to a limited extent in the high and medium elevations of Latin America. In the Andean zone naturalized grazinglands are frequently broken and planted to potatoes–wheat–maize. After several years of cropping the land reverts to grass for grazing. In the cooler regions, *Pennisetum clandestinum* and *Trifolium repens* regenerate naturally but other species are sometimes sown, such as *Dactylis glomerata, Lolium perenne, L. multiflorum, Festuca arundinacea* and *Trifolium pratense*. In the medium elevation, *Melinis minutiflora, Paspalum* and *Axonopus* spp. regenerate. In the upper limits of the warm climate, crops such as cotton, beans and maize follow a naturalized grass fallow, after which grasses such as *Digitaria decumbens, Panicum maximum* and *Hyparrhenia rufa* are established after an undetermined cropping period. Crops are frequently fertilized so that the subsequent grass benefits from the residual nutrients, especially P. Furthermore, the naturalized legume population increases after the cropping period. There is no regular pattern of alternating the arable crops and the grass fallow, but grazing management is practised and attention given to animal health.

Information on the value of the ley for maintaining soil fertility has come from a number of studies where crops follow grazingland (Brockington *et al.*, 1965; Clarke, 1962; Ellis, 1953; Foster, 1971; Jameson and Kerkham, 1960; Peat and Brown, 1962; Rains, 1963; Scaife, 1971; Tiley, 1965). In general, results have shown that grass leys are as effective as naturally regenerated fallows that are not grazed. Compared to continuous cultivation, they reduce surface runoff and soil erosion, enhance the acceptance of rainfall and improve the nutrient and physical status of the soil. This depends to a great extent on stocking rate and intensity of grazing. With overgrazing the grass cover diminishes, bare soil spots appear and soil compaction occurs. These factors arrest the beneficial effects of the ley on physical properties of the soil.

Although economic yields of arable crops may be increased following the grass ley, the total harvest does not usually compensate for the loss of cropping during the ley period (Fauck *et al.*, 1969; Jameson and Kerkham, 1960; Peat and Brown, 1962; Scaife, 1971). It appears, therefore, that mixed agriculture will be satisfactory only if crop yields can be considerably increased during the arable break. In addition, appreciable returns must be realized from animal production on the ley. This probably cannot be achieved without the use of fertilizers, improved managerial skills and practices of crop and animal husbandry, as well as modifications in the land tenure systems of some countries.

Crop yields after fallow

The ultimate aim of alternating arable crops and a fallow period is to maintain or increase economic yields within a suitable and economically viable cropping system. The first crop yield after a long natural fallow in the forest and humid savanna is usually abundant. There is usually a marked drop in the second season with a steady decline in succeeding years. In regions of adequate rainfall a rotation of about 3 years of grass or legume and 3 years of cropping maintains the soil fertility level at a moderate level, provided the soil is not initially infertile or exhausted by intensive cropping (Allan, 1965; Jurion and Henry, 1969; Foster, 1971; Scaife, 1971; Smith, 1962; Stephens, 1960; Nye and Greenland, 1960). The planted fallow does not bring the soil fertility level back to the original level of the long-term forest fallow. It is usually no more effective than the natural fallow in the savanna regions when the latter is not annually burned. *Pennisetum purpureum* when grown during the fallow period has consistently shown advantages over other grass in improving soil fertility and increasing crop yields. Other grasses that have been used include *Panicum maximum, P. coloratum, Chloris gayana, Setaria anceps, Brachiaria ruziziensis, Cynodon, Paspalum* and *Andropogon* species.

A deficiency of available N often occurs in the first crop following a grass fallow, especially if the test crop is a cereal (Berlier *et al.*, 1956; Greenland, 1958; Griffith, 1951; Hosking, 1937; Meyers, 1959; Mills, 1953; Rounce and Milne, 1936; Theron and Haylett, 1953). The incorporation of a large bulk of herbage and roots with a wide carbon–nitrogen (C–N) ratio creates an immediate demand for N by soil organisms that are responsible for decomposition of the residue. The imbalance is further complicated by the low accumulation of nitrates in the soil when the dominant cover is grass. Apparently, a difference exists among species in the repressive effect on nitrate formation. An acute N deficiency occurs during the first cropping year after long-term *Andropogon* and *Hyparrhenia* species, and some deficiency may continue into the second year. The persistency of nitrate shortage depends to some extent on the age of the grass and the species composition of the vegetation cover, as it is more pronounced after long periods of fallow (more than 3 years). It can be offset to some extent by application of N to the ensuing crop. Furthermore, after the mass of herbage and roots have had time to decompose there may be a notable increase in crop yields because of mineralization of the added organic matter. With short-term fallows of *Andropogon* the adverse effect becomes less noticeable, and the problem has not been observed with 3- and 4-year fallows of *Pennisetum purpureum, Panicum maximum* and *Cynodon*.

Under grass cover and natural regeneration fallow, soil fertility increases by the mobilization of soil nutrients through roots and their accumulation in the plant residues left on, and incorporated into, the soil. This is usually reflected by increased crop yields after the fallow period. There is a problem of low nitrate accumulation under some grass species which is associated with fewer nitrifying microbial organisms within the rhizosphere of these species. The reasons for low nitrate formation have not been clearly determined (Meiklejohn, 1962; Nye, 1958; Vine, 1968), but is probably associated with a lack of available carbohydrates needed by the microorganisms.

References

Adil, M. L., Y. V. Kathavate and **A. Sen** (1966) Changes in soils associated with continuous growth of some vegetation, *Pl. Soil* **25**, 73–80.

Allan, W. (1965) *The African Husbandman*, Oliver and Boyd: London.

Barrow, N. J. (1969) The accumulation of soil organic matter under pasture and its effect on soil properties, *Aust. J. Exptl Agric. Anim. Husb.* **9**, 437–44.

Bartholomew, W. V., I. Meyer and **H. Laudelot** (1953) Mineral nutrient immobilization under forest and grass fallow in Yangambi (Belgian Congo) region, *NEAC Série Scientifique* **57**, 1–27.

Berlier, Y., B. Dabin and **N. Leneuf** (1956) Physical, chemical and microbiological comparison of forest and savanna soils on the tertiary sands of the lower Ivory Coast, *Trans 5th Intl. Cong. Soil Sci.* E, 499–502.

Birch, H. F. (1962) Effect of a legume on soil nitrogen mineralization and percentage nitrogen in grasses. In E. L. Gaden, Jr. (ed.), *Global Impacts of Applied Microbiology*, Biotech. Bioenerg. Symp. no. 1, FAO, Addis Ababa, pp. 299–303.

Brockington, N. R., T. H. Stobbs, P. W. Newhouse and **G. A. Wadsworth** (1965) The effect of leys on soil fertility in the annual cropping areas of Uganda, *Afr. Soils* **10**, 473–81.

Bruce, R. C. (1965) Effect of Centrosema pubescens on soil fertility in the humid tropics, *Queensld J. Agr. Sci.* **22**, 221–6.

Bryan, W. W. (1962) The role of the legume in legume/grass pastures. In *A Review of Nitrogen in the Tropics with Particular Reference to Pastures*, Commonw. Bur. Past. Fld Crops, Bull. 46, Farnham Royal, pp. 147–60.

Burkill, I. H. (1935) *A Dictionary of the Economic Products of the Malay Peninsula*, 2 vols, Crown Agents for Colonies: London.

Charreau C. (1969) Influences des techniques culturales sur le développement du ruissellement et de l'érosion en Casamance, *Agron. Trop. (Paris)* **24**, 836–42.

Clarke, R. T. (1962) The effect of some resting treatments on a tropical soil, *Emp. J. Exptl Agr.* **30**, 57–62.

Dove, M. D. (1980) Symbiotic relationships between human population and Imperata cylindrica: The question of ecosystematic succession and preservation in South Kalimantan, *Intl. Symp. Consr.*, Kuala Lumpur (mimeo).

Dugain, F. and **R. Fauck** (1959) Erosion and runoff measurement in middle Guinea, *Third Inter-Afr. Soils Conf.*, Dalaba, pp. 597–600.

Ellis, B. S. (1953) The soil farming systems in S. Rhodesia with special reference to grass leys, *Rhod. Agric. J.* **50**, 379–90.

Fauck, R., C. Moureaux and **C. Thomann** (1969) Bilans de l'évolution de sols de Sefa (Casamance, Senegal) après quinze annes de culture continue, *Agron. Trop. (Paris)* **24**, 263–301.

Foster, K. L. (1971) Crop yields after different elephant grass ley treatments at Kawanda Research Station, Uganda, *E. Afr. Agric. For. J.* **37**, 63–72.

Greenland, D. J. (1958) Nitrate fluctuation in tropical soils, *J. Agric. Sci. (Camb.)* **50**, 82–92.

Griffith, G. (1951) Factors affecting nitrate accumulation in Uganda soils, *Emp. J. Exptl Agric.* **19**, 1–12.

Hosking, N. R. (1937) The unsuitability of certain virgin soils to the growth of grain groups, *E. Afr. Agric. J.* **2**, 313–14.

Hudson, N. W. (1957) Erosion control research, *Rhod. Agric. J.* **54**, 297–323.

Jameson, J. D. and **R. K. Kerkham** (1960) The maintenance of soil fertility in Uganda. I. Soil fertility experiments at Serere, *Emp. J. Exptl Agric.* **29**, 179–92.

Jenny, H. F., F. T. Bingham and **B. Padilla-Saravia** (1948) Nitrogen and organic matter contents in equatorial soils of Colombia, South America, *Soil Sci.* **66**, 173–86.

Jenny, H. F., S. P. Gessel and **F. T. Bingham** (1949) Comparative decomposition rates of organic matter in temperate and tropical regions, *Soil Sci.* **68**, 419–32.

Jurion, F. and **J. Henry** (1969) Can primitive farming be modernized? *INEAC Serie HORS*, Belgian Co-op. Dev. Office: Brussels.

Laudelot, H., R. Germain and **W. Kessler** (1954) Preliminary results on the chemical dynamics of grass fallows and of pastures at Yangambi, *Trans. 5th Intl. Cong. Soil Sci.* **2**, 312–21.

Laudelot, N. (1961) Dynamics of tropical soils in relation to their fallowing techniques, unpubl., FAO (quoted by H. Vine, 1968).

Laudelot, N. (1962) Fallowing techniques of tropical soils, *UN Confr. Appl. Sci and Tech. E/Confr.*, 39/c/26.

Marques, J. A., J. Bertoni and **G. Berreto** (1961) Perdas por erosao no estado de Sâo Paulo, *Bragantia* **120**, 1143–82.

Meiklejohn, J. (1962) Microbiology of the nitrogen cycle in some Ghana soils, *Emp. J. Exptl Agric.* **30**, 115–26.

Meyers, J. A. (1959) Fluctuations of mineral N in soils under food crops, *Third Inter-Afr. Soils Conf.*, Dalaba.

Mills, W. R. (1953) Nitrate accumulation in Uganda soils, *East Afr. Agric. J.* **19**, 53–4.

Moore, A. W. (1962) The influence of a legume on soil fertility under a grazed tropical pasture, *Emp. J. Exptl Agric.* **30**, 239–48.

National Academy Sciences (1977) Leucaena: *Promising Forage and Tree crop of the Tropics*, Washington, D.C.

Nye, P. H. (1958) The relative importance of fallows and soils in storing plant nutrients in Ghana, *J. W. Afr. Sci. Assoc.* **4**, 31–41.

Nye, P. H. and **D. J. Greenland** (1960) *The Soil under Shifting Cultivation*, Commonw. Bur. Soils Tech. Comm. No. 51, Farnham Royal, England.

Nye, P. H. and **R. G. Hutton** (1957) Some preliminary analyses of fallows and cover crops at the West African Institute for Oil Research, Benin, *J. W. Afr. Inst. Oil Palm Res.* **2**, 237–43.

Peat, J. E. and **K. J. Brown** (1962) The yield response of rain grown cotton at Ukiriguru in the Lake Province of Tanganyika, I and II, *Emp. J. Exptl Agric.* **30**, 305–14.

Pereira, H. C., E. M. Chenery and **W. R. Mills** (1954) The transient effects of grasses in the structure of tropical soils, *Emp. J. Exptl Agric.* **22**, 148–60.

Popenoe, H. (1957) The influence of the shifting cultivation cycle on soil properties in Central America, *Ninth Pacif. Sci. Cong.*, Bangkok, pp. 72–7.

Rains, A. B. (1963) *Grasslands Research in Northern Nigeria, 1952–62*, Samaru Misc. Paper No. 1.

Rounce, N. V. and **G. Milne** (1936) The unsuitability of certain virgin soils to the growth of grain crops, *E. Afr. Agric. J.* **2**, 145–8.

Ruthenberg, H. (1980) *Farming Systems in the Tropics*, Clarendon Press: Oxford.

Sah, S. N., C. R. Prasad and **W. Hassan** (1968) Studies on changes in some physiochemical properties of the red loam soils of Chotangpur under different crop rotations, *Indian J. Agric. Chem.* **1**, 43–52.

Saleem, M. A. M., H. R. Chheda and **L. V. Crowder** (1972) Effects of lime on herbage production and chemical composition of Cynodon IB.8 and on some chemical properties of the soils, *E. Afr. Agric. For. J.* **40**, 217–26.

Scaife, M. A. (1971) The long-term effects of fertilizers, farmyard manure and leys at Mwanhala, Western Tanzania, *E. Afr. Agric. For. J.* **37**, 8–14.

Seavoy, R. E. (1975) The origin of tropical grasslands in Kalimantan, Indonesia, *J. Trop. Geogr.* **40**, 48–52.

Simpson, J. R. (1961) The effects of several agricultural treatments on the nitrogen status of a red earth in Uganda, *E. Afr. Agric. For. J.* **26**, 158–63.

Smith, R. M. and **F. Abruña** (1955) *Soil and Water Conservation Research in Puerto Rico 1938 to 1947*, Puerto Rico Agric. Res. Bull. 124.

Smith, C. A. (1962) Tropical grass/legume pastures in Northern Rhodesia, *J. Agric. Sci. (Camb.)* **59**, 14–18.

Spencer, J. E. (1966) *Shifting Cultivation in Southeastern Asia*, Univ. Calif. Press: Berkley.

Sperow, C. B. Jr., and **R. F. Keefer** (1975) *An Introduction to Soil Science Applied to East Africa*, West Virginia Univ. Publ.: Charlotesville.

Stephens, D. (1960) Three rotation experiments with grass fallows and fertilizers, *Emp. J. Exptl Agric.* **28**, 165–78.

Stephens, D. (1967) Effects of grass fallow treatment in restoring fertility of Buganda clay loam in South Uganda, *J. Agric. Sci. (Camb.)* **68**, 391–403.

Suarez de Castro, F. (1952) Perdidas de suelo por erosión, *Agric. Trop. (Colombia)* **8**, 35–7.

Theron, J. J. and **D. Y. Haylett** (1953). The regeneration of soil humus under a grass ley, *Emp. J. Exptl Agric.* **21**, 86–98.

Tiley, G. E. D. (1965) The effects of the grass ley on arable crops with special reference to the elephant grass areas of Uganda, *Afr. Soils* **10**, 409–15.

Tuley, P. (1968) Stylosanthes gracilis, *Herb. Abst.* **38**, 88–94.

Vicente-Chandler, J., R. Caro-Costas, R. W. Pearson, F. Abruña, J. Figarella and **S. Silva** (1964) *The Intensive Management of Tropical Forages in Puerto Rico*, Univ. Puerto Rico Agric. Expt. Sta., Bull 187.

Vine, H. (1968) Developments in the study of soils and shifting agriculture in tropical Africa. In R. P. Moss (ed.), *The Soil Resources of Tropical Africa*, Cambridge Univ. Press, Ch. 5.

Williams, W. A. (1967) The role of Leguminosae in pasture and soil improvement in the neotropics, *Trop. Agric. (Trin.)* **44**, 103–15.

Chapter 16

The role of legumes

Legumes are desirable components of pastures and grazinglands where (a) N is a limiting factor for optimal growth of associated grasses, (b) a need exists for increased crude protein in herbage available for grazing, (c) extended grazing into the dry season is desired and (d) stability of the pasture system is given high priority. Legumes remain green when grasses mature and become dry with the onset of drought. Most legume species have a tap-root that penetrates deeper into the soil than the fibrous roots of grasses. In addition, the enlarged root section just below the crown of many legumes provides a nutrient and water reserve during stress periods. Forage legumes are characterized by a symbiotic relationship with bacteria that infect their roots and transform atmospheric nitrogen into a form utilized directly by the legumes and made available to associated plants. The bacteria depend on the legume for basic nutrients needed to sustain their life functions. The nitrogen-fixing capability of legumes in mixtures stabilizes the soil nitrogen and organic carbon content in the root zone to a greater extent than grasses grown alone.

In evaluating the presence and value of legumes in pastures, consideration must be given to three relationships (a) legume–*Rhizobium* symbiosis, (b) legume–grass symbiosis and (c) plant–animal symbiosis.

Legume–bacteria symbiosis

Fixation of nitrogen by legume–bacterial symbiosis depends upon the presence of an effective strain of *Rhizobium* that infects the legume root, availability of soil and plant nutrients, environmental conditions, competition of associated vegetation and intensity of defoliation or grazing. A majority of tropical legumes are readily infected with a wide range of rhizobial types commonly present in soils of the tropics and subtropics. Some tropical legumes require a specific type of *Rhizobium* for effective nodulation and N fixation.

An effective strain of rhizobium must be readily accessible to the legume seedling for nodulation to occur and effective symbiosis to begin.

The bacteria and nodulation

The legume-infecting bacteria are small, motile, rod-shaped cells found naturally in some soils where they adopt a free existence from their hosts and are capable of prolonged persistence. They prefer the environment of the rhizosphere, however, and multiply rapidly under the influence of root secretions. Usually the more fertile soils contain greater numbers of rhizoba. In soils of low fertility some rhizobial strains have not been introduced or fail to persist (Harris, 1953, Norris, 1966). The rhizobia gain access to the plant by penetrating root hairs at about the time the first true leaf emerges. A classic study of this process was made in *Trifolium* spp. (Nutman, 1948, 1962, 1971). A large number of root hairs may be infected, but only a few develop nodules. Inside the nodule, the bacteria utilize mono- and disaccharide sugars from the host plant and in turn provide nitrogen for direct use by the plant. Essential features of the nodule consist of a central tissue composed of 10 000–40 000 host cells per nodule and a vascular system connection to the roots of the host plant. The central tissue contains up to 100 000 bacteria and is pink due to the red pigment leghaemoglobin that is characteristic of bacterial activity and N fixation (Bergersen, 1971). Near the point of the root meristem nodules are filled with actively multiplying, rod-shaped bacteria. Further along the root the bacteria are transformed into inactive, non-motile bacteroids but still surrounded by the red pigment. Still further back the bacteroids disintegrate and their contents are absorbed by the plant (Norris, 1966).

The shape, number and distribution of nodules vary with the host species. The total number on a single plant ranges from a few to several thousand. Among tropical legumes spherical nodules are most common, although cylindrical and digitate shapes also occur. Generally, they enlarge by increasing in size in all directions. If an effective *Rhizobium* strain is present, the nodule will be filled with red tissue. This can be readily seen on splitting a nodule. If the bacteria are not effective symbionts, nodules are likely to be smaller and will appear white or greenish inside. The lack of pink colour does not necessarily indicate ineffective association unless the plant shows symptoms of N deficiency. Nodules of a green and vigorous-growing plant may be small and white because of a copious supply of nitrate (Diatloff, 1972; Norris, 1966).

Nodule longevity is closely related to the growth habits of the host. Effective nodules on herbaceous plants usually begin to degenerate in 35–70 days, but on woody species they may persist longer, sometimes for over 1 year. As nodules degenerate the tissue becomes brown and soft, along with wrinkling of the surface. At this stage they are usually sloughed off at the narrow connection to the root (Harris, 1953).

Factors affecting nodulation and N fixation

The factors affecting legume–*Rhizobium* symbiosis fall into categories attributable to the bacteria and its host, nutritional and moisture aspects of the soil and effects of the climate (Herridge and Roughley, 1976; Norris and Date, 1976; Philpotts, 1975; Souto and Döbereiner, 1970; Wilkins, 1967). Calcium is important as a nutrient for functioning of the bacteria and the host plant. The bacteria are Ca-insensitive and require only trace amounts, which they usually obtain from the rhizosphere. The host plant, on the other hand, requires a larger supply of this nutrient, especially in nodule formation. Soil acidity does not interfere with rhizobial activity unless extreme. The effect of low pH is largely a function of Ca nutrition. In acid soils Ca may be unavailable due to excess Al or Mn, both of which antagonize Ca uptake. Phosphorus deficiency in many soils will severely limit both the formation of nodules and N fixation. Molybdenum plays an essential role in symbiotic N fixation, since it is required for the formation of the enzyme nitrogenase. Cobalt is essential for the growth of *Rhizobium* and is related to haemoglobin content of the nodule. Sulphur affects the host plant in respect to protein synthesis. In S-deficient soils this process declines, reducing the plant need for N and thus nodulation is indirectly affected. High levels of nitrates in the soil solution tend to inhibit nodulation, due to uptake of N from this source by the legume plant and reduced demand for N produced by the nodule. A strong interaction exists between soil moisture and soil temperature in respect to rhizobial activity. Some strains of rhizobia can tolerate temperatures of 50 °C in dry soil, but are killed by similar exposure to moist soils. Optimum temperatures for most strains range from 25 °C to 35 °C.

Types of Rhizobium and cross-inoculation

An effective symbiosis between the bacteria and legume must be quickly established for the legume to compete with the vigorous tropical grasses. An effective *Rhizobium* is thus a strain of bacteria that infects the host plant roots and rapidly begins to fix N.

The genus *Rhizobium* is composed of several species separated into groups as based on cross-inoculation trials. In early legume–*Rhizobium* studies it was noted that certain closely related groups of legume species would nodulate effectively when bacteria from each was used to inoculate the others. By 1932 eight cross-inoculation groups had been recognized: alfalfa, clover, pea, bean, lupin, soybean, lotus and cowpea (Fred *et al.*, 1932). This concept was readily accepted by agronomists, since it offered a convenient identification of *Rhizobium* types for field inoculation purposes. They are not so neatly separated on basis of morphology, cultural and biochemical characteristics because of integrating types. One biochemical grouping is of significance, however (Norris, 1965, 1967). Some strains produce an alkaline reaction when grown in cultural media due to accumulation of certain end-products. Others produce an acid reaction. The alkali-producers are slow-growing, monoflagellate bacteria associated with tropical legumes, and are classified as the cowpea type. The acid-producers are associated with legumes such as lucerne and clovers that are adapted to the temperate zone. These legumes grow on fertile soils of a much less acidic nature than most tropical soils. If the soils are acidic, i.e. below a pH of 5.3–5.5, then application of lime is needed for their successful establishment and forage production. Research with temperate legumes began much before tropical legumes, so that use of lime in the temperate zone was commonplace. Furthermore, early research with tropical legumes was largely carried out by those trained in the temperate zone who had a preconceived association of lime with legume. Thus, it followed that application of lime was commonly recommended for tropical legumes when these researchers found soils having a pH of 5.0–4.0 or even lower, and before consulting a legume bacteriologist familiar with legume–*Rhizobium* symbiosis in the tropics.

Most tropical legumes are tolerant of acid soils and possess a high efficiency in the extraction of Ca from soils low in this element (Andrew and Norris, 1961). The cowpea types of rhizobia required by most tropical legumes also tolerate acid soil conditions (Norris, 1965). This type of legume–*Rhizobium* symbiosis forms the basis for Ca requirement in the establishment and maintenance of tropical as compared to temperate legumes. Applications of lime are rarely needed, especially if superphosphate dressings are made, except in very infertile and highly acidic soils.

A number of tropical legumes are promiscuous in respect to effectiveness of the common cowpea type of *Rhizobium* (Table 16.1). The major reason for the outstanding success of Townsville stylo and Siratro is their symbiotic promiscuity. Natural nodulation of most legumes occurs, but the rhizobia are not always effective. Certain legume species and selections within species require a specific *Rhizobium* strain for effective symbiosis, e.g. centro, *Desmodium* and stylo (Bowen, 1959; Diatloff, 1968; Halliday, 1978; Norris, 1972). Extreme specificity was found in a South African species of *Lotononis bainesii* by Norris (1958). Herbage production of this species depends on inoculation with a particular red stain of *Rhizobium* obtained originally from Africa. This clearly demonstrated that introduction of specific legumes into a region where they have

Table 16.1 *A guide to inoculum and lime requirements of legumes used in tropical pastures (Norris, 1967, 1970)*

Species	Common name	Lime response	Inoculum
*Cajanus cajan	Pigeon pea	No	Cowpea
*Calopogonium mucunoides	Calopo	No	Cowpea
Centrosema pubescens	Centro	No	Specific
Desmodium intortum	Greenleaf	No	Desmodium
D. uncinatum	Silverleaf	Rarely	Desmodium
Glycine wightii	Glycine	Occasionally	Cowpea
Lablab purpureus	Rongai dolichos	No	Cowpea
Leucaena leucocephala	Leucaena	Yes	Specific
Lotononis bainesii	Miles lotononis	No	Specific
*Macroptilium atropurpureum	Siratro	No	Cowpea
*M. lathyroides	Phasey bean	No	Cowpea
*Macrotyloma axillare	Archer dolichos	No	Cowpea
*M. unifolorum	Leichardt dolichos	No	Cowpea
Medicago sativa	Lucerne	Yes	Medicago
* Pueraria phaseoloides	Kudzu, Puero	No	Cowpea
*Stizolobium deeringianum	Velvet bean	No	Cowpea
Stylosanthes guianensis	Oxey finestem	No	Specific
*S. guianensis	Schofield	No	Cowpea
*S. humilis	Townsville stylo	No	Cowpea
Trifolium repens	White clover	Yes	Clover
Trifolium semipilosum	Kenya white clover	Yes	Specific
*Vigna luteola	Dalrymple vigna	No	Cowpea
*V. radiata	Mung bean	No	Cowpea
*V. unguiculata	Cowpea	No	Cowpea

* Indicates a promiscuous species that will normally nodulate with native cowpea *Rhizobium* even if not inoculated.

not grown before makes it imperative that a supply of reliable inoculant be available.

The selection of a specific strain of *Rhizobium* for high N-fixing capacity in the laboratory is not sufficient to ensure effective legume symbiosis. It must be followed by field trials to check adaptation to soil conditions where the inoculum is to be used. An example of the need for field observation was noted with *Leucaena leucocephala* at the Cunningham Tropical Pastures Laboratory in Queensland (Norris, 1972). This legume is a specialized species that tends to occur naturally on limestone soils in the tropics and responds to application of lime when planted on acid soils. An acid-producing strain of *Rhizobium* isolated in New Guinea was incorporated in commercial inoculants. None the less, serious nodulation problems had been recorded with *Leucaena* on acid soils. A subsequent *Rhizobium* strain (CB 81) with a slightly alkaline reaction isolated at the Cunningham Laboratory was compared with the commercial type (NGR8) in a soil of pH 5.0 (Table 16.2). Seedlings dug 8 weeks after inoculation showed that the acid-type NGR8 produced nodules when seeds

Table 16.2 *Nodulation performance of Leucaena leucocephala using acid- and an alkali-producing strain of Rhizobium in a soil of pH 5.0 (Norris, 1972)*

Rhizobium	Inoculation method	Plants nodulated after storage*		Mean nodules per plant*	
		1 day (%)	28 days (%)	1 day (no.)	28 days (no.)
CB81 alkali-producer	1. No inoculation	0	0	0	0
	2. Inoculated just before sowing	25	43	0.82	1.37
	3. Inoculated with sticker[†] only	19	1	0.36	0.02
	4. Lime-pelleted	33	25	0.81	0.61
NGR8 acid-producer	1. No inoculation	0	0	0	0
	2. Inoculated just before sowing	1	0	0.01	0
	3. Inoculated with sticker[†] only	1	0	0.01	0
	4. Lime-pelleted	31	23	0.77	0.53

* Plants examined at day 28 after inoculating and sowing.
[†] 4% 'methofas' (methyl cellulose).

were lime-pelleted. The alkali-type CB81 produced nodules with or without a sticker for adherence of inoculum to the seeds. This strain was used to replace the NGR8. Useful techniques for selecting, testing and maintaining rhizobia are described by Norris and Date (1976), Date and Norris, (1979) and Vicent (1970). Date (1970) discussed the factors affecting survival of inoculum, production of nodules by the legume plant and methodology of inoculant control service. Roughley (1970) reviewed the preparation and use of modern seed inoculants and Norris (1970) prepared a résumé about nodulation of tropical pastures.

Seed inoculation

Whenever seeds are purchased, the buyer should ascertain that the proper commercial inoculant is available. Legume seeds are usually inoculated just prior to sowing by using a water slurry, i.e. mixing the peat inoculum with water and wetting the seeds with the mixture so that a thin film of the black inoculum covers a portion of each seed. In some instances, the seeds are mixed with a commercial sticker, such as 4 per cent methyl cellulose (diluted to 1 or 2%) or 40 per cent gum arabic (diluted to 15%), then adding the fresh inoculum with further stirring or mixing. Honey diluted with water can be used as the sticker substitute. A word of caution: the inoculum should be fresh. If the peat has dried out it is likely that the *Rhizobium* has died.

Recently seeds have been pelleted on a commercial basis, thus eliminating the need for further inoculation. For most tropical legumes rock phosphate should be used as the base for pelleting (Norris, 1967). The cowpea types of rhizobia are acid tolerant, and their survival is greater with this material as compared to lime used as the pellet base. Lime pelleting is more appropriate for seeds of *Leucaena*, however, since the specific *Rhizobium* for this legume produces an alkaline reaction.

Nitrogen fixation

Cultivation of leguminous plants provides the most economical means of adding N to the soil–plant–animal system. A widespread belief prevailed for some years that tropical legumes did not greatly benefit a mixed pasture or were less efficient in their capability to fix N than temperate legumes. Since the early 1960s evidence to discount this belief has been accumulated from widely dispersed trials of several tropical legumes. Studies showed a wide range in quantities of N fixed, but the amounts were generally comparable under tropical and temperate conditions (reviews by Bryan, 1962, 1969; Jones, 1972; Thomas, 1973). For example, the annual production of 567 kg/ha of N by *Leucaena leucocephala* obtained in Australia (Hutton and Bonner, 1960) approaches that of 604 kg produced by *Trifolium repens* in New Zealand (Sears and Evans, 1953). Such quantities exceed those commonly obtained under experimental conditions, and undoubtedly under pasture conditions of the animal producer. In Hawaii, N fixation varied between 97 kg/ha annually with *Desmodium canum* to 264 kg/ha with *D. intortum* in mixture with Pangola grass (Whitney and Green, 1969–Table 16.3) and 360 kg/ha with *D. intortum* grown with Pangola and elephant grasses in cinder culture (Whitney *et al.*, 1967). In other locations, the values for *D. uncinatum* ranged between 110 and 145 kg/ha (Henzell *et al.*, 1967; Suttie, 1968); *Centrosema pubescens* in pure stand and with various grasses, 123–280 (sources cited by Teitzel and Burt, 1976); *Glycine wightii*, 175 kg (Gethin-Jones, 1942); *Macroptilium atropurpureum* with a grass, 70–150 kg (Jones *et al.*, 1967; Miller and van der List, 1977; Johansen and Kerridge, 1979); *Pueraria phaseoloides* with molasses grass, 145–190 kg (Vicente-Chandler *et al.*, 1953); *Stylosanthes guianensis* mixed with *Heteropogon contortus*, 96 kg (cited by Henzell and Norris, 1962); *S. humilis* with buffel grass, 110 kg (Norman, 1960). The differing abilities of legumes to fix nitrogen are positively correlated with total dry weight production (Erdman and Means, 1962). Thus, the amount of N added to the soil depends to a great extent on herbage yields (Jones, 1972).

Table 16.3 *Performance of grass alone, legume alone and mixtures of grass + legume (Whitney and Green, 1969)**

Treatment	Dry matter yields (kg/ha)			Crude protein yields (kg/ha)		Nitrogen[†] (kg/ha)	
	Grass	Legume	Legume (%)	Grass	Legume	Fixed	Transferred
1. Pangola grass	3 780	–	–	260	–	–	–
2. Pangola + *Desmodium canum*	5 460	2 050	27	460	420	99	32
3. Pangola + *D. intortum*	4 290	7 670	64	490	1 420	264	37
4. *D. intortum*	–	10 630	100	–	2 070	290	–

* Experiment located on a Latosol; plots 2.1 × 6.1 m; basic fertilizer of P, K, micro-nutrients and limestone; data are averages of 2 years.
† N fixation estimated by subtracting N yield of Pangola grass from total N of mixture; apparent transfer of N estimated by subtracting N yield of Pangola from N yield of grass fraction in mixture.

This is illustrated in Fig. 16.1 which shows a near linear relationship of estimated N fixation and dry matter yield of legumes. Studies in Queensland showed that approximately 3 000 kg/ha of legume dry matter was needed for a return of 100 kg/ha of N (Jones *et al.*, 1967; Thairu, 1972). Differences in N fixation by legumes are influenced by the species, cultivar or selection, fertilizer rates, grass–legume combinations or climatic variables. The total legume N yield is the sum accumulation of aerial and root growth, as well as nodules that are sloughed off as the plant develops and matures. Once a legume is firmly established with a dense canopy there is considerable leaf fall and root senescence, resulting in litter on the soil surface and organic residue in the soil. Decay of this material is a useful source of N. Whitney *et al.* (1967) estimated that fallen leaves of *D. intortum* supplied about 2.2 g/ha of N weekly.

The ability of *Rhizobium* strains to fix nitrogen can be determined by growing legume plants in sand culture using a 'Leonard' jar technique (Norris and Date, 1976). A bottle 7.5 cm in diameter and 25 cm in height, with the base cut off, rests in an inverted position in a glass jar 10 × 10 × 15 cm tall. The jar contains 500 ml of a nutrient solution. The neck of the bottle is plugged with cotton wool and the bottle filled with 900 g of washed sand in which plants grow. The cotton plug allows absorption of nutrients and keeps the plants irrigated from below. Bacteria are cultured in yeast–mannitol broth with 1 ml of the broth pipetted on to the sand surface soon after seedling germination. After 8–12 weeks, depending on the legume species, plants are washed free of sand for counting nodules and determining dry weight of the whole plant. Relative N-fixing ability of the *Rhizobium* is determined by correlation curves of dry weight.

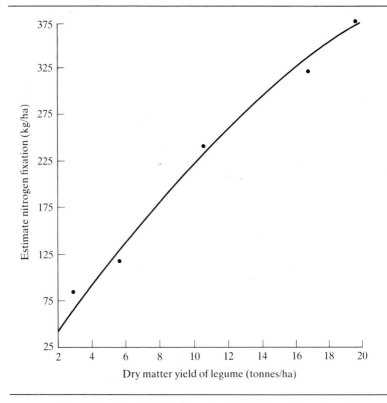

Fig. 16.1   The relationship of N fixation and yield of the aerial portion of legumes (Data of Whitney *et al.*, 1967; redrawn from Jones, 1972).

This technique gives an estimate of *Rhizobium* efficiency, but field-testing is needed to confirm success in nodule formation and N fixation. Haydock, *et al.* (1980) pointed out that the most effective strains can be selected on the basis of dry weight yields of whole plants or tops without the need for N analyses.

N fixation by legume/*Rhizobium* symbiosis is measured by one of the following methods:

1. The most accurate estimates have been obtained from lysimeters, i.e. growing legumes in field plots where N can be measured in the whole plant or aerial portion, in the soil and in drainage water. It is assumed that N taken up by legumes is obtained solely from two sources – fixation from the air or available N in the soil. This implies that other sources of N, e.g. non-symbiotic and rainfall, are negligible or balance out in quantities lost and not measured.
2. Total N produced in aerial and root portions of the legume plant

via Kjeldahl analysis. In some studies only the aerial part is utilized because of complications in removing roots from the soil. A conversion factor is applied for estimating the quantity of roots.

3. Indirect measurements. (a) Comparison of N yield of grass–legume combinations with that of pure grass fertilized with different amounts of N, and taking comparable yields as the amount fixed by the legume. In this method it must be realized that grass species differ in their efficiency of N utilization. Also, management treatments such as cutting interval and height influence N fixation. (b) Comparison of crop yields following grass–legume versus grass alone. Care must be taken to assure that the field crops follow closely the pasture crops and bare soil is reduced to a minimum. (c) Comparison of liveweight production of animals grazing grass–legume mixtures and pure grass pastures.

4. Acetylene-reducing ability based on nitrogenase activity, i.e. the characteristic of this enzyme to reduce C_2 to NH_3 is supplied to an actively N_2–fixing system (Hardy *et al.*, 1973). Ethylene is measured by gas chromatography. With this technique rate of N fixation is determined at a given time rather than over a period of time. Furthermore, it requires expensive equipment.

Formation of stem nodules

The symbiotic relationship between rhizobia and leguminous plants occurs almost exclusively in roots. Aurora (1954) noted an exception in *Aeschynomene indica* which formed nodules in the lower parts of stems which remained under water for some time. Upper stems nodules (up to 50 cm above the rooting zone) were noted by Yatzawa and Yoshida (1979) and the bacterium was found to be identical to that of root nodules. The development of stem nodules was enhanced by soil N deficiency (Yatzawa and Susilo, 1980).

Legume – grass symbiosis

The effects of legume–grass associations can be expressed in terms of dry matter yields, N content of herbage, yield of protein, animal intake of forage and animal productivity. Factors to consider in legume–grass symbiosis include:

(a) Grasses can utilize nitrogenous compounds fixed by the legume.

(b) Effectively nodulated legumes make little or no demand on soil N.

(c) Nitrogen content of the grass herbage in mixtures will be higher than in unfertilized pure stands, while N content of the legume will likely be lower than in pure stands.

(d) Neither yields as much in mixtures as grown separately in pure

stands, assuming grass receives fertilizer N, but the combined yield may exceed that of legumes grown alone.
(e) Yield of legumes is likely to be more depressed in the association than that of grasses.

Dry matter production

Dry matter yields of grass–legume mixtures usually exceed that of grass alone not receiving fertilizer N (Fig. 16.2). This is illustrated by data shown in Table 16.3 where *Desmodium canum* (Kaimi clover) + Pangola about doubled and *D. intortum* (greenleaf desmodium) + Pangola more than tripled the yield of grass alone (Whitney and Green, 1969). Similar results have been reported in other studies (e.g. Asare, 1974; Grof and Harding, 1970; Horrell, 1964; Kretschmer *et al.*, 1974; Ramirez *et al.*, 1976; Skerman, 1977). Yields will vary widely, depending on such factors as species of grasses and legumes, inherent fertility of soil, fertilization (amount and time of application), percentage of legume, available soil moisture, intensity of defoliation (interval and height), light intensity and temperature. In general, legumes become established more slowly than grasses so that the latter may predominate for 1 or 2 years. Application of phosphate to an established grass–legume pasture often causes a response of the legume so that the percentage of this component exceeds that of grass for a period of time, i.e. until the N

Fig. 16.2 *Glycine wightii* provides nitrogen in combination with Pangola grass (left), extends the grazing season and adds protein to the available herbage. Pangola grass alone on the right.

level of the soil increases so as to favour growth of the grass. In fact, the botanical composition of the mixture usually alternates between grass and legume, depending on the N status of the soil and timely application of maintenance P fertilizer. Application of N to the grass–legume association diminishes the legume fraction. High-yielding grasses grown alone and heavily fertilized with N outyield grass–legume combinations or legume grown alone. For example, yields up to 30 tonnes/ha of Pangola grass dry matter are obtained with high levels of applied N (see Ch. 9).

Transfer of nitrogen from legumes

Movement of legume N to the associated grass and forbs primarily occurs in the following ways:

1. The main mechanism is by sloughing-off and decay of nodules, thus releasing N into the soil solution (Butler and Bathurst, 1956). Once nitrates are released from the nodules into the soil solution, they are readily available to roots of the associated species but could also be reabsorbed by roots of the legume.

2. Decomposition of legume residue is an important source of N for associated plants. Leaf and petiole fall adds considerable organic matter to the soil surface and releases N into the soil as decomposition occurs. Older roots also provide N as they are replaced by newer growth. Once the legume plant matures, of course, all of the remaining tissue is converted into organic matter for immediate and delayed release of N.

3. Direct transfer, i.e. excretion from the nodules, of N from legume plants to the associated grass or forbs is usually negligible, for example no more than 0.5 per cent (Henzell, 1962, 1970). Simpson (1965), however, found evidence of greater excretion from lucerne than other legumes and attributed it to the more continuous development of nodules. Where more appreciable amounts of N transfer are involved, they could be derived by passive release from actively growing roots and tops or result from normal senescence, death and decomposition of older tissues.

   Underground transfer during the growth of perennial legumes may not exceed 1 or 2 per cent in the short term (Vallis *et al.*, 1967) but over a longer period, e.g. 2 years, the amounts vary from less than 5 to more than 30 per cent (Henzell, 1962; Miller and van der List, 1977; Whitney *et al.*, 1967; Whitney and Green, 1969; Johansen and Kerridge, 1979). The difficulty of studying underground transfer of N from legume plants is related to the inadequacy of available techniques. A number of studies have been carried out in pot culture, using sand and nutrient solution. Under such conditions the root associations may be different than in the field. Furthermore, under controlled conditions the effects of such things as variable defoliation, water stress, shading, alternating temperatures, animal effects, etc. are eliminated. Undoubtedly, these factors have a significant effect on N release in the field. Transfer of N from legume to grass in the field is often estimated by subtracting the N yield of unfertilized grass from that of the grass fraction comprising the legume–grass combination. The exact pathways for underground transfer in legume–grass symbiosis will remain at best a rough estimate until more precise methodology is devised to measure the quantities of N involved (Whiteman, 1980).

Herbage quality considerations

An outstanding feature of the herbage available for grazing in a legume–grass association is the increased N content as compared to grass alone and the constancy of N found in the legume component. Legumes contain higher amounts of N than grasses whether grown alone or in mixture. *Desmodium intortum* in Hawaii ranged from 2.72 to 3.36 per cent over the season in mixture with Pangola grass (Whitney and Green, 1969). The grass component ranged from 1.22 to 2.18 in association with legume and from 0.94 to 1.63 per cent in pure stand not fertilized with N. The N content of leaves in certain legumes, e.g. Leucaena, may exceed 4.0 per cent N, being almost equal to lucerne (NAS, 1977). Grass heavily fertilized with N will seldom exceed 2.75 per cent N in juvenile herbage.

The sustained N level is legume herbage over a period of time is of pronounced significance. The legume plant when effectively nodulated has a 'built-in' supply of N, so that fluctuation within the plant is largely independent of that in the soil. This reservoir of plant N maintains a rather constant supply of N as plants age (Table 16.4). For example, the decline in N concentration of Siratro with increasing age of leaves and stems was quite small over a 16-week regrowth period (Jones, 1972). A similar stability of N content was noted with several tropical legumes in Trinidad (Guyadeen, 1951). In Nigeria, only small changes occurred in centro over a 5-month

Table 16.4 *Sustained nutritive value of Siratro (Macroptilium atropurpureum) with advancing age (Jones, 1972)*

Regrowth (wks)	Leaflets (%)			Stems (%)			Digestibility whole plant (%)
	N	P	K	N	P	K	
4	4.63	0.26	2.33	2.25	0.27	3.86	69.0
8	4.38	0.25	2.03	2.10	0.24	3.67	63.8
12	4.26	0.24	2.48	2.08	0.23	3.58	65.5
16	4.12	0.26	2.18	2.00	0.28	3.50	65.5

growing period, even though length of plant stems increased from 0.4 to over 4.0 m. This was in marked contrast to 20 grasses examined, in which the N content declined from 2.2 in 4-week-old herbage to near 1.0 per cent in 17-week-old herbage (Oyenuga, 1957). The preserved N value of legume herbage is pertinent to animal production. Voluntary intake of legume–grass forage and digestibility will remain at a higher level than grass alone. Furthermore, this advantage continues into the dry season after the grass component becomes rank in growth.

With annual legumes such as Townsville stylo (Fig. 16.3), the nutrient content diminishes more rapidly as the plants age. This is particularly noticeable during maturation and seed development (Fisher, 1969). At the time of flowering, however, a redistribution of nutrients from the leaves to the seeds takes place, so that seeds contain high concentrations of N, P and S. Despite this process, the nutrient composition of legume herbage remains at a higher level than of grass herbage. Furthermore, animals consume a high percentage of pods and seeds of the mature legumes and thus benefit from this plant portion in their diet.

Botanical composition

The legume–grass composition in the temperate zone is roughly in the ratio of 40 : 60. The percentage of legume that constitutes an

Fig. 16.3 Townsville stylo, the annual legume that revolutionized cattle-growing in northeastern Queensland, Australia, and has potential in areas with a long, well-defined dry season.

'adequate proportion' of the mixture has not been established in the tropics and subtropics. In the Northern Territory of Australia dry-season performance of steers was examined when allowed to graze Townsville stylo pastures with varying proportions of sown perennial *Cenchrus setigerus* (birdwood grass). The data given in Table 16.5 show that liveweight gain at 2.75 steers/ha was related to the percentage of legume in the pasture (Norman and Stewart, 1964). At the time of grazing the herbage of legume and grass had matured and turned brown. In this region of Australia livestock graze native-type pasture in the 6-month rainy season of about 890 mm and make satisfactory liveweight gains. They lose weight in the dry season unless supplemented with annual pastures such as bulrush millet or Townsville stylo.

Table 16.5 *Dry-season performance of steers on sown pastures with varying proportions of grass and legume (Norman and Stewart, 1964)*

Pasture mixture					
Birdwood grass (%)	Annual grass (%)	Townsville stylo (%)	N content of herbage at beginning (%)	Liveweight gain (kg/head)	Period of gain (wks)
51.5	25.7	22.8	0.75	22	8
9.9	45.4	44.7	1.12	109	20
–	37.4	62.6	1.3	216	22

Liveweight gains in southeast Queensland over a 7-year period of study and under annual precipitation of 1 650 mm (70% in October–April) were also related to percentage of legume fraction (Evans, 1970). At a stocking rate of 2.47 animals/ha (250 kg initial liveweight), mean annual liveweight gains of 290, 336 and 545 kg/ha were obtained with average pasture legume contents of 13, 20 and 35 per cent, respectively. The proportion of legume showed seasonal variation, reaching a maximum in the warmer months. A résumé of 3 years' data that shows the effect of legume content on liveweight gains at different stocking rates and two P levels is given in Fig. 16.4. The legume component differed between phosphate levels, being higher with 252 kg than 125 kg/ha annual topdressings of superphosphate. These were the average of several grass–legume mixtures. There was an obvious trend of higher animal output with increasing legume content and a marked difference between phosphate treatments at both stocking rates. A pronounced decline in legume percentage occurred with increased grazing pressure (Table 16.6). Within a 3-year period it dropped by about one-third when 2.47 steers/ha grazed continuously and about one-half when 1.23 steers/ha grazed continuously. The botanical composition data were taken in December during the warmer and wetter season. Note that

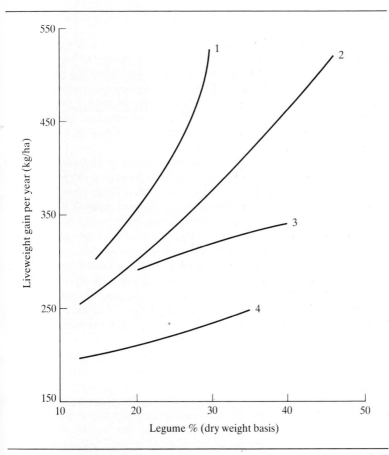

Fig. 16.4 Legume content of pasture and liveweight gains. Lines 1 and 2: 2.47 steers/ha, 252 and 125 kg/ha of superphosphate. Lines 3 and 4: 1.23 steers/ha, 252 and 125 kg/ha of superphosphate (Evans, 1970).

Table 16.6 *Effect of stocking pressure on liveweight gain and legume content of legume–grass pastures in Queensland (Evans, 1970)*

Stocking rate (no.)	1965–66		1966–67		1967–68	
	LWG* (kg/ha)	Legume† (%)	LWG* (kg/ha)	Legume† (%)	LWG* (kg/ha)	Legume† (%)
2.47	488	36	384	26	280	12
1.65	369	39	315	27	294	17
1.23	313	38	319	33	254	20

* LWG = liveweight gain; average of several fertilizer treatments.
† Legume percentage in December of each year during the warm, rainy season; average of several mixtures.

the heavier grazing pressures more adversely affected the legume components.

A decline in the legume fraction affects the overall efficiency of herbage utilization. In a study with Pangola grass mixed with legume, Minson and Milford (1967) found that digestible dry matter intake was linearly related to the legume content in excess of 10 per cent. They recorded an increase in voluntary intake of herbage, an increase in dry matter and crude protein digestibilities and a corresponding rise in liveweight gain as the proportion of legume increased from 0 to 100 per cent.

An increase in liveweight gain as influenced by the legume component is clearly depicted by the graphs in Fig. 16.5. This evidence from grazing experiments strongly suggests that N supply is the main factor affecting animal production in a pasture–animal system. In legume–grass pastures animal output is obviously determined by the amount of legume in the mixture. Chemical analyses generally show that the N content of grass in mixture with legumes is less than 1 per cent until the legume fraction exceeds 40 per cent. This does not necessarily mean that a desirable legume : grass ratio should possess more than this figure since the animal is not dependent on the grass component for its major source of N. It strongly indicates, however, that the animal producer should make an effort to maintain at least 30–40 per cent of legume in a pasture mixture to obtain optimal animal production.

Shrub legumes

In discussing legume–grass symbiosis we tend to think primarily of herbaceous legumes as comprising a part of the mixture. Another means of obtaining benefits of legumes in grazed vegetation is by utilizing shrub or tree legumes that have the capacity to develop an effective symbiosis with N-fixing bacteria. The use of trees and shrubs as sources of herbage in tropical and subtropical regions was reviewed by Gray (1970) and Whyte (1947). Information about *Leucaena leucocephala*, the shrub legume most extensively used as fodder and having far-reaching potential, was given by Gray (1968), Hill (1971), NAS (1977), Jones (1979) and Blom (1980). This legume can be grown in pure stand, usually in rows of about 1.0 m width, and grazed alternatingly with grass pastures or in random plantings as browse, or used as fresh feed, being cut as needed.

In Australia steers of about 130 kg liveweight were used to graze Leucaena pastures for (a) 1 month then N-fertilized Pangola 2 months, (b) Leucaena 2 months and Pangola 1 month and (c) Pangola full time (Blount and Jones, 1977). Over a 308-day grazing period, steers on the three treatments gained about 90 kg/head. Those on Pangola full time gained weight in a linear manner, but

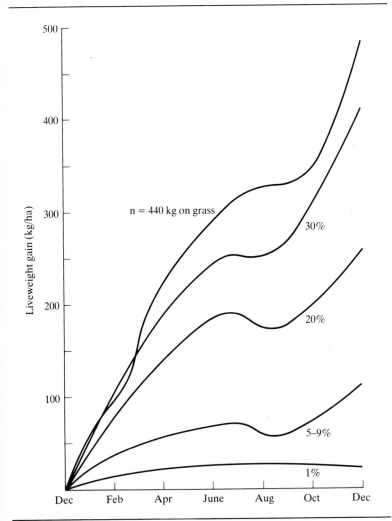

Fig. 16.5 Effect of legume component in grass–legume pastures and N-fertilized grass on beef cattle liveweight gains. Redrawn from data in Australia (Evans, 1968).

those alternated between Leucaena–Pangola gained more when on Leucaena. The reason for alternating between legume and grass was an attempt to reduce the toxicity effect of mimosine when cattle graze Leucaena (Jones *et al.*, 1976). The effect of mimosine is cumulative, so that animals having continuous access to Leucaena make poor liveweight gains and may develop enlarged thyroids. The toxicity effect is overcome by removing the legume from the diet of animals (Blount and Jones, 1977). The rotation of one month

Leucaena and two months Pangola, however, was not sufficient to allow complete dissipation of the toxic substance. In general, mimosine *per se* is not the major problem of toxicity but rather a breakdown product DHP (3-hydroxy-4-1(H)-pyridone) which occurs in the rumen. DHP prevents iodination of tyrosine, results in goitre, and reduced levels of thyroxine in the serum (Jones *et al.*, 1976, 1978).

Leucaena can be interspersed in grass pastures, either as randomly spaced plants or placed in rows separated about 2–3 m apart. In this way sunlight filters through the leaf canopy onto the associated grass. The combination provides a highly productive, two-level pasture. Leucaena grows slowly in the early stages of development and must be protected from grazing. Once established, however, it is compatible with the most vigorous grasses such as Pangola, guinea and Signal grass (*Brachiaria decumbens*). Under a favourable system of grazing management, the combination remains well balanced so that neither the Leucaena nor the grass dominates (NAS, 1977). Once the legume plants reach a height of about 1.0 m they can be browsed, but overstocking reduces forage yields and understocking allows plants to grow too tall for cattle to reach. Overly tall plants can be cut back after which they regenerate new growth.

Leucaena may be cut as daily fresh feed and taken to enclosed or tethered animals. In the Philippines young beef stock are kept near the household and force-fed Leucaena leaves and twigs. Rapid liveweight gains are made over a short period of time. Beef cattle in Queensland were lot-fed sugar-cane supplemented with Leucaena, meat meal or urea + sodium sulphate (Siebert *et al.*, 1976). Leucaena leaves were stripped from shrubs 5.0 m tall in sufficient amounts for 3 days of feeding at 1.5 kg/animal daily. Cattle consumed 71 gm organic matter per kg $LW^{0.75}$ per day and digested 57.9 per cent of the diet with Leucaena as the protein source. Those fed a diet of meat meal and urea + sodium sulphate digested 64.4 and 66.2 per cent, respectively. Differences came about due to the diminished rumen retention of Leucaena. Liveweight gains on all treatments ranged from 0.5 to 0.6 kg/head per day.

Some advantages of shrub legumes in animal feeding as compared to the herbaceous forms include (taken in part from Humphreys, 1974);

1. Long-term persistence due largely to a more extensive root system and to drought tolerance.
2. Retention of leaves further into the dry season.
3. The two-tiered canopy in shrub legume–grass combinations favours more efficient light utilization and increased dry matter production.
4. Plants can be grown along road or pathways, hedges, fences, as

well as for erosion control, wind-breaks and shade and utilized as a feedstuff.
5. Branches can be lopped for a dry season feed reserve.
6. Rapid growth is regenerated after cutting for wood or charcoal and plants tolerate repeated cutting.

Plant–animal symbiosis

Output of the grazing animal provides a sensitive measure of the quality and quantity of available pasture herbage. Legumes in pastures may benefit animals in the following ways:
1. They are essential for maximum health and production, as well as for improved calf crop.
2. They give a more even distribution of high-quality herbage over the year, i.e. extending into the dry season.
3. Voluntary intake of legume–grass mixtures exceeds that of pure grass herbage due to reduce rumen retention, thus increased animal intake with a higher proportion of legume in the diet.
4. The amount and digestibility of crude protein is higher in legumes than associated grasses, thus animals can utilize grasses of low nutritive value.

Increased liveweight gain from legume–grass pastures
The combination with grasses is illustrated by the information given in Fig. 16.5 (Evans, 1968). Data used to construct this graph were taken from several experiments in Queensland:
(a) The curve of 1.0 per cent legume represents a native pasture that produced about 20 kg liveweight gain annually in a rainfall area of 600 mm.
(b) The 5–9 per cent legume curve came from an unfertilized *Paspalum dilatatum–Axonopus affinis–Trifolium repens* pasture that yielded about 100 kg/ha of liveweight gain in a region with slightly more rainfall.
(c) The curves of 20 and 30 per cent legume denote several pastures of different grasses in mixtures with *Desmodium intortum, D. uncinatum, Lotononis bainesii, Macroptilium lathyroides* and *T. repens* and located in an area of about 1 650 mm rainfall.
(d) A Pangola grass pasture fertilized with 440 kg/ha of N and grazed with about 5 animal units per hectare in the same region.

A significant feature of the graph is the increased liveweight gain due to the higher percentages of legume in the available herbage. The upper curves are averages of the entire year and represent a decline in production during the cooler months, but a high rate of gain during the hot, summer season. Presence of the legume increased the supply of crude protein for animal production, along with a greater intake of digestible dry matter. The grasses, however, gained little benefit from N being fixed and transferred by the legume. Generally, the legume component must exceed 40 per cent before the protein content of the associated grass is improved.

Improved liveweight gains were reported in other studies, e.g. stylo pastures in Queensland (Gillard *et al.*, 1980); a 3-year accumulation of 880 kg/ha from steers grazing guinea grass + centro and 660 kg from guinea grass alone in Queensland (Grof and Harding, 1970); 526 kg/ha from centro + stylo + grass and 127 kg from grass alone in Uganda (Stobbs, 1965); 150 kg/ha from stylo + *Hyparrhenia rufa* and 90 kg from *H. rufa* alone in Peru (Santhirasegaram, 1975).

Effect of legumes plus fertilizer

A combination of legume and fertilizer usually permits heavier stocking rates and improved animal productivity as compared to grass alone. This was evidenced from an experiment in the *Heteropogon contortus* grasslands located in Queensland (Shaw and Mannetje, 1970; Shaw, 1978). Applications of fertilizer or oversowing with Townsville stylo gave an animal response, but the combination was more dramatic (Table 16.7). About 50 per cent more animals were carried with a five-fold liveweight gain per hectare as compared to the native pasture. In addition, cattle on the improved pasture were ready for slaughter at an earlier age than those on native pasture.

The impact of Townsville stylo in terms of increased quantity and quality of herbage available to grazing animals was even more striking in the drier area of Northern Australia (Norman, 1970). Under conditions of a prolonged dry season of about 8 months and about 900 mm annual precipitation, animal gain was linearly related to the number of days cattle had access to the legume. Over a 630-day period, including wet and dry seasons, animals gained 60 kg/head on native pasture and no access to Townsville stylo pasture. Those given full access to the legume–native grass pastures gained 280 kg/head. During the dry season of about 112 days animals on native pastures lost nearly 40 kg/head, thus experiencing a 'stop-and-go' pattern of growth. Those on Townsville stylo gained weight throughout the year, including 60 kg/head in the dry season. In the Philippines, liveweight gains on Townsville stylo–native grass pastures, stocked at 10 animal units per hectare, were 126 kg/ha, while that on native pasture was 28 and 17 kg/ha at stocking rates of 0.5 and 1.0 animal units, respectively. Animals on native pasture lost weight in the dry season, but not on legume–grass pasture (Siota *et al.*, 1976).

Table 16.7 *The influence of oversown Townsville stylo and fertilizer on animal production from native Heteropogon contortus grasslands in southeast Queensland, 1959–66 (Shaw and Mannetje, 1970)*

Pasture treatment	Stocking rate (no./ha)	Gain (kg/head)	Gain (kg/ha)
1. *H. contortus*	0.61	47	29
2. *H. contortus* + fertilizer*	0.61	100	62
3. *H. contortus* + Townsville stylo	0.74	121	93
4. *H. contortus* + Townsville + fert.	0.95	149	148

* 125 kg/ha superphosphate, 37.5 g/ha molybdenum and 63 kg/ha potassium choloride topdressed annually.

In the Philippines, grasslands dominated by *Imperata cylindrica* (cogon grass) were grazed with one animal unit per hectare (Magadan *et al.*, 1974). This is the commonly accepted stocking rate of the weedy grass which is fairly nutritive in the juvenile stage. It rapidly becomes stemmy, highly lignified and not well accepted by livestock. Two improved pastures were established in the 'cogonland', one by burning the cogon, preparing the land and transplanting pará grass, and the other by sowing centro with pará (Table 16.8). The legume–grass pasture was superior in terms of animal performance, forage production and suppression of weeds as compared to the others. Utilization of available legume–grass herbage was slightly less than the pará fertilized with 100 kg/ha of N, being 41 and 44 per cent, respectively. This suggested that a heavier stocking rate might have been used but at the risk of greater invasion of weeds. Obviously, the stocking rate of one animal unit per hectare of Cogon was too severe and led to deterioration of the pasture, as noted by the in-

Table 16.8 *Liveweight gains and herbage production from native Imperata cylindrica and improved grass alone and with centro in the Philippines* (Magadan et al., 1974)*

Pasture[†]	Stocking rate (no.)	Liveweight gains[‡] (kg)			Herbage[§] (t/ha)		Weed increase (%)
		Daily/ head	Yearly/ head	Yearly/ ha	Prod/ yr	Consm/ yr	
1. Native *Imperata*	1.0	0.27	100	100	14.5	5.0	59
2. Pará + centro	2.0	0.42	153	305	56.4	23.0	18
3. Pará + N₁	2.0	0.36	130	260	40.2	17.8	26
4. Pará + N₂	3.0	0.28	104	310	41.7	21.9	47

* Rainfall about 2 400 mm/year with a 3–6-month dry season.
† No fertilizer; (2) 50 kg/ha P_2O_5 annually; (3) 100 kg/ha N + 50 kg P_2O_5; (4) 200 kg/ ha N + 75 kg P_2O_5; pará is *Brachiaria mutica*; centro is *Centrosema pubescens*.
‡ Pastures were divided into three paddocks and rotationally grazed with Nellore–Brahmin steers.
§ Forage yields obtained by clipping fifteen 1 m² paired, open and caged quadrats in each replicate after each grazing cycle for calculation of amount produced and consumed. Samples were hand-separated to determine botanical composition. Centro comprised about 24 per cent of the vegetation of pasture 2.

crease of weeds, 59 per cent after 1 year, being up from 7.0 per cent at the start of the trial.

Legumes and milk production

There are fewer data to substantiate the beneficial effects of legumes in mixtures or alone on milk production, as compared to grass alone. In New South Wales, Australia, cows that continuously grazed *Pennisetum clandestinum* + *Glycine wightii* for 10 months produced 12.2 kg of milk per head, as compared to 9.8 kg for those on N-fertilized *P. clandestinum* (Holder, 1967). In southeast Queensland, under commercial conditions on steep hillsides, this grass–legume mixture raised the butter fat per cow from 33 to 66 kg (Luck and Douglas, 1966). At the Samford Experiment Station, Hamilton *et al.* (1970) found that Jersey cows grazing *Lablab purpureus* averaged 10.5 kg/head of milk and those grazing *Setaria anceps* or *Chloris gayana* produced only 7.0 kg daily. Stobbs (1971), in contrast, obtained less milk from *Macroptilium atropurpureum* and *Desmodium intortum* than from fertilized *Digitaria decumbens*, 7.7 and 9.0 kg/head per day, respectively. He attributed the difference to prolonged grazing on these trailing legumes with large leaves.

Legumes and calving

The legume component in the plant–animal symbiosis not only enhances liveweight gain but may also increase the conception rate of cows and decrease calf mortality. A comprehensive study was carried out in Florida to evaluate beef production and reproduction of lactating cows on legume–grass and all grass pastures. Warnick *et al.* (1965) examined the effects of pastures consisting of a mixture of Pangola and bahia grass in programme 1 and these two grasses plus white clover in programme 2. During the winter months the grasses were dormant because of frost but the clover remained green. Under this condition replacement heifers received a low or high protein supplement. These effects were less notable than type of pasture on weight gains and reproduction. Reproductive response was measured by calving and weaning percentages. Heifers on clover–grass pastures averaged 47 kg heavier at 30 months of age than those on pure grass pastures. Furthermore, the former had a 96 per cent calving rate compared to 81 per cent for the latter. Eighty-nine per cent of heifers on legume–grass pastures weaned calves, whereas 76 per cent of those on all-grass pastures weaned calves. In another study Koger *et al.* (1961) obtained an average of 83.5 per cent calves weaned from cows grazing clover–grass pastures and 64.0 per cent from those grazing all grass. Grasses consisted of one-third Coastal bermuda, one-third Pangola and one-third bahia grass. The legume was white clover. Reproduction of cows on all-grass pastures was nil during the first 2 years of the experiment, but improved to equal that of the legume–grass during the last 2 years. The beneficial effects of white clover in the mixture were attributed to improved herbage quality, seasonal distribution of forage and possibly some qualitative influence on the occurrence of oestrus in young cows. Calf weaning weights did not differ between all grass and legume–grass pastures. A second 3-year phase of the same experiment (Koger *et al.*, 1970) showed less dramatic differences between the two pasture types. Birth rate of cows on legume–grass was 97 per cent, and on all grass with 202 kg/ha of N annually, 91 per cent. Calf weaning rates on the pastures averaged 89 and 95 per cent, respectively. Thus, reproductive efficiency of the all-grass pasture improved in the second 3-year phase and reproductive efficiency by including white clover in the grass pastures was less beneficial. Guyton, *et al.* (1980) reported that legume-grass combinations (greenleaf desmodium, Siratro and glycine plus digit grass and buffel grass) gave lamb weaning percentages of 97.5 per cent compared to grass alone of 88.7 per cent.

Application of superphosphate to Townsville stylo pastures in north Queensland improved breeding performance as well as beef production (Edye *et al.*, 1972; Rural Research, 1973). Conception rate increased from 74 per cent with annual topdressing at 110 kg/ha of superphosphate to 85 per cent with 330 kg/ha. This compared to 66 per cent conception rate of cows grazing on an unfertilized pasture mixture of Townsville stylo and spear-grass. Moreover, most cows on the unimproved pasture conceived every second year. The increased conception on fertilized legume pastures resulted from improved nutrition of herbage and of lactating cows during mating. Tropical legumes such as Townsville stylo growing in soils with low P have the capacity to extract sufficient amounts of this element for their development. The P content of legume herbage from such soils, however, frequently drops below 0.10 per cent and is inadequate for maintenance of animal health. The critical plant level of this nutrient and others has been established. For example, optimal forage yields of Townsville stylo occurred when herbage contained 0.17 per cent of P, centro 0.16 per cent, Glycine 0.23 per cent and *Desmodium* 0.24 per cent (Andrew and Robins, 1969). Unfortunately, information about critical herbage levels may not be directly related to standards set for animal performance.

Animal selectivity of herbage and structure of pasture canopy

Legume–grass combinations provide a wider array of selectivity by grazing animals than either component grown alone. The proportion of plant species consumed, however, may be different than the botanical composition of available herbage might indicate. The material selected by grazing animals depends on such factors as species and species combinations, season of grazing, stage of plant growth, sward structure and individual animal preferences. At certain times of the year tropical grasses are preferred over legumes. For example, cattle preferred *Hyparrhenia rufa* to *Stylosanthes guianensis* when moved into a grass–legume pasture in Uganda (Stobbs, 1969). Also, at the Cunningham Pasture Laboratory in Queensland, botanical composition of oesophageal fistula samples showed that cattle preferentially selected leaves of the grass *Urochloa mosambicensis* following early rains at the beginning of the wet season (CSIRO Report, 1974–75). This preference occurred in pastures where the green *Urochloa* leaves comprised only 10 per cent of the dry matter available in a Townsville stylo dominated sward. At the nearby Samford Station, the diet of grazing animals on a legume–grass pasture contained only 3 per cent of Siratro in the spring and early summer, whereas this legume comprised 40 per cent of the herbage consumed in autumn. Further study of penned animals showed that cows preferred autumn-grown compared to summer-grown Siratro when free choice was allowed. This indicated that some change in the legume plant had affected acceptability.

The large leaves of legumes such as Siratro, *Desmodium, Pueraria* and *Lablab* interfere with the biting behaviour of grazing animals (Stobbs, 1973). Cattle draw herbage into their mouths by a swipe of the tongue, tearing material from the plants. Thus, the size arrangement and distribution of leaves in the canopy influence the ease with which they select the ingested material. A measure of herbage yield without describing its physical distribution, therefore, may give no indication of the ease whereby the feed can be harvested by animals. Leaf content of a species may be low, but if readily accessible a relatively high proportion might be eaten. Vertical distribution of herbage within tropical pasture swards show a loose arrangement, as compared to temperate pastures. For example, leaf bulk density of *D. intortum* average only 32 kg/ha/cm, compared to a range of 47–104 kg/ha/cm for white clover (Stobbs and Imrie, 1976). Less quantity of herbage would, therefore, be harvested per bite when grazing a *Desmodium* pasture. Bite size of tropical legumes is generally smaller than that of grasses. For example, the amount of Siratro herbage taken in by one swipe of the tongue was less than Pangola and Setaria due to the greater accessibility and bulk density of the grasses (Stobbs and Hutton, 1974). The amount of stems in the Siratro sward exceeded that of the grasses. This, and the larger leaves, interfered with bite size from the legume. Under conditions of low sward density in which leafy material is difficult to harvest, cattle compensate for small bite size by increasing the time spent grazing and the number of bites taken. Feeding behaviour studies suggested that animal output could be improved by (1) selecting more leafy pasture plant types that are easily harvested, (2) fertilizing pastures to produce dense leafy swards and (3) imposing grazing management practices that maintain sward density near to the optimum so as to maximize bite size (Stobbs, 1973).

Animal disorders and tropical legumes

Certain legumes contain compounds that are potentially toxic to livestock. Mimosine is an amino acid that causes loss of hair in cattle and wool in sheep, breeding problems in swine and horses, leads to enlarged thyroid glands and depressed liveweight gain in cattle and has been implicated in calf mortality (NAS, 1977).

The presence of tannins in some tropical legumes may adversely affect voluntary intake and digestibility (Marshall *et al.*, 1979). The higher quantities of these compounds in *Desmodium intortum* (Rotar, 1965) may be the reason for lower *in vitro* digestibility as compared to *Macroptilium atropurpureum* (Jones, 1969). Stobbs (1971), however, did not encounter differences in milk production or liveweight gains of cows grazing both legumes.

Bloat is uncommon in the tropics, largely because of the low soluble protein content and the presence of precipitating agents in tropical legumes (Jones and Lyttleton, 1971). However, bloat has been recorded when cattle have grazed *Lablab purpureus* (Hamilton and Ruth, 1968).

Oestrogens in temperate legume species appear to interfere with reproduction, but tropical legumes do not contain sufficient amounts to be potentially dangerous (Bindon and Lamond, 1966; Little, 1969). Indospicine, a toxic amino acid in *Indigofera spicata*, causes liver damage and restricts the use of this species unless lines low in this substance can be developed (Hegarty and Pound, 1968).

The elusive tropical legume

The Gramineae predominate as constituents of the vegetation found in native and naturalized grazinglands and in most permanent pastures in the tropics and subtropics. Thus, grasses are the most important component of the diet of grazing animals. This is the prevailing situation in the monsoon regions of Africa, Asia and Australia as evidenced by surveys reported by Whyte (1968). Herbaceous legumes comprise a greater proportion of the vegetation in many South American grazinglands. It is not uncommon to be able to stand and count within the limit of vision five or more legume species in naturalized pastures and in improved pastures sown with grasses alone. Their contribution to total herbage yields, however, is highly variable and fluctuates over the year. At times they may supply 20 per cent or more of the total production in a localized area, but at other times their contribution is nil or insignificant. The difficulty encountered in attempting to modify tropical grazinglands was vividly portrayed by Whyte (1962) in his discussion of 'the myth of tropical grasslands'. He stated that finding and developing legumes suitable for grass–legume mixtures in the tropics would be a formidable task and feasible only for limited areas. It has been estimated that some 50 genera of tropical legumes are likely to contain species of potential value for pasture improvement (Williams *et al.*, 1976). Species, selections and cultivars have been identified and evaluated under many environmental conditions but relatively few are considered important. The list includes *Calopogonium mucunoides, Centrosema pubescens, Desmodium intortum, D. uncinatum, Glycine wightii, Leucaena leucocephala, Lotononis bainesii, Macroptilium atropurpureum, M. lathyroides, Macrotyloma axillare, Medicago sativa, Pueraria phaseoloides, Stylosanthes guianensis, S. hamata, S. humilis, S. capitata, S. scabra, Vigna luteola, Zornia diphylla* (not in order of importance). Several selections or strains exist among some species but only one, Siratro (*Macroptilium atropurpureum*) was developed by breeding. Unfortunately, despite continued research with varied intensity for about half a century, forage legumes make little contribution to total herbage production and overall animal production in the tropics and subtropics. For example, in 1972 it was estimated that in Queensland, Australia, where a major part of legume-based pasture research has been carried out, only about 1.7 per cent of the total 148 million ha in rural holdings had been planted to improved pastures (Commonwealth Statistics, cited by Jones, 1972). In 1978, the replacement of native pastures by improved species represented less than 5 per cent of the potential area (Mannetje, 1978). The percentage would be even smaller in other parts of the tropics. Thus, the animal industry in tropical and subtropical countries relies primarily on native and naturalized pastures for production of meat, milk, wool and hides.

In view of the benefits gained by growing legumes in combination with grasses it is surprising that they have not been more widely used. One can surmise that most livestock producers are not aware of these benefits, have encountered difficulty in management of grass–legume pastures, have experienced low economic returns from use of sown legumes or consider that extensive livestock production is more economic than intensive pasture management. Traditionally, livestock production in the tropics and most of the subtropics has been carried out on an extensive rather than an intensive basis. Often livestock-keeping is relegated to marginal lands unsuited for crops and unproductive in terms of herbage output.

Under such conditions the so-called improved pasture legumes cannot be utilized without soil amendment and better pasture management practices.

Productive legumes may be absent for one or more of the following reasons (taken in part from Humphreys and Jones, 1975):

1. Inadequate supply of nutrients

P is the nutrient most likely to be needed. Many tropical and subtropical soils are deficient in this element and application of P fertilizers is necessary for legume establishment and maintenance. In some instances S and certain micronutrients may be required.

2. Ineffective nodulation

Legumes may become inoculated by rhizobia in the soil. Under some conditions, however, the bacteria may not be present or else a specific *Rhizobium* is required by a given legume. Thus, use of the proper commercial inoculant is recommended. Legume plants that are not effectively nodulated make poor growth and are unlikely to compete successfully with associated grasses.

3. Incorrect management practices

Few tropical legumes develop crown shoots as does lucerne, but new tillers arise from leaf axils. New growth also develops from apices of the main stem as well as from side and axillary branches. The stems of trailing and decumbent legumes such as centro, Siratro, *Desmodium, Glycine*, kudzu, calopo possess long internodes. Poor rooting occurs at the nodes as contrasted to the temperate legume white clover. Thus, when stems of tropical legumes are damaged by hooves of trampling animal there is little chance of regenerated growth. Furthermore, heavy grazing and frequent defoliation results in low yields and eventually reduced plant population. Unlike the twining or trailing types, Townsville stylo appears to thrive under heavy grazing. Other legumes appear to be intermediate in response to grazing pressure.

Of equal or greater importance is the contrasting cycle of growth of legumes and associated grasses. Grass seedlings grow more vigorously than legumes and thus have a competitive advantage for nutrients, water and light during the establishmental stage of development. Furthermore, this competitive advantage of more rapid growth continues throughout the life cycle of grasses, since they possess the more efficient C_4 carbon fixation pathway. Legumes use the C_3 carbon fixation pathway (Hatch and Boardman, 1973). In general, seed-bearing grasses develop flowering stems before most legumes initiate flower primordia. The twining legumes climb on to upright grasses to that new growth arises at some distance above ground level. Most grasses produce crown tillers and new stems arise from the nodes of decumbent stems or from stolons and rhizomes. Thus, grasses have a regrowth advantage when grazed or cut close to the soil. All of these factors add to the complexity of choosing grass–legume combinations.

4. Water requirements

Even though most legumes develop a deep penetrating tap-root and display drought tolerance once they are well established, many of the perennial species appear to have a threshold of about 1 000 mm annual precipitation. The annuals such as Townsville stylo complete their life cycle and are productive with less than this amount. Equally, *Stylosanthes* species, notably *S. hamata* and *S. scabra* from South America, show potential for good herbage production in areas of rainfall below 1 000 mm.

5. Constraint of seed supply and costs

The seed supply of tropical legumes is limited, or if available somewhat expensive in most tropical countries. Most seeds found in commercial channels are produced in Australia, so that time and cost of shipping hinder movement into other countries. Once annual legumes are established regeneration of new stands depends on the supply of volunteer seeds. A number of factors, such as variable rainfall and intensity of grazing, can influence the quantity of seeds produced. Fortunately, Townsville stylo plants have the capability of developing seed heads from axillary branches near soil level even under close grazing.

6. Inability to grow at cool temperatures

This problem occurs largely in the subtropics, mostly beyond latitude 23° and at high elevations where frost occurs.

It has been suggested that a measure of the compatibility of a legume is its capability to invade an established grass sward. Furthermore, the legume should possess a 'weedy' characteristic that allows colonization under adverse conditions. Townsville stylo appears to have developed both traits in tropical and subtropical areas of Australia and the Sudan savanna of West Africa. White clover behaves similarly in parts of the temperate zone and the high elevations of the tropics. It was anticipated that Siratro would have these capabilities, and indeed volunteer plants are often seen along roadways, fence rows and in waste places where seeds were scattered from sown pastures (Jones and Jones, 1978). This is also observed in regions of tropical America where *Macroptilium atropurpureum* is a part of the native or naturalized vegetation. Regardless of the traits deemed to be desirable, legumes in sown pastures generally tend to diminish over time and livestock producers have difficulty stabilizing a legume–grass combination.

The lack of persistence of legumes in tropical pastures is a challenge to the pasture agronomist, the plant breeder and the animal scientists. Plant persistence may occur through many pathways and involve traits such as longevity of the original plant, length of internodes, regeneration of new roots at nodes, cycle of seed production, rate of seedling emergence and seedling vigour. Assessment of these factors and others must be integrated with agronomic, pasture and livestock management practices. Even though interest in tropical legumes has stimulated increased research in recent years, a great need still exists for expansion and intensification of scientific and practical studies. One can only surmise as to the outcome of a unified global programme concentrated on a single tropical legume such as that expended on lucerne. The first problem, of course, might be unity in deciding which tropical legume is of paramount importance.

References

Andrew, C. S. and **D. O. Norris** (1961) Comparative responses to calcium of five tropical and four temperate pasture legume species, *Aust. J. Agric. Res.* **12**, 40–55.

Andrew, C. S. and **M. F. Robins** (1969) Effect of phosphorus on the growth and chemical composition of some tropical pastures. I. Growth and critical percentage of phosphorus, *Aust. J. Agric. Res.* **20**, 665–74.

Asare, E. O. (1974) Dry matter yield, chemical composition and nutritive value of buffel grass grown alone and in mixture with other tropical grasses and legumes, *Proc. 12th Intl. Grassld Cong.* pp. 45–55.

Aurora, N. (1954) Morphological development of the root and stem nodules of Aeschynoneme indica L., *Phytomorph.* **4**, 211–16.

Bergersen, F. J. (1971) The biochemistry of symbiotic nitrogen fixation in plants. In J. Döbereiner, P. A. da Eira, A. A. Franco and A. B. Campelo (eds), *As Leguminosas na Agricultura Tropical*, Seminar Sobre Met. e. Plan. de Pesq. com Leg. Trop., Brazil, pp. 98–107.

Bindon, B. M. and **D. R. Lamond** (1966) Examination of tropical legumes for deleterious effects on animal reproduction, *Proc. Aust. Soc. Anim. Prod.*, **6**, 109–16.

Blom, P. S. (1980) Leucaena a promising versatile leguminous tree for the tropics, *Abst. Trop. Agric.* **6**, 9–17.

Blount, C. G. and **R. J. Jones** (1977) Steer liveweight gains in relation to the proportion of time in Leucaena leucocephala pastures, *Trop. Grassld* **11**, 159–64.

Bowen, G. D. (1959) Specificity and nitrogen fixation in the Rhizobium symbiosis of Centrosema pubescens Benth, *Queensld J. Agric. Sci.* **16**, 267–82.

Bryan, W. W. (1962) The role of the legume in legume/grass pastures. In *A Review of Nitrogen in the Tropics with Particular Reference to Pastures*, Commonw. Bur. Past. Fld Crops, Bull. 46, Hurley: England, pp. 147–60.

Bryan, W. W. (1969) Desmodium intortum and Desmodium uncinatum, *Herb. Abst.* **39**, 183–91.

Butler, G. W. and **N. O. Bathurst** (1956) The underground transference of nitrogen from clover to associated grass, *Proc. 7th Intl. Grassld Cong.*, pp. 168–78.

CSIRO Report (1974–75) *Tropical Agronomy Division*, Commonw. Sci. Ind. Res. Org., Brisbane, Australia. pp. 43, 109.

Date, R. A. (1970) Microbiological problems in the inoculation and nodulation of legumes, *Plant and Soil* **32**, 703–5.

Date, R. and D. O. Norris (1979) Rhizobium screening of Stylosanthes species for effectiveness in nitrogen fixation, *Aust. J. Agric. Res.* **30**, 80–104.

Diatloff, A. (1968) Nodulation and nitrogen fixation in some Desmodium spp., *Queensld J. Agric. Anim. Sci.* **25**, 165–7.

Diatloff, A. (1972) Recognising a nodulation problem, *Trop. Grassld* **6**, 70–1.

Erdman, L. W. and U. M. Means (1962) Use of total yield for predicting nitrogen content of inoculated legumes grown in sand cultures, *Soil Sci.* **73**, 231–5.

Edye, L. A., J. B. Ritson and K. P. Hyadock (1972) Calf production of Droughtmaster cows grazing a Townsville stylo-speargrass pasture, *Aust. J. Exptl Agric. Anim. Husb.* **12**, 7–11.

Evans, T. R. (1968) Sources of nitrogen for beef production in the Wallum, *Trop. Grassld* **2**, 192–5.

Evans, T. R. (1970) Some factors affecting beef production from subtropical pastures in the coastal lowlands of southeast Queensland, *Proc. 11th Intl. Grassld Cong. Husb.* **9**, 196–208, pp. 803–7.

Fisher, M. J. (1969) The growth and development of Townsville lucerne (*Stylosanthes humilis*) in ungrazed swards at Katherine, N.T., Aust. J. Exptl. Agric. Anim. Husb.

Fred, E., I. L. Baldwin and E. McCoy (1932) *Root Nodule Bacteria and Leguminous Plants*, Studies in Science, No. 5, Univ. of Wisconsin.

Gethin-Jones, G. H. (1942) The effect of leguminous cover crop on building up soil fertility, *E. Afr. Agric. J.* **8**, 48–52.

Gillard, P., L. A. Edye and R. L. Hall (1980) Comparison of Stylosanthes humilis, S. hamata and S. subsericea in the Queensland dry tropics. Effects on pasture composition and cattle liveweight gains, *Aust. J. Agric. Res.* **31**, 205–20.

Gray, S. G. (1968) A review of research on Leucaena leucocephala, *Trop. Grassld* **2**, 19–30.

Gray, S. G. (1970) The place of trees and shrubs as sources of forage in tropical and subtropical pastures, *Trop. Grassld* **4**, 57–62.

Grof, B. and W. A. T. Harding (1970) Dry matter yields and animal production of guinea grass (Panicum maximum) on the humid tropical coast of north Queensland, *Trop. Grassld* **4**, 85–95.

Guyadeen, K. D. (1951) A note on some promising tropical legume forages, *Trop. Agric. (Trin.)* **28**, 231–2.

Guyton, R. F., T. E. Cathupoulis and J. E. Baylor (1980) Productivity of digit grass (Digitaria decumbens) and buffel grass (Cenchrus ciliaris) with and without legumes utilized by native ewes in Bahamas, *Turrialba* **30**, 189–95.

Halliday, J. (1978) Field responses by tropical forage legumes to inoculation with Rhizobium. In P. A. Sanchez and L. E. Tergas (eds), *Pasture Production in Acid Soils of the Tropics*, Seminar Proc., CIAT, Colombia, pp. 123–37.

Hamilton, R. I. and G. Ruth (1968) Bloat on Dolichos lablab, *Trop. Grassld* **2**, 135–6.

Hamilton, R. I., L. J. Lambourne, R. Roe and D. J. Minson (1970) Quality of tropical pastures for milk production, *Proc. 11th Intl. Grassld Cong.*, pp. 860–4.

Hardy, R. W. F., R. C. Burns and R. D. Holsten (1973) Applications of the acetylene-ethylene assay for measurement of nitrogen fixation, *Soil Biol. Biochem.* **5**, 47–81.

Harris, J. R. (1953) The significance of symbiotic nitrogen fixation. In R. O. Whyte, G. Nilsson-Leissner and H. C. Trumble (eds), *Legumes in Agriculture*, FAO Agric. Studies No. 21, Ch. 9.

Hatch, M. D. and N. K. Boardman (1973) Biochemistry of photosynthesis. In G. W. Butler and R. W. Bailey (eds), *Chemistry and Biochemistry of Herbages*,

Academic Press: London, pp. 25–56.

Haydock, K. P., D. O. Norris, and **L. 't Mannetje** (1980) The relation between nitrogen percent and dry weight of inoculated legumes. *Plant and Soil* **57**, 353–62.

Hegarty, M. P. and **A. W. Pound** (1968) Indospicine, a new hepatotoxic amino-acid from Indigofera spicata, *Nature (Lond.)* **217**, 354–5.

Henzell, E. F. (1962) Nitrogen fixation and transfer by some tropical and temperate pasture legumes in sand culture, *Aust. J. Exptl Agric. Anim. Husb.* **2**, 132–40.

Henzell, E. F. (1970) Problems in comparing the nitrogen economics of legume-based and nitrogen-fertilized pasture systems, *Proc. 11th Intl. Grassld Cong.* pp. A112–20.

Henzell, E. F. and **D. O. Norris** (1962) Process by which nitrogen is added to the soil/plant system. In *A Review of Nitrogen in the Tropics with Particular Reference to Pastures*, Commonw. Bur. Past. Fld Crops, Bull. **46**, Hurley; England, pp. 1–18.

Henzell, E. F., I. F. Fergus and **A. E. Martin** (1967) Accretion studies of soil organic matter, *J. Aust. Inst. Agric. Sci.* **33**, 35–37.

Herridge, D. F. and **R. J. Roughley** (1976) Influence of temperature and Rhizobium strain on nodulation and growth of two tropical legumes, *Trop. Grassld* **10**, 21–23.

Hill, G. D. (1971) Leucaena leucocephala for pastures in the tropics, *Herb. Abst.* **31**, 111–19.

Holder, J. M. (1967) Milk production from tropical pastures, *Trop. Grassld* **1**, 135–41.

Horrell, C. R. (1964) Effect of two legumes on the yield of unfertilized grass at Serere, *E. Afr. Agric. For. J.* **30**, 94–6.

Humphreys, L. R. (1974) Pasture species, nutritive value and management. In P. C. Whiteman, L. R. Humphreys and N. H. Montieth (eds), *A Course Manual in Tropical Pasture Science*, Aust. Vice-Cham. Comm., Brisbane, pp. 70–130.

Humphreys, L. R. and **R. J. Jones** (1975) The value of economical studies in establishment and management of sown tropical pastures, *Trop. Grassld* **9**, 125–31.

Hutton, E. M. and **I. A. Bonner** (1960) Dry matter and protein yields in four strains of Leucaena glauca Benth., *J. Aust. Inst. Agric. Sci.* **26**, 276–7.

Johansen, C. and **P. C. Kerridge** (1979) Nitrogen fixation and transfer in tropical legume-grass swards in southeastern Queensland, *Trop. Grassld* **13**, 165–70.

Jones, R. J. (1969) A note on the *in vitro* digestibility of two tropical legumes, Phaseolus atropurpureus and Desmodium intortum, *J. Aust. Inst. Agric. Sci.* **35**, 62–3.

Jones, R. J. (1972) *The Place of Legumes in Tropical Pastures*, ASPAC Food Fert. Techn. Center, Tech. Bull. No. 9, Taiwan.

Jones, R. J. (1979) The value of Leucaena leucocephala as a feed for ruminants in the tropics, *World Anim. Rev.* **31**, 13–23.

Jones, R. J., C. G. Blount and **J. H. Holmes** (1976) Enlarged thyroid glands in cattle grazing Leucaena pastures, *Trop. Grassld* **10**, 113–16.

Jones, R. J., J. G. Davies and **R. B. Waite** (1967) The contributions of some tropical legumes to pasture yields of dry matter and nitrogen at Samford, south-eastern Queensland, *Aust. J. Exptl Agric. Anim. Husb.* **7**, 57–65.

Jones R. J. and **R. M. Jones** (1978) The ecology of Siratro-based pastures. In J. R. Wilson (ed.), *Plant Relations in Pastures*, CSIRO: Melbourne, Ch. 23.

Jones, R. J., C. G. Blount and **B. I. Nurnberg** (1978) Toxicity of Leucaena leucocephala: The effect of iodine and mineral supplementation on penned steers fed a sole diet of Leucaena, *Aust. Vet. J.* **54**, 387–92.

Jones, W. T. and **J. W. Lyttleton** (1971) Bloat in cattle. XXXIV. A survey of legumes that do not produce bloat, *N. Z. J. Agric. Res.* **14**, 101–7.

Koger, M., W. G. Blue, G. B. Killinger, R. E. L. Green, H. C. Harris, J. M. Myers, A. C. Warnick and **N. Gammon** (1961) *Beef Production, Soil and Forage Analysis and Economic Returns from Eight Pasture Programs in North Central Florida*, Agric. Expt. Sta., Bull. 631, Univ. Florida, Gainesville.

Koger, M., W. G. Blue, G. B. Killinger, R. E. L. Green, J. M. Myers, N. Gammon,

A. C. Warnick and **J. R. Crockett** (1970) *Production Response and Economic Returns from Five Pasture Programs in North Central Florida*, Agric. Expt. Sta., Bull. 740, Univ. Florida, Gainesville.

Kretschmer, A. E., G. H. Snyder, J. B. Brohman and **G. J. Gasho** (1974) Seasonal distribution of dry matter and crude protein in tropical legume–grass mixtures in south Florida, *Proc. 12th Intl. Grassld Cong.*, pp. 241–8.

Little, D. A. (1969) The examination of Townsville lucerne (Stylosanthes humilis) for oestrogenic activity, *Aust. Vet. J.* **45**, 24–6.

Luck, P. E. and **N. J. Douglas** (1966) Dairy pasture research and development in the near north coast centered on Cooroy, Queensland, *Trop. Grassld Soc. Aust. Proc. No. 6*, pp. 35–49.

Magadan, P. B., E. Q. Javier and **J. C. Madamba** (1974) Beef production on native (Imperata cylindrica (L.) Beauv.) and para grass (Brachiaria mutica (Forsk.) Stapf) pastures in the Philippines, *Proc. 12th Intl. Grassld Cong.*, pp. 293–9.

Mannetje, L. 't (1978) The role of improved pastures in beef production in the tropics, *Trop. Grassld* **12**, 1–9.

Marshall, D. R., P. Broué and **F. Murday** (1979) Tannins in pasture legumes, *Aust. J. Exptl Agric. Anim. Husb.* **19**, 192–7.

Miller, C. P. and **J. T. van der List** (1977) Yield, nitrogen uptake and liveweight gains from irrigated grass–legume pasture on a Queensland tropical highland, *Aust. J. Exptl Agric. Anim. Husb.* **17**, 949–60.

Minson, D. J. and **R. Milford** (1967) The voluntary intake and digestibility of diets containing different proportion of legume and mature Pangola grass, *Aust. J. Exptl Agric. Anim. Husb.* **7**, 546–51.

NAS (1977) Leucaena, *Promising Forage and Tree Crop for the Tropics*, Natl. Acad. Sci.: Washington, D. C.

Norman, M. J. T. (1960) *Performance of Buffel Grass and Buffel Grass-Townsville Lucerne Mixtures at Katherine, N. T.*, CSIRO Aust. Div. Land Res. Reg. Surv., Tech. Paper No. 11.

Norman, M. J. T. (1970) Relationships between liveweight gain of grazing beef steers and availability of Townsville lucerne, *Proc. 11th Intl. Grassld Cong.*, pp. 829–32.

Norman. M. J. T. and **G. A. Stewart** (1964) Investigations on the feeding of beef cattle in Katherine region, N. T., *J. Aust. Inst. Agric. Sci.* **30**, 39–46.

Norris, D. O. (1958) A red strain of Rhizobium from Lotononis bainesii Baker, *Aust. J. Agric. Res.* **9**, 629–32.

Norris, D. O. (1965) Acid production by Rhizobium – a unifying concept, *Plant and Soil* **22**, 143–66.

Norris, D. O. (1966) The legumes and their associated Rhizobium. In W. Davies and C. L. Skidmore (eds), *Tropical Pastures*, Faber and Faber: London, Ch. 6.

Norris, D. O. (1967) The intelligent use of inoculation and lime pelleting for tropical legumes, *Trop. Grassld* **1**, 107–21.

Norris. D. O. (1970) Nodulation of pasture legumes. In R. M. Moore (ed.), *Australian Grasslands*, Aust. Nat. Univ. Press: Canberra, Ch. 22.

Norris, D. O. (1972) Leguminous plants in tropical pastures, *Trop. Grassld* **6**, 159–69.

Norris, D. O. and **R. A. Date** (1976) Legume bacteriology. In N. W. Shaw and W. W. Bryan (eds), *Tropical Pasture Research, Principles and Methods*, Commonw. Bur. Past. Fld Crops., Bull. 51, Hurley: England, Ch. 7.

Nutman, P. S. (1948) Physiological studies on nodule formation. I. The relationship between nodulation and lateral root formation in red clover, *Ann. Bot. N. S.* **12**, 81–96.

Nutman, P. S. (1962) The relationship between roothair infection by Rhizobium and nodulation in Trifolium and Vicia, *Proc. Roy. Soc. B.* **156**, 122–37.

Nutman, P. S. (1971) The physiology of root hair infection, 2.2 Physiology of nodule formation, 2.3. In J. Döbereiner, P. A. da Eira, A. A. Franco and A. B. Campelo (eds), *As Leguminosas na Agricultura Tropical*, Seminar Sobre Met. e Plan. de Pesq. com Leg. Trop., Brazil, pp. 61–81.

Oyenuga, V. A. (1957) The composition and agricultural value of some grass species in Nigeria, *Emp. J. Exptl Agric.* **25**, 237–55.

Philpotts, H. (1975) The effect of lime and Rhizobium strain on the nodulation of Glycine wightii and Macroptilium atropurpureum on acid soils, *Trop. Grassld* **9**, 37–43.

Ramirez, A., A. Michelin and **J. Lotero** (1976) Producción y consumo de la mezcla de tres gramineas y cinco leguminosas forrajeras tropicales bajo condiciones de pastero controlado, *Revista ICA (Colombia)* **11**, 327–38.

Rotar, P. P. (1965) Tannins and crude proteins of tick clovers (Desmodium spp.), *Trop. Agric. (Trin.)* **42**, 333–47.

Roughley, R. (1970) The preparation and use of legume seed inoculants, *Plant Soil* **32**, 675–701

Rural Research (1973) *Superphosphate Lifts cow Fertility*, CSIRO Quarterly No. 80, pp. 30–2.

Santhirasegaram, K. (1975) Management of legume pastures in a tropical rainforest ecosystem of Peru. In E. Bornemisza and A. Alvarado (eds), *Soil Management in Tropical America*, North Carolina University, Ch. 24.

Sears, P. D. and **L. T. Evans** (1953) Pasture growth and soil fertility. 3. The influence of red and white clovers, superphosphate, lime and dung and urine on soil composition and on earthworm and grass-grub populations, *N. Z. J. Sci. Tech.* **35**(*A*), Suppl. **1**, 42–52.

Shaw, N H. and **L. 't Mannetje** (1970) Studies on a spear grass pasture in central coastal Queensland – the effect of fertilizer stocking rate and oversowing with Stylosanthes humilis on beef production and botanical composition, *Trop. Grassld* **4**, 43–56.

Shaw, N H. (1978) Superphosphate and stocking rate on a native pasture sown with Stylosanthes humilis in central coastal Queensland, *Aust. J. Exptl Agric. Anim. Husb.* **18**, 788–807.

Siebert, D. D., R. A. Hunter and **P. N. Jones** (1976) The utilization by beef cattle of sugarcane supplemented wth animal protein, plant protein or non-protein nitrogen and sulphur, *Aust. J. Exptl Agric. Anim. Husb.* **16**, 789–94.

Simpson, J. R. (1965) The tranferance of nitrogen from pasture legumes to an associated grass under several systems of management in pot cultures, *Aust. J. Agric. Res.* **16**, 915–26.

Siota, C. M., A. P. Castillo, F. A. Moog and **E. Q. Javier** (1976) Productivity of native grass and native grass–stylo pastures, *Phil. J. Anim. Husb.* **31**, 101–8.

Skerman, P. J. (1977) *Tropical Forage Legumes*, FAO: Rome.

Souto, S. M. and **J, Döbereiner** (1970) Problems in the establishment of perennial soybean (Glycine javanica L.) in a tropical region, *Proc. 11th Intl. Grassld Congr.*, pp. 127–31.

Stobbs, T. H. (1965) Beef production from Uganda pastures containing Stylosanthes gracilis and Centrosema pubescens, *Proc. 9th Intl. Grassld Cong.*, pp. 938–42.

Stobbs, T. H. (1969) The effect of grazing management upon pasture productivity in Uganda. IV. Selective grazing, *Trop. Agric. (Trin.)* **46**, 303–9.

Stobbs, T. H. (1971) Quality of pasture and forage crops for dairy production in the tropical regions of Australia. I. Review of the Literature, *Trop. Grassld* **5**, 159–70.

Stobbs, T. H. (1973) The effect of plant structure on the intake of tropical pastures. I. Variation in the bite size of grazing cattle. II. Differences in sward structure, nutritive value and bite size of animals grazing Setaria anceps and Chloris gayana at various stages of growth, *J. Agric. Res.* **24**, 809–29.

Stobbs, T. H. and **E. M. Hutton** (1974) Variations in canopy structures of tropical pastures and their effects on the grazing behaviour of cattle, *Proc. 12th Intl. Grassld Cong.*, pp. 510–7.

Stobbs, T. H. and **E. C. Imrie** (1976) Variation in yield canopy structure, chemical composition and in vitro digestibility within and between two Desmodium species and interspecific hybrids, *Trop. Grassld* **10**, 99–106.

Suttie, J. M. (1968) Pasture legume research in Kenya, *E. Afr. Agric. For. J.* **33**, 281–5.

Teitzel, J. K. and **R. L. Burt** (1976) Centrosema pubescens in Australia, *Trop. Grassld* **10**, 5–14.

Thairu, D. M. (1972) The contribution of Desmodium intortum to the yield of Setaria sphacelata, *E. Afr. Agric. For. J.* **37**, 215–19.

Thomas, D. (1973) Nitrogen from tropical legumes on the African continent, *Herb. Abst.* **43**, 33–9.

Vallis, I., K. P. Haydock, P. J. Ross and **E. F. Henzell** (1967) Isotopic studies on the uptake of nitrogen by pasture plants. III. The uptake of small additions of ^{15}N-labelled fertilizer by Rhodes grass and Townsville stylo, *Aust. J. Agric. Res.* **18**, 865.

Vicente-Chandler, J., R. Caro-Castros and **J. Figarella** (1953) The effects of two heights of cutting and three fertility levels on a tropical kudzu and molasses grass pasture, *Agron. J.* **45**, 397–400.

Vicent, J. M. (1970) *A Manual for the Practical Study of Root-nodule Bacteria*, I. B. P. Handbook No. 15, Blackwell Scientific: Oxford and Edinburgh.

Warnick, A. C., M. Koger, A. Martinez and **T. J. Cunha** (1965) *Productivity of Beef Cows as Influenced by Pasture and Winter Supplement during Growth*, Agric. Expt. Sta. Bull 695, Univ. Florida, Gainesville.

Whiteman, P. C. (1980) *Tropical Pasture Science*, Oxford Univ. Press: London.

Whitney, A. S. and **R. E. Green** (1969) Legume contributions to yields and compositions of Desmodium spp. -Pangola grass mixtures, *Agron. J.* **61**, 741–6.

Whitney, A. S., Y. Kanchiro and **G. D. Sherman** (1967) Nitrogen relationships of three tropical forage legumes in pure stands and in grass mixtures, *Agron. J.* **59**, 47–50.

Whyte, R. O. (1947) Africa. In *The Use and Misuse of Shrubs and Trees as Fodder*, Imp. Agric. Bur. Joint Publ. No. 10, pp. 94–109.

Whyte, R. O. (1962) The myth of tropical grasslands, *Trop. Agric. (Trin.)* **39**, 1–11.

Whyte, R. O. (1968) *Grasslands of the Monsoon*, Faber and Faber: London.

Wilkins, J. (1967) The effect of high temperatures on certain root-nodule bacteria, *Aust. J. Agric. Res.* **18**, 299–304.

Williams, R. J., R. L. Burt and **R. W. Strickland** (1976) Plant introduction. In N. H. Shaw and W. W. Bryan (eds), *Tropical Pasture Research, Principles and Methods*, Commonw. Bur. Past. Fld Crops, Bull. 51, Hurley, England, Ch. 5.

Yatzawa, M. and **S. Yoshida** (1979) Stem nodules in Aeschynomene indica and their capacity to fix nitrogen, *Physiol. Plant* **45**, 293–5.

Yatzawa, M. and **H. Susilo** (1980) Development of upper stem nodules in Aeschynomene indica under experimental conditions, *Soil Sci. Plant Nutr.* **26**, 317–19.

Chapter 17

Plant improvement and breeding

Evaluation and testing of introductions and cultivars at various experimental sites in different ecological areas furnish information regarding geographical limits within which a species is adapted. A thorough knowledge of environmental influences will be of considerable value in choosing the appropriate species for genetic improvement. The eventual success in developing cultivars that satisfy the breeder's objectives will primarily depend on the availability of a wide germplasm base. Many pasture species possess a wide genetic diversity and are adapted to a range of climatic and soil conditions. Such forms or ecotypes have been recognized under native and naturalized conditions and will continue to be of value in the search for new cultivars. The intensive plant exploration, collection and evaluation of tropical pasture species carried out in recent years has barely tapped the existing genetic potential. The tremendous benefits already derived from such collections reaffirm the urgent need for further collections of genetic material from native regions. There is overwhelming evidence to suggest that time and effort spent in introducing and evaluating exotic germplasm during the initial stages of a breeding programme is likely to produce greater benefits than a long-term breeding programme using local types with a restricted germplasm base.

Factors to consider in breeding

Even though the basic procedures utilized by a forage breeder in the genetic improvement of pasture crops are similar to those developed and practised by breeders engaged in field crops improvement, the task of the forage breeder is often more difficult and complex. Such difficulties emanate from the following interrelated factors.

Complex soil–plant–animal relationships

The primary objective of pasture crop improvement may not be higher herbage yield per unit area, but rather increased and more efficient animal production (Bray and Hutton, 1976). Consequently, the forage breeder must pay attention to a diversity of factors such as intake, digestibility and presence of toxic substances that influence animal productivity on one hand and actual herbage productivity on the other.

Variations in forage plant culture and utilization

Forage plants are grown and utilized in simple or complex mixtures of grasses and legumes for grazing by various classes of livestock, or preservation in form of hay or silage. Non-forage uses such as soil erosion control or soil improvement may also be important considerations. Therefore, special procedures need to be developed to test and evaluate the performance of improved forage genotypes.

Mode of reproduction and seed production

The vast majority of potentially useful tropical pasture grasses are either cross-pollinated or apomictic and show varying degrees of self-sterility, e.g. *Cynodon* spp., *Paspalum notatum* and *Pennisetum clandestinum*. Seed set in many species is generally poor. Flowers are often small, so that controlled hybridization is difficult. These factors account in part for the lack of fundamental genetic research in many species.

Life span and growth habit

The perennial nature of many important pasture species increases the time period (often up to 15 years) for the development of new genotypes and their evaluation in terms of sustained productivity, persistence and animal performance (Thompson and Wright, 1971). Various growth habits, ranging from stoloniferous to tufted growth in grasses and erect to creeping or twining growth in legumes and relatively large plant sizes, cause maintenance problems of large populations. Collection of data in such populations is difficult, time-consuming and often prone to considerable error.

Breeding and improvement objectives

The rate of progress, and eventual success, of any plant-breeding programme depends on the clarity of breeding objectives and the vision with which the improvement programme is planned. The list of breeding objectives for pasture crop improvement is varied (Hanson and Carnahan, 1956; Poehlman, 1959; McWilliam, 1969; Hutton,

1971; Williams *et al.*, 1976; Crowder and Chheda, 1977). It includes dry matter yield, disease and pest resistance, tolerance to climatic conditions and to adverse soil conditions, time of maturity, seed production and seed characteristics, growth cycle and distribution of herbage yield during growing season, growth habit and compatibility with associated forage species, lodging and root development, per-. sistence or longevity and ability to withstand occasional over- or underutilization by grazing animals, rapid early growth and/or re-growth for ground cover and reduced weed competition, acceptability by livestock, herbage quality and nutritive value, freedom from toxic substances, ease of establishment and eradication, and for legumes, nitrogen-fixing capability and efficient rhizobium symbiosis associated with high dry matter and crude protein yield.

Even though a breeder is interested in many features of a pasture crop, significant success will only be achieved by selecting a limited number of characteristics based on the species under investigation, area of adaptation and intended use of the improved genotypes. Five major objectives that should receive attention in the development of superior genotypes are (a) total dry matter yield and its distribution during the growing season, (b) animal acceptability and nutritive value of the herbage, (c) persistence or plant longevity, (d) ability to grow in mixtures and (e) reproductive potential and ease of management. Bray (1975) listed the following requisites in choosing a characteristic for study and genetic improvement (1) the possibility of growing large populations, (2) the ease and precision of evaluation, (3) the genetic variation in the breeder's plots (plants), (4) the meaningful genetic variation under field conditions, (5) a reasonable correlation between expression of genetic potential in the breeder's plots and in the field.

Yield of dry matter and its distribution

The ultimate aim of pasture improvement through breeding is to develop genotypes capable of increasing animal productivity. Thus, it is necessary to develop varieties that produce high yields of digestible nutrients, i.e. increased dry matter production, with extended availability, and that have high nutritive value. Experimental evidence suggests that heritability of yield in cross-pollinated perennial pasture species on a single plant basis is low and selection for high yielding genotypes is difficult (Burton, 1970). None the less, worthwhile progress can be achieved by giving attention to such factors as disease and insect resistance, tolerance to environmental stresses such as drought and high or low temperature, suitable maturity period, responsiveness to fertilization and management, heterosis and aggressiveness.

Tropical legumes generally possess lower dry matter yield potential than grasses and are less competitive. Selections within some species possess greater drought tolerance, and vary in rate of decline in nutritive value with advancing maturity in terms of digestibility and crude protein content. These traits should receive special attention in tropical forage improvement programmes.

Acceptability and nutritive value

Williams *et al.* (1976) questioned the value of placing emphasis on selection for high dry matter yield in the tropics because of a noticeable inverse relationship between yield and feed quality. Burton *et al.* (1967), however, observed that correlation coefficient values between yield and protein content and between yield and dry matter digestibility in *Cynodon* spp. were insignificant. Digestibility of forage within this genus, therefore, could be improved without reducing herbage yield. Hutton (1968) suggested that breeding less vigorous tropical grass varieties would allow for a greater proportion of legume in the pasture and thus raise its feeding value.

Animals graze pastures selectively by showing preference for some plants and plant parts and refusing others. Acceptability, palatability and preference are three terms used to describe this phenomenon. In a strict sense, palatability and preference influence acceptability of a forage by a given class of livestock. Intra- and interspecific variations in acceptability have been observed among selections of grasses and legumes. Leafiness, growth habit, response to fertilizer application and recuperation after grazing are factors that influence palatability and preference, and therefore acceptability.

A measure of acceptability can be obtained by allowing animals to graze freely various genotypes grown in small replicated plots. Grazing should be restricted to a few hours with adequate numbers of livestock so that all genotypes are tested. Percentage utilization is calculated from herbage yields before and after grazing of individual genotypes. The procedure should be repeated over a full growing season. Using this procedure, Chheda (1971) evaluated 24 promising *Cyndon* introductions and hybrids in Nigeria and observed variations in per cent utilization ranging from 28 to 68 per cent. In Nigeria, Aken'Ova (1975) noted 11 per cent greater forage utilization of *Pennisetum* F_1 hybrids than *P. purpureum* by grazing Zebu steers (46 and 35% of herbage on offer, respectively). The *P. americanum* × *P. purpureum* hybrids produced 20 per cent less dry matter (6.3 vs. 7.56 tonnes/ha, respectively) but showed 10 per cent greater herbage intake per animal (2.2% of bodyweight) and gave 46 per cent higher liveweight gains than *P. purpureum* (267 vs. 152 kg/ha).

Higher acceptability and greater herbage intake will minimize wastage of mature herbage and thus increase animal production on tropical pastures. Milford and Minson (1966) indicated the feasibility of selecting and breeding for higher intake at various stages of maturity for Rhodes grass cv. Callide, Siratro and *Glycine wightii*. Voluntary intake of the leaf fraction was 46 per cent higher than that of the stem fraction of five tropical grasses (Laredo and Minson, 1973), indicating that stem residues were retained for a longer period in the rumen. This led to reduced voluntary intake by animals. Selection of types within a species having densely packed leaves and a minimum quantity of stem should enhance herbage intake (Burton *et al.*, 1969; Stobbs, 1973; Minson, 1976). In *Pennisetum americanum* the incorporation of dwarf genes decreased stem internode length and reduced the stem percentage in herbage. Intake by cattle of this dehydrated forage increased by 20 per cent. Liveweight gains increased by 20 per cent when dwarf material was grazed as compared to normal height materials (Burton *et al.*, 1969).

Selection and breeding for greater forage digestibility hold a distinct possibility for improving tropical forages. Species and variety differences have been observed in *Cynodon* (Chheda, 1971; Burton and Monson, 1972; Fig. 17.1), *Pennistum* (Aken'Ova, 1975), *Digitaria* (Klock *et al.*, 1975) and *Setaria* (Hacker and Minson, 1972) genotypes. Heritability estimates of digestibility range between 0.3 and 0.7 (Burton and Monson, 1972; Bray, 1975). Development of the *in vitro* technique (Tilley and Terry, 1963; Donefer 1970; Goto and Minson, 1977) to estimate digestibility has provided forage breeders with a powerful tool to select suitable genotypes from variable populations. Use of the 'nylon bag technique' (Burton *et al.*, 1967) to estimate digestibility is useful where laboratory facilities are not available or where a small number of genotypes are to be tested.

Control of the lignification process can result in increased digestibility. The brown midrib (*bmr*) mutants observed in maize and sorghum have significantly less lignin than their normal counterparts (Gee *et al.*, 1968; Porter *et al.*, 1978). The lignin content of sorghum mutants was reduced by 5–51 per cent in stems and 5–25 per cent in leaves. In addition, digestibility was increased 33 per cent when *bmr* genes were present. Similarities between the maize and sorghum mutants suggest the existence of similar genes in other grasses.

Protein deficiency of mature herbage, especially during the dry season, limits animal production in the tropics. Lack of nitrogen depresses bacterial activity in the rumen causing a marked decline in forage intake, and digestibility when crude protein content falls below 7 per cent (Milford and Minson, 1966). Although it is desirable to develop forage varieties with high nitrogen content, success may be limited because of small genetic variations within species, along with environmental, seasonal and genotype interactions, as noted in temperate species by Simon (1976).

Considerable varietal differences in chemical composition were reported by Dougall and Bogdan (1958) and Bredon and Horrell

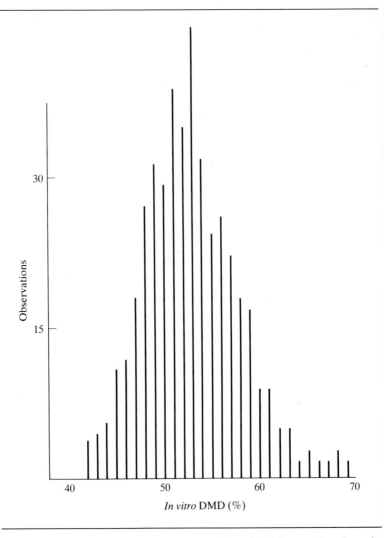

Fig. 17.1 *In vitro* dry matter digestibilities of 5-week-old forage taken from the world collection of bermudagrasses (over 500 genotypes) grown at Tifton, Georgia, USA (Burton and Monson, 1972).

(1962) in East Africa. These authors suggested that such differences would be of importance in forage improvement programmes. Mineral deficiencies can be rectified by appropriate mineral supplementation. In the tropics where animals are maintained under extensive grazing systems, however, plant breeders must give attention to herbage mineral content that approaches the optimum for animal production. Critical percentages of essential elements for maximum pasture growth have been determined for several species. This information could be used by breeders in developing cultivars of tropical pasture crops.

Persistence

Persistence of a forage genotype is influenced by ability to withstand grazing or mowing; tolerance of environmental stresses, such as drought, flooding and frost; disease and pest resistance; and ability to regenerate from self-sown seed or vegetative means. Existence of adequate genetic variability in one or more of these characters when diverse genotypes are screened permits selection of persistent types. Genotypes within certain species that exhibit pronounced stoloniferous growth usually withstand grazing more effectively than those with weak stolons or tufted growth habit (Bogdan, 1965; Chheda, 1971). Cold and drought tolerance influence persistence, and field observations have indicated considerable differences among grass and legume genotypes (Rawal and Harlan, 1971; Clements, 1976; Williams *et al.*, 1976).

Disease and pest incidences apparently have less effect on persistence of tropical pastures than other factors, particularly of grasses. This might be due to the wide genetic base of most tropical pasture species. For example, it was feared that Pangola stunt virus would devastate pastures of this species since it is vegetatively propagated. Screening of *Digitaria* germplasm collections in Florida, USA, however, showed that resistant or tolerant clones existed (Schank and Decker, 1963). *Digitaria decumbens* cv. Transvala is resistant to the Pangola stunt virus. Persistence of Rhodes grass is affected by *Helminthosporium* spp., and some cultivars such as Nzoia are severely affected by dietback of leaves and stems followed by death of affected plants. Lines selected from mixed populations of Nzoia were almost free of disease (Bogdan, 1965, 1969). Braverman and Oakes (1972) reviewed disease resistance in warm-season forage grasses commonly grown in southern USA. The list included species of *Cenchrus*, *Chloris*, *Cynodon*, *Digitaria*, *Panicum*, *Paspalum*, *Pennisetum* and *Setaria*.

Anthracnose caused by *Colletotrichum gloeosporoides* causes extensive losses of *Stylosanthes* spp. The disease is endemic in South and Central America and has been reported in Australia, Africa and Florida. Systematic screening of 850 accessions of *Stylosanthes* spp. revealed a high degree of resistance in a large number of genotypes, particularly those belonging to *S. capitata* (CIAT, 1977, 1978; Grof *et al.*, 1979).

Ability to grow in mixtures

The plant breeder usually observes individual genotypes growing in space-planted field plots and is therefore unable to predict accurately their potential performance in mixed swards. A practical procedure would be to grow and study the performance of promising genotypes in mixtures at a number of locations and over a reasonable period of time. The aggressive growth of grasses, however, may adversely affect the fate of legumes in a mixture. Low-growing prostrate legumes will be less compatible than more upright trailing types when grown in mixture with tall bunch-type grasses. Allelopathic effects, such as those observed in *Cynodon* root exudates (unpublished information by authors), could be a significant factor in retarding the germination, establishment and growth of undesired weed species as well as legumes. Modifications of fertilization schedules, intervals and intensity of grazing, and management practices may alter the performance of genotypes in terms of their competitive ability when grown in mixtures.

Reproductive potential and ease of management

Ability to produce adequate quantities of seed or quickly regenerate by vegetative means after transplanting are important attributes in forage species, both for establishment of new pastures and for persistence. Several important tropical grass species and cultivars, because of sterility or low seed-production ability, are propagated by vegetative means, e.g. *Cynodon dactylon, C. nlemfuensis, Pennisetum purpureum, P. clandestinum, Digitaria decumbens* and *Panicum maximum*. Considerable variations in per cent establishment and early growth have been observed among genotypes of *Cynodon* species when planted by rooted stolon cuttings (unpublished data by authors).

Wide differences exist among genotypes of grasses and legumes in respect to seed production. The forage breeder, however, should not place undue emphasis on this trait and neglect that of herbage yield. Frequently, an inverse relationship exists between seed yield and herbage yield. This can be obviated by giving attention to both characters and by the possibility of producing hybrid seed in commercial quantities, using various genetic systems such as self-incompatibility and male sterility (Burton *et al.*, 1954; Powell and Burton, 1966; McWhirter, 1969; Aken'Ova, 1975).

Time of flowering and seed-setting in the area of intended use need to be considered in order to ensure production of adequate amounts of seed for species regeneration in the subsequent rainy season. Cameron (1967a) observed that the late flowering types of *Stylosanthes humilis* gave higher dry matter yields than the early types in the seeding year, even in areas with a short growing season. Sward performance suffered in the following season, however, because of the inability of late types to flower and set seed.

Selections and cultivars vary in regard to ease of management because of genetic differences and environmental responses. For example, certain types of *Pennisetum purpureum* may be difficult to manage for grazing but not for silage. Other genotypes possess flexibility, so that different management schedules do not significantly alter their productivity, acceptability and persistence under grazing or silage. These can be classified as 'easy to manage'. Aspects of this intangible characteristic should also be given consideration in formulating objectives in forage breeding programmes.

Breeding behaviour

Detailed information on the mode of reproduction of a pasture species is essential in (a) formulating appropriate breeding strategies, (b) producing seeds of improved varieties, (c) maintaining germplasm nurseries and (d) providing guidelines for collecting additional genetic materials.

Mode of reproduction varies with different species of forage crops (Table 17.1). Most of the important tropical pasture grasses are perennial or tend to perenniate. They are (1) cross-pollinated and normally produce highly heterozygous populations that show varying degrees of inbreeding depression and self-incompatibility when selfed or (2) they are obligate apomicts and produce highly heterozygous but uniform progenies. Apomixis is a form of asexual reproduction whereby the embryo develops from an unreduced cell of the ovule without the fusion of the egg and sperm nuclei. The most common form of apomixis in tropical grasses is pseudogamous apospory, in which the diploid embryo sac is formed from a somatic cell by mitotic divisions but requires pollination to stimulate development of the seed.

Most of the tropical legumes are self-pollinating. However, in some, such as *Glycine* spp., *Desmodium* spp. and *Macroptilum atropurpureum*, considerable cross-pollination can occur (Bray and Hutton, 1976). *Medicago sativa, Trifolium mattirolianum* and *T. semipilosum* are examples of cross-pollinated legumes.

Determination of mode of reproduction

Even though reproduction methods are known for many tropical forages (Hanson and Carnahan, 1956; Fryxell, 1957; Hutton, 1970; Bray and Hutton, 1976), environmental conditions can cause modifications. For example, *Chloris gayana* performed as a cross-pollinated species with a high degree of self-incompatibility in East Africa (Bogdan, 1963). In Australia, however, Hutton (1961) found apomictic reproduction based on field observations, as well as cytological and embryological studies. Progenies from single plants pro-

Table 17.1 Principal cultivated grasses and legumes in the tropics and subtropics

Botanical name	Common name	Area of diversity	Chromosome numbers*	Breeding behaviour+	Cultivars and selections (Based on Bogdan, 1977)
Andropogon gayanus	Gamba	Tropical Africa	20,40	CP	
Axonopus compresus	Carpet	Tropical and subtropical America	40	CP	
A. scoparius	Imperial	Central and S. America	20	CP	Clone 60, Clone 72 (Colombia)
Brachiaria brizantha	Palisade	Tropical Africa	36,54	Apo	
B. decumbens	Signal	E. Africa	36	Apo	Basilisk (Australia)
B. mutica	Pará	Tropical Africa and tropical S. America	36	Apo	
B. ruziziensis	Congo	Tropical Africa	18,36	CP; Apo	Kennedy (Australia)
Cenchrus ciliaris	Buffel	Africa, India, Indonesia	36	Apo	Biloela, Molopo, Boorara, Lawes, Nunbank, Tarewinnbar, Gayndah, Western Australian (Australia); Higgins (USA); Pusa Giant (India)
Chloris gayana	Rhodes	Africa	20,40	CP	Pioneer, Katambora. Samford, Callide, Commercial (Australia); Mpwapwa, Kongwa, Mbarara, Masaba, Pokot, Nzoia, Elmba (E. Africa)
Cynodon dactylon	Bermuda	Tropics and subtropics	18,36	CP	Tift bermuda, Coastal, Midland Suwanee, Coastcross-1 (USA); Star grass-1, Star grass-4, Muguga, (E. Africa); IB.8, IBX.7 (Nigeria)
C. nlemfuensis	Giant Star	Tropical Africa	18,36	CP	
C. aethiopicus	Giant Star	E. Africa	18,36	CP	
C. plectostachyus	Giant Star	E. Africa	18	CP	
Dichanthium spp.	Angleton	E. Africa and Asia	40,50,60	CP	Marvel 8, Marvel 40, Marvel 93 (India)
Digitaria decumbens	Pangola	S. Africa	27	Sterile	Pangola, Transvale
D. pentzii	Woolly finger	S. Africa	18,27,36,54	Sterile; CP	
D.		S. Africa	18,36	CP	

					negro, Roxo, Francano, Branco (Brazil)
Panicum antidotale	Blue panic	S.E. Asia	18,36	CP	A–130 (USA); 297 (India); T.15327
P. coloratum	Kleingrass	E. and S. Africa	18,36,54	CP; Apo	Solai (E. Africa); Kabulabula, Zhilo (S. Africa); Bambatsi, Bushman mine (Rhodesia); Pollock, Burnett (Australia)
P. maximum	Guinea	Tropical and subtropical Africa	16,32,48	CP; Apo	Coloniao, Siempre verde, Guinea, Boringuen, Broadleaf, King Ranch (N. and S. America); Sigor, Arusha, Nchisi, Sabi, Makueni, Slender guinea, Embu (E. Africa); Gatton, Hamil, Petrie (Australia)
Paspalum notatum	Bahia	Tropical and subtropical S. America	20,40	CP; Apo	Pensacola, Tifhi-1, Tifhi-2, Common, Argentina, Batatai, Paraguay, Wilmington, Anche da Rocha, Capivari (N. and S. America)
P. dilatatum	Dallis	Tropical and subtropical S. America	40,50	CP; Apo	
P. plicatulum	Plicatulum		40	Apo	Rodds Bay, Hartley (Australia)
Pennisetum clandestinum	Kikuyu	East and Central Africa	36	CP; Apo	Whittet, Breakwell (Australia)
P. pedicilatum	Kaysuwa	W. Africa &	38,54	CP; Apo	G–73, T–15 (India)
P. purpureum	Elephant	Tropical Africa	28	CP	Merkeron (USA); Capricorn (Australia); Minerio (Brazil); Uganda hairless, Cameroons, Gold Coast, French Cameroons, (E. Africa); Cubano, Domira, Panama and (*P. purpureum* × *P. americanum*) F$_1$ hybrid varieties in Asia, Africa and Australia
P. americanum	Pearl millet, Bulrush millet	Tropical Africa and W. Africa	20,40	CP	Tift 23B, Starr, Gahi-1 Millex-22 (USA); Bh-4 (India); Katherine pearl, Ingrid pearl, Tamworth (Australia); Tjolotjo bearded (Rhodesia)

Cont. p. 468

Table 17.1 (cont'd)

Botanical name	Common name	Area of diversity	Chromosome numbers*	Breeding behaviour[+]	Cultivars and selections (Based on Bogdan, 1977)
Setaria anceps	Setaria	East and South	18,36,54	CP	Nandi, Nandi Mark-2, Bua river Toittskraal (E. Africa); Kazungula (S. Africa); Narok (Australia)
Sorghum almum	Columbus	S. America	40	CP	Grooble (Australia); Nunbank (S. Africa); De Soto (USA)
S. sudanense	Sudan	Tropical Africa	20,40	CP	Tift, Piper, Greenleaf (USA) and several varieties originating from interspecific crosses
Calopogonium mucunoides	Calopo	S. America	36	SP	
Centrosema pubescens	Centro	S. America	20	SP	Deodoro (Brazil): Belalto (Australia)
Desmodium intortum	Greenleaf	Tropical America	22	SP; CP	Greenleaf (Australia)
D. uncinatum	Silverleaf	Tropical America	22	SP; CP	Silverleaf (Australia)
Glycine wightii	Glycine	Africa and India	22,44	SP	Tinaroo, Cooper, Clarence (Australia); Kenya white glycine, Kenya violet glycine (Kenya)
Lablab purpureus	Hyacinth bean	Africa and Asia	22	SP	Rongai (Australia)
Leucaena leucocephala	Leucaena, Ipil-ipil	Tropical America	104	SP	Hawaii, Peru, Elsalvador, Guatemala, Australia, Cunningham
Lotononis bainesii	Lotononis	S. Africa	36	SP	Miles (Australia)
Macroptilium atropurpureum	Siratro	Tropical America	22	SP	Siratro (Australia)
Medicago sativa	Alfalfa, Lucerne	Central Asia and southern Europe	32	CP	Hairy Peruvian (Peru); Hunter river (Australia) and various others
Pueraria phaseoloides	Puero, Kudzu	E. Asia	22	SP	
Stylosanthes guianensis	Stylo	Tropical America	20	SP	Schofield, Oxley, Cook, Endeavour (Australia); Deodoro, Deodoro-2, IRI-1022 (Brazil)
S. humilis	Townsville	S. America	20	SP	Gordon, Lawson, Paterson (Australia)
S. hamata		S. America	20,40	SP	Verano

duced uniform populations even though they were grown close to different types of *C. gayana*. Later, Jones and Pritchard (1971), noted that *C. gayana* was cross-pollinating under conditions similar to those utilized by Hutton.

Mode of reproduction and prevalence of natural crossing within a given species should be ascertained at the beginning of a breeding programme in a specific locality. In some instances, mode of reproduction can be determined by examining the flower structure and flowering behaviour, e.g. cleistogamy indicates self-pollination, monoecy and dichogamy indicate cross-pollination. Absence of seed production when individual plants are isolated or individual inflorescences are bagged indicates cross-pollination. Reduced seed set with self-pollination and presence of inbreeding depression among segregating progenies suggest that cross-pollination had previously occurred. On the other hand, good seed set when individual inflorescences are bagged indicates either self-pollination or apomixis. Apomixis is confirmed if crosses involving diverse parental types produce progeny exactly like the female parent. Histological studies will be necessary to determine the exact nature of apomictic development. To determine the amount of natural crossing requires growing plants carrying a recessive marker gene in the proximity of plants carrying the dominant alternative allele. Open-pollinated seeds harvested from the recessive individual plants are grown and the per cent of natural crossing is calculated as the proportion of phenotypically dominant offspring.

Hybridization methods

Development of efficient cross-pollinating techniques requires a detailed understanding of the floral biology of a given species. This includes such factors as induction of flowering and seasonal variations in flowering behaviour, time of anthesis, length of time that pistils remain receptive and pollen grains viable, interval between pollination and fertilization. Many tropical pasture species are short- or intermediate-day plants, or day-neutral. Small seasonal changes in day lengths and temperature variations produce marked effects on inflorescence initiation in certain species. Daily time of flowering and pollen shedding varies within and between species (Foster, 1962), is influenced by climatic conditions and is usually restricted to a few hours per day. For simultaneous flowering of different genotypes it may be necessary to control photoperiod by use of artificial light. Collected pollen of some species may remain viable for several days if stored at low temperature and high humidity in a refrigerator (Aken'Ova and Chheda, 1970). Pollen germination and pollen-tube growth in some tropical legumes, e.g. *Phaseolus* and *Indigofera* spp., may not occur unless high relative humidity is maintained around the stylar tissue. This is accomplished by enclosing the female flowers in clear polythene bags with moistened cotton balls (Hutton, 1964).

Small flower size of some forage species makes emasculation difficult and tedious. Alternative procedures for making crosses have been developed as follows:

1. Some cross-pollinating tropical grasses such as *Cynodon*, *Setaria* and *Digitaria* spp. are often highly self-sterile. Emasculation can then be omitted and inflorescences of the two parents having the same stage of development can be bagged together in vegetable parchment bags or sleeves to produce hybrid seed.

2. Anthesis and seed-setting will occur in some forage species (e.g. *Cynodon* and *Dichanthium* spp.) when culms ready to flower are cut close to the ground and placed in bottles containing water. Thus, hybrid seeds can be produced by placing culms of the desired parents together and isolating them from other genotypes.

3. Emasculation can also be omitted if hybrids resulting from a cross between two parental types can be easily distinguished from selfed progeny, i.e. use of marker genes.

4. Availability of male-sterile lines, for example in *Pennisetum americanum*, greatly facilitates production of F_1 hybrid seed.

5. Hot-water treatment has been applied successfully for bulk emasculation of forage grasses (Keller, 1952; Hutton, 1964). Immersion of panicles 2–3 days before anthesis for 1–10 min and at temperatures ranging from 45 °C to 48 °C provide a guide to determine actual treatment of a particular species.

6. Cross-pollination of legumes that requires tripping by insects to set seeds can be accomplished by enclosing the two parents in a wire or plastic cage with the appropriate pollinating insects.

7. Anthers and pollen of some legumes can be removed with suction, or washed off with a jet of water or pollen grains inactivated by hot water or alcohol solution. Hutton and Gray (1959) recommended the use of weak solution of a non-toxic wetting agent to effect emasculation in *Leucaena*.

Genetic and cytogenetic studies

Existence of wide genetic diversity offers considerable opportunity for development of improved pasture cultivars by assembling, evaluating and selecting ecotypes suitable for a given environment. A number of pasture cultivars now popular in East Africa and subtropical Australia have originated from such efforts (Bogdan, 1965; Hutton, 1970; Crowder, 1977). For sustained and efficient varietal improvement programmes, and for development of suitable breeding strategies capable of effective exploitation of existing variation, genetic and cytogenetic information should be sought from published literature and pursued as the breeding programme proceeds.

Cytogenetics

Many pasture species are polyploids (Table 17.1) and are therefore cytogenetically complex. Several species have variable chromosome numbers and thus display meiotic irregularities such as univalents, multivalents, chromatin bridges, unequal chromosome distribution at anaphase and varying chromosome laggards at anaphase. These irregularities cause varying degrees of sterility. Knowledge of chromosome number and behaviour of diverse genotypes during meiosis is essential for effective planning of a breeding programme.

Information about chromosome numbers of pasture grasses and their behaviour during meiosis was reviewed by Myers (1947) and Carnahan and Hill (1961). Somatic chromosome numbers of indigenous and exotic tropical grasses in South Africa were compiled by Pienaar (1954), of grasses and legumes in India by Whyte (1964) and in Australia by Pritchard and Gould (1964). In the last two decades information about the range of cytological variability, cytogenetical relationships and phylogenetic affinities within related tropical pasture species and species complexes have been made available to plant breeders.

Cynodon species

Cytomorphological studies of *Cynodon* collections, their geographical distribution and ecological adaptation, range of hybridization (Tables 17.2 and 17.3) and cytogenetical analysis of parents and hybrids have furnished valuable information about evolutionary mechanisms in the genus, phylogenetic relationships and mode of chromosome pairing. This information enhanced the development of superior hybrid varieties in West Africa and delineated the regions of introductions for use in future breeding programmes (Chheda, 1974).

Published work on the biosystematics of *Cynodon* (Burton, 1951; Forbes and Burton, 1963; Harlan and de Wet, 1969; Harlan *et al.*, 1969; 1970; Harlan, 1970; Chheda and Rawal, 1971; Rawal and Chheda, 1971; Chheda, 1971) indicates the following conclusions:

1. Both diploid (2n = 18) and tetraploid (4n = 36) forms exist in species of *Cynodon*. Chromosome pairing in diploids and tetraploids is regular, with some accessions showing a few univalents and occasional multivalents at metaphase I.
2. Both intra- and interspecific diploid hybrids are characterized by fairly regular chromosome pairing with 2 univalents and very rarely 4 or 6 univalents.
3. In triploid hybrids involving tetraploid × diploid parents (or 2n

Table 17.2 *Some agro-botanical observations on Cynodon introductions grown at Ibadan, Nigeria (Chheda, 1974)*

Species	No. studied	Origin	Chrs*	Height (cm)†	Tiller density†	Leaf area index†	Rhizome presence	DM yield (t/ha) per yr†	Persistence
C. arcuatus	6	Malagasy, India S.E. Asia, Australia	36	41 (33–48)	18 (15–21)	5.1 (4.4–6.0)	No	2.07 (0.45–3.81)	V. poor
C. barberi	2	India	18	27 (25–28)	31 (26–35)	5.2 (4.8–5.6)	No	0.89 (0.53–1.24)	V. poor
C. dactylon var. dactylon	24	Ubiquitous	18,36	33 (15–71)	21 (7.5–40)	5.50 (4.1–7.3)	Yes	2.09 (0.09–6.04)	Poor to good
var. afghanicus	2	Afghanistan	18,36	46 (41–48)	11 (9–13)	7.1 (6.8–7.4)	Yes	2.86 (0.44–5.28)	V. poor to fair
var. aridus	5	E. and S. Africa, India, Malagasy	18	33 (15–38)	30 (22–37)	7.0 (6.0–8.0)	Yes	4.39 (3.14–7.33)	fair
var. coursii	5	Malagasy	36	48 (33–61)	16 (11–20)	5.90 (5.2–7.2)	No	4.98 (2.87–6.17)	Fair to good
var. elegans	32	S. Africa	36	51 (20–66)	26 (13–48)	8.0 (4.8–11.4)	No	6.60 (0.93–10.22)	Fair to v. good
C. aethiopicus	10	Ethiopia, E. Africa	18,36	51 (30–69)	16 (10–21)	6.6 (6.2–8.1)	No	4.75 (2.95–7.88)	Fair to good
C. incompletus C. var. incompletus	7	S. Africa	18	20 (15–26)	24 (15–28)	6.1 (5.0–7.7)	No	0.59 (0.11–1.04)	Poor
var. hirsutus	2	S. Africa	18	18 (15–20)	28 (24–31)	5.5 (5.1–5.9)	No	0.50 (0.48–0.52)	Poor
C. nlemfuensis var. nlemfuensis	15	E. Africa	18,36	58 (41–74)	13 (12–25)	8.4 (5.6–11.3)	No	6.94 (0.33–11.09)	Fair to good
var. robustus	11	E. Africa	18,36	61 (38–91)	12 (8–15)	8.2 (7.0–9.3)	No	7.00 (5.47–9.25)	v. good
C. plectostachyus	10	E. Africa	18	43 (23–58)	13 (6–18)	6.6 (4.7–10.0)	No	4.57 (3.56–6.28)	Fair to poor
C. transvaalensis	2	S. Africa	18	(Lawn or turf grass not suitable for grazing)					

Table 17.3 *Hybridization range within Cynodon (Chheda, 1974)*

	aethiopicus	dactylon	afghanicus	aridus	coursii	elegans	incompletus	nlemfuensis	robustus	transvaalensis
C. aethiopicus	+									
C. dactylon										
var. dactylon	−	+								
var. afghanicus	−	+	+							
var. aridus	−	+	+aa	+						
var. coursii	−	+	+aa	+aa	+					
var. elegans	−	+	+	−	−	+				
C. incompletus	−	+	+	+	+	+	+			
C. nlemfuensis	+	+	+	+aa	+	+aa	−	+aa		
var. robustus	−	+	+	+	+	+aa	+	+aa	+aa	
C. transvaalensis	−	+	−	+	+	+	−	+	−	−

Note. + and − indicate successful and unsuccessful hybridization attempts, respectively, *C. arcuatus, C. barberi* and *C. plectostachyus* do not hybridize with other species; aa indicates hybrid combinations showing heterosis under Ibadan conditions.

× 2n crosses with an unreduced gamete functioning), the usual configurations at metaphase I are 9_{II} and 9_{I}. Trivalent formation is rare, but Forbes and Burton (1963) reported triploid hybrids with 8 or 9_{III}.

4. In hybrids involving tetraploid parents, or 2n × 4n crosses with unreduced 2n gametes, chromosome behaviour in most cases has been found to be regular at metaphase I. In some cells up to two or three multivalents have been noted.

5. The observed chromosome pairing in 2n, 3n and 4n hybrids suggests (a) considerable homology between genomes of various species that could be crossed, (b) essentially autopolyploid nature of *Cynodon* polyploids, and (c) autosynthetic and preferential pairing of chromosomes.

6. From a plant-breeding point of view these studies indicated that: (a) *C. arcuatus*, *C. barberi* and *C. plectostachyus* are valid species completely isolated from other members of the genus. (b) The remaining members of the genus can be hybridized and appear to share the same, genome. (c) Complete sterility of the inter-specific and some intervarietal hybrids, or lack of seed germination from partially fertile hybrids, may be due to barriers of gene flow accounting for the absence of appreciable amount of introgression within the group. (d) The barriers to gene flow between various taxa are predominantly ecological and geographical, and arose as a result of fragmentation of the gene pool into subpopulations and their divergene evolution under distinct environmental conditions.

7. Selection of suitable ecotypes and production of hybrid varieties are two major ways by which improved cultivars can be developed quickly and efficiently in this genus.

Setaria anceps complex

A cytological analysis of 100 genotypes of the *S. anceps* complex revealed the presence of diploid (2n = 18), tetraploid, pentaploid, hexaploid, octaploid and decaploid forms, as well as aneuploids (Hacker, 1966). Crosses were easily obtained between different ploidy levels except diploid and tetraploid (Hacker, 1967). Many of the hexaploid forms exhibited frost tolerance while tetraploid forms were generally more vigorous. Pentaploids were partially fertile with a high level of chromosome-pairing during meiosis, suggesting the possibility of transferring frost resistance from hexaploid to tetraploid strains by hybridization and a backcrossing programme.

Maximae agamic complex

Combes and Pernes (1970), Pernes (1971) and Pernes *et al.* (1975) reported the presence of fully sexual diploids (2n = 16), as well as apomictic polyploids, in *Panicum maximum*. The diploids were found among collections originating from Korogwe in the Tanga area of Tanzania and around Dar es Salaam. Chromosome-doubling of the diploids produced fully sexual tetraploids. In nature the polyploids of this agamic complex are essentially facultative apomicts (pseudogamous apospory), with progeny tests indicating less than 5 per cent sexual reproduction (Warmke, 1954; Bogdan, 1963). In the Philippines, Javier (1970) made cytological studies of the embryo sacs and estimated that sexual reproduction ranged from 22 to 53 per cent. Interspecific hybridization at the tetraploid level between *P. maximum* and other members of the *maximae* complex such as *P. infestum* and *P. trichocladum*, were obtained and the resultant allotetraploids found to be facultative apomicts. No major differentiation of the *maximae* genomes were found (Pernes *et al.*, 1975). Tetraploidy ($4n = 32$) is most common in this agamic complex and hexaploids ($6n = 48$) are frequently found. Examination of their progenies indicated a higher frequency of plants arising by sexual reproduction than in tetraploid progenies (Combes and Pernes, 1970). Triploids, pentaploids, octaploids, monoploids and aneuploids (31, 36, 37, 38 chromosomes per cell) have been encountered in experimental populations but rarely occur in natural populations.

Pennisetum hybrids

The F_1 hybrids between *P. americanum* ($2n = 14$) and *P. purpureum* ($4n = 28$) are triploid ($3n = 21$), sterile and can only be propagated vegetatively. The hybrids have considerable potential for forage and have been used in pastures in India, Australia and parts of Africa. Fertile amphidiploids ($6n = 42$, having up to 21_{II} at metaphase I) were produced by treating coleoptiles of germinating F_1 seed with 0.2 per cent colchicine solution for 3 h (Aken'Ova and Chheda, unpublished). Seed set was excellent. Seeds were about three times heavier than elephant grass seeds, with germination percentage (80%) comparable to the millet parent and exceeding that of elephant grass. The amphidiploids exhibited high seedling vigour and forage yield was comparable to the F_1 hybrids. Such amphidiploids appear to have potential as a pasture crop, and because of their fertility provide avenues for further genetic improvement.

Bothriochloa – Capillipedium – Dichanthium generic complex

Intensive biosystematic studies of this generic complex carried out in the USA provided information useful for a breeding programme (Harlan *et al.*, 1961, 1962, 1964). The extremely complex cytogenetic background of the materials studied was evidenced by:

1. Variation in chromosome numbers ranging from $2n = 20$ to 180 per cell in multiples of 10.
2. Presence of apomixis (pseudogamous apospory) in the Old World forms where diploids were completely sexual, tetraploids

were largely apomictic and other polyploids were highly to completely apomictic.

3. Occurrence of polyhaploids due to parthenogenetic development of cytologically reduced female gametes.
4. Segmental allopolyploid nature of most polyploids, with several being genomic allopolyploids and few being autopolyploids.
5. Widespread occurrence of introgressive hybridization at intergeneric, interspecific and intraspecific levels in nature, as well as in hybridization between forms differing in their ploidy levels.
6. Union of functional unreduced eggs with normal male gametes resulting in genome-building and production of new chromosome races.
7. Sterility of aneuploids or subvital nature of their progenies.
8. Autosyndesis and preferential pairing of chromosomes and presence of a gene responsible for the predominant bivalent formation commonly observed during meiosis.

These mechanisms have far-reaching implications on the evolution of this highly polymorphic group and on the adaptability of its members in very diverse habitats (Harlan and de Wet, 1963). The members of this generic section are primarily apomictic, both obligate and facultative. This factor, along with intraspecific, interspecific and intergeneric hybridization at various ploidy levels, adds an aspect of dynamic evolution within the group. A plant breeder can direct the evolutionary processes in achieving designated goals. Sexual tetraploids are derived by doubling the chromosome number of a sexual diploid or by crossing two facultative apomictic plants. Desirable genomes from one species can be introduced into another by means of hybridization, then a stable biotype can be ensured through apomixis. For example, a sexual and highly self-sterile *Bothriochloa intermedia* hybrid (clone X-750) was crossed with 8 species of *Bothriochloa*, 6 species of *Dichanthium* and 3 species *Capillipedium*. The hybrids varied greatly in viability, vigour and fertility. Several of the *B. intermedia* × *B. ischaemum* hybrids showed evidence of heterosis and produced strikingly vigorous plants with a high degree of winter hardiness and abundant seed set. The hybrids were apomictic and therefore capable of maintaining their desirable characteristics in later generations. Similarly, drought resistance of some of the *Dichanthium* spp. could be incorporated into *B. intermedia* and the resultant hybrids should have potential in low rainfall tropical and subtropical areas (Chheda and Harlan, 1964).

Glycine wightii

Crosses of many cultivars of *G. wightii* have been unsuccessful because of diploid ($2n = 22$) and tetraploid ecotypes in this species (Pritchard and Wutoh, 1964). Chromosome numbers in some species were noted to be either 2n = 40 or 4n = 80. The chromosomes were much smaller than those of *G. wightii*.

Leucaena leucocephala

Gonzalez *et al.* (1967) studied the cytogenetics of *Leucaena* in relation to breeding of low-mimosine lines and observed that *L. leucocephala* (4n = 104) crossed easily with a related low-mimosine species *L. pulverulenta* (2n = 56). The F_1 hybrids (3n = 80) had low mimosine content and grew vigorously despite meiotic irregularities (26_{II} and 28_{I} at metaphase I). The F_I hybrids exhibited good fertility producing 72 per cent well-filled seeds and only 25 per cent pollen abortion. Chromosome numbers in F_2 progeny ranged from 56 to 88. There is a need for acid tolerant types of *Leucaena leucocephala* which are adapted to the highly acid oxisols and ultisols (pH 4.0–4.5) with high Al saturation (70–90 per cent). At CIAT in Colombia, fertile hybrids with acid tolerance were obtained from crosses between *L. leucocephala* and *L. pulveulenta*, then backcrossed to *L. leucocephala* cv. Cunningham (Hutton, 1980). Large segregating populations gave up to 5 per cent progenies with acid tolerance. At 2–3 months of age, the intolerant seedlings grew only 5 cm high, did not possess nodules, and had small, yellowish leaves. In contrast, the acid tolerant seedlings reached 20 cm in height, were well nodulated, and had thick stems and large green leaves. In addition, some of the progenies contained only 5–6 per cent mimosine in the dry matter of young leaves as compared to 10–11 per cent in leaves of Cunningham. Hybridization of selections and cultivars within *L. leucocephala* also indicated the possibility of decreasing mimosine and increasing edible dry matter (Hutton and Beattie, 1976).

Stylosanthes species

Differences in chromosome numbers, size and morphology were observed among 10 species of *Stylosanthes* (Cameron, 1967b). *Stylosonthes humilis, S. guianensis, S. viscosa, S. hamata, S. macrocarpa* and *S. montevidensis* are diploids (2n = 20), while *S. mucronata, S. tuberculata* and *S. subcericea* are tetraploids (4n = 40) and *S. erecta* is a hexaploid (6n = 60). Colchicine treatment of the sterile *S. humilis* × *S. guianensis* and *S. humilis* × *S. hamata* hybrids resulted in the production of fertile tetraploids (Cameron, 1967b; 1968, quoted by Hutton, 1970).

Genetics

A thorough understanding of the hereditary mechanisms controlling various traits in a pasture species is needed in order to choose methods applicable for varietal improvement or development of more efficient breeding procedures. Lack of basic genetic information, particularly inheritance of quantitative traits, has placed a major constraint on tropical pasture breeding programmes (Burton, 1952; Bray, 1975).

Qualitative and quantitative characters

Some characters are easily quantified, e.g. dwarfism and absence of trichomes in *Pennisetum americanum* (Burton and Fortson, 1966; Powell and Burton, 1971), apomixis in several perennial tropical grasses (Burton and Forbes, 1960; Harlan *et al.*, 1964; Taliaferro and Bashaw, 1966; Hanna *et al.*, 1973), brown midrib mutants associated with lower lignin content and higher digestibility in maize and sorghum (Bula *et al.*, 1977; Porter *et al.*, 1978), rhizome development in *Cynodon* (Rawal and Harlan, 1971), branching habit in *Leucaena leucocephala* (Gray, 1967a) and flowering time in *Stylosanthes humilis* (Cameron, 1968, quoted by Hutton, 1970). Incorporation of such qualitative characters into a new cultivar is relatively easy and can be achieved by simple breeding procedures.

Unfortunately, economic traits such as herbage yield, seed production, seedling vigour, herbage quality, adaptation, etc. show continuous variation, are strongly influenced by environmental and management conditions and are polygenically controlled. They must therefore be studied by suitable statistical methods developed for the analysis of quantitatively inherited traits (Moll and Stuber, 1975).

Components of variance

Many different methods proposed for estimating genetic variances are reviewed by Cockerham (1963), and methods proposed by Stuber (1970) are particularly suitable when dealing with self-pollinated species. The choice of a particular method depends on (1) the mode of reproduction, (2) limitations imposed by crop under study, (3) time and facilities available and (4) assumptions made for simplifying the statistical treatment of the data.

Researchers studying cross-pollinated pasture species have often adopted quantitative methods originally developed for maize. The usefulness of such methods, however has been limited in many pasture species due to flower structure, perennial growth habit, self-incompatibility and polyploidy, such that the forage breeders are confronted with lower selection intensities, longer generation intervals and less effective use of non-additive genetic variance (Hill, 1977). In much of the tropics pasture-breeding is still in its infancy, thus complicated methods may not be necessary in the early stages of a breeding programme. Instead, estimates of genetic variances and breeding value of genotypes can be obtained from crosses routinely made in a breeding programme (Bray and Hutton, 1976;

Latter, 1964). Burton and DeVane (1953) described a method of separating total genetic variation from environmental variation, utilizing clonally propagated materials grown in one or more locations and during one or more seasons. The method is relatively simple and suitable for many perennial pasture species. The method also permits estimates of heritability in the broad sense and calculations of gains from selection.

Burton (1959) reported the results of an 11-year study of 18 single-cross populations of *Pennisetum americanum* involving 296 inbred lines to compare herbage yield and to provide information on additive and non-additive genetic variances. Non-additive genetic variance was a major constituent of total genetic variance in most populations. On the average 55.9 per cent of the total genetic variance for forage yield was non-additive, which indicated the feasibility of rapid yield advance by using the F_1 hybrid directly. A comparison of annual green forage yields of Starr, a stabilized synthetic variety (54.6 tonnes/ha), and Gahi-1, an F_1 hybrid (70.6 tonnes/ha), with the common variety (45.6 tonnes/ha), supported this prediction.

Estimates of combining abilities of materials available for a breeding programme are often made by plant breeders to facilitate the choice of parents in a breeding programme. For a particular quantitative character, the general combining ability (gca) of genotype is the average performance of the genotype in a series of crosses, and differences in gca are related to additive genetic variance. Differences in specific combining ability (sca) which indicate deviation from performance predicted on the basis of gca, are accounted for by the non-additive genetic variance due to dominance and epistasis.

Gray (1967b) used a diallel series of crosses to determine the combining ability of length of the main stem and stem number in four selections of *Leucaena leucocephala*, namely Peru, Hawaii, El Salvador and Guatemala. Significant differences were found, and Peru which had positive gca for each of these traits and a high sca for variance for stem length was considered to be a useful parent in changing the length of the main stem and increasing stem number. Based on experimental data, it was predicted that a cross between Peru and Hawaii should produce progenies that were shorter and more densely branched than Peru. Therefore, they would be more suitable for grazing. In addition, the cross between Peru and Guatemala was predicted to produce highly productive tall lines. These predictions were substantiated by subsequent performance among segregating progenies. The cultivar Cunningham, capable of producing 49 per cent higher forage yield than Peru, was released in 1976 and originated from the Peru × Guatemala cross (Hutton, 1976).

In another diallel cross of five accessions of *Glycine wightii*,

Wutoh *et al.* (1968a, 1968b) measured gca and sca components of variance for nine characters. The gca component of variance was larger than sca with respect to flowering time, maturity data and seed weight, and therefore genetic advance should be possible in these traits by increasing the frequency of desirable additive genes by means of simple selection procedures in the segregating generations. Since the sca component of variance was larger than gca variance component for yield, stolon length and stolon number, and since *G. wightii* is a cleistogamous species not suitable for production of hybrid varieties, genetic progress in these traits would be limited and variance due to dominance and epistasis will be lost during inbreeding. The cultivar Tinaroo used in their studies exhibited high gca and low sca for various traits associated with herbage yield and was considered a suitable parent in breeding for increased yield.

Genotype × environment interactions

Interactions between genotypes and locations, genotypes and years and genotypes × locations × years interactions introduce considerable bias in estimation of genetic variances, and their implications in plant-breeding have been discussed by Allard and Bradshaw (1964). Such interactions affect the reliability of variance estimates, reduce correlation between phenotype and genotype and hinder crop improvement programmes due to discrepancies between predicted and actual response to selection. To develop stable and well-buffered cultivars with broad adaptation requires low genotype × environment interactions. A large genotype × environment interactions suggests the need to develop types adapted to specific environments. As a general rule heterogeneous varieties in self- and cross-pollinated crops are likely to be more stable than homogeneous varieties. Several methods available to measure genotype × environment interactions have been reviewed by Moll and Stuber (1975). Breese and Hayward (1972) suggested that in pasture-breeding repetitive measurements on the same plant and establishment of clonal replicates provide useful information on environmental effects and genotype × environment interactions. Regression technique, originally proposed by Finlay and Wilkinson (1963) and later modified by several researchers, can be employed to characterize responses of genotypes in varying environments and to provide measures of phenotypic stability (Burt and Haydock, 1968; Bray, 1975).

Heritability and gains from selection

Herbitability estimates and expected gains by selecting among individuals in the population are important in order to develop effective and efficient selection procedures applicable to improvement of quantitative traits. In cross-pollinated pasture species parent–progeny regressions are commonly used to provide estimates of heritability particularly for single plant characters such as time of flowering, growth habit, seed weight, seed-shattering and seedling vigour. In many tropical pasture species where plants are normally larger and form loose swards, heritability of most characters including herbage yield, drought resistance, aggressiveness and persistence may also be measured on single plant basis for practical plant-breeding purposes. Several methods available to estimate total genetic variance and heritability in the broad sense (Burton, 1952; Burton and De Vane, 1953) have been commonly used by tropical pasture workers (Burton, 1951; Potts and Holt, 1967; Wutoh *et al.*, 1968a; Hearn and Holt, 1969; Burton and Monson, 1972; Boonman, 1977).

For a particular character, the magnitude of genetic variance, along with heritability, influences the genetic advance that is possible by applying selection pressure. High heritability must be accompanied by high genetic variance to achieve a high rate of progress from selection (Burton, 1952).

For characters that show high heritability, e.g. 50 per cent or more, individual plant or mass selection is preferred to other methods of plant improvement. Aids to selection such as clonal replication, progeny-testing or family selection, generally give minor increases in the rate of progress or may not be beneficial due to lower selection intensity and lengthening of the generation interval (Latter, 1964). Clonal evaluation, i.e. assessment of performance of genotypes selected from space-planted nurseries in a replicated clonal trial, is useful for single plant characters of low heritability, characters which cannot be adequately measured on a single plant, or for studying the response of genotypes to different environmental conditions and management practices. Morley and Heinrichs (1960) found individual selection and polycross progeny-testing were superior to individual selection alone for single plant characters with 10 per cent heritability. Progeny-testing is, however, extremely useful when measurements are made on a sward basis. When a breeder is selecting for two characters of widely differing heritability values, selection could be based on family performance for the poorly inherited character and on individuals within a family for the character with a high heritability.

Burton and Monson (1972) estimated heritabilities for dry matter digestibility (DMD) of forage samples collected from 148 *Cynodon* selections and hybrids grown in three replicated clonal experiments. The DMD ranged from 40.3 to 64.7 per cent. Heritability values calculated on the annual average of two to four clippings ranged from 27 to 78 per cent, indicating the possibility of increasing DMD by breeding. A hybrid (Coastal × Kenya F_1 hybrid No. 14) produced during the course of the breeding programme was observed to be 12.3 per cent more digestible than Coastal bermudagrass

(Burton *et al.*, 1967), and was subsequently released as Coastcross-1 This cultivar produced 30 per cent more liveweight gain but required 20 per cent less dry matter per kilogram of gain than Coastal bermudagrass when cut at comparable stages of early growth, dehydrated and fed to dairy heifers (Lowrey *et al.*, 1968). In *Setaria anceps* Hacker (1972) also found high heritability for *in vitro* digestibility and predicted that improved quality could be obtained by breeding.

Boonman (1977) observed genotypic variances and broad sense heritability values for seeding characters and herbage yield in *Chloris gayana*. Mass selection was considered an appropriate method to identify superior genotypes. Plant populations were developed with different maturity dates. Eight superior clones were identified by selecting for increased seed yield within an early maturing population of the Mbarara cultivar. These were bulked to form a new cultivar named Elmba. It showed over 100 per cent gain in pure germinating seed yield and over 10 per cent gain in dry matter yield as compared to the parental type.

A parent open-pollinated progeny test involving 12 selections of *Panicum coloratum* revealed significant and heritable differences in seed production components such as fertile florets per panicle, seed per panicle and per cent fertility (Hearn and Holt, 1969). An advanced generation of plants selected for individual seed production components produced 47 per cent more seed (65 kg/ha) than unselected plants (45 kg/ha).

Correlation between characters and indirect selection

Practical plant-breeding programmes deal with the improvement in one or a few characters, but it is important to know to what extent other traits are affected during the breeding process. For example, a breeder attempting to increase seed yield or *in vitro* digestibility of a pasture species may wish to know the effects of selection for these traits on dry matter yield or maturity date. This is accomplished by examining correlation coefficients of trait relationships.

Genetic correlations can be utilized to increase the rate of response of a primary character (e.g. seed yield) that may have a low heritability, or to which high selection intensity cannot be applied, by selecting for a secondary character (e.g. date of maturity) that exhibits high heritability, or to which high selection intensity can be applied, and is closely correlated to the primary character. Achieving progress in trait X through the correlated response of trait Y is called indirect selection. Much of pasture-plant breeding where the eventual aim is higher livestock productivity deals with indirect selection. In many pasture-breeding programmes early selection is carried out in space-planted nurseries or small plots. Bray (1975) discussed indirect selection in pasture-breeding, and by a specific example showed the progress to be expected under pasture conditions when selection is carried out in spaced plants for *in vitro* digestibility in *Cenchrus ciliaris*, assuming different heritability and genetic correlation values.

Inheritance studies of agronomic characters in *Centrosema virginianum* in Australia revealed that genetic variance in herbage yield was predominantly additives with a genetic correlation of 0.87 between yield and leaf size in spaced plants (Clements, 1976). Also a high genetic correlation of 0.97 is found for yield among crosses between diverse genotypes when measured at two different experimental sites. These observations indicated considerable scope for yield improvement by indirect selection of leaf size. Furthermore, genotype × environment interaction would unlikely limit breeding progress in the initial stages of the improvement programme.

Breeding methods

Breeding procedures utilized for pasture crops are the same or similar to those developed for other self- and cross-pollinated crop species. Hanson and Carnahan (1956), Whyte *et al.* (1959), Harlan, (1962), Latter, (1964), Bogdan (1966), Burton (1970), Wit (1971), Bray and Hutton (1976) and Bashaw (1975) reviewed pasture-breeding procedures with special reference to tropical pasture species.

Seed propagated sexual species

Tropical pasture crop breeders have utilized mass selection, pure-line selection, controlled hybridization followed by pedigree selection or the backcross method for the improvement of self-pollinated species. For cross-pollinated species that are normally propagated from seed, genetic improvement has been based on mass selection, the backcross method, selection by progeny-testing and the development of synthetic and hybrid cultivars.

Mass selection

In the initial stages of a breeding programme, where the assemblage of germplasm contains products of natural selection from diverse habitats, mass selection offers an approach whereby improved cultivars can be developed rapidly. Further improvements may then be effected by the subsequent use of more elaborate breeding methods.

The method involves establishment of a large (several thousand plants), space-planted nursery which is maintained under good fertility conditions and frequently clipped or grazed. The forces of natural selection in the form of environmental conditions, disease and insect pests, cold, drought, etc. are allowed to operate. In addition, the breeder may remove obviously undesirable plants prior to pollination or further select from the surviving plants those with desired characteristics. Seeds are harvested from the promising plants and bulked to form the next generation. The method is simple and relatively inexpensive, enables rapid concentration of desirable true breeding genotypes in the land races of self-pollinated species and increases the frequency of desired genes in the heterogeneous cross-pollinated species.

Since selection is based on the measured performance of an individual phenotype, the method is effective for improvement of characters that show high heritability and where the measured performance of an individual phenotype is correlated to its progeny performance. Success depends on maintaining a high level of selection intensity for the character in which improvement is sought. For characters such as dry matter yield where heritability on an individual plant basis is generally low, the rate of improvement may be negligible or, at best, slow.

A larger number of tropical pasture cultivars have been derived by mass selecting and multiplying superior, locally collected or introduced ecotypes. For instance, the development of Pokot Rhodes grass was based on five outstanding plants selected for their vigour, persistence and ability to develop fast-growing stolons from a population grown from seed collected in the West Pokot district of Kenya. Similarly, the Usenge cultivar was based on 12 single plants selected for their dense and low-growing herbage, and for their fine leaves, from a population grown from seed collected at Usenge in Kenya (Bogdan, 1965). Boonman (1977) developed Elmba by bulking seed of the eight best clones selected for their seed-yielding ability, vigour and early maturity from a population of Mbarara.

Three cycles of single-plant mass recurrent selection in the Nandi variety of *Setaria anceps* resulted in the development of late-flowering Nandi Setaria Mark-2 with a rapid regeneration capacity after grazing. In each of the generations, 40–50 of the top performing plants were selected from 1 000–2 000 singly spaced plants tested in a polycross block (Bogdan, 1965). The Narok variety of *S. anceps* was developed using mass selection with restricted crossing. Twelve frost-tolerant plants were selected from a population of 10 000. These were intercrossed, then 87 plants selected from progeny rows and again intercrossed to produce the cultivar Narok (Bray and Hutton, 1976).

Modified versions of mass selection have been used to achieve improvement, particularly in characters with low heritability such as dry matter yield. For example, Burton (1974) described a procedure called recurrent restricted phenotypic selection. In Pensacola bahia grass, this procedure resulted in an 18 per cent increase in herbage yield after four 2-year cycles of selection at 20 per cent selection intensity. The procedure involved selection of spaced plants on the basis of herbage yield at the end of the first season plus the spring growth the following year. The following restrictions were imposed to increase the efficiency of selection:

(a) Division of the nursery into 25-plant, square plots and selecting the five best plants from each plot to reduce adverse effects of soil heterogeneity.

(b) Imposition of paternal as well as maternal selection by using intercrossed seed produced on selected phenotypes isolated from the population instead of open-pollinated seed.

(c) Facilitating intermating and gene recombination by isolating two culms ready to flower from each selected phenotype and placing them along with culms from other selected plants in 4 litre cans filled with water.

(d) Avoiding unequal representation of parents in the next cycle by using seed from the two heads of each selected genotype.

(e) Choosing germplasm with a high degree of self-incompatibility to reduce the likelihood of inbreeding.

At the end of the study Burton suggested that the rate of progress could have been doubled by using a 1-year cycle and higher selection intensity.

Neal Wright (1976) studied the effectiveness of recurrent phenotypic selection through various cycles to isolate genotypes of heavy and light seed weight of *Panicum antidotale*. Selection was much more effective for heavy seed weight than for light seed weight, and at the end of the sixth cycle seed weight showed 73 per cent increase and 17 per cent decrease as compared to the average of the original population.

Pure-line selection

This method is useful for self-pollinating species, especially when genetically variable populations are available. The usual procedure is to select carefully a large number of single plants from random populations or ecotypes, harvest their seed and grow the progenies. Lines with obvious defeats are eliminated, and the number is further reduced by suitable aids to selection such as artificially produced disease epiphytotics, heavy grazing pressure, compatibility with associated grass species, performance in diverse environments, etc. Finally, the progenies of the remaining lines are studied in replicated field trials to identify the single most valuable progeny which can form the basis of a new cultivar. For example, the three most prominant cultivars of *Stylosanthes humilis* – Gordon, Lawson and Paterson – developed in Queensland, Australia, are each based on a single plant. The three types are erect and morphologically similar, but differ in maturity dates. Gordon, the late flowering cultivar, is suitable for areas with 1 100 mm rainfall. The early flowering cultivar Paterson is recommended for the low rainfall (600–900 mm) areas. The cultivar Lawson, with an intermediate flowering behaviour, grows well in areas of Queensland receiving around 1 000 mm of precipitation. The cultivars differ in seed colour which facilitates variety identification as well as seed certification.

When selection in the cross-pollinated species is based on progeny tests and when a group of superior progeny lines are bulked to form a new cultivar, the procedure is called line-breeding.

Pedigree method

Improvement of some pasture species has involved hybridization of two or more parental genotypes with a view to combine their desirable traits into a new cultivar. In self-pollinated pasture species, pedigree selection in the segregating generations following hybridization has produced worthwhile results as exemplified by the development of Siratro (Hutton, 1962; Hutton and Beall 1977; Shaw and Whiteman, 1977; Hutton *et al.*, 1978). This cultivar was developed from a cross between two Mexican introductions of *Macroptilium atropurpureum* (C.P.I. 16877 and 16879). The two introductions were morphologically similar but one had greater stoloniferous development than the other. Both were highly resistant to root-knot nematode and the virus disease legume little leaf.

An F_2 population of 1 240 plants, 50 plants of each parent and the F_1, were grown in a field and rated for degree of rooting from stems used for transplanting, herbage yield and seed yield. The plants were grown at a spacing of 1.8×1.8 m to allow for lateral spread. Several of the F_2 hybrid plants had higher stem-rooting than the parents and were considered transgressive segregates for the stoloniferous character. Thirty-four of the superior recombinants with the highest herbage rating were selected for further evaluation.

The F_3 and F_4 families were grown in replicated plots of 49 plants, each transplanted at 1.6×1.6 m spacing. In order to simulate pasture conditions and favour seed production of the strongly stoloniferous plant, the plots were oversown with *Chloris gayana* and subjected to intensive, but intermittent, grazing with cattle. An area of 3.6 to 1.2 m in each plot was used to estimate dry matter yield and crude protein content prior to each grazing. At the end of the first season, six F_3 families showed high rating for yield and percentage of plot covered by the legume. Bulk seed was collected from them. By the third season *Digitaria didactyla*, an undesirable grass, had invaded most plots. Three of the families, however, maintained a high percentage of *Chloris gayana* and had lower content of the *Digitaria* invasion.

The most promising families and parents were compared in the F_4 generation in sown and transplanted plots. The three selections having high competitive ability with *Digitaria* were significantly greater in dry matter yield and stolon development. Further field-testing of these selections showed them to be similar in most plant morphological characters. The F_5 seeds of the three (i.e. 0.24% of the original population of 1 240 F_2 plants) were bulked in equal proportion to produce the cultivar Siratro. The composite nature of Siratro ensures sufficient variability for adaptation to widely differing environments.

Hutton and Beall (1977) Hutton *et al.* (1978) and Jones, *et al.* (1980) reviewed their breeding efforts with *Macroptilium atropurpureum* subsequent to the development of Siratro. Three series of crosses were made in 1959–63, 1964–66 and 1972–73 using a wide range of introductions, the cultivar Siratro and promising selections derived from the first Siratro cross. Selection criteria among progenies from the various crosses included herbage yield, superior nodulating ability, ease of establishment, stoloniferous habit, persistence, root-knot nematode resistance, disease resistance – particularly halo blight resistance, tolerance to acid soil conditions and wide adaptability. Series II and Series III crosses produced eight superior lines with herbage yields exceeding Siratro by 25 per cent.

Although the pedigree method permits the plant breeder to exercise skill in selection, it also places limitation on the amount of material that can be evaluated. For example, field evaluation of Series II Siratro cross-bred progenies arising from 20 crosses required about 30 ha of land in four field stations.

Brim (1966) developed a modified pedigree method called single-seed descent method for improvement of self-pollinated species. The method is potentially useful for breeding tropical pasture legumes, and has been recommended by Bray and Hutton (1976) and Hutton and Beale (1977). It consists of advancing each selected F_2 plant by single-seed descent. In the F_2 and succeeding generations only one seed is used from each plant in the population selected as a parent for the next generation. The method enables rapid advancement of generations to homozygosity. When the desired level of inbreeding is achieved after about five generations, each progeny that traces to a different F_2 parent is bulked, multiplied and used for subsequent field-testing. Thus, repeated field evaluations and associated field-work with the F_3 to F_6 generations are completely eliminated, saving human and land resources. By using glasshouse and irrigation facilities several generations can be grown in 1 year (Imrie, 1976).

Clements (1976) suggested a multi-parent approach to breeding self-pollinated tropical pasture legumes. He indicated that the 30 accessions of *Centrosema virginianum* contained ecotypes of little value in themselves, but possessed useful genes. Broadly based populations derived from many parents (Fig. 17.2) would provide

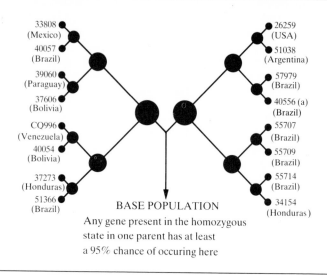

33808 (Mexico)
40057 (Brazil)
39060 (Paraguay)
37606 (Bolivia)
CQ996 (Venezuela)
40054 (Bolivia)
37273 (Honduras)
51366 (Brazil)

26259 (USA)
51038 (Argentina)
57979 (Brazil)
40556 (a) (Brazil)
55707 (Brazil)
55709 (Brazil)
55714 (Brazil)
34154 (Honduras)

BASE POPULATION
Any gene present in the homozygous
state in one parent has at least
a 95% chance of occuring here

Fig. 17.2 Development of a multi-parent gene pool in *Centrosoma virginianum*. The highly heterogeneous base population offers the opportunity to select desirable lines for further field-testing and eventual production of superior cultivar(s) (Clements, 1977).

long-term selection responses and offer the possibility of incorporating many useful genes into a single cultivar. Two populations were developed by first crossing the parents and then successive generations, so that a gene present in the homozygous state in only one parent had at least a 95 per cent chance of occurring in the base population. For the development of superior cultivar(s) from such a base population the author proposed the use of the single-seed descent method to carry forward about 1 000 lines, following a single cycle of selfing and intense selection for frost resistance and seedling characteristics. Seed of these lines could then be multiplied for field-testing.

Backcross method

This method is used to incorporate one or a few highly heritable, desirable characteristics into a cultivar or selection. Plants of the adapted type (recurrent parent) are crossed with those of another type (donor parent) possessing the desirable trait. The progenies are subsequently backcrossed to the recurrent parent with selection of the desirable trait among progenies. The procedure is repeated five or six times to reconstitute the recurrent parent. Finally a number of lines similar to the recurrent parent are bulked and increased to form the basis of a new cultivar. The method is useful for both self- and cross-pollinated species.

Aken'Ova (1975) transferred sterility factors from the cytoplasmic male-sterile Tift 23A *Pennisetum americanum* to the West African photoperiod-sensitive Maiwa cultivar by the backcross method. This permitted large-scale commercial production of F_1 hybrid seeds using *P. purpureum* as the pollen parent.

Synthetic varieties

In cross-pollinated pasture species, population improvement through the production of synthetic varieties is a common method of breeding in the temperate countries and is being increasingly utilized by the breeders in the tropics. A synthetic variety is derived by hybridizing a number of selected genotypes in all possible combinations. The combination is maintained and multiplied by random mating in isolation. The method involves (a) a large source population consisting of inbred lines, clones or mass-selected materials from which selections can be made, (b) identifying phenotypically outstanding plants in the source population as potential parents, (c) testing their genotypes for combining ability, (d) intercrossing parent plants to synthesize the variety and (e) stabilizing it under random mating in isolation for two to three generations (Fig. 17.3). The ultimate aim is to increase the frequency of desirable genes in the population while avoiding inbreeding depression.

Pasture crop breeders have tested genotypes for combining ability by one of the following four tests:

1. Open-pollinated progeny test

Seed is collected from promising plants in the source nursery wherein the pollen source was non-restricted. Relative performance of progenies is derived by evaluating progenies of the selected parental plants.

2. Top-cross test

Selected clones or inbred lines are grown in rows alternating with a tester variety to produce top-cross seed. When the tester variety is an inbred line or clone the top-cross test gives a measure of specific combining ability. When an open-pollinated variety is used as the tester a measure of general combining ability can be obtained.

3. Single-cross test

Each selected clone or inbred line is crossed with a number of other selected clones or inbred lines. Performance of any one single cross provides an estimate of specific combining ability. The mean performance of an entry in its crosses with all the other selected entries

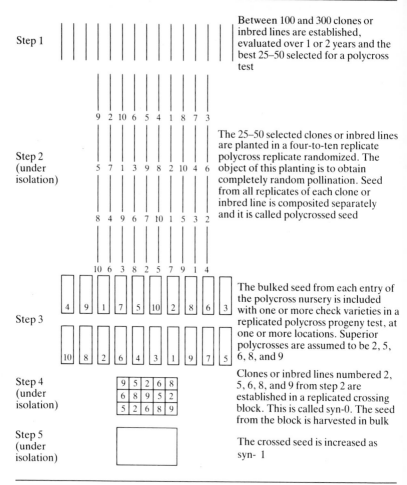

Fig. 17.3 Steps in the development of a synthetic variety, using a polycross test. With perennial pasture species each step may require more than 1 year. The synthetic variety is usually evaluated as Syn-2 and Syn-3 in comparison with other commercial varieties (Briggs and Knowles, 1967).

with which it was crossed reflects general combining ability. When the single crosses are made in all possible combinations of lines or clones, the resulting set of crosses is called a diallel cross. This scheme is appropriate when a few genotypes need to be tested.

4. Polycross test

In the early stages of a breeding programme suitable tester varieties may not be available for top crosses, leading to the use of a poly-cross test. For the production of seed, propagules of each selected clone are transplanted at random into small compact plots in a replicated (usually 4–10) and isolated polycross nursery. The aim is to obtain seed from a clone or line that is subject to outcrossing with other selected materials growing in the same nursery. Volume of pollen produced, time of anthesis, differential cross compatibilities, level of inbreeding, plant height and lodging affect random pollination. Equal amounts of seed from each entry harvested from all replications are composited to give the polycross seed.

The polycross progeny test provides information on the general combining ability of each entry. Identification of superior clones is based on herbage yield data as well as observations of traits such as disease resistance, maturity period, seed production, etc. A polycross test provides material that can be used in the subsequent cycles of selection. Advanced cycles reduce variation among progenies tested, however, and may hinder full exploitation of the diversity existing in the original parental materials.

To obtain a reliable measure of combining ability, it is important that progenies be grown under conditions approximating a sward, i.e. close spacing, oversowing with grass or legume and intermittent heavy grazing. It is of value to use several replications at different locations.

Synthesis of a variety

Based on the combining ability test, 4 to 10 superior clones or inbred lines that are reasonably uniform in their flowering behaviour are established in a replicated crossing block to permit random and, as far as possible, equal crossing of all components. These materials which form the basis of a synthetic variety are called Syn-O. Equal amounts of seed of each component harvested from each replication are bulked to grow the next generation called Syn-1, under isolation and with adequate replication. The Syn-1 derived from diverse selected genotypes comprises a large proportion of single crosses and is likely to yield more herbage than later generations. Thus, it is necessary to stabilize yield and other characteristics by multiplying seeds in isolation for two to three generations before final evaluation. The final evaluation should be carried out over a number of years at several locations and under conditions closely approximating commercial production. If found superior to other existing cultivars, the advanced generation is released as a new synthetic variety. It is important to maintain the basic components of a synthetic variety so that it can be resynthesized or modified in the future.

Two synthetic varieties of *Pennisetum americanum*, Starr and Tiflate, were developed in Georgia, USA (Burton and Powell, 1968). Several synthetic varieties of *Setaria anceps* produced in Australia are reportedly superior to the existing Nandi variety. Synthetics have also been produced from a cross between *S. splendida* × *S.*

anceps in an attempt to combine plant vigour and high seed production potential in one variety (Hacker, 1976).

Hybrid varieties

A hybrid variety can be produced by crossing inbred lines, clones or open-pollinated varieties that are genetically dissimilar and which combine with each other to give maximum expression of heterosis. The various avenues by which hybrid varieties can be produced in sexually propagated pasture species are as follows (Burton and Sprague, 1961; Burton, 1970):

1. First-generation chance hybrids

Gahi-1 variety of *Pennisetum americanum* is a first-generation chance hybrid capable of producing up to 50 per cent more herbage than the common variety and 25–30 per cent more than the synthetic variety Starr. It was produced by allowing natural open-pollination of four inbred lines randomly sown under field conditions. The harvested seed (over 2 tonnes/ha) contained 68–75 per cent hybrids. The advantage gained by chance hybrids is based on the greater competitive ability of hybrid seedlings as compared to self-pollinated seedlings. The former predominated in the established plant population.

2. Self-incompatibility hybrids

Several tropical pasture species contain a significant proportion of plants that are highly self-incompatible but cross-compatible with other genotypes. For example, over 20 per cent of plants in Pensacola bahia grass (*Paspalum notatum*) set less than 2 per cent seed when selfed. However, when grouped in pairs to permit cross-pollination over 90 per cent seed set was recovered (Burton, 1955). A similar situation was observed in selected *Cynodon* clones by Burton and Hart (1967). By estimating the herbage production potential of hybrids between such selected cross-compatible clones it may be possible to identify superior hybrids. Two such selected clones can then be transplanted vegetatively in alternate rows in isolated seed fields to produce commercial hybrid seed. Proper maintenance should ensure hybrid seed production for several years from the same field by preventing seedlings to develop. Commercial production of Tifhi-1 bahia grass F_1 hybrid seed is based on such a procedure. Two clones, 14 and 108, selected for their ability to produce highly vigorous hybrid progeny, are used as the parental materials. Beef production on Tifhi-1 pastures averaged 514 kg/ha per yr which was 77 kg/ha per yr above that obtained from the open-pollinated parental Pensacola variety (Hein, 1958).

3. Cytoplasmic male-sterile hybrids

Cytoplasmic male-sterility mechanisms are known for many crops, including the pasture species *Pennisetum americanum, Sorghum sudanense* and *Desmodium sandwicense* (Burton, 1959; Craigmiles, 1966; McWhirter, 1969). A thorough search in other species is likely to reveal cytoplasmic sterility. The mechanisms offer one of the most practical ways of producing F_1 hybrid seeds for sexual species. Suitable pollen parents that combine well with the male-sterile parent need to be identified. The sterility of F_1 hybrids in the pasture species is not a disadvantage.

With the development of male-sterile Sudan grass the height of Rhodesian Sudan grass was reduced from 12 feet to 4 feet (Craigmiles, 1966). The pollinator Tift Sudan grass restored full height and yield potential of the F_1 hybrid. These hybrids, when grown in Georgia, USA produced 31 per cent more herbage than the average Sudan grass and 22 per cent more than the highest-yielding commercial variety. Many of the successful sorghum-Sudan grass hybrids are produced by crossing selected Sudan grass males with cytoplasmic male-sterile sorghums.

Commercial F_1 hybrid seeds of *Pennisetum americanum* pasture varieties such as Millex 22 are produced by crossing selected pollen parents with the male-sterile Tift 23A and harvesting the seed produced on the male sterile parent. Commercial F_1 hybrid seed of *P. americanum* × *P. purpureum* is also feasible by using cytoplasmic male-sterile stocks of *P. americanum* (Powell and Burton, 1966; Aken'Ova, 1975).

McWhirter (1969) proposed a scheme for the commercial production of a hybrid cultivar based on interspecific hybrids in *Desmodium* using male sterility (Fig. 17.4). The scheme attempts to obviate, in part, the technical difficulties arising due to the procumbent, trailing growth habit and negligible amounts of wind pollination in the species. The author suggested that commercial seed could be produced from F_2 hybrid plants that carry the self-fertile gene from the R-line.

Asexually propagated species

Some tropical grass species (e.g. *Cynodon* spp., *Pennisetum purpureum, P. clandestinum*) are readily propagated vegetatively, although sexual reproduction is normal. Others (e.g. *Brachiaria* spp., *Bothrichloa* spp., *Paspalum* spp., *Panicum maximum, Cenchrus ciliaris*) are primarily apomictic but are sometimes propagated by vegetative cuttings. Asexual reproduction offers several advantages to the breeder (Burton, 1970):

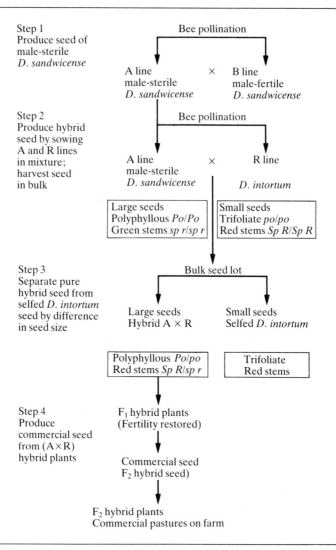

Fig. 17.4 A hybrid *Desmodium* seed production scheme (McWhirter, 1969).

(a) An improved variety can be based on the discovery or creation of one superior genotype.

(b) The benefits of heterosis and favourable agronomic traits can be maintained indefinitely either by clonal propagation or by obligate apomixis.

(c) Progeny-testing which involves time and effort is not necessary.

(d) Wide crosses aimed at combining the desirable traits of two or more species or genera can be attempted.

(e) Variety identification and maintenance is simple.

The main disadvantages are:

(a) Vegetative propagation is generally more expensive and less convenient than seeding.

(b) Search or development of sexual types in apomictic species for use in a breeding programme can be laborious, time-consuming and may not always be fruitful.

Vegetatively propagated species

Breeding procedures of vegetatively propagated species aim to isolate a superior genotype that can be released as an improved cultivar. This requires the collection or creation of a population containing variable genotypes.

Cynodon improvement in the USA and in West Africa led to the development of cultivars such as Coastal, Midland, Coastcross-1, IB.8, IBX. 7 and illustrates various aspects of breeding for vegetatively propagated types.

At Ibadan, Nigeria, over 200 introductions, representing eight species and various cultivars of *Cynodon*, were grown in duplicate plots (1 m × 2 m with cultivated 0.75 m alleys) and evaluated over two growing seasons (Table 17.2 p. 472). Fourteen clones were selected for additional study during one growing season. The five top-performing genotypes, along with a local selection, were further tested in another replicated field trial over 2 years. Individual plot size was 9 m × 15 m. Observations were made on rate of spread, drought tolerance, herbage yield, nutritive value and animal acceptability. Two superior clones, IB.8 (*C. nlemfuensis* var. *nlemfuensis*) and IB.12 (*C. aethiopicus*), along with the local cultivar *C. nlemfuensis* var. *robustus*, were each established in large paddocks to observe animal performance. These and other highly ranked clones were also evaluated in three regional trials and grown in observation plots in different parts of the country. One clone, IB.8 was released as a commercial cultivar in 1968. This genotype is tetraploid, well adapted to the forest and savanna areas of southern Nigeria, high yielding (15–20 tonnes/ha per yr of dry matter), capable of growing well into the dry season, and highly responsive to improved cultural and management practices. Up to 650 kg/ha per yr liveweight gains by Zebu steers have been recorded on 3–5-year-old IB.8 pastures, as compared to around 300 kg/ha per yr gain on the local cultivar (Chheda, 1974).

All potentially useful species of *Cynodon* are cross-pollinated. Seed set is generally poor and highly self-sterile clones are commonly encountered. Controlled hybridization within and between species is possible, however. A number of hybrids were produced and established in a nursery at Ibadan along with material obtained from Stillwater, Oklahoma. Several genotypes in a collection of 150 hybrids displayed heterosis under Ibadan conditions (Table 17.3). Studies that involved herbage for yields, chemical composition, *in vitro* digestibility, drought tolerance, persistence and animal acceptability at Ibadan and several regional research stations resulted in the release of IBX.7 (Ademosun and Chheda, 1974). IBx.7 is an interspecific hybrid between *C. nlemfuensis* var. *nlemfuensis* and *C. dactylon* var. aridus. It is comparable in herbage yield and quality to IB.8, possesses more compact herbage and is more fully utilized under grazing.

The development of Coastal bermudagrass by Burton (1943, 1947) is considered a landmark in pasture and livestock development in the USA and in the annals of plant-breeding. A species long considered a weed was converted into an extremely useful pasture crop by breeding. Two strains of local *C. dactylon* were interplanted with two, tall-growing South African strains with a view to obtaining a large number of natural hybrids. Open-pollinated seeds were collected and 5 000 seedlings established on 1.5 m × 1.5 m squares in a field nursery. After 1 year of careful observations 147 promising clones were selected for critical evaluation. Ten centimetre sod plugs from each were placed in the centre of cultivated plots measuring 1.25 m × 7.50 m. Three plots of each genotype were established. Over a 7-year period records were kept on rate of spread, sod density, winter hardiness, disease resistance, yield, percentage weeds, percentage cover, seeding habits, compatibility with oversown legumes, root-knot nematode resistance and persistence. Simultaneous studies carried out on the top-performing clones provided information regarding their responsiveness to fertilizer applications, chemical composition, palatability, yield potential and persistence under simulated close-grazing conditions. Stolons of the highest ranked genotypes were evaluated in regional trials. Clone 35 was described and released in 1943 as Coastal. By 1970 it had been planted on more than 4 million ha from Texas to the Atlantic coast in the USA and contributed over US$200 million to grazers as compared to the common type *Cynodon*. Coastal yields two to four times more herbage than common strains of *Cynodon*, responds efficiently to nitrogen fertilization, is tolerant of *Helminthosporium* leaf spot and root-knot nematode, flowers sparsely and remains vegetative over the growing season. In regions of adaptation Coastal produces 100 kg/ha or more beef than the naturalized stains at equivalent levels of fertility and management.

Though frost-tolerant, Coastal did not possess sufficient winter hardiness for some areas. A hybrid between it and a *C. dactylon* strain from Indiana was selected for hardiness and called Midland (Harlan *et al.*, 1954).

Subsequent to the release of Coastal many data clearly indicated

its high herbage yield per unit area, but moderate output per grazing animal primarily due to relatively low dry matter digestibility.
Using the nylon bag technique, Burton *et al.* (1967) compared the
dry matter digestibility of 23 uniformly managed genotypes including a number of high-yielding F_1 interspecific hybrids with Coastal.
A hybrid derived from Coastal × *C. nlemfuensis* var. *robustus* averaged 12.3 per cent more digestible dry matter than Coastal and was
subsequently released as Coastcross-1. Feeding trials have shown
higher herbage intake and daily gains by cattle when fed Coastcross-
1 as compared to other *Cynodon* cultivars.

Using methods similar to those described for *Cynodon*, breeders
in various parts of the tropics have released cultivars of other vegetatively propagated pasture species such as *Pennisetum purpureum*,
P. purpureum × *P. americanum* F_1 hybrids (Aken'Ova, 1975; Muldoon and Pearson, 1979).

Apomictic species

In species with obligate apomixis, development of superior cultivars
depends on the availability of desirable mutants within a large collection of diverse ecotypes. A single plant can thus form the basis of
a true breeding cultivar (Bashaw, 1962; Sherwood *et al.*, 1980). A
number of cultivars of *Panicum maximum*, *Paspalum* spp., *Cen-
churs ciliaris*, *Brachiaria* spp., *Melinis minutiflora* and *Dichanthium*
spp. have been identified in this manner.

Discovery or creation of sexual types in apomictic species, or in
the agamic complex to which the species belongs, permits improvement by breeding. Sexual forms can be obtained as follows:

1. Screening a germplasm collection

A careful morphological study of plants derived from open- and
self-pollinated seeds collected from individual plants helps to identify parents that are obligate apomicts (no segregation among progenies), facultative apomicts (several off-type plants in the progeny),
or completely sexual (diverse segregation among progenies). Using
this procedure, Smith (1972) and Burton *et al.* (1973) discovered
completely sexual plants of *Panicum maximum* (Table 17.4).

2. Doubling the chromosome number of a sexual diploid

A typical agamic complex in the grasses is based upon one or more
sexual diploid species, with a complex polyploid superstructure of
apomictic species as observed in *Bothriochloinineae* (Harlan and de
Wet, 1963), *Maximae* agamic complex, and *Paspalum notatum*.
Doubling the chromosome numbers of such sexual diploid species
can produce a sexual tetraploid (Harlan *et al.*, 1962; Burton and
Forbes, 1960).

Table 17.4 *Reproductive system in 295 plants of Panicum maximum, based on a 10 to 20-plant progeny test at Tifton, Ga., in 1971 (Burton et al, 1973)*

Reproductive system	Parent plants		Plants studied	
	(no.)	(%)	Total	Offtype
Obligate apomixis	259	87.8	3 751	0
Facultative apoximis	34	11.6	568	48
Obligate sexuality	2	0.6	27	27

3. Function of unreduced egg of a sexual tetraploid

When sexual tetraploids are hybridized with apomictic tetraploid forms, a higher polyploid (6n) with the sexual mode of reproduction may arise due to the union of an unreduced egg from the sexual parent with a reduced male gamete from the apomictic parent, as in *Bothriochloa* (Harlan and de Wet, 1963).

4. Crossing of facultative apomictic types

Facultative apomictic plants may be heterozygous for a gene(s) controlling apomixis. An occasional progeny shows sexuality. Harlan *et al.* (1962) crossed two facultative apomictic plants of *Bothriochloa intermedia* and obtained a semi-sterile but fully sexual plant. Most of the F_2 progenies, however, were sterile or apomictic.

Genetics of apomixis

Inheritance studies of apomixis indicate that the complex phenomenon is controlled by relatively few genes. In *Paspalum notatum* apomixis is recessive to sexuality and controlled by a few genes that give autotetraploid ratios (Burton and Forbes, 1960). Genetic evidence in *Bothriochloininae* suggests dominance for apomixis, with two genes involved at the tetraploid level (one gene per genome). Most of the apomictic plants appear to be heterozygous at both loci (Harlan *et al.*, 1964). In *Cenchrus ciliaris*, Taliaferro and Bashaw (1966) noted that the method of reproduction is conditioned by two epistatic genes. The inheritance data, based on progeny of selfed sexual plants and sexual × apomictic hybrids, led Hanna *et al.* (1973) to conclude that apomixis is controlled by two recessive genes in *Panicum maximum*, but conditioned by modifying genes.

Utilization of sexual forms encountered in apomictic species

Taliaferro and Bashaw (1966) developed a model in *Cenchrus ciliaris* for breeding apomictic species in which heterozygous sexual plants are identified (Fig. 17.5). Apomictic plants appear among the selfed progenies of a sexual plant and among progenies derived from crossing a sexual plant with apomictic types. Using the former

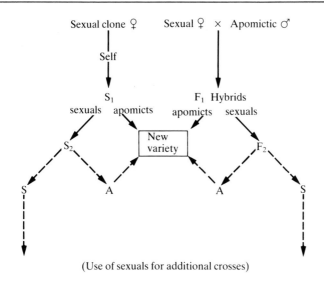

Fig. 17.5 Diagram of methods of developing new apomictic types of buffel grass using heterozygous sexual plants (modified from Taliaferro and Bashaw, 1966).

approach Bashaw (1968) developed the Higgins cultivar by increasing the seed of an apomictic segregate. The cultivar combines high herbage and seed yields and persists under grazing.

Burton *et al.* (1973) observed considerable variation in important agronomic characters of *Panicum maximum*, e.g. plant height (53–248 cm), *in vitro* dry matter digestibility (41–70%), green plant yield (0.09–6.13 kg), selfed seed set per head (3–580 mg), hairiness of plant parts, maturity date and winter hardiness. Sexual plants have been isolated in this species and their use as females in hybrids will release variability through genetic recombination. Apomictic plants recovered in progenies of segregating progenies could be exploited to fix heterosis in the superior genotypes.

Team approach

During the course of a breeding programme the breeder may wish to screen germplasm material for disease and pest resistance, nutritive value, drought or cold tolerance, nodulation in case of legumes, along with other characteristics. Improved animal performance is the ultimate goal in development of new cultivars. It is therefore prudent for a breeder to seek the cooperation of agronomists, entomologists, pathologists, virologists, nematologists, physiologists, and biometricians and other specialists in order to achieve worthwhile results.

Registration, maintenance and distribution of improved cultivars

A new cultivar is usually registered before release to the general public. The International Code of Nomenclature of Cultivated Plants provides guidelines for selecting names for identification. In Australia a Herbage Plant Registration Authority was established in 1965 for review of new cultivars before release (Barnard, 1972). The registry contains origin, description and agronomic characteristics of tropical grasses and legumes. Whiteman (1978) reviewed the improved cultivars utilized in Australia. In many other countries registration is accomplished by publication in a leading national agricultural or botanical journal. For example, the origin and brief descriptions of grass and legume cultivars grown or developed in Kenya are published in the East African Agricultural and Forestry Journal.

The responsibility of the breeder does not end with development and release of the new cultivar. Another duty is to ascertain that the genetic integrity of the cultivar is maintained. In countries where seed certification schemes operate, the breeder provides basic seed stocks on a continuing basis to the appropriate authorities for increase and subsequent distribution to the farmers. Standards for seed certification have been recommended by the OECD (1966).

References

Ademosun, A. A. and **H. R. Chheda** (1974) Regional evaluation of Cynodon genotypes, Ile-Ife area, *Niger. Agric. J.* **11**, 25–30.

Aken'Ova, M. E. (1975) Improvement of Pennisetum purpureum Schum. for forage in the low-altitude humid tropics, Ph.D. Thesis, Univ. Ibadan, Nigeria.

Aken'Ova, M. E. and **H. R. Chheda** (1970) Effects of storage on viability of elephant grass (Pennisetum purpureum Schum.) pollen, *Niger. Agric. J.* **7**, 111–14.

Allard, R. W. and **A. D. Bradshaw** (1964) Implications of genotype–environmental interactions in applied plant breeding, *Crop Sci.* **4**, 503–8.

Barnard, C. (1972) *Register of Australian Herbage Plant Cultivars*, CSIRO, Canberra: Australia.

Bashaw, E. C. (1962) Apomixis and sexuality in buffelgrass, *Crop Sci.* **2**, 412–15.

Bashaw, E. C. (1968) Registration of Higgins buffelgrass, *Crop Sci.* **8**, 397–8.

Bashaw, E. C. (1975) Problems and possibilities of apomixis in the improvement of tropical forage grasses. In E. C. Doll and G. O. Mott (eds), *Tropical Forages in Livestock Production Systems*, ASA Special Pub., No. 24, Wisconsin, pp. 23–30.

Bogdan, A. V. (1963) Chloris gayana without anthocyanin colouration, *Heredity* **18**, 364–8.

Bogdan, A. V. (1965) Cultivated varieties of tropical and subtropical herbage plants in Kenya, *E. Afr. Agric. For. J.* **30**, 330–8.

Bogdan, A. V. (1966) Plant introduction, selection, breeding and multiplication. In W. Davies and C. L. Skidmore (eds), *Tropical Pastures*, Faber and Faber: London, Ch. 5.

Bogdan, A. V. (1969) Rhodes grass, *Herb. Abst.* **39**, 1–13.

Bogdan, A. V. (1977) *Tropical Pasture and Fodder Plants*, Longman: London.

Boonman, J. G. (1977) *On Rhodes Grass Breeding, Seed Yield and Herbage Quality*, Min. Agric. Kenya Grass Br. Proj., Tech. Rep. GBP/June/77/1, pp. 1–61.

Braverman, S. W. and A. J. Oakes (1972) Disease resistance in warm season forage, range and turf grasses, *Bot. Rev.* **38**, 491–544.

Bray, R. A. (1975) Genetic adaptation of grasses and legumes to tropical environments, *Trop. Grasslds* **9**, 109–16.

Bray, R. A. and E. M. Hutton (1976) Plant breeding and genetics. In N. H. Shaw and W. W. Bryan (eds), *Tropical Pasture Research Principles and Methods*, Commonw. Bur. Past. and Fld Crops Bull. 51, Hurley, England, Ch. 14.

Bredon, R. M. and C. R. Horrell (1962) The chemical composition and nutritive value of some common grasses in Uganda, *E. Afr. Agric. For. J.* **39**, 13–17.

Breese, E. L. and M. D. Hayward (1972) The genetic basis of present breeding methods in forage crops, *Euphytica* **21**, 324–36.

Briggs, F. N. and P. F. Knowles (1967) *Introduction to Plant Breeding*, Rheinhold: New York, Ch. 20.

Brim, C. A. (1966) A modified pedigree method of selection in soybeans, *Crop Sci.* **6**, 220.

Bula, T. J., V. L. Lechtenberg and D. A. Holt (1977) Potential of temperate zone cultivated forages. In *Potential of the World's Forages for Ruminant Animal Production*, Winrock Intl. Livestock Res. Trg Center, Arkansas, pp. 7–28.

Burt, R. L. and K. P. Haydock (1968) Yield stability as an aid to the preliminary assessment of introduced forage plants, *J. Aust. Inst. Agric. Sci.* **34**, 228–30.

Burton, G. W. (1943) *Coastal Bermuda Grass*, Georgia Coastal Plains Expt. Sta., Circ. 10.

Burton, G. W. (1947) Breeding bermuda grass for the Southeastern United States, *J. Amer. Soc. Agron.* **39**, 551–69.

Burton, G. W. (1951) Intra and inter-specific hybrids in bermuda grass, *J. Hered.* **42**, 152–6.

Burton, G. W. (1952) Quantitative inheritance in grasses, *Proc. 6th Intl. Grassld Cong.*, pp. 277–83.

Burton, G. W. (1955) Breeding Pensacola bahiagrass, Paspalum notatum. 1. Method of reproduction, *Agron. J.* **47**, 311–14.

Burton, G. W. (1959) Breeding methods for pearl millet (Pennisetum glaucum) indicated by genetic variance component studies, *Agron. J.* **51**, 479–81.

Burton, G. W. (1970) Breeding subtropical species for increased animal production, *Proc. 11th Intl. Grassld Cong.*, pp. A56–63.

Burton, G. W. (1974) Recurrent restricted phenotypic selection increases forage yields of Pensacola bahiagrass, *Crop Sci.* **14**, 831–5.

Burton, G. W. and E. H. DeVane (1953) Estimating heritability in tall fescue (Festuca arundinacea) from replicated clonal material, *Agron. J.* **45**, 478–81.

Burton, G. W., E. H. DeVane and J. P. Trimble (1954) Polycross performance in Sudan grass and its possible significance, *Agron. J.* **46**, 223–6.

Burton, G. W. and I. Forbes, Jr. (1960) The genetics and manipulation of obligate apomixis in common bahia grass (Pasaplum notatum Flugge), *Proc. 8th Intl. Grassld Cong.*, pp. 66–71.

Burton, G. W. and J. C. Fortson (1966) Inheritance and utilization of five dwarfs in pearl millet (Pennisetum typhoides) breeding, *Crop Sci.* **6**, 69–72.

Burton, G. W. and R. H. Hart (1967) Use of self-incompatibility to produce commercial seed propagated F_1 bermudagrass hybrids, *Crop Sci.* **7**, 524–6.

Burton, G. W., R. H. Hart and R. S. Lowrey (1967) Improving forage quality in bermudagrass by breeding, *Crop Sci.* **7**, 329–32.

Burton, G. W., J. C. Millot and **W. G. Monson** (1973) Breeding procedures for Panicum maximum Jacq. suggested by plant variability and mode of reproduction, *Crop Sci.* **13**, 717–20.

Burton, G. W. and **W. G. Monson** (1972) Inheritance of dry matter digestibility in bermudagrass, *Crop Sci.* **12**, 375–8.

Burton, G. W., W. G. Monson, J. C. Johnson, Jr., R. S. Lowrey, H. D. Chapman and **W. H. Marchant** (1969) Effect of the d_2 dwarf gene on the forage yield and quality of pearl millet, *Agron. J.* **61**, 607–12.

Burton, G. W. and **J. B. Powell** (1968) Pearl millet breeding and cytogenetics, *Adv. Agron.* **20**, 1–20.

Burton, G. W. and **G. F. Sprague** (1961) Use of hybrid vigor in plant improvement. In *Germ Plasm Resources*, AAAS: Washington, D.C., pp. 191–203.

Cameron, D. F. (1967a) Flowering in Townsville lucerne (Stylosanthes humilis), 2. The effect of latitude and time of sowing on the flowering time of single plants, *Aust. J. Exptl Agric. Anim. Husb.* **7**, 489–94.

Cameron, D. F. (1967b) Chromosome number and morphology of some introduced Stylosanthes species, *Aust. J. Agric. Res.* **18**, 375–9.

Cameron, D. F. (1968) Studies of the ecology and genetics of Townsville lucerne (Stylosanthes humilis HBK), Ph. D. Thesis, Univ. Queensland.

Carnahan, H. L. and **H. D. Hill** (1961) Cytology and genetics of forage grasses, *Bot. Rev.* **27**, 1–162.

Chheda, H. R. (1971) Cynodon *Improvement in Nigeria*, Seminar on For. Crop Research W. Africa, University Ibadan, Nigeria (Mimeo).

Chheda, H. R. (1974) Forage crops research at Ibadan. 1. Cynodon spp. In J. K. Loosli, V. A. Oyenuga and G. M. Babatunde (eds), *Animal Production in the Tropics*, Heinemann (Nigeria): Ibadan, pp. 79–94.

Chheda, H. R. and **J. R. Harlan** (1964) Chromosomal evolution and genome building in Bothriochloinae, *Niger. Agric. J.* **1**, 18–21.

Chheda, H. R. and **K. M. Rawal** (1971) Phylogenetic relationships in Cynodon. I. C. aetheiopicus, C. nlemfuensis and C. transvaalensis, *Phyton* **28**, 15–21.

CIAT (1977 and 1978) *Annual Reports*, Centro Intl. Agric. Trop.: Cali, Colombia.

Clements, R. J. (1977) Plant breeding strategies for Centrosema virgianum, *CSIRO Trop. Crops and Past. Div. Rept. (1975–1976)*, pp. 95–9.

Cockerham, C. C. (1963) Estimation of genetic variance. In W. D. Hanson and H. F. Robinson (eds), *Statistical Genetics and Plant Breeding*, NSA-NRC Publ. 982, pp. 53–84.

Combes, D. and **J. Pernes** (1970) Variations dans les nombres chromosomiques du Panicum maximum Jacq. en relation avec de la mode de reproduction, *C. R. Acad. Sci. (Paris)*, 270, 782, 785.

Craigmiles, J. P. (1966) Utilization of heterosis in Sudan grass breeding, *Proc. 10th Intl. Grassld Cong.*, pp. 801–3.

Crowder, L. V. (1977) Potential of tropical zone cultivated forages. In *Potential of the World's Forages for Ruminant Animal Production*. Winrock Intl. Livestock Res. Trg Center, Arkansas, pp. 49–78.

Crowder, L. V. and **H. R. Chheda** (1977) Forage and fodder crops. In C. L. A. Leaky and J. B. Wills (eds), *Food Crops of the Lowland Tropics*, Oxford Univ. Press, pp. 127–59.

Donefer, E. (1970) Forage solubility measurements in relation to nutritive value, *Proc. Nat. Conf. For. Qual. Eval. Util.*, Nebraska Center Cont. Ed., Lincoln, Nebraska.

Dougall, H. W. and **A. V. Bogdan** (1958) The chemical composition of the grasses of Kenya, *E. Afr. Agric. For. J.* **24**, 17–23.

Finlay, K. W. and **G. N. Wilikinson** (1963) The analysis of adaptation in a plant breeding program, *Aust. J. Agric. Res.* **14**, 742–54.

Forbes, I. and **G. W. Burton** (1963) Chromosome numbers and meiosis in some Cynodon species and hybrids, *Crop Sci.* **3**, 75–9.

Foster, W. H. (1962) Investigations preliminary to the production of cultivars of Andropogon gayanus, *Euphytica* **11**, 47–52.

Fryxell, P. A. (1957) Mode of reproduction of higher plants, *Bot. Rev.* **23**, 135–233.

Gee, M. S., O. E. Nelson and **J. Kuc** (1968) Abnormal lignins produced by the brown midrib mutants of maize. II. Comparative studies on normal and brown midrib-1 dimethyl formamide lignins, *Arch. Biochm. Biophys.* **123**, 403–8.

Gonzalez, V., J. L. Brewbaker and **D. E. Hamill** (1967) Leucaena cytogenetics in relation to the breeding of low mimosine lines, *Crop Sci.* **7**, 140–3.

Goto, I. and **D. J. Minson** (1977) Prediction of the dry matter digestibility of tropical grasses using a pepsin–cellulase assay, *Anim. Feed Sci. Tech.* **2**, 247–53.

Gray, S. G. (1967a) Inheritance of growth habit and quantitative characters in intervarietal crosses in Leucaena leucocephala (Lam.) De Wit, *Aust. J. Agric. Res.* **18**, 63–70.

Gray, S. G. (1967b) General and specific combining ability in varieties of Leucaena leucocephala, *Aust. J. Agric. Res.* **18**, 71–6.

Grof, B., R. Schultze-Kraft and **F. Muller** (1979) Stylosanthes capitata Vog., some agronomic attributes and resistance to anthracnose (Colletotrichum gloesporiodes Penz.), *Trop. Grassld* **13**, 28–37.

Hacker, J. B. (1966) Cytological investigations in the Setaria sphacelata complex, *Aust. J. Agric. Res.* **17**, 297–301.

Hacker, J. B. (1967) The maintenance of strain purity in the Setaria sphacelata complex, *J. Aust. Inst. Agric. Sci.* **33**, 265–7.

Hacker, J. B. (1972) Setaria species, *CSIRO, Division of Trop. Past., Ann. Rpt* *(1971–1972)*, pp. 41–2.

Hacker, J. B. (1976) Evaluation of Setaria synthetics and Digitaria hybrids, *CSIRO, Trop. Crops and Past. Div. Rept (1975–1976)*, p. 44.

Hacker, J. B. and **D. J. Minson** (1972) Varietal differences in in vitro dry matter digestibility in Setaria and the effects of site, age and season, *Aust. J. Agric. Res.* **23**, 959–67.

Hanna, W. W., J. B. Powell, J. C. Millot and **G. W. Burton** (1973) Cytology of obligate sexual plants in Panicum maximum Jacq. and their use in controlled hybrids, *Crop Sci.* **13**, 695–7.

Hanson, A. A. and **H. L. Carnahan** (1956) *Breeding Perennial Forage Grasses*, USDA Tech. Bull. No. 1155, Washington, D.C., pp. 1–116.

Harlan, J. R. (1962) Breeding superior forage plants for the great plains, *J. Range Mgmt* **13**, 86–9.

Harlan, J. R. (1970) Cynodon species and their value for grazing and hay, *Herb. Abst.* **40**, 233–8.

Harlan, J. R., M. H. Brooks, D. S. Borgaonkar and **J. M. J. de Wet** (1964) The nature and inheritance of apomixis in Bothriochloa and Dichanthium, *Bot. Gaz.* **125**, 41–6.

Harlan, J. R., G. W. Burton, and **W. C. Elder** (1954) *Midland Bermudagrass, A New Variety for Oklahoma Pastures*, Oklahoma Agric. Expt. Sta., Bull. B-416.

Harlan, J. R., H. R. Chheda and **W. L. Richardson** (1962) Range of hybridization with Bothriochloa intermedia (R. Br.) A. Camus, *Crop Sci.* **2**, 480–3.

Harlan, J. R., J. M. J. de Wet, W. L. Richardson and **H. R. Chheda** (1961) *Studies on Old World Bluestems III*, Oklahoma State Univ., Tech. Bull. T-92, pp. 1–30.

Harlan, J. R. and **J. M. J. de Wet** (1963) Role of apomixis in the evolution of Bothriochloa-Dichanthium complex, *Crop Sci.* **3**, 314–16.

Harlan, J. R. and **J. M. J. de Wet** (1969) Sources of variation in Cynodon dactylon, *Crop Sci.* **9**, 774–8.

Harlan, J. R., J. M. J. de Wet and **W. L. Richardson** (1969) Hybridization studies with species of Cynodon from East Africa and Malagasy, *Amer. J. Bot.* **56**, 944–50.

Harlan, J. R., J. M. J. de Wet, K. M. Rawal, N. R. Felder and **W. L. Richardson** (1970) Cytogenetic studies in Cynodon (L.) C. Rich, *Crop Sci.* **10**, 288–91.

Hearn, C. J. and E. C. Holt (1969) Variability in components of seed production in Panicum coloratum L., *Crop Sci.* **9**, 38–40.

Hein, M. A. (1958) Registration of varieties and strains of grass Paspalum (Paspalum spp.) I. Tifhi (Bahiagrass), *Agron. J.* **50**, 401.

Hill, R. R. Jr. (1977) Quantitative genetics of forage-potentials and pitfalls, *Agron. Abst.*, Madison, Wisconsin, p. 58.

Hutton, E. M. (1961) Inter-variety variation in Rhodes grass (Chloris gayana Kunth), *J. Brit. Grassld Soc.* **16**, 23–9.

Hutton, E. M. (1962) Siratro – a tropical pasture legume bred from Phaseolus atropurpureus, *Aust. J. Exptl Agric. Anim. Husb.* **2**, 117–25.

Hutton, E. M. (1964) Plant breeding and genetics. In *Some Concepts and Methods in Sub-tropical Pasture Research*, Commonw. Bur. Past. Fld Crops Bull. **47**, Hurley England, Ch. 7.

Hutton, E. M. (1968) Australia's pasture legumes, *J. Aust. Inst. Agric. Sci.* **34**, 203–18.

Hutton, E. M. (1970) Tropical pastures, *Adv. Agron.* **22**, 1–73.

Hutton, E. M. (1971) Plant improvement for increased animal production, *J. Aust. Inst. Agric. Sci.* **37**, 212–25.

Hutton, E. M. (1976) Leucaena leucocephala. *CSIRO Trop. Crops and Past. Div. Rept. (1975–1976)*, p. 40.

Hutton, E. M. (1980) Breeding Leucaena for acid tropical soils, *Leucaena Newsletter*, Univ. Hawaii, p. 7.

Hutton, E. M. and L. B. Beall (1977) Breeding of Macroptilium atropurpureum, *Trop. Grasslds* **11**, 15–31.

Hutton, E. M., L. B. Beall, and W. T. Williams (1978) Evaluation of a bred line of Macroptilium, *Aust. J. Exptl Agric. Anim. Husb.* **18**, 702–7.

Hutton, E. M. and W. M. Beattie (1976) Yield characteristics in three bred lines of the legume Leucaena leucocephala, *Trop. Grassld* **10**, 187–94.

Hutton, E. M. and S. G. Gray (1959) Problems of adapting Leucaena glauca as a forage for the Australian Tropics, *Emp. J. Exptl Agric.* **27**, 187–96.

Imrie, B. C. (1976) Breeding methods research, *CSIRO Trop. Crops and Past. Div. Rept (1975–1976)*, p. 45.

Javier, E. Q. (1970) The flowering habits and mode of reproduction of Guinea grass (Panicum maximum Jacq.), *Proc. 11th Intl. Grassld Cong.*, pp. 284–9.

Jones, R. J. and A. J. Pritchard (1971) The method of reproduction in Rhodes grass (Chloris gayana Kunth), *Trop. Agric. (Trin.)* **48**, 301–7.

Jones, R. M., R. J. Jones and E. M. Hutton (1980) A method for advanced evaluation of pasture species. A case study with bred lines of Macroptilium atropurpureum, *Aust. J. Exptl Agric. Anim. Husb.* **20**, 703–9.

Keller, W. (1952) Emasculation and pollination technics, *Proc. 6th Intl. Grassld. Cong.*, pp. 1613–19.

Klock, M. A., S. C. Schank and J. E. Moore (1975) Laboratory evaluation of quality in subtropical grasses. III. Genetic variation among Digitaria species in in vitro digestion and its relationship to plant morphology, *Agron. J.* **67**, 672–5.

Laredo, M. A. and D. J. Minson (1973) The voluntary intake, digestibility and retention time by sheep of leaf and stem fraction of five grasses, *Aust. J. Agric. Res.* **24**, 875–88.

Latter, B. D. H. (1964) Selection methods in the breeding of cross-fertilized pasture species. In C. Barnard (ed.), *Grasses and Grasslands*, Macmillan: London, Ch. 10.

Lowrey, R. S., J. C. Johnson, G. W. Burton, W. H. Marchant and W. C. McCormick (1968) In vivo *Studies with Coastcross-1 and Other Bermudas*, Georgia Agric. Expt. Sta., Res. Bull. 55.

McWhirter, K. S. (1969) Cytoplasmic male sterility in Desmodium, *Aust. J. Agric. Res.* **20**, 227–41.

McWilliam, J. R. (1969) Introduction, evaluation and breeding of new pasture species, *J. Aust. Inst. Agric. Sci.* **35**, 90–8.

Milford, R. and D. J. Minson (1966) The feeding value of tropical pastures. In W. Davies and C. L. Skidmore (eds), *Tropical Pastures*, Faber and Faber: London, Ch. 7.

Minson, D. J. (1976) Plant factors involved in digestibility and intake, *CSIRO Trop. Crops and Past. Div. Rept (1975–1976)*, pp. 100–4.

Moll, R. H. and C. W. Stuber (1975) Quantitative genetics – empirical results relevant to plant breeding, *Adv. Agron.* **27**, 277–313.

Morley, F. H. W. and D. H. Heinrichs (1960) Breeding for creeping root in alfalfa, *Can. J. Pl. Sci.* **40**, 424–33.

Myers, W. M. (1947) Cytology and genetics of forage grasses, *Bot. Rev.* **13**, 319–421.

Muldoon, D. K. and C. J. Pearson (1979) The hybrid between Pennisetum americanum and Pennisetum purpureum, *Herb. Abst.* **49**, 189–99.

Neal Wright, L. (1976) Recurrent selection for shifting gene frequency of seed weight in Panicum antidotale Retz, *Crop. Sci.* **16**, 647–9.

OECD (1966) *OECD Scheme for the Varietal Certification of Herbage Seed Moving in International Trade*, Document Agric. Food 76.

Pernes, J. (1971) Problems posed by the improvement of the tropical forage species: *Panicum maximum* (Jacq.), *Seminar on Forage Crops in W. Africa*, Univ. Ibadan, Nigeria (mimeo).

Pernes, J., Y. Savidan and R. René-Chaume (1975) Panicum: Structures genetiques complexes des "Maximae" et organisation de les populations naturelles en relation avec la speciation, *Boissiera* **24**, 383–402.

Pienaar, R. de V. (1954) The chromosome numbers of some indigenous South African and introduced Gramineae. In D. Meredith (ed.), *The Grasses and Pastures of South Africa*, Central News Agency: Cape Town, Ch. 2(b).

Poehlman, J. M. (1959) *Breeding Field Crops*, Henry Holt: New York.

Porter, K. S., J. D. Axtell, V. L. Lechtenberg and V. F. Colenbrander (1978) Phenotype, fiber composition, and in vitro dry matter disappearance of chemically induced brown midrib (*bmr*) mutants of sorghum, *Crop Sci.* **18**, 205–8.

Potts, H. C. and E. C. Holt (1967) Parent-progeny offspring relationships in Kleingrass (Panicum coloratum L.), *Crop Sci.* **7**, 145–8.

Powell, J. B. and G. W. Burton (1966) A suggested commercial method of producing an interspecific hybrid forage in Pennisetum, *Crop Sci.* **6**, 378–9.

Powell, J. B. and G. W. Burton (1971) Genetic suppression of shoot-trichomes in pearl millet (Pennisetum typhoides), *Crop Sci.* **11**, 763–5.

Pritchard, A. J. and K. F. Gould (1964) Chromosome numbers in some introduced and indigenous legumes and grasses, *CSIRO Div. Trop. Past., Tech. Paper No. 2*.

Pritchard, A. J. and J. G. Wutoh (1964) Chromosome numbers in the genus Glycine L., *Nature (Lond.)* **202**, 322.

Rawal, K. M. and H. R. Chheda (1971) Phylogenetic relationships in Cynodon. II. C. nlemfuensis, C. dactylon vars. afghanicus and aridus, *Phyton* **28**, 121–30.

Rawal, K. M. and J. R. Harlan (1971) The evolution of growth habit in Cynodon (L.) C. Rich, *Trans. Illinois State Acad. Sci.* **64**, 110–18.

Schank, S. C. and H. F. Decker (1963) The Florida garden of Digitaria introductions, *Sunshine State Agric.* (April), Gainesville, Florida, pp. 8–9.

Shaw, N. H. and P. C. Whiteman (1977) Siratro – a success story in plant breeding, *Trop. Grassld* **11**, 7–14.

Sherwood, R. T., B. A. Young and E. C. Bashaw (1980) Facultative apomixis in buffelgrass, *Crop Sci.* **20**, 375–8.

Simon, U. (1976) Improvement of the protein content in forage plants by plant cultural and breeding measures, *Plant Res. Dev.* **3**, 90–100.

Smith, R. L. (1972) Sexual reproduction in Panicum maximum Jacq., *Crop Sci.* **12**, 624–7.

Stobbs, T. H. (1973) The effect of plant structure on the intake of tropical pastures. II. Differences in sward structure, nutritive value and bite size of animals grazing Setaria anceps and *Chloris gayana* at various stages of growth, *Aust. J. Agric. Res.* **24**, 821–9.

Stuber, C. W. (1970) Estimation of genetic variance using inbred relatives, *Crop Sci.* **10**, 129–35.

Taliaferro, C. M. and **E. C. Bashaw** (1966) Inheritance and control of obligate apomixis in breeding buffelgrass (Pennisetum ciliare), *Crop Sci.* **6**, 473–6.

Thompson, A. J. and **A. J. Wright** (1971) Principles and problems in grass breeding, *Plant Breeding Institute, Ann. Report*, Cambridge, pp. 31–67.

Tilley, J. M. A. and **R. A. Terry** (1963) A two-stage technique for in vitro digestion of forage crops, *J. Brit. Grassld Soc.* **18**, 104–11.

Warmke, H. E. (1954) Apomixis in Panicum maximum, Amer. J. Bot. **41**, 5–11.

Whiteman, P. C. (1978) *Tropical pastures – Australia's Contribution to Regional Pasture Research and Development*, ASPAC Food Fert. Techn. Cent., Extn. Bull. No. 118, Taiwan.

Whyte, R. O. (1964) *The Grassland and Fodder Resources of India.* ICAR: New Delhi, Ch. 10.

Whyte, R. O., T. R. G. Moir and **J. P. Cooper** (1959) *Grasses in Agriculture*, FAO Agric. Stud. 21, Rome.

Williams, R. J., R. L. Burt and **R. W. Strickland** (1976) Plant introduction. In N. H. Shaw and W. W. Bryan (eds), *Tropical Pasture Research Principles and Methods*, Commonw. Bur. Past. and Fld Crops Bull. 51, Hurley: England, Ch. 5.

Wit, F. (1971) General problems and methods in breeding herbage crops, *Ist Intl. Course Pl. Br. Wageningen*, Netherlands (mimeo).

Wutoh, J. G., E. M. Hutton and **A. J. Pritchard** (1968a) Combining ability in Glycine javanica, *Aust. J. Agric. Res.* **19**, 411–18.

Wutoh, J. G., E. M. Hutton and **A. J. Pritchard** (1968b) Inheritance of flowering time, yield and stolon development in Glycine javanica, *Aust. J. Exptl Agric. Anim. Husb.* **8**, 317–22.

Chapter 18

Seed production, multiplication and processing

Adequate seed supplies of acceptable standards in terms of purity and germination of the important grasses and legumes are practically non-existent or very limited, in most tropical countries. This lack of seed places a serious constraint on various national efforts to raise animal productivity quickly and effectively by increasing the land area under productive sown pastures and improving the species composition of existing rangelands. The need for information leading to economic and efficient seed production and processing is acute. In recent years detailed seed investigations have been carried out in East Africa, Australia, Colombia, the Philippines and to some extent by pasture agronomists in other tropical countries.

Status of forage seed production in the tropics

Seed production of potentially useful genotypes is generally not fully assessed during the initial stages of introduction and field-testing. As a result, wide inter- and intraspecific variation exists in the seed production potential of important tropical pasture species (Okigbo and Chheda, 1966; Javier, 1970; Boonman and van Wijk, 1973; Boonman, 1977).

Seed yields of grasses

Seed yields range from nil in species such as *Digitaria decumbens*, some cultivars and clones of *Cynodon* spp., *Brachiaria mutica*, *Pennisetum purpureum* and *P. clandestinum*; low (less than 60 kg/ha of harvested seeds) in *Brachiaria mutica*, *Panicum coloratum* var. *makarikariense*, *Andropogon gayanus*, *Pennisetum purpureum* and *Hyparrhenia rufa*; to moderate (60–250 kg/ha) in *Brachiaria brizantha*, *B. ruziziensis*, *Cenchrus ciliaris*, *Panicum coloratum* var. *coloratum*, *P. maximum*, *Chloris gayana*, *Melinis minutiflora* and *Setaria anceps* (Strickland, 1971; Boonman, 1972a, 1972b; Javier and Mendoza, 1976; Jones and Roe, 1976; Bogdan, 1977; Ferguson,

1978). Proper agronomic practices, particularly nitrogen fertilization and efficient harvesting methods, increase yields substantially. Even so, they do not compare favourably with the 500–1 000 kg/ha obtained with many grasses and legumes grown in temperate environments. *Paspalum plicatulum* is one exception, with yields exceeding 500 kg/ha (Chadhokar and Humphreys, 1970, 1973). Boonman (1973b) found that potential yields of *Setaria anceps* approach those of temperate grasses grown in Europe.

The purity of many tropical grass seeds rarely exceeds 25 per cent when cleaned and processed (Combes and Verburgt, 1971; Boonman, 1977) and may be no more than 1–10 per cent in hand-harvested seeds of grasses containing empty florets, shrivelled seeds, pieces of stems, leaves, etc. (Javier, 1970; Crowder, 1977; Bogdan, 1977). Thus, the effective seed yields of tropical grasses may range from less than 10 kg/ha to no more than 50 kg/ha annually.

Low seed yields and high requirements of labour or machinery for growing and processing seed crops make pasture seed a high priced commodity. Boonman (1971a) and Combes and Verburgt (1971) estimated seed costs at 20–25 per cent of the initial cost of establishment of tropical grass leys in Kenya.

Seed yields of legumes

The potential seed yields of tropical legumes exceeds that of grasses (Hopkinson and Loch, 1973, 1977; Wickham *et al.*, 1977), but actual yields are lower due to irregular flowering and seed-shattering. Machine-harvested yields range from 75 to more than 100 kg/ha for *Desmodium intortum* and *Lotononis bainesii*, 100 to 200 kg/ha for *Stylosanthes guianensis*, *Centrosema pubescens* and *Macroptilium atropurpureum*, and 200 to 300 kg/ha for *Desmodium uncinatum*, *Glycine wightii* and *Stylosanthes humilis* (Strickland, 1971; Javier and Mendoza, 1976; Bogdan, 1977; Ferguson, 1978). Hand-harvesting at regular intervals can reduce seed losses and is considered economic in areas where labour is plentiful and relatively cheap. Hard seededness and dormancy (Fig. 18.1) are responsible for poor germination of fresh seeds of many grass and legume species (Prodonoff, 1967; Cameron, 1967; McLean and Grof, 1968; Strickland, 1971; Bogdan, 1977).

Sources of planting materials

The development of forage seed production enterprises in most tropical countries has been hindered by (1) low seed yields, (2) lack of adequate technology needed for growing and processing forage seed crops and (3) non-existent or inefficient seed marketing and distribution networks. Seeds are not regularly available for large-scale

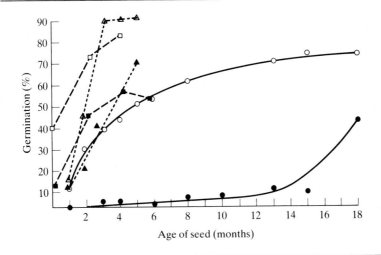

Fig. 18.1 Hard-seededness and dormancy in tropical grasses and legumes. *Panicum coloratum* var. *kabulabula*: ●——● control, ○——○ scarified; *Setaria anceps*: ■——■ control, □——□ glumes removed; and *Stylosanthes humilis*: ▲——▲ control, △——△ scarified (redrawn from Strickland, 1971).

pasture establishment or range improvement except in Australia and East Africa. As a result, grassland improvement in much of the tropics has depended on labour-intensive vegetative propagation, locally produced seed of variable quality and expensive imported seed.

Vegetative propagation

Despite the fact that a number of tropical grass species and cultivars are sterile, or possess low seed production potential, they are recommended and accepted by growers in many areas because of higher herbage yield, ability to withstand grazing or mowing, persistence and higher nutritive value than presently used types. Their stoloniferous growth habit or large tuft size enables rapid ground cover when propagated vegetatively from splits, stem cuttings, stolons or rhizomes. In the humid tropics of Central and South America, the Pacific Islands and Southeast Asia (Osbourn, 1975) and the lowland humid tropics of West Africa, pasture improvement is based primarily on vegetative propagation. Grasses such as *Brachiaria mutica*, *Setaria anceps* and *Panicum maximum* which could be grown from seed are propagated vegetatively when seeds are scarce or expensive and when cheap labour and adequate vegetative planting materials are available.

Vegetative propagation is labour-intensive and more dependent on weather than planting by seed. Long-distance shipment of large quantities of planting materials is difficult. Criste *et al.* (1973) in the

Philippines estimated the cost of establishing 1 ha of *Panicum maximum* from root stocks at 150–200 pesos, and from seeds (2–3 kg/ha) at 10–15 pesos. With some grasses, such as *Cynodon dactylon* in southeastern USA (Burton, 1966) and *Pennisetum clandestinum* in Australia (Wilson, 1970), transplanting operations on cultivated soils have been successfully mechanized.

Legumes are seldom established vegetatively. Whyte *et al.* (1953) referred to an occasional practice of propagating *Calopogonium mucunoides* by cuttings in Malaya. *Lotononis bainesii* can be established by stem cuttings (Bogdan, 1977). In some areas where seed-setting is poor *Pueraria phaseoloides* has been established vegetatively using crown divisions or rooted vine cuttings.

Seed production

Pasture seeds in much of the tropics are still an uncertain by-product of agricultural operations rather than a specialized farming enterprise as in most temperate countries. Seeds are often harvested by hand from natural stands. For example, *Andropogon gayanus* in West Africa (Haggar, 1966), *Hyparrhenia rufa* in tropical South America (Crowder *et al.*, 1970) and the Old World bluestems in India are hand-harvested from native grasslands not subjected to specific cultural or management practices. The harvested material is rarely processed and may contain less than 5 per cent viable seed. For satisfactory establishment, large quantities must be planted to obtain 1–2 kg/ha of pure germinating seeds. Thus, the recommended sowing rate for the hand-harvested, unthreshed seed of *Andropogon gayanus* in West Africa is 40–50 kg/ha, and for *Melinis minutiflora* in South America 20–25 kg/ha.

Seed is also collected from cultivated pastures grown and managed primarily for animal production. Grazing is discontinued to encourage flowering during the later part of the rainy season. Harvesting is done mostly by hand and occasionally with a reaper-binder when labour is scarce. The sheaves are stooked in the field and threshed some 10–12 days later. Seed yields are generally low and few farmers have satisfactory cleaning facilities to deal with the various kinds of pasture seeds. Ministries or departments of agriculture, pasture research stations and extension services often maintain source nurseries of certain recommended pasture species and cultivars, and produce small quantities of clean seed for distribution to farmers and ranchers in surrounding areas.

Small quantities of pasture legume seeds are harvested locally for sowing as cover crops in plantations and for mixed grass–legume pastures (Javier and Mendoza, 1976; Crowder and Chheda, 1977). Ferguson (1978) described several systems of seed production in Latin America that included harvests from natural populations, pastures and commercial plantings. Large-scale needs, however, are generally met by imported commercial seed, mostly from Australia.

In East Africa and Australia superior strains of pasture species have been developed. Widespread acceptance of these genotypes by farmers and ranchers led to an ever-increasing demand for seed that was not satisfied by government agencies. Consequently, successful seed production enterprises developed in these areas primarily to satisfy national needs. In recent years exports have been made to other countries and the demand is steadily increasing.

In Kenya, seeds of the important recommended cultivars of *Setaria anceps, Chloris gayana, Panicum coloratum, P. maximum, Brachiaria ruziziensis, Melinis minutiflora* and *Cenchrus ciliaris* are produced on a commercial scale by farmers and sold to a seed company for cleaning, processing and marketing. The Rhodesian Pasture Seed Growers' Association produces certified seeds of these same species, as well as cultivars of *Cynodon nlemfuensis* and *Pennisetum purpureum* (Whyte *et al.*, 1959). Seed growers in East Africa are instructed with regard to selection site, land preparation, planting procedures, fertilizing, harvesting, threshing and drying of seeds.

Pasture seed-growing, harvesting and processing have attained high levels of sophistication in Australia. Field operations are usually mechanized and efficient machinery is used to harvest, clean and process the seeds. Linnett (1977) listed 34 major tropical pasture grasses and legumes grown commercially for seed in Queensland with more than 75 per cent having purities of 99 per cent (Table 18.1).

The Kenyan example

The history of pasture crop seed production in Kenya clearly illustrates that great strides can be made in overcoming restraints and developing a viable industry in a relatively short time (15–20 years – Combes and Verburgt, 1971). Prior to 1950 grass seeds were hand-harvested from natural stands. With the development and release of selected cultivars, government agencies arranged for harvesting and bulking of initial seed stocks. To meet the increased demand for seed of improved varieties a group of local farmers established the Kenya Seed Company in 1956. The company obtained seeds from domestic contract growers, cleaned and processed them and sold certified seeds through the Kenya Farmers' Association (Co-op) Ltd. The processing plant and most of the contract growers were situated in the Trans Nzoia district of western Kenya. This area has a mean altitude of 1 800 m, mild climate, adequate and well-distributed rainfall during the growing season and relative humidity of less than 60 per cent. It is ideally suited for profitable seed production of many pasture species. From a modest beginning of less than 100 ha

Table 18.1 *Tropical pasture seeds processed in Queensland, Australia (Linnett, 1977)*

Botanical name	Common names and cultivars	Purity (%)
Astrebla spp.	Mitchell grasses	99.5
Brachiaria mutica	Pará grass	80.0
B. decumbens	Signal grass	99.0
Calopogonium mucunoides	Calopo	99.5
Cenchrus ciliaris	Buffel grass (several cultivars)	99.0
Centrosema pubescens	Centro	99.5
Chloris gayana	Rhodes grass	99.0
Desmodium intortum	Greenleaf Desmodium	99.5
D. uncinatum	Silverleaf Desmodium	99.5
Dichanthium aristatum	Angleton grass	75.0
Glycine wightii	Glycine, 'Cooper', 'Tinaroo'	99.9
Lablab purpureus	Lablab, Dolichos	99.9
Leucaena leucocephala	Leucaena	99.9
Lotononis bainesii	Lotononis	99.9
Macroptilium atropurpureum	Siratro	99.7
Macrotyloma axillare	Axillaris Dolichos	99.7
M. uniflorum	Leichhardt Dolichos	99.1
Melinis minutiflora	Molasses grass	89.0
Panicum antidotale	Blue panic, Giant panic	99.9
P. coloratum	Makarikari grass, Bambatsi grass	99.6
P. maximum	Guinea grass, Hamil grass, Gatton panic, Coloniao grass	89.0
P. maximum var. *trichoglume*	Green panic, Sabi panic	98.0
Paspalum dilatatum	Paspalum, Dallis grass	85.0
P. plicatulum	Plicatulum, 'Hartley', 'Rodds Bay'	99.5
P. notatum	Pensacola, Bahia grass	99.9
Pennisetum clandestinum	Kikuyu	98.5
Pueraria phaseoloides	Puero	99.9
Setaria anceps	Setaria, 'Kazungula', 'Nandi,' 'Narok'	99.0
Sorghum spp.	Pasture types	99.9
Stylosanthes guianensis	Stylo, 'Cook', 'Endeavour', 'Schofield', 'Oxley Fine-stem'	99.8
S. hamata	Caribbean stylo, 'Verano'	99.0
St. humilis	Townsville stylo, 'Paterson', 'Gordon', 'Lawson'	99.7
Urochloa mosambicensis	Sabi grass	99.0
Vigna spp.	Mung bean, Cowpeas	99.8

before 1960, more than 1 800 ha were contracted annually to produce more than 140 000 kg of processed seeds in 1971. Annual seed production is expected to exceed 1 million kg of certified seed by 1980.

Based on projected seed requirements, the contract growers are selected after due consideration of (a) farmer's financial position, (b) geographic situation of the farm, (c) type of farming operation, (d) farmer's experience in seed production and (e) his motivation. The crop is either grown primarily for seed with two harvests annually, a combination of forage and seed with one harvest per year, or primarily for livestock production and one seed crop in 2–3 years. Constant guidance and some supervision is provided to farmers so as to overcome technical difficulties in producing certified seeds in terms of varietal purity, isolation distances and contamination sources.

A new seed certification scheme with rules and regulations as set forth by OECD was initiated in 1969. The Kenya Seed Company leased a 730 ha farm in 1970 to produce the necessary basic seed and to conduct applied research aimed towards increased seed production.

Factors limiting seed production of tropical pasture species

The yield of pure live seed in pasture crops is a function of inflorescence density, numbers of flowers differentiated on each inflorescence, number of seeds formed and harvested from each inflorescence, individual seed weight and the percentage of viable seeds. Genetic and environmental factors exert varying degrees of influence on each of these yield components.

The pioneering research of Boonman (1973b) in Kenya contributed greatly to an understanding of the physiological and genetic constraints of tropical grass-seed production. He demonstrated that yields of pure germinating seeds can be improved by agronomic and management practices and by varietal improvement without sacrificing herbage yield and nutritive value. Strickland (1971), Jones and Roe (1976) and Javier and Mendoza (1976) reviewed seed-production problems in tropical grass and legume species and Humphreys (1979) and Bogdan (1977) compiled information regarding seed production of individual pasture species.

Boonman (1973b) concluded that the potential seed yields of *Setaria anceps* approach the 800 kg/ha yields of temperate grasses such as *Phleum pratense* grown in Europe. Even though tropical and temperate grasses may have similar yield components, actual yields of cleaned seed of existing tropical cultivars are usually no more than 15–20 per cent of the potential, and much lower in terms of pure germinating seeds, as compared to temperate cultivars. Low inflorescence density, poor seed-setting, poor synchronization of yield components and suboptimal management of seed crops appear to be the primary causes of low seed yields of tropical grasses. Strickland (1971) and Javier and Mendoza (1976) considered diffuse flowering, pod-shattering and shading of mature pods to be the main factors limiting seed production in many tropical legumes.

Inflorescence density

Seed yields are positively correlated with inflorescence density (Hacker and Jones, 1971). In grasses inflorescence density is dependent on the rate of appearance and growth of tillers as well as their survival and fertility (Chadhoker and Humphreys, 1970). Variations occur due to species and varietal differences, weather conditions and agronomic practices. The maximum number of tillers in established swards of tropical grasses such as *Andropogon gayanus, Brachiaria ruziziensis, Chloris gayana, Cynodon nlemfuensis, Melinis minutiflora, Panicum coloratum* and *P. maximum* range from 1 000 to 2 000 tillers/m² (Boonman, 1971a, 1971b; Haggar, 1966; Singh and Chatterjee, 1966; Akinola *et al.*, 1971). With intensive management the numbers may approach 3 000. These figures represent only one-half to one-third of those normally observed in temperate grasses. The attainment of maximum tiller number roughly coincides with the initial flowering date, followed by a steady decline and levelling off at about 50–60 per cent of the maximum. The percentage of flowering tillers when computed on the basis of surviving tillers at the time of harvest is variable. Depending on the genotype and on growing conditions, it can range from 20 to 90 per cent in the sparsely tillering tropical grasses. The numbers of seed tillers at the optimum time of harvest usually do not exceed 10–20 per cent of the total because of irregular flowering (Haggar, 1966; Boonman, 1971a). Occasionally it reaches 50 per cent (Chadhokar and Humphreys, 1970). Prolonged flowering and selection for leafiness and late maturity rather than low tillering ability *per se* contribute to low inflorescence density at optimum harvesting time and consequently low seed yields in many tropical grasses (Boonman, 1971a).

Jones and Roe (1976) compared published data on inflorescence density in temperate and tropical grasses and noted that fertilized temperate grasses produced from 300 to over 1 000 inflorescences/m² and yielded 400–900 kg/ha of seed. Most tropical grasses produced fewer than 300 inflorescences/m² with less than 160 kg/ha of seed. With an inadequate nitrogen supply the inflorescence density of tropical grasses rarely exceeded 70 inflorescences/m², resulting in less than 50 kg/ha of seed. Wide variation in inflorescence density within and between tropical legume species and cultivars also occurred.

Prolonged flowering and seed-shedding

Inflorescence initiation in temperate grasses is primarily dependent on photoperiod and in some cases on exposure to low temperature followed by long days. Under favourable conditions, tiller elongation is rapid and inflorescence emergence is concentrated over a very short period. A vast majority of tropical grasses, on the other hand, are indifferent to day length or may respond to short-day conditions by flowering earlier. Most perennial tropical grasses commence flowering within 1–3 months after cutting and continue to flower over an extended period of 3–4 months (Fig. 18.2).

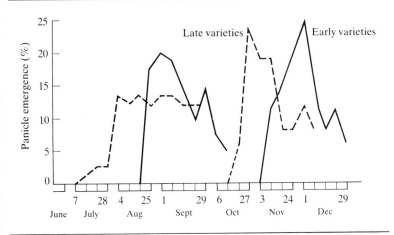

Fig. 18.2   Panicle emergence of Guinea grass at weekly intervals for two flowering periods (Javier, 1970; from Javier and Mendoza, 1976).

Boonman (1971a) noted that inflorescence production in many tropical grasses follows a sigmoid pattern and prolonged flowering in tropical grasses is due to the combined effects of (1) extended anthesis and stigma exsertion within an individual inflorescence (e.g. up to 7 weeks in the early emerged heads of *Setaria anceps*), (2) prolonged inflorescence emergence within individual plants (up to 12 weeks or longer) and (3) interplant variation in days to flower within a cultivar (e.g. up to 7 weeks in *Chloris gayana*).

Prolonged flowering results in uneven maturity of inflorescences and spikelets within an individual inflorescence. Late-emerging inflorescences produce few seeds and these are of low quality. Seed production of tropical grasses is therefore spread over a period of several weeks (Table 18.2). Spikelets of *Panicum maximum* begin shedding within 2–4 weeks of inflorescence emergence (Table 18.3). As a result, yields of adequately mature seeds on any particular date are low. When harvested too early most of the spikelets are immature. If harvested late, large seed losses occur due to shattering. The highest quantity of seed that could be recovered from an inflorescence was only 19 per cent of the total seeds produced. This occurred 12–14 days after inflorescence emergence when 40–60 per cent of the spikelets had already shed. Hacker and Jones (1971) calculated that the harvested seed of *Setaria anceps* came from only 5–7 per cent of the spikelets produced.

Table 18.2 *Seed yield of Panicum coloratum harvested by shaking ripe seed from inflorescences at seven different dates and the proportion that each harvest contributed to the total yield (Roe, 1972)*

Date of harvest	Yield at each date (kg/ha)	Per cent of total collection
Dec. 15	84	33
Dec. 16	26	10
Dec. 17	23	9
Dec. 18	29	11
Dec. 21	64	25
Dec. 24	23	9
Dec. 30	7	3

Table 18.3 *Days from inflorescence emergence to different stages of maturity in Panicum maximum (Javier, 1970)*

Date to	Late varieties	Early varieties
Complete inflorescence exsertion	3	3
25% bloom	6	6
50% bloom	9	8
50% shatter	14	12
100% shatter	31	20

The relationship between seed yields and flowering characteristics of eight cultivars of tropical grasses belonging to five different species was studied by Boonman (1971b). Mean weekly increase in inflorescence number prior to optimal harvesting time (i.e. degree of concentrated flowering) showed the highest correlation with seed yields (Table 18.4). Species and cultivars differ in concentrated flowering which can be modified by weather conditions, N fertilization and management practices. Under East African conditions, *Panicum coloratum* shows a greater degree of concentration in inflorescence emergence and produces significantly higher yields than other grasses with a more diffuse pattern of flowering. In the Philippines early flowering selections of *Panicum maximum* exhibited a peak in inflorescence production followed by a sharp decline during both growing seasons of the year. The late selections produced a comparable flowering peak only in the late growing season (Fig. 18.2). The seed yields of early-flowering types were substantially higher than late-flowering types. *Paspalum plicatulum* in Australia produced high seed yields with highly concentrated flowering in autumn.

Many of the important tropical pasture legume species such as *Stylosanthes guianensis, S. humilis, Desmodium intortum, D. uncinatum,* and *Glycine wightii* are short-day plants, while a few such as *S. hamata, Macroptilium lathyroides* and some of the African *Trifolium* spp. may be day-neutral (Hutton, 1970; Mannetje, 1966).

Table 18.4 *The correlation coefficients between seed yield and tillering and flowering characteristics of eight tropical grasses* (Boonman, 1971b)*

Between seed yield and:	Correlation coefficient(r)	
	1968	1969
Maximum tiller number	−0.22	−0.07
Tiller number at optimal harvesting time	0.03	−0.11
Inflorescence number at optimal harvesting time	0.80[†]	0.82[†]
Percentage of tillers flowering at optimal harvesting time	0.81[†]	0.84[‡]
Tillers flowering at optimal harvesting time as percentage of the maximum number	0.76[†]	0.77[†]
Mean weekly increase in inflorescence number prior to optimal harvesting time	0.85[‡]	0.88[‡]

* The grasses studied were: *Setaria anceps* cv. Nandi I, cv. Nandi III; *Chloris gayana* cv. Mbarara, cv. Masaba, cv. Pokoti; *Panicum coloratum* cv. Solai; *P. maximum* cv. Makueni; *Brachiaria ruziziensis.*
[†] Significant at 5% level.
[‡] Significant at 10% level.

Some species, e.g. *Indigofera spicata*, may contain both day-neutral and short-day forms (Hutton, 1960), while *S. guianensis* contains long-day forms. Near the equator flowering of most legumes is little affected by photoperiod of around 12 h. At some distance from the equator, as in the subtropical environment of Australia, the short-day types flower in the months of autumn and early winter. Day and night temperatures may also influence flowering in some species (Hill, 1967; Cameron, 1967; Mwakha, 1969; Skerman and Humphreys, 1973; Wutoh *et al.*, 1968), while moisture stress plays an important role in producing flushes of flowering in *Macroptilium atropurpureum* (Hopkinson, 1977).

Flowering of legumes is usually spread over a considerable period and pod-shedding or seed-shattering are common (Javier and Mendoza, 1976; Bogdan, 1977). Hopkinson and Loch (1973) estimated total seed production *Macroptilium atropurpureum* at 1 600 kg/ha with only 550 kg ripe at harvest. The remainder were either unripe or shattered. In contrast, *Stylosanthes guianensis*, which has a relatively concentrated flowering habit under northern Queensland conditions in Australia and relatively good seed retention, produced 700–800 kg/ha of seed from a closed canopy (Fig. 18.3). At optimum harvest time (about 3 weeks after peak flowering), up to 600 kg/ha of seeds were retained and losses in the form of fallen seed remained below 10 per cent.

Low seed-setting

Seed formation in tropical grasses is often low with large numbers of

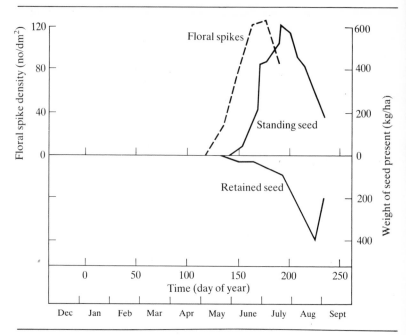

Fig. 18.3 Pattern of flowering and seed production in *Stylosanthes guianensis* (re-drawn from Loch *et al.*, 1976).

empty spikelets (Burton, 1943; Bogdan, 1977). Gildenhuys (1951) reported fertility values ranging between 18 and 29 per cent for cultivars of *Setaria anceps*. *Brachiaria ruziziensis* clones when cross-pollinated averaged 30 per cent of spikelets with caryopses, but only 0.5 per cent when selfed (CIAT, 1972). In West Africa up to 60 per cent of mature spikelets of *Andropogon gayanus* contained caryopses, but a large portion of seed shattered during ripening and harvesting (Bowden, 1964). Sterility occurs in *Cynodon nlemfuensis* and *C. aethiopicus*, with less than 15 per cent seed set in many genotypes. A few, however, displayed greater fertility and one clone had over 80 per cent seed set (Okigbo and Chheda, 1966).

Sterility and low fertility in tropical grasses have been associated with triploidy (*Digitaria decumbens* and *Pennisetum purpureum* × *P. americanum* F_1 hybrids), aneuploidy (*Bothriochloa–Dichan-thium–Capillipedium* agamic complex), polyploidy (*Setaria anceps*) and apomixis (*Panicum maximum*). In addition, the presence of various genetic systems control self-incompatibility (*Paspalum notatum* and *Cynodon* spp.) and male sterility (*Bothriochloa intermedia*, *Pennisetum clandestinum* and *P. americanum*).

Environmental conditions, particularly day and night temperatures and relative humidity, affect seed formation in some pasture legumes. *Glycine wightii* fails to form seed when day temperatures exceed 27 °C (Wutoh *et al.*, 1968). Night temperatures of 24 °C and 31 °C do not affect seed yields of *Stylosanthes humilis*, but very little seed set occurs at 10 °C nyctotemperature (Skerman and Humphreys, 1973). This adverse effect of cool nights on seed-setting, and therefore on self-regeneration, places a limitation on the distribution of *S. humilis* in some parts of the tropics. Cool day and night temperatures restrict activity of bees during periods of flowering and cause low seed set in cross-pollinated legumes such as *Desmodium uncinatum* (Nicholls *et al.*, 1973). A number of cleistogamous legumes require high relative humidity for pollen tube development, and seed-setting is poor when the relative humidity drops below 75 per cent (Hutton, 1960).

Diseases and pests

Many grasses are susceptible to smut, bunt and ergot infections. Seed production of legumes is generally more affected by diseases and pests than grasses. *Botrytis* spp. cause aborting of florets in *S. guianensis* (Loch *et al.*, 1976). Insects (e.g. *Zonabris tristrigina* beetles in East Africa) attack flowers and pods, reduce seed yields and often result in complete failure of the seed crop. Seed production of stylo severely declines when *Colletotrichum* spp. develop late in the growing season (Irwin and Cameron, 1978).

Migrating weaver birds such as the Sudan Dioch (*Quelea quelea*) are major pests of grass seed heads and consume large quantities of seed, especially in East Africa (Wiggs *et al.*, 1973). Other birds alight on the seed heads of some grasses (e.g. *Panicum maximum*) and pick seeds from the florets.

Plant characteristics

Intermingling of inflorescences with vegetative parts occurs as a result of lodging in grasses and indeterminate growth habit of many legumes. This causes serious harvesting problems particularly in *Centrosema pubescens*, *Pueraria phaseoloides*, *Macrotyloma axillare* and *Glycine wightii*. Presence of bristles, hooks, awns, involucres and other seed parts present difficulties in harvesting and cleaning (Javier and Mendoza, 1976), and result in considerable seed loss during these operations.

Genetic influence

In Kenyan genotypes of *Setaria anceps*, Boonman (1973b, 1977) suggested that mass selection could be used for varietal improvement in terms of seed yield. In a space-planted nursery vigorous plants within a selection should be grouped according to days required for flowering to obtain relatively homogeneous populations with higher seed yield potential. Additional improvement should be sought by further selection, utilizing existing variations in seed weight, seed-setting and inflorescence number within populations arising from the first cycle of selection. Data presented in Tables 18.5 and 18.6 indicate the feasibility of this approach. Average pure live seed yields of three selected clones of *Setaria anceps* were increased by almost 200 per cent, due largely to increase in inflorescence density and improved seed-setting. In *Chloris gayana*, average seed yields of all selections (early medium, late and very late) were 24 per cent higher than the standard varieties, with the early selections averaging 71 per cent higher pure germinating seed than the parental varieties. The early selection developed from Mbarara showed 127 per cent increase in pure live seed yield (67.1 kg/ha), when compared to the standard Mbarara cultivar (29.5 kg/ha).

Large variations in seed yields of individual plants of different ecotypes of *Stylosanthes guianensis* (ranging from 0.7 to 28.9 g of dehulled seed per plant) were observed in Colombia (CIAT, 1973, 1978). This indicated a strong possibility of raising the genetic potential for seed yields in self-pollinating legume species.

Agronomic and management practices for seed production

Seed yields of pasture crops can be significantly increased with appropriate crop husbandry practices. These practices are similar to those essential for the satisfactory establishment and maintenance of productive sown pastures. For profitable seed production, however, special attention must be given to location, isolation requirements, weed control, plant density, time and rate of fertilizer application, disease and pest control, defoliation effects and time and method of seed harvesting.

Choice of location
Seed yields are influenced by weather and climatic conditions, both of which are essentially beyond human control. Regions where a species is adapted as a pasture crop may not necessarily be ideal for its seed production. As a general rule, high rainfall, low radiation and high humidity restricted flowering in some species and prolonged it in others. Incidences of diseases and pests are also likely to prevail. Under such conditions, consistently high seed yields of most tropical pasture species are difficult to obtain. Subtropical and high-

Table 18.5 *Genetic variation in seed yield and yield component in 18 early-flowering Setaria anceps cv. Nandi clones (Boonman and van Wijk, 1973)*

	Seed yield per plant (g)		Pure live seed (%)	Inflorescence density (no./m²)	Inflorescence length (cm)	Germinating seeds/inflorescence (no.)	100 seed weight (mg)
	Pure live seed	Clean seed					
Mean of 18 clones	0.45	1.66	27.1	72	19.7	9.8	520
Mean of 3 selected clones	1.39	2.79	49.8	148	17.8	17.9	540
Advantage (%)	+208.8	+68		+105.6	−10.7	+82.7	3.8

Table 18.6 *Effect of selection on pure live seed yield and yield components in Chloris gayana (Boonman, 1977)*

	Pure live seed yield (mean of 4 harvests) (kg/ha)	Inflorescence density (no./m²)	Germinating seeds/inflorescence (no.)	1000 seed weight (mg)
Mean of all selections*	27.5	179	50.4	305
Mean of standard varieties	22.1	164	42.9	314
Advantage (%)	+24.4	+9.1	+17.5	−2.8

* Selections based on days to flower and vigour, and designated early, medium, late and very late, were made from each of the three standard varieties, viz. Mbarara, Masaba and Pokot.

altitude areas with frequent frosts and long periods of cold weather should also be avoided.

In order to achieve satisfactory seed yields of a particular pasture species the site chosen should provide (a) adequate length of growing season, with radiation, rainfall and temperature conditions optimum for a healthy and vigorous vegetative growth prior to flowering, (b) favourable photoperiod and day and night temperatures for floral induction and relatively concentrated flowering, (c) less frequent rainfall during the late reproductive phase of the crop for satisfactory flowering, pollinating and seed-setting, (d) calm weather with little or no precipitation during seed-ripening and harvesting and (e) supplementary irrigation facilities, particularly in dry regions with prolonged periods of drought during the growing season (Kernick, 1961; Humphreys, 1979; Hopkinson, 1977).

A number of districts in Queensland, Australia, are endowed with climatic conditions ideal for tropical pasture seed production. The areas are almost frost-free with brief periods of restricted vegetative growth caused by low temperatures. The wet season of 4–5 months (December–April) and the dry season are reasonably consistent and reliable (Hopkinson, 1977).

Near the equator the less humid, medium-altitude areas have good potential for pasture seed production. Kitale (1° N) in the Trans Nzoia district of Kenya and situated at 1 900 m altitude is a good example. Mean monthly temperatures during the warm months (February–April) are about 19.5 °C and 17 °C in the cool months (July–August). Maximum and minimum temperatures rarely exceed 30 °C and 9 °C, respectively. Precipitation is approximately 1 200 mm during 150–160 days, extending from late March to November. Drought periods of more than 10 days during the growing season are rare, except during June. Profitable seed production is possible in such areas and two full seed crops a year without supplementary irrigation are feasible for most grasses (Combes and Verburgt, 1971; Boonman, 1973b). *Panicum maximum, P. coloratum* and *Brachiaria ruziziensis* are among the species grown in the area for seed, even though cultivars of these are not ideally suited as pasture crops in the region. Seed grown in this area is exported to regions within the country and to other tropical countries with warmer climates (Boonman and van Wijk, 1973).

The history of Siratro (*Macroptilium atropurpureum*) seed production was reviewed by Hopkinson (1977). Soon after the release of this cultivar, farmers in parts of Northern Australia attempted to produce seed and found that production at an economic level was difficult. At present districts extending discontinuously from Walkamin through Lakeland Downs to the Endeavour and McIvor Rivers, north of Cooktown in Northern Queensland, produce the bulk of Siratro seed. In this area Siratro is considered a poor pasture legume, but seed yields exceed 150 kg/ha, almost four times greater than those obtained in the Bundaberg–Maryborough districts (35 kg/ha) where the cultivar is well suited as a pasture legume.

Townsville stylo (*Stylosanthes humilis*) is well adapted as a seed crop in northeastern Thailand. Under experimental conditions Robertson *et al.* (1976) and Wickham *et al.* (1977) obtained yields of 1 550 kg/ha, two to three times higher than those reported from the Northern Territory of Australia (Fisher, 1969) and from Queensland (Shelton and Humphreys, 1971), and almost 13 times those reported by Javier and Mendoza (1976) in the Philippines. Wickham *et al.*, (1977) attributed the attainment of high yields to the favourable environmental conditions in this area of Thailand, i.e. 14–19° N latitude, 100–300 m above sea-level, 1 000–1 100 mm of rainfall from mid-April to mid-October followed by a long reliable dry season.

A study of *Desmodium uncinatum* by Nicholls *et al.* (1973) revealed the unreliability of the southeastern Queensland climate for producing seeds of this species. Seed yields averaged 4.7, 22.9 and 50.0 kg/ha during 1968, 1969 and 1970, respectively. The authors pointed out that the unfavourable weather conditions (wet, windy and cold) at peak flowering in May of 1968 and 1969 were responsible for low seed-setting. The summer growing season in 1969 was abnormally dry, while that of 1970 was abnormally wet.

Simpson (1972) noted that cold weather during the autumn and winter months resulted in low seed set of *Setaria anceps* in southern Queensland. Furthermore, heavy rains in the spring led to flooding and lodging.

Once a region with climatic conditions favourable for seed production of a species or cultivar is identified, it is usually possible to locate soils appropriate for that species. Moderately fertile soils of adequate depth and good drainage are generally preferable. Highly fertile soils tend to produce excessive vegetative growth that causes harvesting problems, particularly in the wet seasons, and weed control is difficult. For Siratro and Townsville stylo, where suction-harvesting of the seed crop is contemplated, an even land surface and light-textured, compact soil cause fewer problems during harvesting. Sandy soils and imperfections such as looseness or presence of stones greatly reduce efficiency of seed recovery (Hopkinson and Vicary, 1974).

Choice of varieties and isolation

When seed production is aimed at satisfying local demand, superior cultivars adapted to the region should be planted for profitable returns. When the seed produced in a region is sold elsewhere, however, it is important to realize that the genetic purity of the cultivar could be significantly altered by growing for several generations in an environment other than that of intended use as a pasture crop. This occurs because of differences in rainfall, length of growing season, temperature, photoperiod, disease and pest incidences, etc. To minimize the danger of such genetic shift, it is advisable to multiply the basic seed originating from the environment of intended pasture use for no more than three to four generations in the different environment of the seed-growing region. Arrangements should be made to obtain authentic basic seed from suitable areas on a continuing basis. Genetic shift is likely to be more pronounced in cross-pollinating species, while self-pollinated types will be comparatively more stable.

To maintain the genetic purity of a cultivar, the seed crop should be isolated from possible sources of contamination which occurs due to mechanical mixing and cross-pollinating with other cultivars. As a general rule cross-pollinating grasses such as *Setaria anceps* and *Chloris gayana* and legumes such as *Desmodium* spp. should be separated from adjacent fields of different cultivars of the same species by at least 200 m or more. Isolation distances of 20–30 m are adequate to prevent contamination due to mechanical mixing in self-pollinated legumes such as *Stylosanthes* spp., *Macroptilium atropurpureum*, *Lotononis bainesii* and apomictic grasses like *Panicum maximum*, *Melinis minutiflora* and *Paspalum* spp.

Adequate roguing, i.e. removal of off-type plants, is also important to maintain genetic purity of a cultivar and should be carried out before flowering.

Weed control

Weeds pose serious problems in pasture seed production. They reduce seed yields and adversely affect seed quality if not controlled effectively. Regulations to control seed quality exist in several tropical countries. Presence of excessive amounts of weed seeds, or even mere presence of certain noxious weed seeds in pasture seeds offered for sale, seriously limit their marketability in many regions (Jong, 1961; Purcell, 1969).

Some weed seeds are similar in size, shape and density to grass or legume seeds, and their physical separation during seed-cleaning and processing is difficult. Some pasture species are inherently more difficult to process than others. Thus, removal of significant quantities of weed seeds cannot be achieved without incurring considerable losses of actual pasture seed. For instance, Bogdan (1966) examined samples of *Chloris gayana*, *Melinis minutiflora*, *Setaria anceps* and *Trifolium semipilosum* seeds originating from the Trans Nzoia and Uasin Gishu districts of Kenya. He found that *C. gayana* and *T. semiplosum* were often contaminated with a number of weed seeds (Table 18.7). Similarly, species of *Sida*, *Stachytarphaeta* and

Table 18.7 *Weed seeds in some herbage seeds in Kenya (Bogdan, 1966)*

	Chloris gayana	*Setaria anceps*	*Melinis minutiflora**	*Trifolium semipilosum*
Number of samples examined	32	9	5	6
Average number of weed seeds in a 10 g sample:				
Before cleaning	90	20	6	236
After cleaning	64	5.5	1	108
Per cent weed seeds retained after cleaning	71	27	16	46
Important weed seeds difficult to separate	*Digitaria velutina, D. ternata, Ageratum conyzoides, Tagetes minuta, Bidens pilosa*	*Paspalum commersonii, B. pilosa*		*Commelina benghalensis C. subulata, Amaranthus spp.·, Nicondra physaloides, Eleusine africana*

* The low weed seed content was ascribed to hand-picking of ripe panicles.

Crotolaria commonly occur among the weeds in *Stylosanthes guianensis* fields and their seeds are difficult to separate from the legume seeds during normal processing. It is important that seed growers make a concentrated effort to keep their seed crops free of weeds in order to reduce contamination at harvest time. Chemical weed control as an adjunct to cultural methods is often an economic proposition because of the high value of pasture seed crops.

A combination of measures usually provides a reasonable control of weeds in seed fields. These include careful selection of the site, such as virgin grassland; clean fallow or use of crop rotation; thorough and repeated cultivation prior to planting; pre-emergence herbicidal sprays, such as dalapon and Atrazine, or mixing herbicidal compounds such as Trifluralin with the soil 3–4 weeks before planting; row-sowing with appropriate seed rates and regular mechanical or hand cultivation during early stages of crop growth; foliar applications of herbicides such as 2,4-D, 2,4,5-T, Dicamba and Picloram and their combinations to control broad-leaved weeds in grasses. When adequate measures are taken to control weeds in the year of establishment, competition from the grass crop can provide an effective check on weed growth in subsequent years. With legumes, however, weed problems increase with age of pasture and efficient selective herbicides are not yet available. Consequently, the decline in seed yield and quality occurs so that many perennial legumes can be grown as profitable seed crops for only 1 or 2 years (Jones and Roe, 1976). On the other hand, perennial tropical grasses provide economic seed returns for 3–5 years when properly managed.

Sowing rates and plant density

Sowing rates are relatively less important in perennial pasture species that tiller or branch freely and quickly to produce a dense canopy than in annual species (Boonman, 1972a, 1972b; Javier *et al.*, 1976; Mendoza *et al.*, 1976). Rates of 0.5–2 kg/ha of pure live seed are generally adequate for most perennial tropical grasses, and 2–4 kg/ha for perennial legumes when grown as seed crops. Rates of 0.2–1.8 kg of pure live seed/ha did not significantly alter seed yields of *Setaria anceps*, a bunch-type grass, or *Chloris gayana*, a stoloniferous-type grass (Boonman, 1972a, 1972b). Seed yields tended to decline in both, however, by sowing more than 1 kg/ha of seed. This quantity of *C. gayana* seeds sown in rows 50 cm apart, and assuming 50 per cent failure, would still produce one seedling per centimetre of row. Plant development thereafter is more dependent on environmental conditions than sowing rates.

The effects of row-spacing and N fertilization on seed production of *Setaria anceps* were studied in Australia by Hacker and Jones (1971) and in Kenya by Boonman (1972a). The Australian workers noted absence of significant differences in seed yield due to 50 or 100 cm row-spacing when the grass was fertilized at varying rates of nitrogen (42–336 kg/ha per yr). Boonman, on the other hand, observed that 130 kg/ha of N increased pure live seed yield by 33 per cent with 30 cm row-spacing as compared to 90 cm row-spacing (Table 18.8), due largely to more concentrated inflorescence emergence and flowering. At close row-spacing more N was required to improve yields but less reponse to N occurred at wide row-spacing.

In a study of *Chloris gayana*, seed yields increased with N application but differences due to row-spacing were not significant (Boonman, 1972b). This indicated that stoloniferous grasses can be sown in widely spaced rows without appreciable reduction in seed yields. With bunch-type grasses such as *Setaria anceps*, however, there is an advantage by sowing in closely spaced rows (35–50 cm), particularly under conditions of adequate moisture and nutrient supply.

In the Philippines differences in seed yields of Siratro did not occur with sown at 3, 6 and 9 kg viable seed per hectare in rows spaced 20, 40 and 80 cm apart (Javier *et al.*, 1976). The recommended sowing rate for seed production was 3 kg/ha in rows 80–100 cm apart. Sowing rates of 2, 4 and 6 kg/ha and row-spacings of 20, 40 and 80 cm did not significantly alter seed yield of *Stylosanthes guianensis* (Mendoza *et al.*, 1976).

The annual species usually yield more seed when sown at rates about the same or higher than those optimum for forage production (Rogler *et al.*, 1961). Shelton and Humphreys (1971) observed seed production of *Stylosanthes humilis* when grown at densities of 10, 50, 250, 850 and 3 800 plants/m^2 in frequently irrigated wooden boxes (Fig. 18.4). Maximum seed yield of 69 g/m^2 (690 kg/ha) was

Table 18.8 *Effects of row-spacing and N fertilization on seed production of Setaria anceps (Boonman, 1972a)*

Width between rows:		Effect of N application (kg/ha)		
		0	130	260
30 cm	Clean seed yield (kg/ha)	133.0	49.0	57.0
	Pure live seed yield (kg/ha)	7.6	11.7	13.1
	Pure live seed (%)	23.0	23.9	23.0
	Inflorescences/m^2	56.0	82.0	76.0
60 cm	Clean seed yield (kg/ha)	92.0	94.0	84.0
	Pure live seed yield (kg/ha)	19.5	18.9	14.9
	Pure live seed (%)	21.1	20.7	17.7
	Inflorescenes/m^2	176.0	180.0	188.0
90 cm	Clean seed yield (kg/ha)	99.0	88.0	67.0
	Pure live seed yield (kg/ha)	19.0	14.0	11.0
	Pure live seed (%)	19.2	15.9	11.0
	Inflorescences/m^2	278.0	278.0	152.0

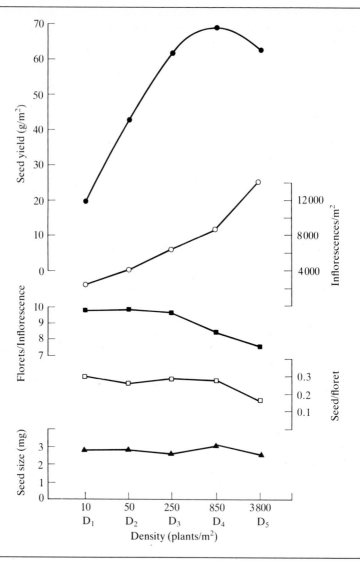

Fig. 18.4 Effect of plant density of *Stylosanthes humilis* on seed yield and its components (Shelton and Humphreys, 1971).

recorded in treatments with 850 plants/m^2. About 90 per cent of maximum yield was obtained with 250 plants/m^2 and 3 800 plants/m^2. A linear relationship existed between seed yield and the logarithm of plant density up to 250 plants/m^2. Inflorescence density was the most important seed-yield component controlling seed yield. A sharp decrease in the proportion of florets that set seed and a reduction in florets per inflorescence accounted for reduced seed yields with 3 800 plants/m^2, as compared to 850 plants/m^2. Extrapolation of these results indicated that a sowing rate of 7 kg/ha of seed pods (assuming 10% establishment) would provide a density of 25 plants/m^2 and yield less than 50 per cent of the maximum. Thus, sowing rates of 35–50 kg/ha will be necessary to achieve high seed yields from first year stands.

Fertilizer application

Research findings, from both temperate and tropical environments, clearly indicate that N fertilization is the most effective way of substantially increasing seed yields of grasses when moisture and other nutrients are not limiting (Foster, 1956; Haggar, 1966; Humphreys and Davidson, 1967; Mishra and Chatterjee, 1968; Cameron and Mullaly, 1969; Grof, 1969; Chadhokar and Humphreys, 1970, 1973; Hacker and Jones, 1971; Boonman, 1972a, 1972b, 1972c; Stillman and Tapsall, 1976; Bahnisch and Humphreys, 1977).

The magnitude of seed yield response to N application varies (Table 18.9) along with soil fertility, form of fertilizer and time of application and availability of moisture.

Increased seed yields due to N fertilization are primarily brought about by greater numbers of fertile tillers and by increased length or branching of the inflorescences. The porportions of fertile tillers at peak flowering in *Paspalum plicatulum*, *Setaria anceps* and *Chloris gayana* were 51, 47 and 57 per cent, respectively, of the total in N treatments, as compared to 41, 4 and 7 per cent, respectively, in control plots with no N (Henzell and Oxenham, 1964). Seed set, however, may be reduced, particularly at high levels of N applications. Seed weight is not affected, or only marginally altered, by additions of N. Effects on germination per cent, which influences the pure live seed yield, appear to be inconsistent (Humphreys, 1979). While some researchers (e.g. Stillman and Tapsall, 1976) reported increased germination percentage with increasing N applications, others (e.g. Wiggs *et al.*, 1973) noted a decrease. Still others (e.g. Cameron and Mullaly, 1969) did not observe a clear relationship between germination per cent and N rates. Studies by Boonman (1972a, 1972b) are presented in Table 18.10 to illustrate the effects of N application on seed yield and its components in tropical grasses.

Published information suggests that the optimum levels of N application for most grass seed crops lies within the range of 100–200 kg/ha per seed crop. At these levels, split application of fertilizer is unlikely to increase seed yields (Hacker and Jones, 1971; Stillman and Tapsall, 1976) and, in fact, may be disadvantageous

Table 18.9 *Effect of nitrogen on the seed yield (kg/ha) of some tropical grasses*

Grass species	Nitrogen application (kg/ha)												
	0	21	28	42	56	84	100	112	168	200	224	336	672
Andropogon gayanus													
Haggar (1966)	25	—	24	—	39	—	58	—	—	—	—	—	—
Mishra and Chatterjee (1968)	69	—	90	—	—	—	—	75	—	—	—	—	—
Brachiaria mutica													
Grof (1969)	13	—	—	—	25	—	—	30	—	—	—	—	—
Cenchrus ciliaris													
Cameron and Mullaly (1969)	6	—	—	—	—	47	—	—	158	—	—	261	503
Wiggs et al., (1973)	67	—	—	—	—	77	—	—	97	—	—	—	—
Chloris gayana													
Boonman (1972b)	16	—	—	—	—	—	97	—	—	110	—	—	—
Setaria anceps													
Hacker and Jones (1971)	—	14	—	17	—	23	—	—	28	—	—	—	—
Boonman (1972a)	13	—	—	—	—	—	90	—	—	87	—	—	—
Stillman and Tapsall (1976)	44	—	—	—	—	—	—	137	—	—	131	—	—

Table 18.10 *Effect of N on mean yield characteristics of Setaria anceps and Chloris gayana sown at row width of 50 cm (Boonman, 1972a 1972b)*

	Setaria anceps (5 crops)			Chloris gayana (6 crops)		
Nitrogen rate (kg/ha)	0	100	200	0	100	200
Clean seed yield (kg/ha)	13	90	87	16	97.7	110
Pure live seed yield (kg/ha)	2.4	18.2	16.3	6.2	41.0	42.9
Pure live seed (%)	18.5	20.2	18.8	38.7	42.4	39.0
Inflorescence/m^2	40	210	240	67	240	293
Percentage flowering tillers	11	28	32	15	35	47
Inflorescence or raceme length (cm)	13.7	14.7	16.7	8.3	10.2	10.6
Raceme number	—	—	—	6.9	6.9	7.3
1 000 seed weight (mg)	340	310	320	290	270	290

(Boonman, 1972c). On light soils (Humphreys, 1979), or with much higher applications of N, the seed crop is likely to benefit from split applications of N. Best results are generally obtained when N is applied as soon as possible at the beginning of each season's growth. A delay of 4 weeks after the onset of the rainy season decreased pure live seed yield of *Setaria anceps* early crop by more than 60 per cent, mainly as a result of poor seed set (Boonman, 1972c). A delay of no more than 2 weeks in the late-season crop gave best results. Chadhokar and Humphreys (1970) reported *Paspalum plicatulum* seed production to be independent of N application after inflorescence emergence. Nitrogen deficiency during vegetative growth, however, decreased tillering and inflorescence branching but increased the percentage of seed set.

The effect of P fertilizer alone did not affect seed yields of *Andropogon gayanus* (Haggar, 1966). Application of 33 kg/ha of P along with 88 and 112 kg/ha of N increased seed yield by 28 and 18 per cent, respectively. Raising the level of P to 67 kg/ha gave lower seed yields than nil P treatments, possibly due to abundant vegetative growth resulting in shading and reduced inflorescence production. Seed yields of *Brachiaria mutica* in the 'wet belt' of north Queensland (Grof, 1969), *Setaria anceps* and *Chloris gayana* in the Trans Nzoia district of Kenya (Boonman, 1972a, 1972b), and *Cenchrus ciliaris* in Tanzania (Wiggs *et al.*, 1973) were not significantly affected by P fertilization.

Nitrogen application to *Desmodium uncinatum* raised seed yields by increasing inflorescence density (Nicholls *et al.*, 1973). This treatment favoured growth of weeds, however, and reduced the vigour of legume plants in subsequent years (Redrup, 1965). Thus, for legume seed production the aim should be to provide adequately the major and minor nutrient elements necessary to promote an efficient legume–*Rhizobium* symbiosis rather than N fertilizer.

Defoliation

Removal of plant shoots by grazing or cutting may increase, decrease or not alter seed yields of pasture grasses and legumes, depending on the species, time and extent of defoliation and length of the growing season. Increased, or more efficient, seed production by defoliation can be explained as follows (Wilaipon and Humphreys, 1976; Fisher, 1973; Loch *et al.*, 1976, Wilson, 1970): (1) removal of apical dominance, (2) greater tillering or branching, (3) increased inflorescence density and/or better synchronization of inflorescence emergence and peak flowering, (4) improved efficiency due to reduced vegetative material and less lodging and (5) weed control and increased availability of light, particularly to the low growing, shade-intolerant legumes.

Harvested seed yields of *Pennisetum clandestinum* usually do not exceed 100 kg/ha. Repeated mowing at intervals of 4–6 weeks at varying heights during different periods of the year (13 mm in late summer and autumn, 50 mm in winter and early spring, 64 mm at peak flowering in mid-September to mid-October) gave seed yields up to 482 kg/ha. Close mowing hindered apical dominance, stimulated lateral shoot development and promoted flowering.

Tillers present early in the rainy season produce the bulk of inflorescences in grasses such as *Andropogon gayanus, Chloris gayana* and *Pennisetum polystachion*. Removal of these tillers by grazing or cutting significantly reduces inflorescence density and therefore seed yields (Haggar, 1966; Foster, 1956; Mishra and Chatterjee, 1968). Addition of nitrogen can reduce, to some extent, the harmful effects of late grazing.

Seed yields of *Stylosanthes hamata* varied with defoliation treatment (Wilaipon and Humphreys, 1976). Undefoliated treatments gave 221 kg/ha. A single heavy grazing followed by mowing at 10 cm height at the early flowering stage boosted production to 335 kg/ha. Intermittent grazing and mowing at a later stage of flowering reduced production, due to removal of inflorescences and reduction in leaf surface.

Defoliation about 4 weeks before first flower initiation in *Stylosanthes guianensis* permitted vegetative regrowth and a return to a closed canopy by the time of a second flower initiation. This allowed improved synchronization of individual shoot development and did not reduce harvestable seed yields (Loch *et al.*, 1976). This practice also produced a more uniform canopy level suitable for mechanical harvesting. Later defoliation resulted in poorly synchronized shoot development and reduced seed yields. Early growth of twining legumes such as *Macroptilium atropurpureum* and *Glycine wightii* can be removed by mowing or grazing without significant reduction in seed yields.

Harvesting

Increases in seed yields of many pasture species brought about by raising their genetic potential for seed yield and by improved agronomic and management practices can be easily lost if sufficient attention is not paid to the time and method of seed harvest.

Time of harvest

Determination of optimum harvest date of a pasture species is difficult because of seasonal variations, prolonged flowering, uneven maturity of inflorescences and seed-shedding. When harvested too early, the pure live seed yield suffers due to a high proportion of immature seeds. When delayed too long, large quantities of seed are shed on the ground. The capacity to germinate deteriorates rapidly during storage in many species when the seeds are immature when harvested (Brzostowski and Owen, 1966). Timing of harvest is more critical when machines are used, particularly in combine-harvesting as compared to a reaper-binder, where some degree of post-harvest maturation occurs in the sheaf.

Days from 'cleaning out', removal of residue from the stubble (Gordin-Sharir and Gelmond, 1966; Criste *et al.*, 1973), initial flowering date (Grof, 1969; Loch *et al.*, 1976) as well as percentage of shatter, are criteria used to determine optimum harvesting time. As a general rule, optimum harvest time is reached well before maximum inflorescence density. Prolonged flowering allows an adequate margin of safety so that harvest in many species may be carried out over a period of 10–15 days. Thus, the seed grower has an option to adjust for weather conditions without fear of significant reduction in seed yields. During this period there is a balance between loss of seed due to shattering and further increase of ripe seeds in later-emerging inflorescences. Boonman (1973a) observed a 6–7-week interval between initial flowering date and optimum harvest date in *Setaria anceps, Chloris gayana* and *Panicum coloratum*. He also noted a maximum percentage of pure live seed at a later stage than maximum yield of clean seed. Contrary to common belief, considerable shattering of spikelets can be tolerated in many grass species (normally 20–30% but up to 40–50% in species such as *P. maximum* and *P. coloratum*) without significant reduction in the pure live seed yield.

Harvest date is critical in species that have a defined peak of flowering and in those that shatter easily. Such species should generally be harvested about 3–4 weeks after peak flowering. A delay in harvesting by 1 week from the optimum date (3 weeks after peak flowering) reduced seed yield of *Stylosanthes guianensis* by 25 per cent (Loch *et al.*, 1976). Seed-shattering in *Brachiaria mutica* is markedly pronounced and the crop must be harvested soon after general anthesis is completed (21 days from emergence of first inflorescences). Delay of 1 week reduced seed yields by 50 per cent or more in Australia (Grof, 1969). Similarly, Haggar (1966) reported linear reduction in seed yield of *Andropogon gayanus* as harvest time was delayed from 9 November (79 kg/ha), when only a few racemes at the distal end of the inflorescences had dropped, to 25 November (12 kg/ha) when about 50 per cent had dropped.

Harvesting any time after 15 weeks from the previous cut during the wet season has been recommended for optimum seed yields in *Panicum maximum* in the Philippines (Criste *et al.*, 1973). Earlier harvest is preferable, however, to allow the succeeding late-season seed crop to flower and mature under more favourable moisture conditions.

Methods of harvest

The bulk of tropical pasture seeds is harvested by combine or header in Australia. Hand-harvesting is more common in other parts of the tropics where labour is relatively cheap and easily available (Fig. 18.5). Reaper-binders are used extensively in some countries, e.g. Kenya during peak periods when labour is scarce. Specialized mechanical procedures such as a repeated beating action for harvesting of some grasses (e.g. *Cenchrus ciliaris* and *Paspalum dilatatum*) and suction-harvesting for some legumes (e.g. *Macroptilium atropurpureum* and *Stylosanthes humilis*) have been developed to maximize seed yields.

Seed yields from four different methods of harvesting were compared with the total production of *Panicum coloratum* to give an estimate of the potential seed losses during harvesting in Australia (Roe, 1972). Total yield was obtained by collecting all seeds as they were shed. Cutting and field-drying gave the highest yield of 317 kg/ha of clean seed, but only 213 kg were retrieved because of seed shed. Hand-shaking allowed the greatest yield of pure live seed (136 kg/ha). Direct heading was recommended, however, despite its low efficiency (30% of potential gross yield and 19% of potential pure live seed yield). Seeds that are shed during field-drying in the

Fig. 18.5   Seed harvest of *Cenchrus ciliaris* in Kenya. Seeds are frequently hand-harvested in the tropics.

wind-row after mowing and during drying in shocks after harvesting with a reaper-binder cannot be retrieved. Additional seed will be lost during the threshing of material cut by mowing and reaper-binder. Therefore, the actual seed yields by these methods are unlikely to greatly exceed those obtained by direct heading.

Hand-harvesting

The effects of prolonged flowering and seed-shedding in terms of seed losses can be minimized by hand-harvesting at regular intervals (Tables 18.11 and 18.12). This is achieved in grasses by (a) shaking ripe seeds into a container, (b) gently stripping the inflorescences to collect relatively mature seeds and (c) pulling out mature inflorescences or cutting them with 50–60 cm of stem, tying into small loose bundles and drying, preferably in the shade, for post-harvest maturation of seeds. Ripe or nearly ripe pods of legumes are hand-picked and dried.

Along with increased seed yields, hand-harvesting results in higher quality of seed with improved germination percentage and fewer weed seeds. Subsequent threshing and cleaning is relatively easier than when harvested by mowing. Hand-harvested legume seeds have a high proportion of hard seeds, however, and require some form of scarification for improved germination (Fig. 18.1, p. 509).

Hand-harvested grasses comprise mature and immature seeds, empty glumes and other impurities. They can be delivered to a seed processing plant or further cleaned by the farmer for local use. Removal of lighter materials is effected by winnowing or with the use of a seed blower. Adequate care should be taken during winnowing and blowing to prevent excessive seed loss. Hand-picked legume pods are dried on a clean surface under cover and allowed to shatter or dry in the sun in porous bags, then flailed or trampled to thresh the pods. Appropriate wire screens and sieves are used for further cleaning.

Reaper-binder

Grass seeds are frequently harvested by a reaper-binder which cuts the plants and binds them into sheaves. Special catch pans placed below the junction of elevator and platform canvas, and below the lower edge of the tying deck, collect some of the seeds that shatter during harvesting. Seed losses are minimized by keeping the shocks small, tying the heads together and covering the shocks with a turned-up sheaf for protection against rain and bird damage (Bogdan, 1966). Where feasible, sheaves should be dried on racks under cover to promote a higher degree of post-harvest maturation and to facilitate the retrieval of shed seed. The sheaves can be threshed on tarpaulins in the field or on a cement floor under cover, with a grain thresher, or by beating the sheaves against an appropriate hard sur-

Table 18.11 *Effects of different methods of harvesting on seed yield, germination and seed weight of Panicum coloratum (Roe, 1972)*

Method of harvest	Clean seed			Pure live seed			1 000–seed weight (g)
	Yield (kg/ha)	Total collection (%)	Germination (%)	Yield (kg/ha)	Total collection (%)		
Total collection	410	–	54	221	–		0.977
Cutting and drying in field	317	77		108	49		
(a) threshed seed	104	25	39	40			0.956
(b) shed seed	213	52	32	68			0.846
Hand-shaking	256	62	53	136	62		1.030
Reaper-binder	179	44		93	42		
(a) threshed seed	31	8	29	9			0.760
(b) shed seed	148*	36	57	84			0.794
Direct heading	123	30	35	43	19		0.737

* Portion of this fraction consisted of seed shaken from panicles before they were put through the threshing machine. It was, therefore, not all shed seed in the normal sense.

Table 18.12 *Seed yields (kg/ha) of some tropical pasture legumes using different methods of harvesting (Javier and Mendoza, 1976; Wickham et al., 1977)*

	Centrosema pubescens	*Macroptilium atropurpureum*		*Stylosanthes guianensis*	*S. humilis*
Hand-picking	488	550	460	254	119
Cutting whole plants	–	–	–	224	60
Cutting mature inflorescences	–	–	–	165	88
Hand-sweeping	–	–	–	–	110
Machine harvest	436	134	300	–	–

face such as wooden slats. Combines can also be used as stationary threshers.

Combine-harvesting

The combine harvester, primarily developed to harvest cereal grains, is available with appropriate modifications to harvest small-seeded pasture species either by direct or indirect heading (Harmond *et al.*, 1961; Jones and Roe, 1976). Direct heading involves cutting the standing crop and simultaneous threshing. This method is commonly used in Australia to harvest most grasses as well as legumes that have a determinate flowering habit or shatter readily (e.g. Siratro), or possess a twining growth habit (e.g. *Glycine wightii*). Seed losses with this method are generally quite large and reach 80 per cent of the total potential production (Roe, 1972; Hopkinson and Loch, 1973).

Indirect heading involves mowing the plants, drying in wind-rows, then picking up the dried crop with a combine harvester for threshing. This technique is suitable with species that do not shatter readily, such as *Setaria anceps, Desmodium* spp. and *Lotononis bainesii*. Indirect heading reduces the amount of herbage that has to pass through the combine during threshing and also favours post-harvest maturation of immature seeds (Jones and Roe, 1976). Rethreshing recovers significant amount of seeds that would otherwise be left in the field.

Factors that limit combine-harvesting in many tropical countries include high initial cost of a combine harvester, generally poor maintenance and repair facilities, a high degree of operator skill, low overall efficiency of seed harvest, requirement of large areas under seed crops and unreliable weather conditions at harvest time. Boonman (1971a) estimated that cost of hand-harvesting grass seeds in Kenya was 30–50 per cent of combine-harvesting.

Repeated beater harvesting

This method permits successive harvests of ripe seed of grasses from

a standing crop. The harvesting machines are often fabricated locally and several designs are available. A simple model described by
Purcell (1969) consists of a box with tensioned wires that is mounted
on the front of a vehicle. In other models (Brzostowski and Owen,
1966; Purcell, 1969) open-front, iron boxes are fitted with revolving,
power-driven beaters (brushes) in front of the box. When driven
through a field of *Cenchrus ciliaris* or *Paspalum dilatatum* at an
appropriate speed (usually 20–30 km/h), the wires, beaters or
brushes knock the ripe seeds back into the box. To avoid emptying
the boxes manually at regular intervals, more sophisticated
machines have suction or air-blast devices that direct seeds from the
boxes into bins or sacks. The efficiency of these methods depends
on the grass species, planting method and clearance of the tractor or
vehicle. After drying, seeds can be processed in the same way as
hand-harvested seed.

Suction-harvesting

Seeds of Siratro and Townsville stylo are harvested in Australia by
using a suction device originally developed to harvest the temperate
legume subterranean clover. Mature fields of Townsville stylo when
mown and raked into wind-rows shed most of their pods on the
ground. The suction harvester picks up these pods. The wind-rows
are then moved on one side and the remaining fallen pods are harvested.

Siratro when conventionally managed during the growing season
usually produces two seed crops. These are harvested by direct
heading, but large quantities of seed fall to the ground because of
shedding. The field can then be closely mowed and wind-rowed for
suction-harvesting. By operating at low ground speed and maximum
vacuum at the mouth of the pick-up duct, up to 350 kg/ha of seed
can be recovered from the near 1 000 kg/ha of fallen seed (Hopkinson and Vicary, 1974). Thus, the use of a suction harvester as an adjunct to direct heading more than doubles Siratro seed yields. On
the irrigated farms in Queensland it is possible to collect from 500 to
800 kg/ha of Siratro seed by inducing multiple-seed crops with successive wet and dry cycles. A single harvest is made at the end of the
season using a combine with the cutter bar set close to the ground.
This is followed by suction-harvesting (Hopkinson, 1977).

Harvesting by suction is slow and the harvested material contains
large quantities of soil particles. This increases subsequent cleaning
and processing costs.

Cleaning and processing

Immediately after harvest, particularly machine harvest, the seeds
contain considerable quantities of impurities in the form of leaf,

stem, pod and inflorescence fragments, weed seeds, immature seeds, insects, inert matter, etc. and generally have a high moisture content.

Seed processing aims to (a) concentrate seeds of the desired species, (b) remove all extraneous matter, (c) obtain the best possible germination and (d) produce a homogeneous bulk (Oomen, 1969). Seed-bearing material that is high in moisture content is dried and then threshed. Cleaning to a reasonable standard of purity and quality may occur on the farm for local sale. In many instances the threshed seeds are delivered to a commercial plant for further cleaning and processing.

Seed drying

The high moisture content of seed causes difficulties in threshing, cleaning and processing, and reduces seed germination because of heating and fungal growth. The seed moisture content should be reduced to 18–20 per cent soon after harvest and to 10–12 per cent after processing. Moisture content should be higher during cleaning and processing than for subsequent storage (Harrington, 1972).

Seeds can be dried in the sun by spreading them on a hard ground surface, tarpaulin or cement floor during periods when rain is unlikely and when the relative humidity is lower than 75–80 per cent. In areas where the days are extremely hot, drying in sheds or on screen-bottom trays is advisable. Seeds can be dried efficiently by blowing ambient or heated air (about 35 °C) through them. Brandenburg *et al.* (1961) discussed the principles and procedures of seed-drying.

Cleaning on the farm

In the absence of an organized seed-processing industry, pasture seeds produced in most tropical countries are cleaned by local growers. With care it is possible to clean pasture seeds to reasonable standards of purity and quality with little equipment and facilities.

Legumes are generally easier to clean than grasses. Townsville stylo, a difficult species to process, was cleaned by farmers in Thailand with seed purity ranging from 77 to 96 per cent (Wickham *et al.*, 1977). When plants are harvested and rolled into wind-rows most of the mature seed, leaves and some stem pieces drop to the ground and form a dense litter. This mass is gathered into small heaps by means of stick brooms, rakes or hoes, then loaded into baskets and transported to a sieving–winnowing area. It is passed through a 3 mm sieve to remove large stems and leaf material and then through a fly-screen gauge sieve to remove small soil particles. A final tossing in bamboo winnowing pans facilitates removal of additional impurities.

A typical farm cleaner described by Linnett (1977) consists of one or two oscillating flat screens of woven wire with variable speed control to remove large particles. An electric fan produces a winnowing action to separate the seeds and inert matter according to density and size. Verhoeven (1964) explained the construction of a simple cleaner made from 24- or 26-gauge galvanized flat iron, 5.0 × 2.5 cm dressed pipe and steel straps. The source of air is derived from a 40 cm household fan with three speeds. For difficult species additional precleaning equipment, such as debearders and huller-scarifiers, can also be improvised on the farm.

Commercial cleaning and processing

To process seeds of a pasture cultivar originating from properly inspected seed multiplication fields, so as to comply with existing regulations for seed certification, requires a range of equipment and a high degree of operator skill (Combes and Verburgt, 1971; Linnett, 1977). The machines effect cleaning and separation due to differences in the physical properties (e.g. size, shape, density, pubescence, texture, specific gravity, etc.) between the seed and the contaminants. Seed-processing can be separated into definite steps that follow a specific sequence (Fig. 18.6). The choice of machines de-

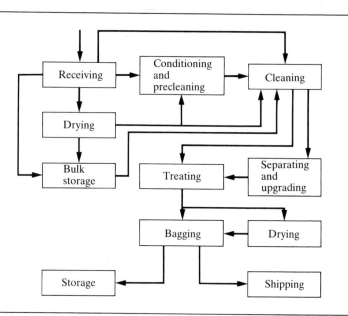

Fig. 18.6 The basic sequence of operations in commercial seed-processing plants (redrawn from Greg, 1977).

pends on the species being processed, the amount and type of contaminants present, and the desired quality standards (Greg, 1973, 1977).

1. Receiving

Seeds delivered to the processing plant are processed immediately or dried and stored for later processing.

2. Conditioning and precleaning

Most legumes and some grasses (*Setaria anceps, Panicum maximum, Brachiaria mutica*) are relatively easy to condition and clean. But certain legumes (*Stylosanthes humilis*) and several grasses (*Chloris gayana, Melinis minutiflora, Cenchrus ciliaris, Hyparrhenia rufa*) are difficult to process. Seed flow and problems in air separation are caused by peculiarities of their seed such as presence of hooks, awns, bristles, hairs, etc. Thus, seeds must be pretreated for easier processing.

A common treatment is scalping by oscillating flat, large-aperture screens or centrifugal screen scalpers to remove excessive amounts of large waste materials. Machines called debearders, oat clippers, brush scarifiers and huller-scarifiers may be used to further improve seed physical condition by removing appendages, husks, glumes or pod fragments (Linnett, 1977; Greg, 1977; Vaughan *et al.*, 1968).

3. Basic cleaning

A screen/air machine has varying numbers of screens and air chambers that improve seed purity and germination by removing large, small and light-weight materials.

4. Separating or upgrading

After screen/air cleaning, the seeds may require further processing to remove specific contaminants. Two commonly used machines include (1) an indented cylinder separator that operates on the principle of differences in seed length and (2) a gravity separator table that operates on the principle of gravity and density variations. Linnett (1977) described the operation of special separators available in Australia for processing Townsville stylo and *Cenchrus ciliaris*, a vacuum cleaner for grass seeds, and a flotation tank for cleaning suction-harvested seeds of Siratro.

5. Treating

Seed protectants, e.g. a fungicide or insecticide, may be applied by dry (dusting), wet (slurry) or spray treatment. These are used to improve storage and to minimize field losses after sowing and prior to germination.

6. Bagging

Seeds are accurately weighed and placed into appropriate containers for storage or shipping.

7. Storage

Life of the seed is doubled with (1) each 1 per cent decrease in seed moisture content between 14 and 4 per cent and (2) each 5 °C decrease in seed temperature between 50 °C and 0 °C (Harrington, 1963). Both rules are logarithmic and act independently. For normal short-term commercial seed storage, or carry-over stocks for not more than 2 years, seeds should be dried to a moisture content of at least 12 per cent and placed in steel bins with tight-fitting lids or in moisture-proof bags (e.g. high-density polyethylene plastic bags) for storage in reasonably cool warehouses.

   Foundation or basic seeds, germplasm seeds and occasionally carry-over seeds require long-term storage. These seeds should be dried to a moisture content below 8 per cent, preferably 4–6 per cent, sealed in airtight containers and stored in an environment where the numerical sum of percentage relative humidity (RH) and the temperature in °F does not exceed 100 (Harrington, 1959). Most seeds can be stored for 3–5 years without significant loss in viability, with RH of about 25 per cent at 30 °C or RH of about 45 per cent at 20 °C or less (Harrington, 1972). These conditions can be achieved by making the storage rooms moisture-proof and using dehumidifiers along with refrigeration units. Good germplasm storage units generally are capable of maintaining 5 °C to −10 °C temperature and 30 per cent relative humidity.

8. Shipping

As needed the seed are released into the local marketing channels for sale to cultivators or exported.

Economic considerations

In tropical countries production of home-grown seeds would be a worthwhile strategy. This approach minimizes the need for expensive imported seed, accelerates tropical pasture research and development and enables establishment of a nucleus of small-scale producers who specialize in pasture seed growing and processing. Substantial quantities of grass and legume seed can be produced on small farms using available family and village labour. In fact, the production of pasture crop seed by manual labour is likely to be more economical than by mechanized means. An economic analysis of Townsville stylo seed production in Thailand (Wickham *et al.*,

1977) by hand labour revealed 37 per cent lower cost per kilogram of seed as compared to mechanical harvesting (Table 18.13). In the Philippines, seed production costs of *Stylosanthes guianensis*, *S. humilis* and *Macroptilium atropurpureum* were 5.00, 5.00 and 1.50 pesos/kg of seed, respectively (Javier and Marasigan, 1976), and of *Panicum maximum* 5.00 pesos/kg of seed (Criste *et al.*, 1973) by use of manual labour. These costs compared favourably with those incurred in Australia using machinery. Home-grown seeds can be sold at prices significantly lower than imported seeds and still provide profitable returns to the small-scale seed producers.

The purchase of a particular seed lot and a decision regarding sowing rate should be based on pure live seed content. For example, suppose seed lot A has 90 per cent purity and 50 per cent germination and seed lot B has 70 per cent purity and 40 per cent germination. The former sells at US$2.00 per kilogram and the latter at US$3.00. The effective price per kilogram of pure live seed of seed lot A will be US$6.97 (3.00/0.9 × 0.5), while that of seed lot B will be US$7.14 (2.00/0.7 × 0.4). To obtain a sowing rate of 2.0 kg of pure live seed per hectare would require 4.44 kg/ha of A and 7.14 kg/ha of B.

Table 18.13 *Comparison of yields, labour use and costs between hand and mechanical harvesting of Stylosanthes humilis (Wickham et al., 1977)*

	Hand-harvesting and cleaning	Mechanical harvesting and cleaning
Clean seed yield (kg/ha)	460	300
Labour use		
hr/ha	2208	55
hr/kg clean seed	4.8	0.2
Cash cost		
US$/kg clean seed	0.51	0.31
Non-cash cost		
US$/kg clean seed	0.20	0.80
Total cost		
US$/ha	324	332
US$/kg clean seed	0.70	1.11

Note. Labour use excludes that used for land preparation, fertilizing and sowing seed. Costs exclude those associated with seed and land preparation. Fertilizer costs have been included. Non-cash costs for hand-harvesting and cleaning include that of family labour.

References

Akinola, J. O., H. R. Chheda and **J. A. Mackenzie** (1971) Effects of cutting frequency and level of applied nitrogen on productivity, chemical composition, growth components and regrowth potential of three Cynodon strains. 2. Growth components: tillering and leaf area, *Niger. Agric. J.* **8**, 63–76.

Bahnisch, L. M. and **L. R. Humphreys** (1977) Urea application and time of harvest effects on seed production of Setaria anceps cv. Narok, *Aust. J. Exptl Agric. Anim. Husb.* **17**, 621–8.

Bogdan, A. V. (1966) Weeds in herbage seeds in Kenya, *E. Afr. Agric. For. J.* **32**, 63–6.

Bogdan, A. V. (1977) *Tropical Pasture and Fodder Plants*, Longman: London.

Boonman, J. G. (1971a) Experimental studies on seed production of tropical grasses in Kenya. General introduction and analysis of problems, *Neth. J. Agric. Sci.* **19**, 26–36.

Boonman, J. G. (1971b) Tillering and heading in seed crops of eight grasses, *Neth. J. Agric. Sci.* **19**, 237–49.

Boonman, J. G. (1972a) The effect of nitrogen and row width on seed crops of Setaria sphacelata cv. Nandi, *Neth. J. Agric. Sci.* **20**, 22–34.

Boonman, J. G. (1972b) The effect of fertilizer and planting density on Chloris gayana cv. Mbarara, *Neth. J. Agric. Sci.* **20**, 218–24.

Boonman, J. G. (1972c) The effect of time of nitrogen top dressing on seed crops of Setaria sphacelata cv. Nandi, *Neth. J. Agric. Sci.* **20**, 225–31.

Boonman, J. G. (1973a) The effect of harvest date on seed yield in varieties of Setaria sphacelata, Chloris gayana and Panicum coloratum, *Neth. J. Agric. Sci.* **21**, 3–11.

Boonman, J. G. (1973b). *On the Seed Production of Tropical Grasses in Kenya*, Center Agric. Publ. Doc.: Wageningen.

Boonman, J. G. (1977) *On Rhodes Grass Breeding, Seed Yield and Herbage Quality*, Min. of Agri. Kenya Grass Br. Proj., Tech. Rep. GBP/June/77/1, Nairobi.

Boonman, J. G. and **A. P. J. van Wijk** (1973) Experimental studies on seed production of tropical grasses in Kenya. 7. The breeding for improved seed and herbage productivity, *Neth. J. Agric. Sci.* **21**, 12–23.

Bowden, B. N. (1964) Studies on Adropogon gayanus Kunth. 3. An outline of its biology, *J. Ecol.* **52**, 255–71.

Brandenburg, N. R., J. W. Simons and **L. L. Smith** (1961) Why and how seeds are dried. In *Seeds, The Yearbook of Agriculture*, USDA: Washington, D. C., pp. 295–306.

Brzostowski, H. W. and **M. A. Owen** (1966) Production and germination capacity of buffel grass (Cenchrus ciliaris) seeds, *Trop. Agric. (Trin.)* **43**, 1–10.

Burton, G. W. (1943) Factors affecting seed setting in several southern grasses, *J. Amer. Soc. Agron.* **35**, 465–74.

Burton, G. W. (1966) Breeding better bermudagrasses, *Proc. 9th Intl. Grassld Cong.*, pp. 93–96.

Cameron, D. F. (1967) Flowering in Townsville lucerne (Stylosanthes humilis). 1. Studies in controlled environments, *Aust. J. Exptl Anim. Husb.* **7**, 489–94.

Cameron, D. F. and **J. D. Mullaly** (1969) Effect of nitrogen fertilization and limited irrigation on seed production of Molopo buffel grass, *Queensld J. Agric. Anim. Sci.* **26**, 41–7.

Chadhokar, P. A. and **L. R. Humphreys** (1970) Effects of time of nitrogen deficiency on seed production of Paspalum plicatulum Michx., *Proc. 11th Intl. Grassld Cong.*, pp. 315–19.

Chadhokar, P. A. and **L. R. Humphreys** (1973) Influence of time and level of urea application on seed production of Paspalum plicatulum at Mt. Cotton, southeastern Queensland, *Aust. J. Exptl Agric. Anim. Husb.* **13**, 275–83.

CIAT (1972, 1973, 1978) *Annual Reports*, Centro Int. Agric. Trop.: Cali, Colombia.

Combes, R. J. and **W. H. Verburgt** (1971) *The Commercial Growing, Cleaning and Marketing of Tropical and Subtropical Pasture Seeds*, Seminar For. Crops. W. Africa, Univ. Ibadan, Nigeria (mimeo).

Criste, P. Q., R. C. Mendoza and **E. Q. Javier** (1973) Seed production of Guinea grass (Panicum maximum Jacq.): Effect of season and date of harvest, *Proc. 4th Ann. Mtg. Crop. Sci. Soc. Phil.*, pp. 313a–i.

Crowder, L. V. (1977) Potential of tropical zone cultivated forages. In *Potential of the World's Forages for Ruminant Animal Production*, Winrock Intl. Livestock Res. Trg Center, Arkansas, pp. 49–78.

Crowder, L. V., H. Chaverra and **J. Lotero** (1970) Productive improved grasses in Colombia, *Proc. 11th Intl. Grassld Cong.*, pp. 147–9.

Crowder, L. V. and **H. R. Chheda** (1977) Forage and fodder crops. In C.L.A. Leaky and J. B. Wills (eds), *Food Crops of the Lowland Tropics*, Oxford Univ. Press: London, pp. 127–59.

Ferguson, J. E. (1978) Systems of pasture seed production in Latin America. In P. A. Sanchez and L. E. Tergas (eds), *Pasture Production in Acid Soils of the Tropics*, Seminar Proc., CIAT, Colombia, pp. 385–95.

Fisher, M. J. (1969) The growth and development of Townsville lucerne (Stylosanthes humilis) in ungrazed swards at Katherine, N. T., *Aust. J. Exptl Agric. Anim. Husb.* **9**, 196–208.

Fisher, M. J. (1973) Effect of time, height and frequency of defoliation on the growth and development of Townsville stylo in pure ungrazed swards at Katherine, N. T., *Aust. J. Exptl Agric. Anim. Husb.* **13**, 389–97.

Foster, W. H. (1956) Factors effecting the production of seed in Rhodes grass, *Rep. Dept. Agric. N. Nigeria Pt. II*, pp. 172–6.

Gildenhuys, P. J. (1951) *Fertility Studies in* Setaria sphacelata *(Schum.) Stapf and Hubbard*, Sci. Bull. 314, Government Printer: Pretoria.

Gordin-Sharir, A. and **H. Gelmond** (1966) Seed setting, production and viability of Rhodes grass in Israel, *E. Afr. Agric. For. J.* **31**, 365–7.

Greg, B. R. (1973) Seed processing, *Seed Research* **1**, 27–39.

Greg, B. R. (1977) Seed processing plant design, with special reference to developing seed industries, *Seed Sci. and Technol.* **5**, 287–336.

Grof, B. (1969) Viability of para grass (Brachiaria mutica) seed and the effect of fertilizer nitrogen on seed yield, *Queensld J. Agric. Anim. Sci.* **26**, 271–6.

Hacker, J. B. and **R. J. Jones** (1971) The effect of nitrogen fertilizer and row spacing on speed production in Setaria sphacelata *Trop. Grasslds* **5**, 61–73.

Haggar, R. J. (1966) The production of seed from Andropogon gayanus, *Proc. Intl. Seed Test. Assoc.* **31**, 251–9.

Harmond, J. E., J. E. Smith Jr. and **J. K. Park** (1961) Harvesting the seeds of grasses and legumes, In *Seeds, The Yearbook of Agriculture*, USDA: Washington, D.C., pp. 181–7.

Harrington, J. F. (1959) Drying, storing and packaging seeds to maintain germination and vigour. In *Proceedings of the 1959 Short Course for Seedsmen*, Mississippi State College.

Harrington, J. F. (1963) Practical advice and instructions on seed storage, *Proc. Intl. Seed Test. Assoc.* **28**, 989–94.

Harrington, J. F. (1972) Seed storage and longevity. In T.T. Kozlowski (ed.), *Seed Biology*, vol. 3, Academic Press: New York, pp. 145–55.

Henzell, D. F. and **D. J. Oxenham** (1964) Seasonal changes in the nitrogen content of three warm climate pasture grasses, *Aust. J. Exptl Agric. Anim. Husb.* **4**, 336–44.

Hill, G. O. (1967) A requirement for chilling to induce flowering in Dolichos lablab cv. Rongai, *Papua New Guinea Agric. J.* **19**, 16–17.

Hopkinson, J. M. (1977) Siratro seed production, *Trop. Grasslds* **11**, 33–9.

Hopkinson, J. M. and **D. S. Loch** (1973) Improvement in seed yields of Siratro (Macroptilium atropurpureum). 1. Production and loss of seed in the crop, *Trop. Grasslds* **7**, 255–68.

Hopkinson, J. M. and **D. S. Loch** (1977) Seed production of stylo in North Queensland, *Queensld Agric. J.* **103**, 116–25.

Hopkinson, J. M. and **C. P. Vicary** (1974) Improvement in seed yield of Siratro (Macroptilium atropurpureum). 2. Recovery of fallen seed by suction harvester, *Trop. Grasslds* **8**, 103–6.

Humphreys, L. R. (1979) *Tropical Pasture Seed Production*, Plant Prod. Prot. Paper 8, FAO: Rome.

Humphreys, L. R. and **D. E. Davidson** (1967) Some aspects of pasture seed production, *Trop. Grasslds* **1**, 84–7.

Hutton, E. M. (1960) Flowering and pollination in Indigofera spicata, Phaseolus lathyroides, Desmodium uncinatum, and some other tropical pasture legumes, *Emp. J. Exptl Agric.* **28**, 235–43.

Hutton, E. M. (1970) Tropical pastures, *Adv. Agron.* **22**, 1–73.

Irwin, J. A. G. and **D. F. Cameron** (1978) Two diseases of Stylosanthes spp. caused by Colletotrichum gloesoporioides in Australia and the pathogenic specialization of one of the causal organisms, *Aust. J. Agric. Res.* **39**, 305–17.

Javier, E. Q. (1970) The flowering habits and mode of reproduction of Guinea grass (Panicum maximum Jacq.), *Proc. 11th Intl. Grassld Cong.*, pp. 284–9.

Javier, E. Q. and **N. M. Marasigan** (1976) Overseeding of legumes on Imperata grassland. Research Papers and Abstracts of Unpublished Theses on Pasture and Forage Crops, *C.R.D. Phil. Council Agric. Res., Los Banos*, pp. 27–42.

Javier, E. Q. and **R. C. Mendoza** (1976) *The Harvesting, Cleaning and Storage of Home Grown Tropical Pasture Seeds*, ASPAC Food Fert. Techn. Cent., Extn. Bull. 65, Taiwan.

Javier, E. Q., G. T. Sasis and **R. C. Mendoza** (1976) Seed production of tropical pasture legumes. II. Phaseolus atropurpureus cv. Siratro. Research Papers and Abstracts of Unpublished Theses on Pastures and Forage Crops, *C.R.D. Phil. Council Agric. Res., Los Banos*, pp. 152–9.

Jones, R. J. and **R. Roe** (1976) Seed production, harvesting and storage. In N. H. Shaw and W. W. Bryan (eds), *Tropical Pasture Research Principles and Methods*, Commonw. Bur. Past. and Fld Crops., Bull 51, Hurley: England, Ch. 16.

Jong, J. J. de (1961) Seed distribution and trade. In *Agricultural and Horticultural Seeds*, Agric. Studies No. 55, FAO: Rome, pp. 121–55.

Kernick, M. D. (1961) Ecology. In *Agricultural and Horticultural Seeds*, Agric. Studies No. 55, FAO: Rome, pp. 32–61.

Linnett, B. (1977) Processing seeds of tropical pasture plants, *Seed Sci. and Technol.* **5**, 199–224.

Loch, D. S., J. M. Hopkinson and **B. H. English** (1976) Seed production of Stylosanthes guyanensis. 1. Growth and development; 2. The consequences of defoliation, *Aust. J. Exptl Agric. Anim. Husb.* **16**, 218–30.

Mannetje, L. 't (1966) Stylosanthes species, *CSIRO Div. of Trop. Past. Ann. Report*, p. 45.

McLean, D. and **B. Grof** (1968) Effect of seed treatments on Brachiaria mutica and B. ruziziensis, *Queensld J. Agric. Anim. Sci.* **25**, 81–3.

Mendoza, R. C., G. T. Sasis and **E. Q. Javier** (1976) Seed production of tropical forage legumes. III. Stylosanthes guyanensis (Aubl.) Sw. cv. Schofield. Research Papers and Abstracts of Unpublished Theses on Pastures and Forage Crops, *C. R. D. Phil. Council Agric. Res., Los Banos*, pp. 191–7.

Mishra, M. L. and **B. N. Chatterjee** (1968) Seed production in the forage grasses Pennisetum polystachion and Andropogon gayanus in the Indian tropics, *Trop. Grasslds* **2**, 51–6.

Mwakha, E. (1969) Observations on the effect of temperature on the growth of Trifolium semipilosum Fres., *E. Afr. Agric. For. J.* **34**, 289–92.

Nicholls, D. F., J. A. Gibson, L. R. Humphreys, G. D. Hunter and **L. M. Bahnisch** (1973) Nitrogen and phosphorus response of Desmodium uncinatum on seed production at Mt Cotton, south-eastern Queensland, *Trop. Grasslds* **7**, 243–8.

Okigbo, B. N. and **H. R. Chheda** (1966) Natural fertility and chromosome numbers in several strains of stargrasses, *Niger. Agric. J.* **3**, 72–5.

Oomen, W. W. A. (1969) Experimental seed cleaning equipment, its use in Wageningen Seed Testing Station, *Proc. Intl. Seed Test. Assoc.* **34**, 15–74.

Osbourn, D. F. (1975) Beef production from improved pastures in the tropics, *World. Rev. Anim. Prod.* **11**, 23–31.

Prodonoff, E. T. (1967) The determination and maintenance of seed quality, *Trop. Grasslds* **1**, 91–8.

Purcell, D. L. (1969) Grass seed harvesting in the Roma Region, *Queensld Agric. J.* **95**, 646–53.

Redrup, J. (1965) An approach to the commercial production of seeds of tropical pasture plants, *Proc. 9th Intl. Grassld Cong.*, pp. 521–6.

Robertson, A. D., L. R. Humphreys and **D. G. Edwards** (1976) Influence of cutting frequency and phosphorus supply on the production of Stylosanthes humilis and Arundinaria pusilla at Khan Kaen, Northeast Thailand, *Trop. Grasslds* **10**, 33–9.

Roe, R. (1972) Seed losses with different methods of harvesting Panicum coloratum, *Trop. Grasslds* **6**, 113–18.

Rogler, G. A., H. H. Rampton and **M. D. Atkins** (1961) The production of grass seeds. In *Seeds, The Yearbook of Agriculture*, USDA: Washington, D. C., pp. 163–71.

Shelton, H. M. and **L. R. Humphreys** (1971) Effect of variation in density and phosphate supply on seed production of Stylosanthes humilis, *J. Agric. Sci. (Camb.)* **76**, 325–8.

Simpson, G. L. (1972) Problems of seed production of Narok Setaria, *Trop. Grasslds* **6**, 72.

Singh, R. D. and **B. N. Chatterjee** (1966) Tillering of perennial grasses in the tropics of India, *Proc. 9th Intl. Grassld Cong.*, pp. 1075–9.

Skerman, R. H. and **L. R. Humphreys** (1973) The effect of temperature during flowering on seed formation of Stylosanthes humilis, *Aust. J. Agric. Res.* **24**, 317–24.

Stillman, S. L. and **W. R. Tapsall** (1976) Some effect of nitrogen on seed formation of Setaria anceps cv. Nandi, *Queensld J. Agric. Anim. Sci.* **33**, 173–6.

Strickland, R. W. (1971) Seed production and testing problems in tropical and subtropical pasture species, *Proc. Int. Seed Test. Assoc.* **36**, 189–99.

Vaughan, C. E., B. R. Gregg and **J. C. Delouche** (eds) (1968) *Seed Processing and Handling*, Handbook No. 1, Seed Tech. Lab., Mississippi State College.

Verhoeven, G. (1964) A simple seed cleaner, *Queensld Agric. J.* **90**, 106–7.

Whyte, R. O., G. Nilsson-Leissner and **M. C. Trumble** (1953) *Legumes in Agriculture*, Agric. Studies No. 21, FAO: Rome.

Whyte, R. O., T. R. Moir and **J. P. Cooper** (1959) *Grasses in Agriculture*. Agric. Studies No. 42, FAO: Rome.

Wickman, B., H. M. Shelton, M. D. Hare and **A. J. de Boer** (1977) Townsville stylo seed production in Northeastern Thailand, *Trop. Grasslds* **11**, 177–87.

Wiggs, P. M., M. A. Owen and **N. J. Mukurasi** (1973) Influence of farmyard manure and nitrogen fertilizers on sown pastures, seed yield and quality of Cenchrus ciliaris L. at Kongwa Tanzania, *E. Afr. Agric. For. J.* **38**, 367–74.

Wilaipon, P. and **L. R. Humphreys** (1976) Grazing and mowing effects on the seed production of Stylosanthes hamata cv. Verano, *Trop. Grasslds* **10**, 107–11.

Wilson, G. P. M. (1970) Method and practicability of kikuyu grass seed production, *Proc. 11th Intl. Grassld Cong.*, pp. 312–15.

Wutoh, J. G., E. M. Hutton and **A. J. Pritchard** (1968) The effects of photoperiod and temperature on flowering in Glycine javanica, *Aust. J. Exptl Agric. Anim. Husb.* **8**, 544–7.

Index